D1552723

18. $\int u\,dv = uv - \int v\,du$ (Integration by Pa

19. $\int \sin^n x\,dx = -\frac{\sin^{n-1}x\cos x}{n} + \frac{n-1}{n}\int \sin^{n-2}x\,dx$

20. $\int \cos^n x\,dx = \frac{1}{n}\cos^{n-1}x\sin x + \frac{n-1}{n}\int \cos^{n-2}x\,dx$

21. $\int \tan^n x\,dx = \frac{\tan^{n-1}x}{n-1} - \int \tan^{n-2}x\,dx$

22. $\int x^m e^{ax}\,dx = \frac{x^m e^{ax}}{a} - \frac{m}{a}\int x^{m-1}e^{ax}\,dx$

Identities

Trigonometric

1. $\sin(x \pm y) = \sin x\cos y \pm \cos x\sin y$
2. $\cos(x \pm y) = \cos x\cos y \mp \sin x\sin y$
3. $\sin x\sin y = \frac{1}{2}\cos(x-y) - \frac{1}{2}\cos(x+y)$
4. $\cos x\cos y = \frac{1}{2}\cos(x-y) + \frac{1}{2}\cos(x+y)$
5. $\sin x\cos y = \frac{1}{2}\sin(x-y) + \frac{1}{2}\sin(x+y)$
6. $\sinh x = \frac{e^x - e^{-x}}{2}$ 7. $\cosh x = \frac{e^x + e^{-x}}{2}$
8. $e^{ibx} = \cos bx + i\sin bx$

Infinite Series

1. $e^x = 1 + x + \frac{x^2}{2!} + \frac{x^3}{3!} + \cdots = \sum_{n=0}^{\infty}\frac{x^n}{n!}$
2. $\sin x = x - \frac{x^3}{3!} + \frac{x^5}{5!} - \cdots = \sum_{n=0}^{\infty}\frac{(-1)^n}{(2n+1)!}x^{2n+1}$
3. $\cos x = 1 - \frac{x^2}{2!} + \frac{x^4}{4!} - \cdots = \sum_{n=0}^{\infty}\frac{(-1)^n}{(2n)!}x^{2n}$
4. $f(x) = \sum_{n=0}^{\infty}\frac{f^{(n)}(a)}{n!}(x-a)^n$ (Taylor Series)

parviz Khajeh-Khalili

Ordinary Differential Equations with Numerical Techniques

Ordinary Differential Equations with Numerical Techniques

John L. Van Iwaarden
Hope College

HARCOURT BRACE JOVANOVICH, PUBLISHERS HBJ
San Diego New York Chicago Atlanta Washington, D.C.
London Sydney Toronto

Copyright © 1985 by Harcourt Brace Jovanovich, Inc.

All rights reserved. No part of this publication may be reproduced or transmitted in any form or by any means, electronic or mechanical, including photocopy, recording, or any information storage and retrieval system, without permission in writing from the publisher.

Requests for permission to make copies of any part of the work should be mailed to: Permissions, Harcourt Brace Jovanovich, Publishers, Orlando, Florida 32887.

ISBN: 0-15-567550-8

Library of Congress Catalog Card Number: 84-81494

Printed in the United States of America

Portions of Chapters 2, 4, 5, 7, and 10 were originally copyrighted in 1980 by the Trustees of Dartmouth College. Permission to use this material has been granted by COMPress, a division of Wadsworth, Inc. Such permission does not imply endorsement of this publication by the Trustees of Dartmouth College or by the National Science Foundation, which supported the original Project COMPUTe.

Preface

Over the years, the methods used in studying differential equations have undergone significant changes. The present availability of computers has led to an increased emphasis on numerical techniques for finding very accurate approximate solutions to those problems—and real-world applications—not formerly considered in a standard differential equations course. Thus, it seems essential in our computer age to incorporate numerical techniques throughout the entire study rather than to relegate them to a single chapter.

This book provides an introduction to the fundamental concepts of differential equations, a wealth of realistic problems that will motivate the study of solution techniques, and a logical and orderly approach that integrates the numerical methods into the overall flow of topics. Numerical techniques are introduced at a number of places as logical extensions of the analytical techniques being studied. This approach will prepare students for the multitude of types of differential equations that may be encountered outside the classroom in a variety of fields.

The background required for this course is a standard elementary calculus course, including only minor references to multivariable calculus. The emphasis throughout is to make the material readable and to provide a sufficient number of fully worked examples.

In addition to exercise sets for each section, Chapters 1, 3, 6, 8, and 9 end with collections of miscellaneous exercises. Chapters 2, 4, 5, 7, and 10 end with a set of problems and a separate set of applications problems, which are taken from a variety of scientific areas to display the widespread applicability of numerical solutions for differential equations and to motivate their study.

The appendixes contain documentation and program listings for the computer programs used in this book. A microcomputer diskette for these programs is produced and marketed by CONDUIT (University of Iowa, Oakdale Campus, Iowa City, Iowa 52242).

The text material has been used successfully for more than eight semesters at Hope College by sophomore science majors. The entire text is covered in a three-semester-hour course that meets 42 times, with nine of the 50-minute class sessions devoted specifically to numerical techniques. A suggested outline appears in the Instructor's Manual. Semester-end questionnaires have verified the usefulness to students of in-depth treatment of numerical methods; returning graduates have echoed these comments. Preparation for today's scientific world requires a thorough working knowledge of how the computer may be used in solving all types of problems, and this book aims at assisting in such preparation.

I wish to thank many persons for their valuable contributions to this work. Gratitude goes to NSF Project COMPUTe at Dartmouth College and, especially, to A. Kent Morton for aid in starting the writing project; to Herbert Dershem and John Stoughton of Hope College and numerous other colleagues for their help and suggestions; to students Christine Brouwer and Deb Van Iwaarden for their problem checking; to Mike Ely and Joellyn Shull for programming modifications; and to expert typists Chris McDowell and Lori McDowell. For their valuable feedback I thank the many students at Hope College who studied from the preliminary materials.

Notable contributions to the completion of this book were made by the staff at Harcourt Brace Jovanovich, Publishers. I am particularly grateful to the acquisitions editor, Richard Wallis, and the manuscript editor, Mary Kitzmiller, for their excellent work. Thanks also go to the following manuscript reviewers for their helpful suggestions: Richard Koch (University of Oregon) and William L. Reddy (Wesleyan University).

A final acknowledgment is made to my wife, Mary, whose patient support, understanding, and encouragement were a valuable contribution in completing this project.

John L. Van Iwaarden

Contents

Ordinary Differential Equations with Numerical Techniques

Chapter 1

Analytical Methods for First-Order Equations

1.1 Introduction

Students enrolled in this course are majoring in various subjects, and you may wonder how important differential equations are to your particular discipline. Since differential equations play a major role in areas such as chemistry, physics, engineering, and medicine, we have built this course around practical applications of differential equations.

Many of the basic laws of the above-mentioned disciplines are defined in mathematical terms that contain unknown functions and their derivatives. Any equation containing an unknown function and its derivatives is called a **differential equation**. Frequently mathematical models are needed for solving problems that include a variable that has a continuous pattern of variation (e.g., time and motion), and usually differential equations serve as the model.

Section 1.1 introduces some basic definitions and concepts regarding elementary differential equations, and subsequent sections present various techniques for finding solutions. The text uses a two-pronged approach: (1) analytical solutions and (2) numerical solutions. When possible, we will find a solution by using only algebra and calculus. Such a solution is called an **analytical solution**. If this cannot be obtained, we will look for a **numerical solution** by using approximations and computer algorithms. Today's scientists find both of these methods indispensable, and an ultimate goal is to develop the expertise to solve almost every differential equation encountered in future scientific work by at least one of these methods.

Definition: An **ordinary differential equation** is an equation that contains an independent variable x, an unknown function $y(x)$, and certain derivatives of y such as $y'(x), y''(x), y'''(x), \ldots, y^{(n)}(x)$. A functional expression that is a combination of the variable x, the unknown function $y(x)$, and the derivatives of y may be written as

$$F(x, y', y'', \ldots, y^{(n)}) = 0 \tag{1}$$

When written in this form, it is called an **ordinary differential equation of order** n, where n is a positive integer.

Definition: The highest order of a derivative (of the unknown function) that appears in a differential equation is also the **order** of that equation.

The following example illustrates differential equations of orders 1, 2, and 3.

Example 1

(a) $y' = 1 + x + y$ is an ordinary differential equation of order 1. In the functional form (1), this differential equation would be written as

$$F(x, y, y') = x + y + 1 - y' = 0$$

(b) $2y'' + 3xy' + y = e^x$, which is an ordinary differential equation of order 2, could be written

$$F(x, y, y', y'') = e^x - y - 3xy' - 2y'' = 0$$

(c) $xy''' - 2y' + e^x y = \ln|x|$ is a third-order differential equation.

Definition: For a differential equation, as in (1), a **solution** is a function $\varphi(x)$ defined on an open interval I such that

$$F\big(x, \varphi(x), \varphi'(x), \ldots, \varphi^{(n)}(x)\big) \equiv 0$$

for every x in the interval I.

In Example 1(a), the function $y = e^x - x - 2$ is an explicit solution on $(-\infty, \infty)$ since $F(x, y, y')$ becomes $x + (e^x - x - 2) + 1 - (e^x - 1) \equiv 0$.

Differential equations are also classified in other ways.

Definition: An equation of order n is said to be **linear** if it has the special form

$$a_0(x)y^{(n)} + a_1(x)y^{(n-1)} + \cdots + a_{n-1}(x)y' + a_n(x)y = f(x) \tag{2}$$

where the $a_i(x)$ are arbitrary functions of x only. Also note that in this form the unknown function y and all of its derivatives appear linearly, (i.e., none may have a power higher than 1 and none may be multiplied by any of the others). The coefficient functions $a_0(x)$, $a_1(x), \ldots, a_n(x)$, and $f(x)$ must not contain y or any of its derivatives. The equation is **nonlinear** if it cannot be put into the form of equation (2).

Example 2

The first equation of Example 1 is linear since it can be written as

$$y' - y = 1 + x$$

The second equation is linear, as may be seen in its present form. The third equation is also linear, even though its coefficient functions are somewhat complicated. It still has the form given in (2).

Example 3

The equation $2xy'' + 3yy' + e^x y = 2x^2$ is nonlinear, because the second term contains the product of y and y'.

The equation $x^2y' + y^2 = e^x + 2$ is nonlinear, because one of the terms is the square of the unknown function.

The equation $2x^2y' - x \sin y = 0$ is nonlinear, because in the second term the unknown function y does not appear linearly.

Definition: If all the coefficient functions $a_0(x), a_1(x), \ldots, a_n(x)$ are constant functions, as in $3y'' + 6y' + y = e^x$, the equation is said to have **constant coefficients.** If any one of the coefficient functions is not a constant function, then the equation is said to have **variable coefficients**.

Example 4

The differential equation $xy'' - 5y' + 2e^x y = \cos 3x$ is second order, is linear, and has variable coefficients.

The differential equation $yy'' - 2xy' = 17$ is second order and nonlinear.

The differential equation $2y''' - y' + 4y = 0$ is third order, is linear, and has constant coefficients.

The use of the word *linear* for differential equations coincides with the use of the same word to describe an algebraic equation such as $3x_1 + 4x_2 = 5$ or $2u - 5v = 8$, where the variables appear linearly. In a quadratic equation, such as $4x_1^2 + 6x_1 - 7 = x_2$, the variable x_1 does not appear linearly; and in a trigonometric equation, such as $z = 3 \sin t$, or an exponential equation, such as $r = 5e^\theta$, the nonlinearity is again apparent.

Definition: A linear equation is called **homogeneous** if $f(x) \equiv 0$ and **nonhomogeneous** if $f(x) \neq 0$. A nonlinear equation is homogeneous if it contains no terms that are functions of x alone. Otherwise, it is nonhomogeneous. (The three linear differential equations of Example 1 are all nonhomogeneous.)

Example 5

The equation $y' + 2y^2 = 0$ is first order, nonlinear, and homogeneous.

The equation $x^2y'' + 2yy' = e^x$ is second order, nonlinear, and nonhomogeneous.

The equation $3y''' - xy'' + 7y' + 9x^2y = \cos 2x$ is third order, linear, and nonhomogeneous.

We can now make our earlier definition more rigorous by the following statement.

Definition: A **solution** of an ordinary differential equation of order n is a function that possesses at least n derivatives on some interval I and that satisfies the equation.

Example 6

For the first-order linear homogeneous equation $y' - 3y = 0$, the solution $y(x) = e^{3x}$ is valid for all real x, since the substitution back into the differential equation produces identical sides.

$$(3e^{3x}) - 3(e^{3x}) \equiv 0$$

Example 7

The second-order linear nonhomogeneous equation $y'' + y' - 2y = 4$ has the solution $y(x) = e^x - 2$. To verify this we calculate $y' = e^x$ and $y'' = e^x$; then we substitute these values in the equation and get $e^x + e^x - 2(e^x - 2) = 4$ or the identity $4 = 4$. Again the solution is valid on the entire real axis.

Example 8

The second-order linear homogeneous equation $x^2y'' + 2xy' - 2y = 0$ has $y(x) = x^{-2}$ as a solution. To verify this, we find $y' = -2x^{-3}$ and $y'' = 6x^{-4}$ and substitute the values to find

$$x^2(6x^{-4}) + 2x(-2x^{-3}) - 2(x^{-2}) = 6x^{-2} - 4x^{-2} - 2x^{-2} \equiv 0$$

This solution is valid on the interval $(0, \infty)$ or on $(-\infty, 0)$.

Example 9

The second-order linear nonhomogeneous equation $y'' + 4y = 12$ has $y(x) = (\sin 2x) + 3$ as a solution. Here $y' = 2\cos 2x$ and $y'' = -4\sin 2x$ so that the differential equation becomes $-4\sin 2x + 4(\sin 2x + 3) = 12$, which is the required identity. The solution is valid for all real x.

Generally, differential equations of lower orders are easier to solve than are those of higher orders; those that are linear are easier than the nonlinear ones; and homogeneous equations are easier than nonhomogeneous ones. This text presents equations in a systematic order, proceeding from the easiest cases to the more difficult ones.

As is shown in most elementary calculus courses, the expression $y' = 2x$ can be solved easily by simple integration and has as its solution(s) the complete family of functions $y(x) = x^2 + C$. This collection is called a **one-parameter family of functions**. When substituted into the differential equation, each of its members changes the equation into an identity. The parameter is the constant C, which can take on all possible real values. From this family of parabolas, an individual member may be inspected by assigning a value to C. To find the particular family member that passes through the point $(3, 1)$, for example, set $x = 3$ and $y = 1$ into the family equation to find $1 = 9 + C$ or $C = -8$. That unique parabola is then $y(x) = x^2 - 8$. Graphs of such individual members would be parabolas with their vertices on the y-axis and opening upward. If one used a black pen to draw this entire family on sheet of white paper, the result would be a black sheet of paper, because through every point of the plane passes at least one curve of the family of parabolas.

In many equations, especially those that are linear and homogeneous, the function $y = 0$ is an automatic solution. Note, for example, that in $x^2y'' + e^xy' + 2y = 0$, it is clear that $y = 0$ satisfies the equation since every term becomes zero. This kind of solution is called a **trivial solution**. In analyzing a differential equation, we must look for the nontrivial solution. The trivial solution gives little or no information.

In many application problems that involve a differential equation, the function to be determined must satisfy a set of additional conditions generally posed at one specific point. A well-known problem involving antidifferentiation in elementary calculus is that of determining the distance that a falling object drops in a specified time interval. The gravitational acceleration equation $d^2y/dt^2 = -g$ is coupled with the initial conditions $y(0) = y_0$ and $\frac{dy}{dt}(0) = v_0$. This is a familiar example of **initial value problem**. An nth order differential equation coupled with n additional conditions. These conditions give the value of the unknown function and all of its derivatives up to the $(n - 1)$st at the same initial point. A solution to an initial value problem is a unique function that, upon substitution, changes the differential equation into an identity and simultaneously satisfies each of the initial conditions. The procedure for incorporating the initial conditions into the solution process will be thoroughly treated in subsequent sections.

The next five sections of this chapter are limited to techniques for finding solutions of a variety of first-order differential equations. The main tool is integration. When one can perform an exact integration and obtain a solution, the solution is called, as inferred earlier, an analytical solution. In certain instances when the integration cannot be done, one must defer to a numerical technique in which an approximate solution is obtained. Such instances are presented in Chapter 2. An alert student will soon discover that both techniques are very important. In most instances, the first attempt will be to find an analytical solution. When coupled with initial conditions, this exact solution may be graphed, analyzed for maximum and minimum points, etc. When an analytical solution cannot be found, an attempt should be made to find a numerical approximate solution. The distinguishing features of each method will be emphasized as we proceed through the text.

Exercises 1.1

1. For each of the differential equations, identify the following attributes: order, linearity, coefficient form, and homogeneity.
   (a) $xy'' + 2y' = 3y$
   (b) $(y')^2 = 2y + 3$
   (c) $3y''' - y' + 4y = 0$
   (d) $e^y y'' + 2xy = 4$
   (e) $\dfrac{dy}{dx} - x^2y^2 = 1$
   (f) $\dfrac{d^2y}{dx^2} - 4y\dfrac{dy}{dx} + 2 = 0$
   (g) $\left(\dfrac{dy}{dx}\right)^2 + 2y^2 = 3x$
   (h) $\cos(y') = 5y + 7$
   (i) $(\cos)y'' + 9e^x y' - 37y = 12x^2$
   (j) $4y'''' - 2y'' + x^6 = 0$
2. For each equation in Exercise 1, follow the definition for initial value problem and pose a valid set of initial conditions.
3. Verify that the function $y = e^{-x}\sin 2x$ is a solution for $y'' + 2y' + 5y = 0$.
4. Verify that the function $y = e^{(2/3)x} + 3e^{-6x}$ is a solution of $3y'' + 16y' - 12y = 0$.
5. Verify that $y = 2x^2e^x$ is a solution of $y'' - 2y' + y = 4e^x$.
6. Verify that the function $y = 3x + xe^{2/x}$ is a solution of $x^3y'' + 2xy' - 2y = 0$.
7. Verify that $y = c_1e^x + c_2xe^x$ is a solution of $y'' - 2y' + y = 0$ for all values of c_1 and c_2.
8. Find the value(s) of k for which e^{kx} is a solution of:
   (a) $y' + 3y = 0$
   (b) $y'' - 3y' - 4y = 0$
   (c) $y'' - 4y = 0$

(d) $y'' + y = 0$
(e) $y''' - y'' - 14y' + 24y = 0$
(f) $y''' - 2y'' - y' + 2y = 0$

9. Determine the value(s) of k for which x^k, $(x > 0)$, is a solution of
(a) $x^2y'' - 3xy' + 3y = 0$
(b) $2x^2y'' + 5xy' + y = 0$
(c) $x^2y'' + 2xy' - 2y = 0$
(d) $x^3y''' + 6x^2y'' + 7xy' + y = 0$

10. Show that the polynomial functions $T_1(x) = x$; $T_2(x) = 2x^2 - 1$; and $T_3(x) = 4x^3 - 3x$ are solutions of the parameterized differential equation

$$(1 - x^2)y'' - xy' + k^2y = 0 \qquad [k \text{ is the integer in the subscript of } T_k(x)]$$

known as **Tchebycheff's equation***, which is used in studies in numerical analysis.

11. Show that the Hermite polynomials $H_1(x) = 2x$; $H_2(x) = 4x^2 - 2$; and $H_3(x) = 8x^3 - 12x$ are solutions of

$$H_k''(x) - 2xH_k'(x) + 2kH_k(x) = 0$$

known as **Hermite's equation****.

12. Near the year 1800, the Reverend T. R. Malthus, a political economist, was studying the population of Europe and noted that it doubled at regular intervals. Continued study indicated that the rate of increase of the population was proportional to the present population. He then formulated the Malthusian Law of Population Growth:

$$\frac{dN}{dt} = kN, \; k > 0$$

If we specify $N(0)$, the number of people at a time determined to be the starting year, we can—by the technique in Section 1.3—easily obtain the result for $N(t)$, the population at any later time. Verify that the function $N(t) = N(0)e^{kt}$ solves this equation.

13. Show that the functions $u_1(x) = 1$ and $u_2(x) = \arctan x$ are solutions of the differential equation $(x^2 + 1)y'' + 2xy' = 0$ on $-\infty < x < \infty$.

14. Verify that $u(x) = 3xe^{2/x}$ is a solution to the differential equation $x^3y'' + 2xy' - 2y = 0$.

15. If $y(0) = 0$ and $y'(0) = 1$, find the value of C_1 and C_2 such that $y = C_1e^x + C_2e^{-x}$ is a solution of $y'' - y = 0$ and both initial conditions are satisfied.

16. If $y(0) = 1$ and $y'(0) = -3$, find the values of C_1 and C_2 such that $y = C_1e^{3x} + C_2e^x$ is a solution of $y'' - 4y' + 3y = 0$ and both initial conditions are satisfied.

---

* Named for Pafnuti L. Tchebycheff, Russian mathematician, 1821–1894.
** Named for Charles Hermite, French mathematician, 1822–1901.

1.2 Solutions by Integration

The previous section cited a simple example in which a single integration immediately produced the general solution. Consider the example $y' = x + 1$. By integrating each side we obtain $y = x^2/2 + x + C$. This collection of functions is the one-parameter family which is the general solution. This is easily verified by substituting the values back into the differential equation.

Suppose now that we have

$$y'' = 6x^2 + 2x \tag{3}$$

This is a second-order example of the more general nth order equation

$$y^{(n)}(x) = f(x) \tag{4}$$

In equations of the type shown in (3) and (4), only one derivative of the unknown function is present; the rest of the equation involves only the independent variable. A common approach is to use the differential notation established in calculus, (i.e., $dy = y'dx$ and its extension $dy' = y''dx$) and to rewrite (3) as $d(y') = (6x^2 + 2x)\,dx$.

Then integrating term by term, we get

$$y' = 2x^3 + x^2 + C_1 \tag{5}$$

This may again be rewritten in differential notation as $d(y) = (2x^3 + x^2 + C_1)\,dx$ and integrated to obtain

$$y = \frac{x^4}{2} + \frac{x^3}{3} + C_1x + C_2 \tag{6}$$

This general solution to (3) is a two-parameter family of functions, each member of which has the following property: If the values of the function and its appropriate derivatives are substituted into the original differential equation, the equation becomes an identity. The interval of validity is all real numbers.

If we add a pair of initial conditions to (3), for example, $y(0) = 2$ and $y'(0) = 3$, we obtain the initial value problem

$$\left\{\begin{array}{l} y'' = 6x^2 + 2x \\ y(0) = 2,\ y'(0) = 3 \end{array}\right\} \tag{7}$$

These initial conditions, applied at the appropriate time, will allow us to determine unique values for the two parameters C_1 and C_2 in (6). After the first integration, we obtained (5): $y'(x) = 2x^3 + x^2 + C_1$. Since this equation is valid for all real numbers, we let x take on its initial value, $x = 0$. The equation then becomes

$$y'(0) = 2(0)^3 + (0)^2 + C_1 = C_1 \tag{8}$$

From the second initial condition, we also know that

$$y'(0) = 3 \tag{9}$$

Equating (8) and (9) we find $C_1 = 3$, and (5) becomes $y'(x) = 2x^3 + x^2 + 3$. After the second integration we obtained (6), which can now be written as

$$y(x) = \frac{x^4}{2} + \frac{x^3}{3} + 3x + C_2$$

Again, because this equation holds for all x, we substitute the value of x given in the initial conditions, $x = 0$, and (6) becomes

$$y(0) = \frac{0}{2} + \frac{0}{3} + 3 \cdot 0 + C_2 = C_2 \tag{10}$$

From the first given initial condition we know that

$$y(0) = 2 \tag{11}$$

Equating (10) and (11) we find $C_2 = 2$, and (6) now becomes

$$y(x) = \frac{x^4}{2} + \frac{x^3}{3} + 3x + 2$$

This is the unique solution to the initial value problem (7). It is the one and only function that simultaneously satisfies the given differential equation and the initial conditions.

As another example of an initial value problem we now are able to solve, consider

$$\begin{Bmatrix} y'' = \sin x \\ y(\frac{\pi}{2}) = 2,\ y'(\frac{\pi}{2}) = 1 \end{Bmatrix} \tag{12}$$

Equation (12) may be written in differential notation as $d(y') = \sin x\,dx$ and integrated to

$$y'(x) = -\cos x + C_1 \tag{13}$$

Because the initial point of x posed in the problem is $x = \frac{\pi}{2}$, we let $x = \frac{\pi}{2}$ to obtain

$$y'(\tfrac{\pi}{2}) = -\cos(\tfrac{\pi}{2}) + C_1 = C_1$$

Since a given initial condition indicated $y'(\frac{\pi}{2}) = 1$, then $C_1 = 1$. Using this value in (13) we get

$$y'(x) = -\cos x + 1 \tag{14}$$

Integrating (14) provides

$$y(x) = -\sin x + x + C_2 \tag{15}$$

Setting $x = \frac{\pi}{2}$ again gives

$$y(\tfrac{\pi}{2}) = -\sin(\tfrac{\pi}{2}) + \tfrac{\pi}{2} + C_2 = -1 + \tfrac{\pi}{2} + C_2 \tag{16}$$

The other initial condition indicates that

$$y(\tfrac{\pi}{2}) = 2 \tag{17}$$

Equating (16) and (17) gives

$$-1 + \frac{\pi}{2} + C_2 = 2$$

or

$$C_2 = 3 - \frac{\pi}{2}$$

Thus (15) becomes

$$y(x) = -\sin x + x + 3 - \frac{\pi}{2}$$

which is the unique solution function solving the initial value problem (12).

A geometrical interpretation of these equations will provide further insight. Let's look again at (7). We are already familiar with several aspects of this equation: its general second derivative has the value $6x^2 + 2x$; its graph passes through the point $x = 0$, $y = 2$ and has a slope value of 3 at $x = 0$; and the unique function which satisfies these criteria is $y(x) = x^4/2 + x^3/3 + 3x + 2$. This graph in Figure 1.1 shows the shape of the curve.

Now let's look back at (12). We were seeking a function that passed through $(\frac{\pi}{2}, 2)$ with slope 1 and whose second derivative value was equal to the value of $\sin x$ at every point. The graph of that function, $y(x) = -\sin x + x + 3 - \frac{\pi}{2}$, is shown in Figure 1.2.

Our next example illustrates a practical problem. All information needed to solve the problem will be given.

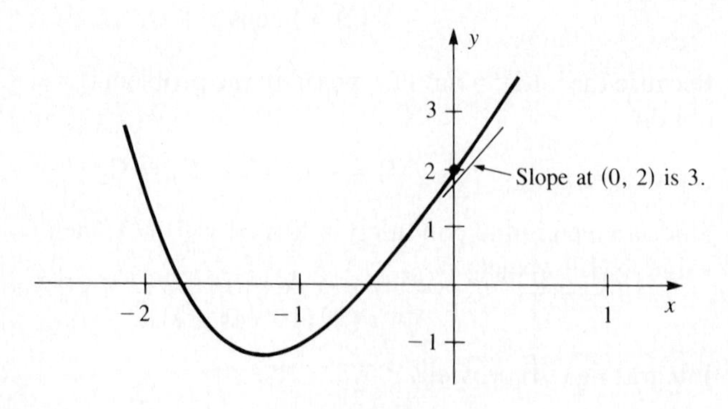

Figure 1.1

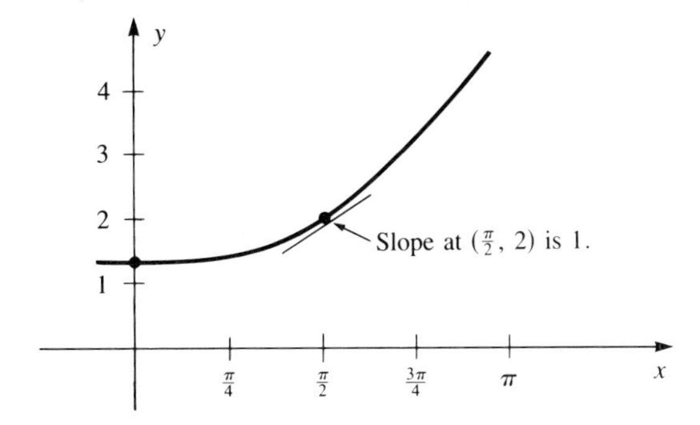

Figure 1.2

Example 10

In the construction of large buildings a long beam is often needed to span a given distance. In deciding what size beam is required, the architect needs to calculate the amount of bending the beam will undergo due to its own weight. If E (the modulus of elasticity of the beam material) and I (the moment of inertia of a cross section about its center axis) are given constants, then the differential equation used to find the sag curve for a beam supported at both ends is

$$EIy'' = Lwx - \frac{wx^2}{2}$$

where the beam has length $2L$ and w is the weight per unit length.

If y denotes the vertical sag distance per horizontal x unit, the natural initial conditions are $y = 0$ when $x = 0$; and $y' = 0$ when $x = L$ (see Figure 1.3).
Since E, I, w, and L are constants, we integrate the differential equation directly:

$$EIy' = Lw\left(\frac{x^2}{2}\right) - \frac{w}{2}\left(\frac{x^3}{3}\right) + C_1$$

Using $y' = 0$ when $x = L$ gives

$$0 = Lw\left(\frac{L^2}{2}\right) - \frac{w}{6}(L^3) + C_1$$

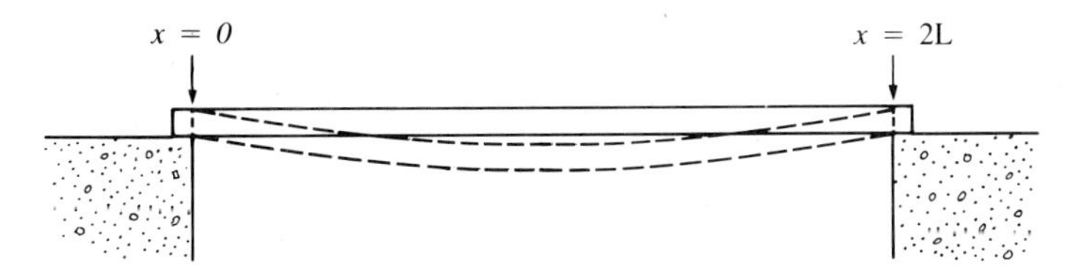

Figure 1.3

or $$C_1 = -\frac{1}{3} L^3 w$$

Then $$EIy' = \frac{Lw}{2} x^2 - \frac{w}{6} x^3 - \frac{1}{3} L^3 w$$

Integrating again gives

$$EIy = \frac{Lw}{2}\left(\frac{x^3}{3}\right) - \frac{w}{6}\left(\frac{x^4}{4}\right) - \frac{1}{3} L^3 wx + C_2$$

Using $y = 0$ when $x = 0$ gives $0 = C_2$.

$$\therefore \quad y = \frac{1}{EI}\left[\frac{Lw}{6} x^3 - \frac{w}{24} x^4 - \frac{L^3 w}{3} x\right]$$

This calculates the sag amount for each x in $0 \le x \le 2L$. An inspection of Figure 1.3 indicates that y should equal zero also when $x = 2L$. If you followed the preceding calculations closely, you will now be able to determine whether y does indeed equal zero when $x = 2L$.

Now consider a more complicated type of differential equation, in which the unknown function appears, either as itself or as a derivative, more than once. For example, consider $y' + ay = 0$. In an attempt to obtain a solution set by the straightforward integration technique, we might discover (quite by accident now but surely by design later) that if we multiply both sides by e^{ax}, obtaining $e^{ax}y' + ae^{ax}y = 0$, that this may be rewritten as $(e^{ax}y)' = 0$. This may immediately be integrated to $e^{ax}y = C$, and by the simple algebraic step of multiplying both sides by e^{-ax} the solution is $y = Ce^{-ax}$. This is a one-parameter family of functions with the property that each member solves the given equation.

Example 11

Consider $y' - 5y = 0$. Multiplication by e^{-5x} produces

$$e^{-5x}y' - 5e^{-5x}y = 0$$

The left side is the same as $d/dx(e^{-5x} \cdot y(x))$, so we have $(e^{-5x}y)' = 0$. Integration now gives $e^{-5x}y = C$. After multiplication by e^{5x} we have $y = Ce^{5x}$ as the general solution.

Example 12

When an object is dropped from rest in a medium that offers no resistance, the equation of motion is easily derived from Newton's law of motion, $F = ma$, where F is the external force, m is the mass of the object, and a is the acceleration in the direction of F.

If the mass is constant, if the object falls in a vacuum, and if the object is near the earth's surface so that the only force on the object is its weight w

(i.e., the gravitational force), then $F = w = -mg$, where g is the acceleration due to gravity (approximately 9.8 meters/second^2). Therefore, the motion equation is

$$-mg = F = ma$$

or

$$a = -g$$

If y denotes the vertical distance positively directed upward and t denotes time, then the equation becomes

$$\frac{d^2y}{dt^2} = -g$$

Natural initial conditions are

$$y(0) = y_0 \quad \text{(the initial height above the origin)}$$

$$\frac{dy}{dt}(0) = 0 \quad \text{(released from rest)}$$

To solve this second-order initial value problem we change to the differential notation

$$d(y'(t)) = -g\,dt$$

and integrate to

$$y'(t) = -gt + c$$

But $y'(0) = -g \cdot 0 + c_1 = c_1 = 0$ from the initial condition. Thus $y'(t) = -gt$. Changing this to

$$d(y(t)) = -gt\,dt$$

and integrating, we obtain

$$y(t) = -g\frac{t^2}{2} + c_2$$

From the initial condition $y(0) = y_0$ we get

$$y(0) = -g \cdot 0 + c_2 = y_0$$

Thus the motion equation is

$$y(t) = -\frac{1}{2}gt^2 + y_0$$

We can illustrate Example 12 with an object that falls from rest in a vacuum from a height of 4.9 meters (see Figure 1.4). The motion equation would be

$$y(t) = -\frac{1}{2}(9.8 \text{ m/sec}^2)(t^2) + 4.9$$

and the position occupied by the ball at any time t could be calculated from this equation. Note that after one second the ball passes through the origin and continues downward as the first term of the equation thereafter $(t > 1)$

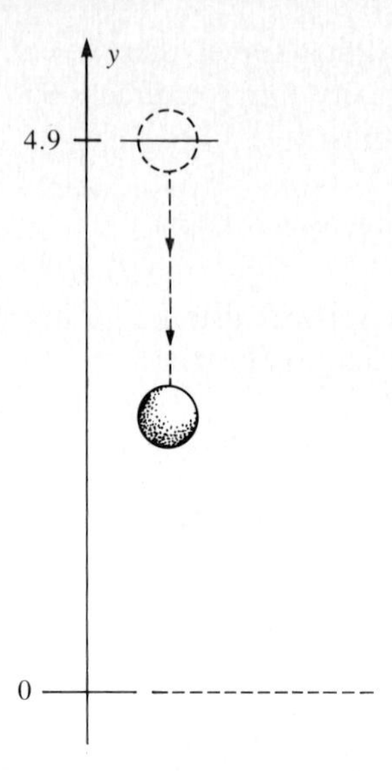

Figure 1.4

dominates the constant term. The earlier equation $y'(t) = -gt$ indicates the velocity at any time $t \geq 0$. At $t = 0$ we get, verifying the initial condition, $y'(0) = 0$. For $t > 0$ the velocity increases linearly with time. If a solid barrier were placed at $y = 0$ and if the object were elastic, then the object would change direction at $y = 0$. We would then have a problem of following the bouncing ball.

Exercises 1.2

Find the general solution for Exercises 1–16. Where initial conditions are given, find the unique solution.

1. $y' = 4x^3$
2. $y' = x^2 - 4x$
3. $y' = e^{2x}$; $y(0) = 2$
4. $y'' = x + 2$
5. $y'' = \cos x$; $y(0) = 1$, $y'(0) = 2$
6. $y' - 2x = 4$; $y(1) = 4$
7. $(y' + x)^2 = 4$
8. $y' + 4y = 0$
9. $y''' = 3x^2$; $y(0) = -1$, $y'(0) = 2$, $y''(0) = 4$
10. $y' = xe^{-x}$; $y(0) = 0$
11. $y' = \frac{1}{2}\sin 3t$; $y(\pi) = -1$

12. $y'' = x^{-2}$; $y(1) = 0$, $y'(1) = 1$

13. $y' = \dfrac{1}{x(x^2 - 4)}$; $y(1) = 0$

14. $y' = \dfrac{2x + xy}{2 + y}$

15. $\dfrac{dx}{dt} = \dfrac{e^t + x^2 e^t}{4xe^t}$

16. $(4 + x^2)y' = 9x$
17. Find the general function whose third derivative vanishes.
18. Find a function that passes through $(0, 2)$ for which the slope is equal to four times the y-coordinate at all times.
19. At each point (x, y) of a certain graph, the slope is $\sec x \tan x$. The graph also contains the point $(\frac{\pi}{4}, \sqrt{2})$. Find the equation of this graph.
20. For a simple electric circuit with a given value of inductance L, resistance R, and applied electromotive force E, the differential equation for the current $i(t)$ is

$$L\frac{di}{dt} + Ri = E \sin kt \quad (\text{constant } k)$$

By inserting the function

$$i(t) = \frac{R(E \sin kt) - kL(E \cos kt)}{R^2 + k^2L^2} + Ce^{-(R/L)t}$$

and its derivative di/dt into the given equation, show that this function gives the general solution family for the variable current. Note the sinusoidal part and the decaying exponential part.
21. As a farmer's wheat crop matures, it is estimated that after t days the crop is increasing at a rate of $\frac{9}{10}t^2 + \frac{3}{5}t + \frac{17}{10}$ bushels per day. If the market price remains stable at \$3.50 per bushel, how much will the value of the crop increase over the next five days?
22. During an Indy race, a driver sees danger ahead and applies the brakes. The car decelerates at a constant rate of 42 ft/sec^2. At the moment of breaking the car has speed v_0. Express the distance the car travels as a function of time and its speed v_0. Also compute the time that will elapse until the car stops. Use

$$\frac{d^2s}{dt^2} = -42$$

Also find the stopping distance in feet if $v_0 = 90$ miles/hour $=$ 132 feet/second.
23. In California's San Fernando Valley an air pollution reading at 6:00 A.M. shows the pollution index to be .30 parts per million. If t

expresses time in hours, the equation

$$\frac{dP}{dt} = \frac{0.21 - 0.03t}{\sqrt{27 + 14t - t^2}}$$

expresses the rate at which the pollution level will change over the next 14-hour period. Find the function $P(t)$ for the pollution level at any time t. When (in hours after 6:00 A.M.) is the pollution reading at its maximum?

1.3 Separable Equations and Producing Separability by Substitution

Separable Equations

The first-order equations we will study in this chapter will usually be expressed in the derivative form $dy/dx = f(x, y)$. It will sometimes be helpful in the solution process to express the differential equation in the equivalent differential form $M(x, y)\,dx + N(x, y)\,dy = 0$. An equation in one form may be easily changed to the other form by simple algebra, as illustrated in Example 13.

Example 13

(a) If we substitute dy/dx for y' in the equation $y' = 3x^2y^3$, and use simple algebra, we get $3x^2\,dx - y^{-3}\,dy = 0$.

(b) The equation $(2y - x)\,dx - x^2\,dy = 0$ is easily changed to $x^2\,dy = (2y - x)\,dx$. Then we can simply divide both sides by $x^2\,dx$ to obtain

$$\frac{dy}{dx} = \frac{2y - x}{x^2}$$

(c) The equation $y' = 5xy^2 + 2y$ is equivalent to

$$\frac{dy}{dx} = 5xy^2 + 2y \Leftrightarrow dy = (5xy^2 + 2y)\,dx \Leftrightarrow (5xy^2 + 2y)\,dx - dy = 0$$

In each case, regardless of the form, we will consider x as the independent variable and y as the dependent variable and will be seeking the collection of functions that constitute the solution family in which y is a function of x.

In the general first-order equation $dy/dx = f(x, y)$, if the function f can be written in such a way that it does not involved the dependent variable y, then the equation may be solved by integration as described in Section 1.2. Even when $f(x, y)$ is more complicated, if it can be factored into distinct parts that involve only one variable per part, it can be written in a special form. The simplest form would be

$$\frac{dy}{dx} = f(x, y) = A(x)B(y)$$

or

$$\frac{dy}{dx} = f(x, y) = \frac{D(x)}{E(y)}$$

A more general form would be

$$\frac{dy}{dx} = \frac{F(x)G(y)}{H(x)J(y)}$$

An equation of this form is called a **separable differential equation**. Its equivalent form is

$$F(x)G(y)\,dx = H(x)J(y)\,dy$$

If we now divide both sides by $H(x)G(y)$ we obtain

$$\frac{F(x)}{H(x)}\,dx = \frac{J(y)}{G(y)}\,dy$$

We have thus physically separated the x variable from the y variable. We may now integrate both sides to the solution family:

$$\int \frac{F(x)}{H(x)}\,dx - \int \frac{J(y)}{G(y)}\,dy = C$$

Example 14

Consider $dy/dx = 2xy^2$. The right side is already factored, so we cross multiply to obtain $2x\,dx = y^{-2}\,dy$. Integrating both sides produces $x^2 + C = -1/y$. Then $y = -1/(x^2 + C)$ is the one-parameter family of functions providing the general solution.

If we add an initial condition, such as $y(1) = -\frac{1}{2}$, and insert this into the general solution, we find $1^2 + C = -1/-\frac{1}{2} = 2$ or $C = 1$. Then the unique solution is $y = -1/(x^2 + 1)$. This solution is easily verified by differentiating the result and noting that the initial condition is satisfied.

Example 15

Consider the first order equation

$$\frac{dy}{dx} = \frac{3x + xy^2}{y + x^2y}$$

If we factor, we obtain

$$\frac{dy}{dx} = \frac{x(3 + y^2)}{y(1 + x^2)}$$

Cross multiplying produces $x(3 + y^2)\,dx = y(1 + x^2)\,dy$ and dividing appropriately gives $x/(1 + x^2)\,dx = y/(3 + y^2)\,dy$. If we integrate now, we get $\frac{1}{2}\ln(1 + x^2) = \frac{1}{2}\ln(3 + y^2) + C$. By algebra and rules of logarithms we obtain

$$\frac{1}{2}\ln\left(\frac{1 + x^2}{3 + y^2}\right) = C$$

or

$$\frac{1 + x^2}{3 + y^2} = e^{2C}$$

Additional algebra would allow us to solve for y explicitly. Since C takes all real values $(-\infty, \infty)$, the term e^{2C} takes the values $(0, \infty)$ so that e^{2C} could be replaced by a new constant $K > 0$.

Example 16

Consider the first-order linear equation

$$x\frac{dy}{dx} = y - xy$$

By factoring the right side and cross multiplying we obtain $x\,dy = y(1 - x)\,dx$. To separate the variable x from the variable y we divide both sides by xy to obtain

$$\frac{dy}{y} = \frac{1 - x}{x}\,dx = (x^{-1} - 1)\,dx$$

This is now easily integrated to $\ln|y| = \ln|x| - x + C$. Rewrite this as $\ln|y| - \ln|x| = C - x$ or $\ln|y/x| = C - x$. We can now "exponentiate" to obtain $|y/x| = e^{C-x} = e^C \cdot e^{-x}$ or $y/x = \pm e^C \cdot e^{-x}$. Since C takes on all possible values, we may replace $\pm e^C$ by C. We then have the explicit solution $y = Cxe^{-x}$. This may be verified as the correct result by substitution.

Example 17

The equation $y' = e^{x+y}$ is easily separated if we write it as $dy/dx = e^x \cdot e^y$ and then as $e^{-y}\,dy = e^x\,dx$. Again an easy integration gives $-e^{-y} = e^x + C$. By taking logarithms of both sides we obtain $\ln(e^{-y}) = \ln(-e^x - C)$ or $-y = \ln|C - e^x|$ and thus the explicit solution $y = -\ln|C - e^x|$.

Example 18

Consider the initial value problem

$$\begin{Bmatrix} \dfrac{dy}{dx} + x = \dfrac{x}{y} \\ y(2) = 0 \end{Bmatrix}$$

Rewriting the differential equation, we find

$$\frac{dy}{dx} = x\left(\frac{1}{y} - 1\right) = \frac{x}{y}(1 - y)$$

or

$$\frac{y}{1 - y}\, dy = x\, dx$$

Now that the equation is separated, we can integrate it:

$$\int \frac{y}{1 - y}\, dy = \int \left(-1 - \frac{1}{y - 1}\right) dy = \int x\, dx$$

obtaining

$$-\ln|y - 1| - y = \frac{x^2}{2} + C$$

Applying the initial condition, we find $-\ln|-1| - 0 = 2 + C$ or $C = -2$. Thus, the unique solution is $-\ln|y - 1| - y = (x^2 - 4)/2$.

The unique solution in Example 18 is expressed in **implicit form**, which is used when solving explicity for y as a function of x is difficult or impossible.

There is an alternate way to solve an initial value problem of the form

$$\left\{\begin{aligned} &\frac{dy}{dx} = f(x, y) \\ &y(x_0) = y_0 \end{aligned}\right\}$$

when the differential equation is separable. If the equation is written as

$$A(x)\, dx + B(y)\, dy = 0$$

and if the initial condition prescribes that the solution curve begins at the initial point (x_0, y_0), we integrate the differential equation between the initial point (x_0, y_0) and the general point (x, y) to obtain the solution

$$\int_{x_0}^{x} A(x)\, dx + \int_{y_0}^{y} B(y)\, dy = 0$$

This solution, however, may not be unique.

Example 19

Consider

$$\left\{\begin{aligned} &e^x\, dx - y^2\, dy = 0 \\ &y(0) = 1 \end{aligned}\right\}$$

We get

$$\int_0^x e^x\, dx - \int_1^y y^2\, dy = 0$$

or

$$e^x\Big|_0^x - \frac{y^3}{3}\Big|_1^y = 0$$

or
$$e^x - 1 - \frac{y^3}{3} + \frac{1}{3} = 0$$

This becomes
$$e^x - \frac{1}{3}y^3 = \frac{2}{3}$$

or
$$y^3 = 3e^x - 2$$

and the unique solution is
$$y = \sqrt[3]{3e^x - 2}, \ x > \ln\frac{2}{3}$$

For additional practice, use simple algebraic manipulation to convince yourself that the following are separable equations:

1. $(1 + y^2)\,dx + (1 + x^2)\,dy = 0$
2. $\dfrac{dy}{dx} + e^y \cos x = 0$
3. $\dfrac{dx}{dt} = \dfrac{\cos x \cot t}{\sin t}$
4. $\dfrac{dy}{dt} = \dfrac{ty + y}{y^2 + 1}$
5. $\dfrac{dx}{dt} = xe^{-t+x}$

The next few examples illustrate the way differential equations can be used to solve actual problems.

Example 20

A rocket with constant mass m will be projected straight up from the surface of the earth with an initial velocity v_0. You have been asked to calculate the rocket's escape velocity. Since the earth's gravitational field varies with altitude, the general expression for the weight w of such a body is derived from the inverse square law of gravitational attraction. If y represents the altitude above sea level, and R the earth's radius, then

$$w(y) = \frac{c}{(y + R)^2} \quad \text{where } c \text{ is constant.}$$

Note that at sea level ($y = 0$), $w = mg$. Thus

$$mg = \frac{c}{(0 + R)^2}$$

or

$$c = mgR^2$$

Thus

$$w(y) = \frac{mgR^2}{(y + R)^2}$$

If we assume negligible air resistance (i.e., a carefully designed rocket!) the motion equation $F = ma$ becomes

$$\frac{-mgR^2}{(y + R)^2} = ma = m\frac{dv}{dt}$$

with the initial condition that $v(0) = v_0$.

The chain rule of differentiation from calculus allows us to write

$$\frac{dv}{dt} = \frac{dv}{dy} \cdot \frac{dy}{dt} = \frac{dv}{dy} \cdot v$$

We then have

$$mv\frac{dv}{dy} = \frac{-mgR^2}{(y + R)^2}$$

which can be separated as

$$mv\,dv = \frac{-mgR^2}{(y + R)^2}\,dy$$

Integration gives

$$\frac{v^2}{2} = \frac{gR^2}{y + R} + K$$

When $t = 0$, then $y = 0$; also the initial condition states that $v = v_0$ when $y = 0$, so $K = \frac{1}{2}v_0^2 - gR$. Thus the solution is

$$v^2 = v_0^2 - 2gR + \frac{2gR^2}{y + R}$$

In order for the rocket to escape the gravitational pull of the earth, the velocity v must remain positive for all y. This occurs if $v_0^2 \geq 2gR$, since the third term is always positive. Hence the equation for the escape velocity is

$$v_e = (2gR)^{1/2}$$

This value is roughly 7 miles per second.

Example 21

A workman was supposed to pour 50 pounds of sugar into a tank that contained 200 gallons of water. By mistake he poured 100 pounds. To correct the error, he thoroughly mixed the solution and then opened a valve at the bottom of the tank. Since the valve allowed 4 gallons of the mixture to escape per minute, the workman simultaneously added 4 gallons of pure water per minute and mixed the solution continuously. If we assume that the tank contained a homogeneous solution at all times, how many minutes were required for the tank to contain the desired amount of sugar?

Solution: Let the variable x represent the number of pounds of sugar in the solution at any time t. As sugar is removed from the mixture, the change in x per unit time is the fractional part of x of the sugar-in-solution flowing out each minute. Thus the rate of change is 4 gallons out of the 200 gallons flowing out of the tank per minute. So

$$\frac{dx}{dt} = -\frac{4}{200}x$$

with the initial condition that the sugar amount $x = 100$ at time $t = 0$. We want to find the t value for which the sugar amount $x = 50$.

Separating the variables produces

$$\frac{dx}{x} = -\frac{1}{50}dt$$

Integration provides

$$\ln|x| = -\frac{1}{50}t + C_1$$

or

$$x = Ce^{-(1/50)t}$$

Inserting the initial condition $x = 100$ when $t = 0$ we obtain $C = 100$. Therefore

$$x = 100e^{-.02t}$$

This equation indicates the exact amount of sugar at any time $t \geq 0$. Putting in $x = 50$ and solving for t gives

$$e^{-.02t} = \frac{1}{2}$$

Taking the logarithm of both sides gives

$$.02t = \log_e 2$$

and thus the time needed is $t = 50 \log_e 2 = 34.65$ minutes.

For additional practice, determine whether or not the following statement is true: If the drawn-off syrup is saved, it will also contain the desired sugar-water concentration.

Example 22

One of the most important distributions in statistics is the normal distribution, which many people have used (e.g., in grading on the curve)

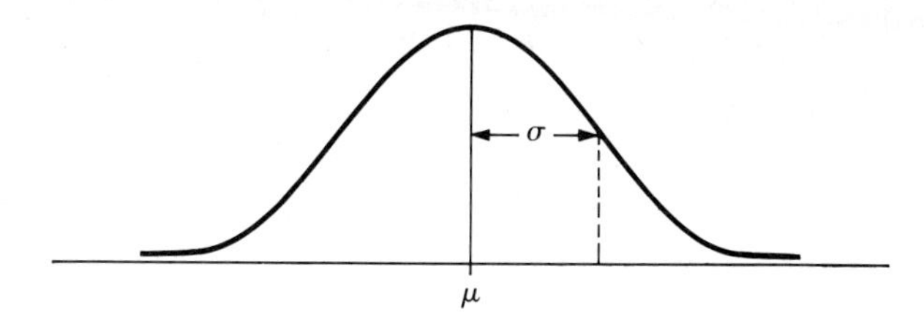

Figure 1.5

without realizing all its properties. The function $N(x)$ which describes this distribution is the solution to the separable differential equation

$$N' = \left(\frac{a-x}{b}\right)N$$

where the constants a and b are, respectively, the mean μ and the variance σ^2 of the distribution. Separating the variables produces

$$\frac{dN}{N} = \left(\frac{\mu - x}{\sigma^2}\right)dx$$

which is integrated to

$$\ln|N| = -\frac{\sigma^2}{2}\left(\frac{\mu - x}{\sigma^2}\right)^2 + C$$

or

$$N(x) = Ke^{-1/2(\mu - x/\sigma)^2}$$

With the constant K given the normalizing value of $1/\sigma\sqrt{2\pi}$, the curve appears as in Figure 1.5.

Example 23

Whenever a viscous fluid flows through a tube, fluid pressure along the tube drops to overcome the frictional resistance at the walls. As a result of this pressure gradient and the elasticity of the tube wall, the radius of the tube R will decrease with distance along the axis. Using cylindrical coordinates r, θ, z, in which r is radial and z is the distance along the axis of the tube, researchers have considered the flow across any tube section of radius $R(z)$. If $R(z)$ is constant, the tube is rigid. For a nonrigid tube, such as a blood vessel, some complicated expressions used in the study of flowing fluids (such as the equation of elasticity, the equation of continuity, and the Navier–Stokes equation for axially symmetric, steady, laminar flow of an incompressible fluid) have been combined to give us a solvable differential equation

$$\frac{dR}{dz} = \frac{-8\mu Q}{E\,\delta\pi\rho\,R^2}$$

where five of the quantities are constant:

E is Young's modulus;

Q is total mass flow across a tube section;

ρ is density;

μ is coefficient of viscosity;

δ is wall thickness.

This equation is easily separated to produce

$$R^2\,dR = \frac{-8\mu Q}{E\,\delta\pi\rho}\,dz \Rightarrow \frac{R^3}{3} = \frac{-8\mu Q}{E\,\delta\pi\rho}\,z + C$$

or

$$R(z) = \sqrt[3]{\frac{-24\mu Qz}{E\,\delta\pi\rho} + 3C}$$

Note that the radius decreases in proportion to the cube root of the distance z along the tube.

Example 24

Although male walruses live in a colony with many others, each dominates a certain subterritory of the land occupied by the whole colony. Hence, a male member of the colony has under his control a packaged resource. Every new incoming male member (even one born in the colony) must either find an unused package or contend with a present member for his package.

Competition of this type restricts the growth of the colony's population, and a reduction in growth rate results. The rate of change of the population is positive for natural propagation but negative for the competition. Hence, if W denotes the male walrus population, the governing differential equation is

$$\frac{dW}{dt} = rW - cW(W-1)$$

where r is the natural unconstrained growth rate and c is a positive constant called the competition parameter.

With this equation goes the natural initial condition that at some starting time t_0 the population contains W_0 members.

$$W(t_0) = W_0$$

With some algebra, the differential equation may be rewritten as

$$\frac{dW}{dt} = kW - cW^2 \quad \text{where } k = r + c$$

Now let $R = k/c$ and obtain

$$\frac{dW}{dt} = c(R - W)W$$

The equation is now separable and becomes $dW/(R - W)W = c\,dt$. Integrating by partial fractions decomposition, with the assumption that $R - W(t) > 0$ for $t > t_0$, we obtain

$$\frac{1}{R}[\ln W - \ln(R - W)] = ct + k_1$$

or

$$\frac{W}{R - W} = k_2 e^{cRt}$$

Solving this equation for $W(t)$ we obtain

$$W(t) = \frac{Rk_2 e^{cRt}}{1 + k_2 e^{cRt}}$$

Using the initial condition that $W(t_0) = W_0$ and some algebraic reorganization, we obtain

$$W(t) = \frac{RW_0}{W_0 + (R - W_0)e^{-cR(t - t_0)}}$$

for $t \geq t_0$. This formula gives us the size of the male walrus population W at any time t. We can easily verify that $W(t_0) = W_0$. We also note that the eventual size after a long time is obtained by allowing t to approach infinity. The exponential term will disappear and $W(t) \to RW_0/W_0 = R$. But $R = (r + c)/c$, so the eventual population size is a function of the growth and competition parameters and independent of the original population size W_0.

Producing Separability by Substitution

Some differential equations that are not separable as they stand can become separable with a change of variable. The change-of-variables process is often used in Calculus (e.g., indefinite integration by trigonometric substitution) and is equally useful in solving differential equations. Figure 1.6 illustrates how a change of variables can often ease the difficulties of solving particular problems.

A test is available to help determine whether a change of variables on a given differential equation will put it into separable form. This test involves a concept that applies to a function $g(x, y)$, and this concept is defined as follows.

Definition: The function $g(x, y)$ is called **homogeneous to degree** p if

$$g(tx, ty) = t^p g(x, y)$$

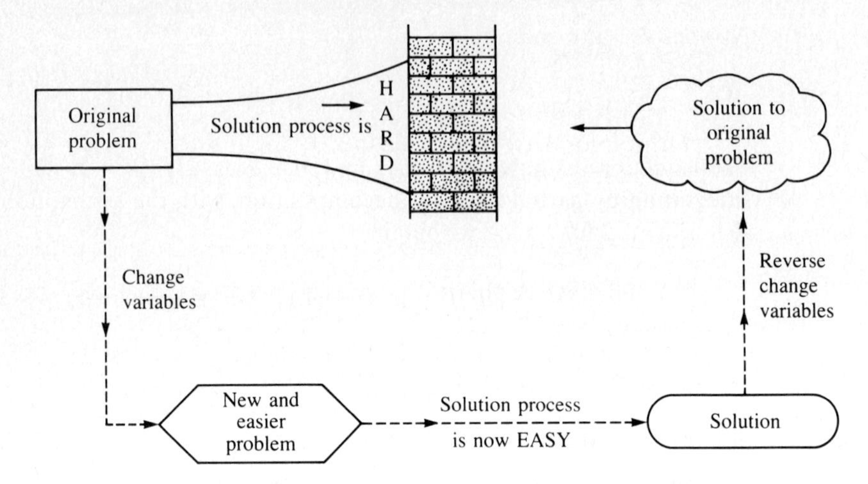

Figure 1.6

In other words, if tx and ty are substituted for x and y, respectively, in $g(x, y)$ and if t^p is then factored out, the remaining expression is the original expression $g(x, y)$. This concept is illustrated in Examples 25–27.

Example 25

The function $g(x, y) = x^2 + y^2$ is homogeneous to degree 2 since $g(tx, ty) = (tx)^2 + (ty)^2 = t^2x^2 + t^2y^2 = t^2(x^2 + y^2) = t^2g(x, y)$.

Example 26

The function $g(x, y) = 4x^4y^2 - 3xy^5$ is homogeneous to degree 6 since

$$\begin{aligned} g(tx, ty) &= 4(tx)^4(ty)^2 - 3(tx)(ty)^5 \\ &= 4(t^4x^4)(t^2y^2) - 3(tx)(t^5y^5) \\ &= 4t^6x^4y^2 - 3t^6xy^5 \\ &= t^6[4x^4y^2 - 3xy^5] = t^6g(x, y) \end{aligned}$$

Example 27

The equation

$$g(x, y) = e^{y/x}$$

is homogeneous to degree 0, whereas

$$g(x, y) = \frac{1}{\sqrt{x^2 + y^2}}$$

is homogeneous to degree -1.

Note: It is important to observe that this definition of "homogeneous" refers to a *function*, not to a *differential equation* as in Section 1.1. The word is the same, but the meaning is different.

The test for determining whether or not an equation is separable can be stated as follows:

> From the differential form $M(x, y)\,dx + N(x, y)\,dy = 0$ consider the functions $M(x, y)$ and $N(x, y)$. If both $M(x, y)$ and $N(x, y)$ are functions homogeneous to the same degree, then the change of variables $v = y/x$ will make the differential equation separable.

Remember that this test-and-procedure instruction provides only a start toward solving the problem. Nevertheless, when an equation passes this test, the change of variables *will* make it separable and thus it is theoretically solvable. However, that solvability depends on the integrability of the separated sides. More will be said on this subject at the end of this section.

Example 28

In $x^2y' = y^2 + 2xy$, algebraic rearranging produces $x^2\,dy = (y^2 + 2xy)\,dx$ or $(y^2 + 2xy)\,dx - x^2\,dy = 0$ and thus $M(x, y) = y^2 + 2xy$ and $N(x, y) = -x^2$. Now

$$M(tx, ty) = t^2y^2 + 2(tx)(ty) = t^2(y^2 + 2xy) = t^2M(x, y)$$

and $$N(tx, ty) = -(tx)^2 = -t^2x^2 = t^2N(x, y)$$

so they both are homogeneous to degree 2 and we should use $v = y/x$ as a change of variables.

Following this process through this example, we will write the substitution in an equivalent expression $y = vx$. Note that we are retaining x as the independent variable and replacing y by v as the dependent variable. Hence, we will also need the differential of y, denoted by dy, for our substitution. In calculus we learned the product rule for differentiation, and the same rule applies for differentials (i.e., if $y = vx$ then $d(y) = d(vx) = v\,dx + x\,dv$). Thus, the original differential equation $(y^2 + 2xy)\,dx - x^2\,dy = 0$ with these substitutions becomes

$$[v^2x^2 + 2x(vx)]\,dx - x^2[v\,dx + x\,dv] = 0$$

and simplifies to

$$x^2(v^2 + v)\,dx - x^3\,dv = 0$$

which is separable to

$$\frac{dx}{x} = \frac{dv}{v^2 + v}$$

This integrates to

$$\ln|x| + \ln C = \int \frac{dv}{v^2 + v} = \int \frac{dv}{v} - \int \frac{dv}{v + 1} = \ln|v| - \ln|v + 1|$$

or
$$\ln C|x| = \ln\left|\frac{v}{v + 1}\right|$$

We must now solve for the dependent variable v. "Exponentiating" both sides gives

$$Cx = \frac{v}{v + 1}$$

Cross multiply to get

$$(v + 1)Cx = v$$

$$vCx + Cx = v$$

$$Cx = v - vCx = v(1 - Cx)$$

or
$$v = \frac{Cx}{1 - Cx}$$

But $v = y/x$, so reverse substitution provides

$$\frac{y}{x} = \frac{Cx}{1 - Cx}$$

and finally the solution is

$$y = \frac{Cx^2}{1 - Cx}$$

Example 29

In a study of population growth, we encounter the initial value problem

$$\begin{Bmatrix} 2xyy' - 3y^2 + x^2 = 0 \\ y(1) = 2 \end{Bmatrix}$$

This may be expressed in the equivalent form

$$\begin{Bmatrix} (x^2 - 3y^2)\,dx + 2xy\,dy = 0 \\ y(1) = 2 \end{Bmatrix}$$

Observing first that the differential equation is not separable as it stands (why?), we test the $M(x, y) = x^2 - 3y^2$ and $N(x, y) = 2xy$ for homogeneity.

$$M(tx, ty) = (tx)^2 - 3(ty)^2 = t^2(x^2 - 3y^2)$$

and
$$N(tx, ty) = 2(tx)(ty) = t^2(2xy)$$

They are both homogeneous to degree 2, so we use the change of variable $v = y/x$.

An interesting case has now appeared. If we use the original derivative form and divide by $2xy$, we get y' as a function of y/x (i.e., a function of v):

$$y' = \frac{-x}{2y} + \frac{3y}{2x}$$

When this happens we can save a little work as we substitute. Since $y = vx$, we get $y' = v + xv'$. The differential equation becomes

$$v + xv' = \frac{-1}{2v} + \frac{3v}{2}$$

which can be simplified to

$$xv' = -\frac{1}{2v} + \frac{v}{2} = \frac{v^2 - 1}{2v}$$

This equation is separable to

$$\frac{2v}{v^2 - 1}\,dv = \frac{dx}{x}$$

and we integrate to get

$$\ln|v^2 - 1| = \ln|x| + \ln C$$

Note that $\ln C$ is equally as valid an integration constant as C since both assume all real values. Then $|v^2 - 1| = C|x|$; and the absolute-value signs may be removed, because C takes on all possible values both positive and negative. Thus $v^2 = Cx + 1$. Since $v = y/x$, we get $y^2/x^2 = Cx + 1$ or $y^2 = Cx^3 + x^2$. The initial condition that $y = 2$ when $x = 1$ means that $4 = C + 1$ or $C = 3$. Therefore, the unique solution to the given initial value problem is then

$$y^2 = 3x^3 + x^2$$

The graph of this solution is presented in Figure 1.7.

Note: In some situations the substitution $v = x/y$ will produce an easier algebraic process. We can recognize this type of situation by noticing that the multiplier of dx is considerably simpler than that of dy. However, the theory and procedure remain the same, as indicated in the following diagram.

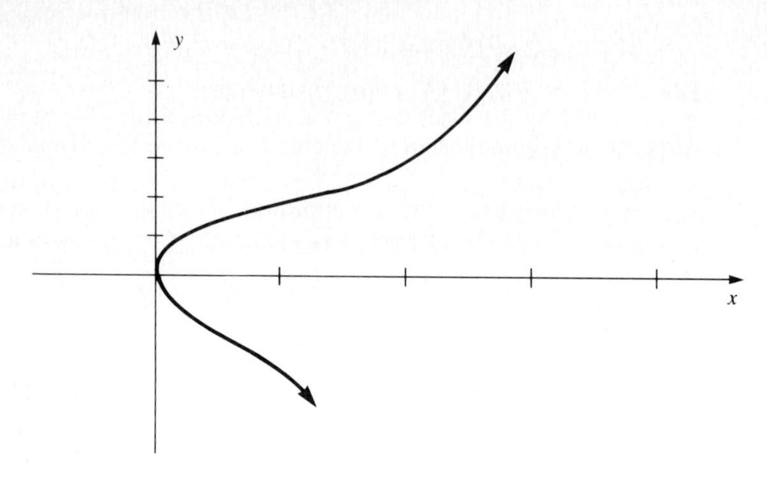

Figure 1.7

When considering the differential equation $M(x, y)\,dx + N(x, y)\,dy = 0$, if M and N are homogeneous to the same degree, we use:

$v = \dfrac{y}{x} \Leftrightarrow y = vx$ .. or .. $v = \dfrac{x}{y} \Leftrightarrow x = vy \Leftrightarrow y = \dfrac{x}{v}$

$dy = v\,dx + x\,dv$ $\qquad \dfrac{dy}{dx} = v + x\dfrac{dv}{dx}$ **OR** $dx = v\,dy + y\,dv$

when $N(x, y)$ is simpler than $M(x, y)$ $\qquad$ when $M(x, y)$ is simpler than $N(x, y)$

As mentioned earlier in this section, the solvability of a separable differential equation depends on the integrability of the separated sides. Surely not all functional combinations are integrable. In a simple-looking equation such as $xy' - \sin x = 0$, a separation gives

$$dy = \frac{\sin x}{x}\,dx$$

However, the right side is not integrable by elementary calculus techniques. To be preposterous but to make a point, we will suppose that a separation produced an equation such as $dy/(y^4 + y - 3) = x^5 e^{3x} \csc(4x^3 - 7)\,dx$. No one would even consider an integration attempt.

But let's not forget about other methods! This is a point at which one might turn to the second solution alternative suggested earlier. We would use a computer to do a numerical approximate integration and obtain a valid result. As we will see in Chapter 2, either of the preceding problems is easily solved by the machine, and the accuracy with which the computer works is astounding. If we could imagine all the conceivable first-order differential equations that might be generated by applications situations in science and engineering, we would surely realize that knowledge of computer-solving techniques is required. Chapter 2 addresses that topic.

Exercises 1.3

Find the general solution for Exercises 1–19. Where initial conditions are given, find the unique solution.

1. $\dfrac{dy}{dx} = \dfrac{3y}{x}$
2. $-2\sin x \cos y\, dy + \cos x \sin y\, dx = 0$
3. $y(x^2\, dy - y^2\, dx) = x^2\, dy$
4. $x\dfrac{dy}{dx} = y - xy;\ y(1) = 2$
5. $\dfrac{dx}{dt} = k(18 - 9x + x^2);\ x(0) = 0,\ x(10) = 2$
6. $x\dfrac{dy}{dx} = xe^{-y/x} + y$
7. $y' - y\tan x = 0$
8. $y' = \dfrac{4y - 3x}{2x - y}$
9. $s' = \dfrac{3s^2}{st + t^2}$
10. $y' = \dfrac{x^2 + 3y^2}{2xy}$
11. $y' - \dfrac{y}{x} = 1$
12. $(3t + x)\, dx = (t - 3x)\, dt$
13. $y' = \dfrac{t - 2y}{2t + y}$
14. $(1 + x)\, dy + 3y\, dx = 0$
15. $x^2(1 - v^2)\, dx - 2vx^3\, dv = 0$

16. $x^2\,dy = \dfrac{1+y^2}{1+x^2}\,dx$

17. $2x\,dy + y\,dx = 0$; $y(3) = 1$
18. $4\,dy + y\,dx = x^2\,dy$; $y(4) = -1$
19. $3e^x \tan y\,dx + (1 + e^x)\sec^2 y\,dy = 0$
20. Find the function $y(x)$ whose slope is always the difference of the y and x coordinates divided by their sum.
21. A particle moves so that its slope is $-\frac{3}{2}x^2y^{-1}$ at all points for which $y \neq 0$. If it passes through (2, 3), find its path.
22. Find the particular solution to $y' = 4x/(3y^2 + 2)$ that passes through the point (0, 4).
23. If p, q, and r are constants with $p > q$, show that the general solution to $y' = r(y - p)(y - q)$ is

$$y = \frac{p + qce^{(p-q)rx}}{1 + ce^{(p-q)rx}}$$

24. A small boy and his sled have a combined mass of 2 slugs, (i.e., their total weight is 64 pounds). They are being pushed in a straight line against the wind by a friend exerting a 10-pound force. Assume that friction is negligible but that the magnitude (in pounds) of air resistance to the motion is equal to twice the velocity of the sled in feet per second. If the sled starts from rest, find the velocity and distance traveled at t seconds. What is the limiting velocity attainable? (Hint: Equate forces and use $F = ma$.)
25. (a) A chemical reaction is called a second-order reaction if the rate of change of x, the amount of produced substance C present at time t, is proportional to the product of the amounts of raw materials A and B remaining at any given time. The differential equation representing this relationship is

$$\frac{dx}{dt} = k(a - x)(b - x)$$

Suppose $a = 4$; $b = 3$; and $x = 1$ when $t = 3$. Find the value of t when $x = 2$.

(b) More generally, suppose that $a = b$ and that $x = a/n$ when $t = t_1$. Find x when $t = 2t_1$.

26. In problems involving intercontinental ballistic missile flights we use the differential equation $dp/dh = -\rho$, which connects the atmospheric pressure p(lb./ft.^{2}), the density of the atmosphere ρ(lb./ft.^{3}), and the height h (in feet) above the earth. If a relationship involving two or three of the variables p, ρ, and h is known, we may eliminate one of the variables and solve the problem. Assume that $p = 192\rho(144 - .001h)$ and the sea level pressure is 2,120 lb./ft.^{2}. Find the pressure at 10,000 feet.

27. Suppose that the derivative dx/dt is proportional to x, that $x = 4$ when $t = 0$, and that $x = 8$ when $t = 5$. What is the value of x when $t = 20$?
28. Given $(dy/dx)^2 = y$, find (a) the general solution and (b) two curves through the point $(1, 4)$ satisfying the differential equation.
29. A woman and her small boat together weigh 160 pounds. The boat's battery-powered electric motor produces a thrust equivalent to a constant force of 5 pounds in the forward direction, but the water resists the motion of the boat with a force equal to 1.5 times the speed. If the woman begins from the dock and flips the switch to full power, what is her speed at any time $t > 0$?
30. In calculating the velocity of water flowing through an opening in the bottom of a tank, we usually assume that the velocity is proportional to the product of the area of the opening and the square root of the depth h of the water; that is,

$$A\frac{dh}{dt} = -kB\sqrt{h}$$

where A is the area of the water surface and B is the area of the opening. Calculate the time t needed to empty a cubical tank whose edge is 5 feet if the tank is originally full of water and has a 2-inch-diameter hole in the bottom (let $k = 4.8$).
31. It is common practice for banks to advertise that interest on your savings is compounded continuously. They use the differential equation $dA/dt = rA$, where A is the amount deposited, r is the interest rate, and t is time in days. If you deposit \$100 at 6% interest, how long will it take to double your money?
32. A complicated but separable differential equation resulted from a study on accumulation processes in the primitive solar nebula:

$$\frac{dx}{dt} = \frac{Ax^{5/6}}{(B - kt)^{3/2}}$$

where A, B, k are constants. Solve for $x(t)$.
33. In the design of a magnetohydrodynamic power generator, an elementary differential equation that arises is

$$[(k + 1)\varphi']' = 0 \qquad k\text{, a positive constant}$$

Solve for $\varphi(x)$.
34. Compute functions $y(x)$ satisfying the following:

(a) $y(x) = k_0 + \int_0^x y(t)\,dt$

(b) $y(x) = k_1 + \int_0^x y^2(t)\,dt$

1.4 Integrable Combinations

An additional solution method is available to solve a differential equation in which the variables are not immediately separable, but which can be arranged into integrable combinations. The process is illustrated in Examples 30–32.

Example 30

Solve $(3x^2 - y)\,dx - x\,dy = 0$.

This equation is not separable, because the $M(x, y)$ that multiplies dx is not separable into a function of x times a function of y. Also, because of the nonhomogeneity of $M(x, y)$, it is not homogeneous to any degree. By transposing we obtain $x\,dy + y\,dx = 3x^2\,dx$. The left side is now in a special form. From our earlier work with differentials, we see that this expression is the differential of xy; that is, $d(x \cdot y) = x\,dy + y\,dx$. Now the equation can be rewritten as $d(x \cdot y) = 3x^2\,dx$. By integrating this equation we get $xy = x^3 + C$, and thus the general solution is $y = x^2 + Cx^{-1}$.

Some of the simpler frequently occurring integrable combinations include:

$$x\,dy + y\,dx = d(xy) \qquad \frac{y\,dx - x\,dy}{y^2} = d\left(\frac{x}{y}\right)$$

$$\frac{x\,dy - y\,dx}{x^2 + y^2} = d\left(\text{Arctan}\,\frac{y}{x}\right) \qquad \frac{x\,dy - y\,dx}{x^2} = d\left(\frac{y}{x}\right)$$

$$\frac{x\,dy - y\,dx}{xy} = d\left(\ln\left|\frac{y}{x}\right|\right)$$

The technique used in this section rearranges the differential equation into a form in which an expression from the equations illustrated above may be inserted to simplify it and prepare it for integration.

Example 31

Solve $(y - x^3)\,dx - x\,dy = 0$. This equation is not separable, but it may be written as $y\,dx - x\,dy = x^3\,dx$. Multiply both sides by $-1/x^2$ to obtain

$$\frac{x\,dy - y\,dx}{x^2} = -x\,dx$$

(Do you see why multiplying by $1/y^2$ would have been a mistake?) Referring to the above listed combinations, we see that the left side is equivalent to $d(y/x)$. Therefore the equation becomes $d(y/x) = -x\,dx$. Integration gives $y/x = -x^2/2 + C$ or $y = -x^3/2 + Cx$.

The frequently occurring combinations should be carefully inspected and some should be memorized. This method provides an easy way to solve a problem that otherwise could be very puzzling.

Example 32

Consider the first-order nonlinear equation $xy' - y = y^2$. Converting this to differential form gives $x\,dy - y\,dx = y^2\,dx$. Since the left side may be converted to an integrable combination by dividing by either x^2 or y^2, we inspect the right side and find that y^2 is both available and needs to be removed before the right-side integration can be performed. Hence, we obtain

$$\frac{x\,dy - y\,dx}{y^2} = dx$$

or

$$-d\left(\frac{x}{y}\right) = dx \Rightarrow -\frac{x}{y} = x + C$$

Then

$$y = \frac{-x}{x + C}$$

is the one-parameter family of functions that solves the given differential equation.

This section closes with a warning. An incorrect method is illustrated here to show how it is sometimes employed on these problems with dire results. Some students have a tendency to integrate a combination such as dy/x into y/x or $2y\,dx$ into $(2y)x$. These, of course, are wrong since y is an unknown function of x.

The following was an attempt to use this incorrect method to solve $xy\,dx - (x^2 + y^2)\,dy = 0$.

$$xy\,dx - (x^2 + y^2)\,dy = 0$$

$$y\,dx - (x + y^2/x)\,dy = 0$$

$$y\,dx - x\,dy = y^2/x\,dy$$

$\dfrac{y\,dx - x\,dy}{y^2} = \dfrac{dy}{x}$ ⇨ $\dfrac{x}{y} = \dfrac{y}{x} + C$, the wrong answer by "integrating"

Integrable to (x/y) NOT integrable

A correct method uses $v = y/x$.

Exercises 1.4

Solve the following equations. Any of the available methods may be used.

1. $\dfrac{dy}{dx} = 1 + \dfrac{y}{x}$
2. $x\,dy - y\,dx = y\,dy$
3. $x\,dx + y\,dy = \sqrt{x^2 + y^2}\,dx$
4. $y' - \dfrac{1}{x}y = -\dfrac{1}{2y}$ $\left(\textit{Hint:}\text{ Write as } \left(\dfrac{x}{y}\right)^{-3} d\left(\dfrac{x}{y}\right)\right)$
5. The slope of what function is always $\dfrac{y^2 + y}{x}$?
6. $y\,dx - 2x\,dy = xy\,dy$
7. $x\,dy - y\,dx + \ln x\,dx = 0,\ (x > 0)$
8. $y(y - x^2)\,dx + x^3\,dy = 0$
9. $y\,dx - (x + y^3)\,dy = 0$
10. $(\frac{1}{x})\,dx - (1 + xy^2)\,dy = 0$ (*Hint:* Divide by x, multiply by e^y.)
11. $x\,dy + y\,dx = 2x^2y^3\,dy$ (*Hint:* Check $d(xy)/(xy)^n$ for some n.)
12. $xy' + 2y = 0$
13. $y\,dx = x(x^2y - 1)\,dy$
14. $(xy + y^2)\,dx + (x^2 - xy)\,dy = 0$
15. Show that an integrable combination may be produced from the differential equation $(x + y)\,dx + (y - x)\,dy = 0$ if we use the factor $1/(x^2 + y^2)$.
16. For the integrable combination listed, verify that the given solution satisfies the differential equation
    (a) $2xy\,dx + x^2\,dy = 0;\ x^2y = C$
    (b) $\sin x\,dy + y\cos x\,dx = 0;\ y = \dfrac{C}{\sin x}$
    (c) $\dfrac{1}{x}\,dx + \dfrac{1}{y}\,dy = 0;\ \ln|xy| = C$
    (d) $\dfrac{y\,dx + x\,dy}{-x^2y^2} = 0;\ \dfrac{1}{xy} = C$

1.5 Exact Differential Equations

In the process of developing solution techniques for first-order differential equations, we will now study a special procedure that applies to a wide class of linear and nonlinear equations. This class (exact differential equations) is defined later in this section.

One of the topics studied in a multivariable calculus course is that of partial differentiation of a function of more than one independent variable. We will see how this topic relates to the special procedure for solving exact equations.

Let $F(x, y)$ be a function of two variables, then fix one of the variables. If we fix y at value y_0, the function whose values are $F(x, y_0)$ is a function of x alone. If at the point $x = x_0$ we calculate the derivative of $F(x, y_0)$ and that derivative exists, we call it the **partial derivative** of F with respect to x at (x_0, y_0). This is designated as

$$\left.\frac{\partial F(x, y)}{\partial x}\right|_{(x_0, y_0)}$$

At any general point x, it is often written

$$\frac{\partial F(x, y)}{\partial x}$$

Similarly, if we first fix x at x_0 and consider the resulting function of y, $F(x_0, y)$, its derivative with respect to y is called the partial derivative of F with respect to y and designated as

$$\frac{\partial F(x, y)}{\partial y}$$

Example 33 illustrates calculations of partial derivatives.

Example 33

(a) Given $F(x, y) = x^2y + 2xy^3$, to calculate

$$\frac{\partial F(x, y)}{\partial x}$$

we fix y and differentiate the function on x. This gives $\partial F/\partial x = (2x)y + (2)y^3$. To get

$$\frac{\partial F(x, y)}{\partial y}$$

we fix x and differentiate on y. This gives $\partial F/\partial y = x^2(1) + 2x(3y^2)$.

(b) For the function $G(x, y) = x^2 \cos y$, the partial derivatives are $\partial G/\partial x = (2x)\cos y$ and $\partial G/\partial y = x^2(-\sin y)$.

(c) For the function $H(x, y) = y^3e^{2x}$, the partial derivatives are $\partial H/\partial x = y^3(2e^{2x})$ and $\partial H/\partial y = (3y^2)e^{2x}$.

Let F, a function of two real variables, be continuous and have continuous first partial derivatives in some simple domain D. The **total differential** dF of the function F is defined as

$$dF = \frac{\partial F(x, y)}{\partial x}\,dx + \frac{\partial F(x, y)}{\partial y}\,dy$$

for all $(x, y) \in D$. Example 34 illustrates the way the total differential is computed.

Example 34

If $F(x, y) = x^2 \cos y + 2xy^3$, then $\partial F/\partial x = 2x \cos y + 2y^3$ and $\partial F/\partial y = -x^2 \sin y + 6xy^2$. Therefore $dF = (2x \cos y + 2y^3)\,dx + (-x^2 \sin y + 6xy^2)\,dy$.

Also consider $G(x, y) = xe^{2y} + 4y \sin x$. In this case, $\partial G/\partial x = e^{2y} + 4y \cos x$ and $\partial G/\partial y = 2xe^{2y} + 4 \sin x$. Then $dG = (e^{2y} + 4y \cos x)\,dx + (2xe^{2y} + 4 \sin x)\,dy$.

Definition: A differential equation $M(x, y)\,dx + N(x, y)\,dy = 0$ is said to be an **exact differential equation** if there exists a function $F(x, y)$ such that

$$d(F(x, y)) = M(x, y)\,dx + N(x, y)\,dy$$

In other words, if there exists a function $F(x, y)$ such that $\partial F/\partial x = M(x, y)$ and $\partial F/\partial y = N(x, y)$.

Example 35 illustrates a differential equation that is exact and one that is not exact.

Example 35

The differential equation $(3x^2 + y^2)\,dx + 2xy\,dy = 0$ is exact, because the partial derivatives of the expression $F(x, y) = x^3 + xy^2$ are $\partial F/\partial x = 3x^2 + y^2 = M(x, y)$ and $\partial F/\partial y = 2xy = N(x, y)$.

From the differential equation

$$(2x + y + 3)\,dx + (5y^2 - x + 2)\,dy = 0$$

we find that $M(x, y) = 2x + y + 3$ and thus $\partial M/\partial y = 1$, while $N(x, y) = 5y^2 - x + 2$ and $\partial N/\partial x = -1$. Therefore, the equation is not exact.

It is our intent to discover how we may identify those differential equations that are exact and how, when they are exact, we may find the general solution.

The following theorem expresses a fundamental property.

Theorem 1.1 (Test for Exactness):

If the functions $M(x, y)$ and $N(x, y)$ and the partial derivatives $\partial M/\partial y$ and $\partial N/\partial x$ are continuous in some domain D containing no holes, then a necessary and sufficient condition for the differential equation $M(x, y)\,dx + N(x, y)\,dy = 0$ to be exact is that $\partial M/\partial y \equiv \partial N/\partial x$.

Proof:

Suppose the differential equation is exact, that is, there exists a function $F(x, y)$ such that $\partial F/\partial x = M(x, y)$ and $\partial F/\partial y = N(x, y)$. Since the mixed partials of a continuous function with continuous partial derivatives are equal, we have

$$\frac{\partial}{\partial y}\left(\frac{\partial F}{\partial x}\right) = \frac{\partial}{\partial x}\left(\frac{\partial F}{\partial y}\right) \quad \text{or} \quad \frac{\partial M}{\partial y} = \frac{\partial N}{\partial x}$$

Conversely, we assume that $\partial M/\partial y = \partial N/\partial x$ for all (x, y) in D and show that there exists a function $F(x, y)$ such that $\partial F/\partial x = M(x, y)$ and $\partial F/\partial y = N(x, y)$ for all (x, y) in D. If we take these last two equations as a pair of simple partial differential equations whose solution $F(x, y)$ we are seeking, we will describe a solution process. This new process, called partial integration, is actually the reverse of partial differentiation. Starting with either equation, we integrate with respect to the differentiation variable. Doing this on $\partial F/\partial x = M(x, y)$ we get $F(x, y) = \int M(x, y)\partial x + f(y)$. In this case the integration of M is on x with y held fixed, and also the function $f(y)$ is the function of integration rather than the constant of integration to which we are accustomed. This "trial" F is now the general solution to one of the two partial differential equations, and we must force it to satisfy the other simultaneously. Hence we differentiate this trial F with respect to the other variable, this time y, and obtain

$$\frac{\partial F}{\partial y} = \frac{\partial}{\partial y}\int M(x, y)\,\partial x + f'(y)$$

Thus we get

$$N(x, y) = \frac{\partial F}{\partial y}$$

or

$$f'(y) = N(x, y) - \frac{\partial}{\partial y}\int M(x, y)\partial x$$

Since f is a function of y only, the right side is independent of x. This also follows if we write

$$f'(y) = N(x, y) - \int \frac{\partial M}{\partial y}\,\partial x$$

and use the hypothesis that $\partial M/\partial y = \partial N/\partial x$ to get

$$f'(y) = N(x, y) - \int \frac{\partial N}{\partial x}\,\partial x$$

Since the partial derivative of the right side with respect to x vanishes, $f'(y)$ is independent of x. We are thus led to a "final" form of the function

$$F(x, y) = \int M(x, y)\partial x + \int [N(x, y) - \frac{\partial}{\partial y}\int M(x, y)\partial x]\partial y$$

This F satisfies both $\partial F/\partial x = M(x, y)$ and $\partial F/\partial y = N(x, y)$, and so $M(x, y)\,dx + N(x, y)\,dy = 0$ is exact. ■

This *test for exactness* is merely that; it is only a test to see if the given differential equation may be solved by the new method. Nevertheless the preceding proof outlines the scheme for finding the desired $F(x, y)$. Once we have found the $F(x, y)$, the general solution to the exact differential equation is easy to find.

Theorem 1.2:

If the differential equation $M(x, y)\,dx + N(x, y)\,dy = 0$ *is exact, then* $F(x, y) = C$ *is the one-parameter family of functions that represents the general solution where* $F(x, y)$ *is a function such that* $\partial F/\partial x = M(x, y)$ *and* $\partial F/\partial y = N(x, y)$.

Proof:

Since the differential equation has passed the test of exactness, we know that $M(x, y) = \partial F/\partial x$ and $N(x, y) = \partial F/\partial y$, so the differential equation may be written as $(\partial F/\partial x)\,dx + (\partial F/\partial y)\,dy = 0$ or as $d(F(x, y)) = 0$. This is integrable immediately to $F(x, y) = C$. ■

At this point, illustrative Examples 36–38 are surely in order.

Example 36

Consider the same differential equation we examined in Example 35:

$$(3x^2 + y^2)\,dx + 2xy\,dy = 0$$

The test for exactness provides $\partial M/\partial y = 2y = \partial N/\partial x$. Thus we may proceed to set up $\partial F/\partial x = 3x^2 + y^2$ and $\partial F/\partial y = 2xy$. Integrating the first of the two preceding equations with respect to x gives the trial F:

$$F(x, y) = x^3 + xy^2 + f(y)$$

Differentiating this gives

$$\frac{\partial F}{\partial y} = 0 + 2xy + f'(y)$$

Comparing this with the second given equation (i.e., $\partial F/\partial y = 2xy$) shows $f'(y) = 0$ and, thus, that $f(y) = \text{constant}$. Hence our final F is $F(x, y) = x^3 + xy^2 + C_1$, and the general solution is $x^3 + xy^2 + C_1 = K$ or $x^3 + xy^2 = C$ where $C = K - C_1$.

Reworking this same example by the reverse integration-differentiation scheme, we begin once again with

$$\frac{\partial F}{\partial x} = 3x^2 + y^2 \quad \text{and} \quad \frac{\partial F}{\partial y} = 2xy$$

but we first integrate the second equation with respect to y. This gives the trial F, $F(x, y) = xy^2 + g(x)$ where $g(x)$ is now the function of integration. Differentiation of this with respect to x provides $\partial F/\partial x = y^2 + g'(x)$, and comparing this with the first equation gives $g'(x) = 3x^2$. Easy integration of

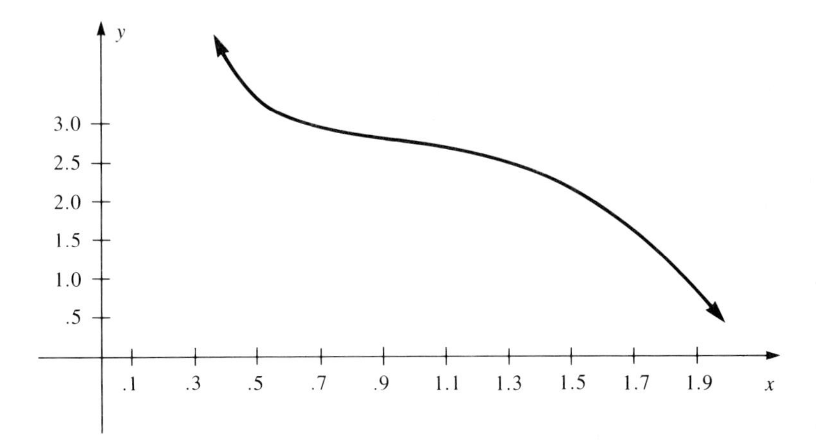

Figure 1.8

this gives $g(x) = x^3 + C_1$ and thus the final F is $F(x, y) = xy^2 + x^3 + C_1$ and the general solution is again $x^3 + xy^2 = C$.

Suppose now we desire the particular curve of this family which passes through the point $(2, -1)$. Set $x = 2$ and $y = -1$ in the family description and obtain $2^3 + 2(-1)^2 = C$ or $C = 10$. Thus the particular curve we seek is $x^3 + xy^2 = 10$ or $y = \pm\sqrt{10 - x^3/x}$. A portion of this curve is shown in Figure 1.8.

The scheme to be followed to solve any exact equation is diagramed in Figure 1.9.

Example 37

In designing a suspension system an engineer might encounter the equation $(y^2 + 6x^2y + x^2)\,dx + (2xy + 2x^3)\,dy = 0$. The test for exactness provides $\partial M/\partial y = 2y + 6x^2 = \partial N/\partial x$, so we may proceed by setting $\partial F/\partial x = y^2 + 6x^2y + x^2$ and $\partial F/\partial y = 2xy + 2x^3$. Arbitrarily we begin with the second equation, integrating with respect to y, and obtaining the trial F

$$F(x, y) = xy^2 + 2x^3y + g(x)$$

Differentiating with respect to x gives

$$\frac{\partial F}{\partial x} = y^2 + 6x^2y + g'(x)$$

Comparing, we find $g'(x) = x^2$, so $g(x) = \frac{1}{3}x^3 + C_1$. Hence the final F is

$$F(x, y) = xy^2 + 2x^3y + \frac{1}{3}x^3 + C_1$$

and the general solution to the given differential equation is $xy^2 + 2x^3y + \frac{1}{3}x^3 = C$.

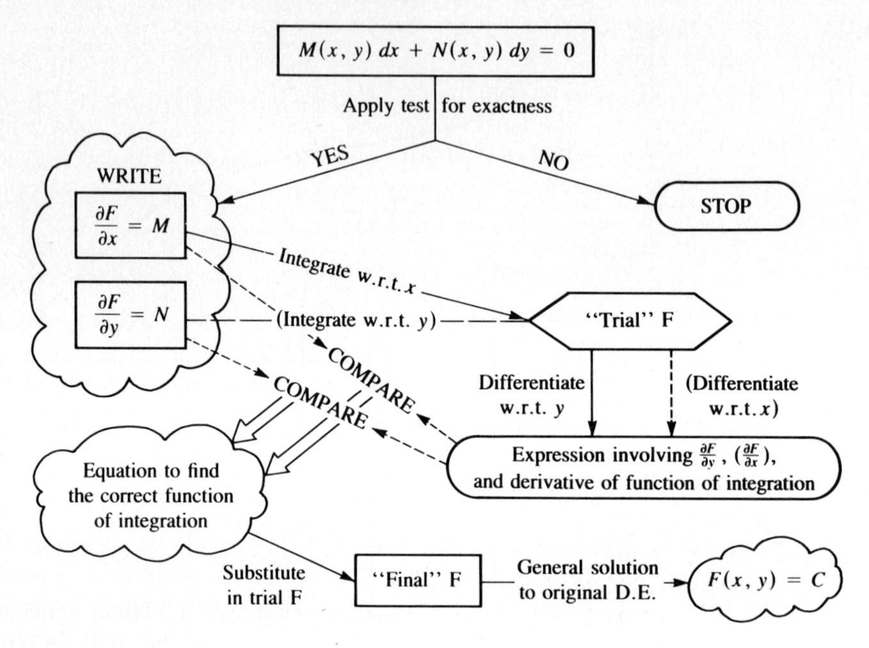

Note: w.r.t. means with respect to.

Figure 1.9 Scheme for Solving Exact Equations

There are times when the solution method must be chosen from among many available. Example 38 shows how two appropriate methods produce the same solution.

Example 38

At first glance, $(3 - x\cos y)\,dy/dx = \sin y + 2x$ might indicate that the exactness test is easy to apply, and $\partial M/\partial y = \cos y = \partial N/\partial x$ shows exactness. If we follow this method of solution, our calculations will be as follows:

$$\frac{\partial F}{\partial x} = \sin y + 2x \Rightarrow F(x, y) = x\sin y + x^2 + f(y)$$

$$\Downarrow$$

$$\frac{\partial F}{\partial y} = x\cos y - 3 \qquad \frac{\partial F}{\partial y} = x\cos y + 0 + f'(y)$$

compare

$$\downarrow$$

$$f'(y) = -3 \Rightarrow f(y) = -3y + C$$

$$\therefore\ F(x, y) = x\sin y + x^2 - 3y + C$$

and the solution is $$x\sin y + x^2 - 3y = K.$$

However, another person might look at the problem as an integrable combination:

$$3\,dy - x\cos y\,dy = \sin y\,dx + 2x\,dx$$
$$\sin y\,dx + x\cos y\,dy = 3\,dy - 2x\,dx$$
$$d(x\sin y) = 3\,dy - 2x\,dx$$

Integrating gives

$$(x\sin y) = 3y - x^2 + C$$

or

$$x\sin y + x^2 - 3y = C$$

Exercises 1.5

Test the differential equations in Exercises 1–22 for exactness and solve those that are exact.

1. $(2x + y)\,dx + (x + 2y)\,dy = 0$
2. $(x^2 + y)\,dx + (x + e^y)\,dy = 0$
3. $y' = \dfrac{y}{y^3 - x}$
4. $\sin y\,dx + (x\cos y + 3y^2)\,dy = 0$
5. $(3x - y)\,dx + (x + 3y)\,dy = 0$
6. $(3x^2y + xy^2 + e^x)\,dx + (x^3 + x^2y + \sin y)\,dy = 0$
7. $\left(\dfrac{x + 3y}{y^2}\right)y' + \left(\dfrac{2xy - 1}{y}\right) = 0$
8. $(x\sin y - y^2)y' = \cos y$
9. $(\cot x\tan y)y' + 1 = 0$
10. $y' + y - 2xe^{-x} = 0$
11. $2yy' = 3(x - 1)^2$
12. $\cos x\sec y\,dx + \sin x\sin y\sec^2 y\,dy = 0$
13. $(2x\cos y - e^x)\,dx - x^2\sin y\,dy = 0$
14. $(x - y) + (2y - x)y' = 0;\ y(0) = 1$
15. $(x^2 + y^2)\,dx + 2xy\,dy = 0;\ y(3) = 1$
16. $(3x^2 - 2xy)\,dx + (2y - x^2)\,dy = 0;\ y(1) = 4$
17. $2ty - (y - t^2)\dfrac{dy}{dt} = 0$
18. $x\cos(xy)y' + y\cos(xy) = -1$
19. $(3x^2 - 3 + t)\dfrac{dx}{dt} + (2 + x + 2t) = 0$
20. $(e^{x+y} + 3)\,dy + (e^{x+y} - 2)\,dx = 0$

21. $(e^y + ye^x)\,dx + (e^x + xe^y)\,dy = 0;\ y(1) = 0$
22. $y\cosh(xy)\,dx + \big(x\cosh(xy) - y\big)\,dy = 0;\ y(0) = \sqrt{20}$
23. Find the value of v that makes the equation exact. Then with the value v find the general solution:
(a) $vx^5y^{v+1}\,dx + 2x^6y^v\,dy = 0$

(b) $4vxe^{x^2}\sin 3y + e^{x^2}\cos(3y)\dfrac{dy}{dx} = 0$

24. Even though an equation $M(x, y)\,dx + N(x, y)\,dy = 0$ may not be exact, we may sometimes find a nonzero function $u(x, y)$ such that $u(x, y)M(x, y)\,dx + u(x, y)N(x, y)\,dy = 0$ is exact. Such a function is called an integrating factor. For example, in section 1.4 we looked at equations of the type $y\,dx - x\,dy = 0$, which is not exact (because it fails the test); but multiplication by $1/y^2$ makes it exact and integrable to the solution $y = Cx$.

Find an integrating factor for the following equations and solve them:
(a) $\cos x \cos y\,dx - 2\sin x \sin y\,dy = 0$
(b) $y\,dx - (x + y^3)\,dy = 0$

(c) $\dfrac{1}{x}\,dx \quad (1 + xy^2)\,dy = 0$ (*Hint:* Check $-e^y/x$.)

(d) $(x^2 + y^2)\,dx + 3xy\,dy = 0$ (*Hint:* Use the method shown in Exercise 25.)

25. In an extended case, we may find that an integrating factor $u(x, y)$ need not be a function of both x and y but only of one. If we find that

$$\left[\frac{\dfrac{\partial M}{\partial y} - \dfrac{\partial N}{\partial x}}{N}\right]$$

depends on x only, then

$$e^{\int((My - Nx)/N)\,dx}$$

is an integrating factor for $M(x, y)\,dx + N(x, y)\,dy = 0$. Similarly, if

$$\left[\frac{\dfrac{\partial N}{\partial x} - \dfrac{\partial M}{\partial y}}{M}\right]$$

depends on y only, then

$$e^{\int((Nx - My)/M)\,dy}$$

is an integrating factor.

Test the following for such an integrating factor. If it works, solve the equation.

(a) $(y + 2x^2)\,dx - (x - x^2y)\,dy = 0$
(b) $y(1 + xy)\,dx - x\,dy = 0$
(c) $ye^{y/x}\,dx - (xe^{y/x} + 2x^2y)\,dy = 0$
(d) $(5xy + 4y^2 + 1)\,dx + (x^2 + 2xy)\,dy = 0$

1.6 First-Order Linear Equations

The general form of an nth-order linear equation was defined in Section 1.1. Using $n = 1$ in that form, we obtain the **general first-order linear equation** $a_0(x)y' + a_1(x)y = f(x)$. If we assume $a_0(x) \neq 0$ in some interval (if it were identically zero we would have no differential equation), then we can divide by $a_0(x)$ and rearrange this equation into the form

$$y' + p(x)y = q(x)$$

where

$$p(x) = \frac{a_1(x)}{a_0(x)} \quad \text{and} \quad q(x) = \frac{f(x)}{a_0(x)}$$

or into

$$[p(x)y - q(x)]\,dx + dy = 0$$

The following derivation will produce the technique for solving this general first-order linear equation.

In an attempt to put this equation into the form of an exact equation, we will seek a factor, which is called an **integrating factor**, that we can multiply through the equation and force the exactness. For simplicity, we will attempt to find an integrating factor that is a function of x alone, and we will call it $m(x)$. We then have

$$m(x)[p(x)y - q(x)]\,dx + m(x)\,dy = 0$$

Using the test for exactness now produces

$$\frac{\partial}{\partial y}[m(x)p(x)y - m(x)q(x)] = \frac{\partial}{\partial x}[m(x)] \quad \text{or} \quad m(x)p(x) = m'(x)$$

This is a simple first-order separable equation in $m(x)$, that is, $m'(x)/m(x) = p(x)$, which integrates to

$$\ln|m(x)| = \int p(x)\,dx + C$$

Making each of these an exponent gives

$$e^{\ln|m(x)|} = e^{\int p(x)\,dx + C} = e^C \cdot e^{\int p(x)\,dx}$$

or

$$|m(x)| = e^C \cdot e^{\int p(x)\,dx}$$

or

$$m(x) = \pm e^C \cdot e^{\int p(x)\,dx}$$

However, we need only one integrating factor $m(x)$, not an entire family, so we choose C such that $m(x) = e^{\int p(x)\,dx}$. Multiplying this integrating factor on both sides of the first-order linear equation produces

$$e^{\int p(x)\,dx}y' + e^{\int p(x)\,dx}p(x)y = e^{\int p(x)\,dx}q(x)$$

which is the same as

$$\frac{d}{dx}[e^{\int p(x)\,dx} \cdot y] = e^{\int p(x)\,dx}q(x)$$

(Verify that these two equations are equivalent by differentiating the product on the left.) We have, by multiplying by the appropriate integrating factor, created an integrable equation, whose integration gives

$$e^{\int p(x)\,dx} \cdot y(x) = \int e^{\int p(x)\,dx}q(x)\,dx + C$$

or

$$y(x) = e^{-\int p(x)\,dx} \int e^{\int p(x)\,dx}q(x)\,dx + Ce^{-\int p(x)\,dx}$$

This is the one-parameter family of functions that is the general solution to $y' + p(x)y = q(x)$.

This solution is in a very general form since $p(x)$ and $q(x)$ need to be specified. Rather than attempting to memorize this formula for the general solution, it is wise to work through the few simple steps needed in each new problem of this type.

Example 39

Consider $xy' - 3y = x^3$. We first rewrite this equation in the standard form $y' + p(x)y = q(x)$ to correctly identify the functions $p(x)$ and $q(x)$. Hence we get

$$y' - \frac{3}{x}y = x^2$$

Since $p(x) = -3/x$ (note that the sign *must* be included) the integrating factor is

$$e^{\int p(x)\,dx} = e^{\int -3/x\,dx} = e^{-3\ln|x|} = e^{\ln|x|^{-3}} = x^{-3}$$

Multiplying by this factor gives $x^{-3}y' - 3x^{-4}y = x^{-1}$. Notice now that the left side is the derivative of the product $ye^{\int p(x)\,dx} = yx^{-3}$, which gives us a check on our computations so far. Thus

$$\frac{d}{dx}(x^{-3}y) = x^{-1}$$

Integrating we obtain $x^{-3}y = \ln|x| + C$ or $y = x^3\ln|x| + Cx^3$. This is the solution family for the original problem.

Example 40

Consider $y' = y + 3x^2e^x$; $y(0) = 4$. By writing this as $y' - y = 3x^2e^x$ we identify $p(x) = -1$; $q(x) = 3x^2e^x$. The integrating factor is now

$$e^{\int p(x)\,dx} = e^{\int -dx} = e^{-x}$$

Multiplying by this factor gives $e^{-x}(y' - y) = 3x^2$. The left side is now the derivative of

$$ye^{\int p(x)\,dx} \quad \text{or} \quad (ye^{-x})' = 3x^2$$

The integration of this produces $ye^{-x} = x^3 + C$, and then the explicit solution to the differential equation is $y = e^x x^3 + Ce^x$. Inserting the initial condition that $y = 4$ when $x = 0$ gives $4 = 0 + C \cdot 1$ or $C = 4$. Thus the unique solution to the initial value problem is $y = (x^3 + 4)e^x$, as illustrated in Figure 1.10.

We must never forget that a solution to any differential equation may be checked by simply substituting the solution and its derivative(s) back into the differential equation and simplifying down to an identity.

For Example 40 we differentiate the result to get

$$y' = (x^3 + 4)e^x + 3x^2e^x$$

When we substitute the solution and its derivative, the differential equation becomes $(x^3 + 4)e^x + 3x^2e^x = (x^3 + 3x^2 + 4)e^x$, which is an identity. The initial condition also checks: $y(0) = (0 + 4)e^0 = 4$.

Example 41

Consider $y' + 2xy = 2x$. This first-order equation is linear, and the integrating factor is

$$e^{\int 2x\,dx} = e^{x^2}$$

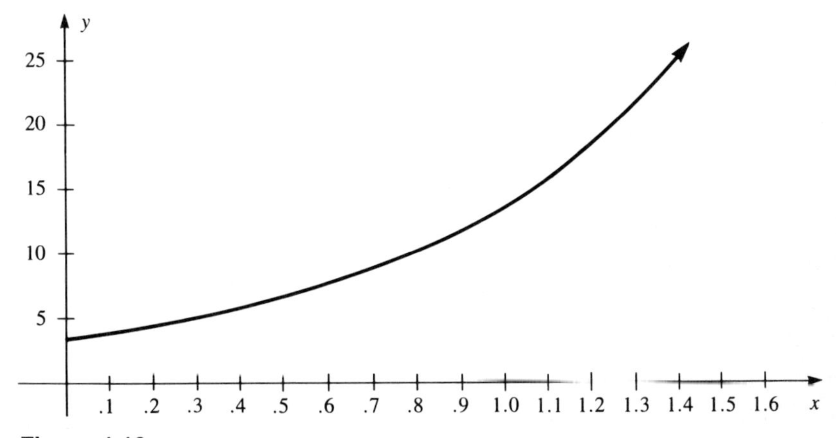

Figure 1.10

Then $e^{x^2}(y' + 2xy) = e^{x^2}(2x)$ integrates to

$$e^{x^2} \cdot y = \int e^{x^2} \cdot 2x \, dx = e^{x^2} + C$$

Then
$$y = \frac{e^{x^2} + C}{e^{x^2}} = 1 + Ce^{-x^2}$$

An alternative method would reorganize the equation to $dy/dx = 2x(1 - y)$. When the variables are separated, we have $dy/(1 - y) = 2x \, dx$. Integrating this gives

$$-\ln|1 - y| = x^2 + C$$

$$(1 - y)^{-1} = e^{x^2 + C} = Ke^{x^2}$$

$$\frac{1}{1 - y} = Ke^{x^2}$$

$$1 - y = -Ce^{-x^2}$$

$$y = 1 + Ce^{-x^2}$$

Example 42

In researching patterns of learning among medical school students, psychologists found that the rate of change of the educational level of an individual at time t is a function that is being diminished by a rate at which information is being forgotten (i.e., it is no longer the student's knowledge) or becoming obsolete (i.e., the student no longer needs it). Thus, if we begin with a unit value of education (or knowledge), the loss of level induces a function proportional to the education. Thus the rate of change of education E is the beginning unit minus the proportional amount. Therefore the governing differential equation is

$$\frac{dE}{dt} = 1 - kE; \; E(0) = E_0 \quad \text{(the starting educational level)}$$

The integration factor for this first-order linear equation is e^{kt}. Thus

$$e^{kt}\left[\frac{dE}{dt} + kE\right] = e^{kt}$$

or
$$(e^{kt} \cdot E)' = e^{kt} \Rightarrow e^{kt} \cdot E\Big|_0^t = \int_0^t e^{kt} \, dt \Rightarrow e^{kt}E(t) - E(0) = \frac{1}{k}(e^{kt} - 1)$$

Thus
$$E(t) = e^{-kt} \cdot E_0 + \frac{1}{k}(1 - e^{-kt})$$

Note that for large t, $E(t) \to 1/k$. You become educated faster if you forget more slowly!

Some nonlinear first-order equations may be transformed by a change of variable into linear equations. Those known as **Bernoulli equations*** make up a class of such equations. These are defined as differential equations in the form $y' + p(x)y = q(x)y^n$. We exclude the cases $n = 0$, because it already is first-order linear, and $n = 1$, because transposing the right term gives a separable equation.

To solve such an equation we multiply both members by y^{-n} and obtain $y^{-n}y' + p(x)y^{1-n} = q(x)$, which is linear in y^{1-n}. After multiplication by the constant $(1 - n)$ we make the variable change $u = y^{1-n}$. Then $u' = (1 - n)y^{-n}y'$. The equation now becomes $u' + p(x)(1 - n)u = q(x)(1 - n)$, which is first order and linear in $u(x)$. This is now solved for $u(x)$, the reverse substitution $u = y^{1-n}$ is made, and the resulting family gives the general solution. This method was discovered by German mathematician Gottfried Wilhelm Leibniz (1646–1716) in 1696.

Example 43

Consider $y' + y = e^{2x}y^2$, a Bernoulli equation. By following the above outline we obtain $y^{-2}y' + y^{-1} = e^{2x}$. Then $-y^{-2}y' - y^{-1} = -e^{2x}$. If we let $u = y^{-1}$, then $u' = -1y^{-2}y'$. When we substitute these into the differential equation we obtain $u' - u = -e^{2x}$. The integrating factor is now $e^{\int -1\,dx} = e^{-x}$

Thus
$$e^{-x}(u' - u) = -e^x$$

or
$$\frac{d}{dx}(e^{-x}u) = -e^x$$

Integration gives

$$e^{-x}u = -\int e^x\,dx = -e^x + C$$

Then $u = -e^{2x} + Ce^x$. However, $u = y^{-1}$, so $y^{-1} = Ce^x - e^{2x}$ or $y = 1/(Ce^x - e^{2x})$. If the initial condition $y(0) = 1$ is added, then $1 = 1/(Ce^0 - e^0) = 1/(C - 1)$ so that $C = 2$. Thus the function satisfying the differential equation and the initial condition is

$$y = \frac{1}{2e^x - e^{2x}} = \frac{e^{-x}}{2 - e^x}$$

The graph of this solution is shown in Figure 1.11.

---

* Named for Swiss mathematician James Bernoulli (1654–1705).

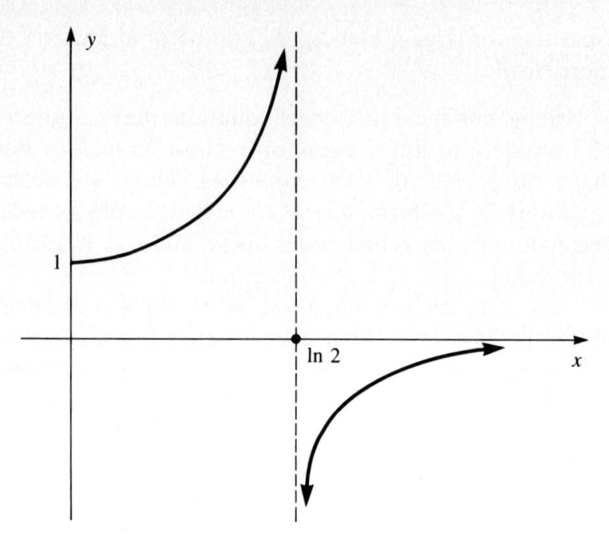

Figure 1.11

Example 44

The differential equation

$$\frac{dy}{dx} = \left(\frac{y}{x}\right)^2 + \frac{2y}{x}$$

arises in the design of exhaust ports in rocket engines. It can be solved by a variety of methods, including the Bernoulli procedure.

$$\frac{dy}{dx} - \left(\frac{2}{x}\right)y = \frac{y^2}{x^2}$$

Dividing by y^2 gives

$$y^{-2}y' - \left(\frac{2}{x}\right)y^{-1} = \frac{1}{x^2}$$

Multiplying by -1 gives

$$-y^{-2}y' + \left(\frac{2}{x}\right)y^{-1} = \frac{-1}{x^2}$$

Now let $u = y^{-1}$ and $u' = -y^{-2}y'$. Then we have

$$u' + \left(\frac{2}{x}\right)u = -\frac{1}{x^2}$$

This is a first-order linear equation for which the integrating factor is

$$e^{\int 2/x\,dx} = e^{2\ln x} = x^2$$

Thus $x^2u' + (2x)u = -1$ can be written as $(x^2u)' = -1$. This is integrated to $x^2u = -x + C$ or $u = (C - x)/x^2$. But since $u = y^{-1} = 1/y$, we have the general solution

$$y = \frac{x^2}{C - x}$$

Exercises 1.6

Solve the differential equations in Exercises 1–27.

1. $\dfrac{dy}{dx} = e^x + y$
2. $xy' + 3y - 1 = x^2$
3. $2x^2y' = xy - 6$
4. $3(x^2y + 1)\,dx + 2x^3\,dy = 0$
5. $(x^3 - xy + x)\,dx - (1 + x^2)\,dy = 0$
6. $y' = \dfrac{xy + 1}{1 - x^2}$
7. $x(y' + y) = x - y$
8. $y' + x - \dfrac{y}{x} = 0$
9. $y' - (x + y) = 1$
10. $y' - \dfrac{y}{x} = \dfrac{x + y}{x - 1}$
11. $y' = \dfrac{y + \sin^3 x}{\sin x \cos x}$; $y\left(\dfrac{\pi}{6}\right) = -\dfrac{1}{2}$
12. $y' - y = xy^2$
13. $xy' + y = y^2 \ln x$, $x > 0$
14. $x^2y' - y^2 = 2xy$
15. $y' - xy = xy^{1/2}e^{x^2}$
16. $t\,dx + x(1 - x^2t^4)\,dt = 0$
17. $xy' + 2y = 3x^3y^{4/3}$
18. $y' + y\cos x = y^3 \sin 2x$
19. $v' - 2uv = uv^2$
20. $\dfrac{x}{y}y' + 1 = 3y^2$
21. $\dfrac{dx}{dy} + \dfrac{y}{x} = \dfrac{x}{y}$

22. $y' + \left(\dfrac{2x}{x^2 + 1}\right)y - x = 0$

23. $y' + y = y^4$

24. $y' + (\cot x)y + 1 = 0;\ y\left(\dfrac{\pi}{4}\right) = 9$

25. $(x + 1)y' = 5 - (x + 2)y;\ y(0) = 4$
26. $y' + (\tan x)y = \sec x;\ y(0) = 1$
27. $(\csc x)y' + (\sec x)y = \cot x$
28. Fifty pounds of salt were added to a tank partially filled with water, resulting in 100 gallons of the salt-and-water mixture. The liquid is constantly stirred and you should assume that a uniform mixture is maintained. Brine containing 2 pounds of salt per gallon runs into the tank at the rate of 3 gal/min, and simultaneously 2 gallons of mixture per minute run out of the tank. Find the amount of salt in the tank after 30 minutes.

    The differential equation should be set up according to the following relationship:

$$\begin{Bmatrix}\text{Rate of increase} \\ \text{of salt (lb/min)}\end{Bmatrix} = \begin{Bmatrix}\text{Rate of inflow} \\ \text{of salt (lb/min)}\end{Bmatrix} - \begin{Bmatrix}\text{Rate of outflow} \\ \text{of salt (lb/min)}\end{Bmatrix}$$

Since the volume of liquid in the tank increases by one gallon each minute, the total amount of liquid after t minutes is $100 + t$. If x is the number of pounds of salt in the tank at t minutes, we have

$$\frac{dx}{dt} = [(3\ \text{gal/min})(2\ \text{lb/gal})] - \left[(2\ \text{gal/min})\left(\frac{x}{100 + t}\ \text{lb/gal}\right)\right]$$

Remembering that $x = 50$ when $t = 0$, solve this for $x(t)$ at $t = 30$.
29. If Exercise 28 had stated that the rate of discharge was 3 gal/min, then the amount of brine in the tank would remain constant (100 gallons). Assume these conditions are true and calculate the amount of salt in the tank after 30 minutes. The differential equation would be

$$\frac{dx}{dt} = 6 - \frac{3x}{100}$$

30. Assume the conditions of Exercise 28 except that pure water is running in. Again solve for the amount of salt in the tank after 30 minutes. Since no salt is entering, the differential equation reduces to

$$\frac{dx}{dt} = -\frac{2x}{100 + t}$$

31. For a simple circuit containing a resistance R and an inductance L in series with a source of emf (electromotive force) e, the governing

differential equation is

$$L\frac{di}{dt} + Ri = e$$

where i is the current. If e is a function of time t, the equation is linear. If e is a constant, it is separable and linear.
(a) Find the general solution if e is constant.
(b) Find the general solution if $e = \sin t$.
(c) If we have a resistance and capacitance C in series with an emf, then

$$R\frac{di}{dt} + \frac{1}{C}i = \frac{de}{dt}$$

Suppose a simple electric circuit contains a resistance of 10 ohms and an inductance of 4 henrys in series with an emf given by $e = 100 \sin(200t)$ volts. If $i = 0$ where $t = 0$, find the current i when $t = .01$ seconds.

32. If we attempt to cool a hot object by conduction, such as immersing a small piece of hot iron into water that is held at a constant temperature, we encounter the classical problem of Newton. To determine how long it would take the iron to cool, we could use Newton's law of cooling: The rate of change in the object's temperature is proportional to the difference between the object's temperature and the constant temperature of the object's environment. The differential equation would be

$$\frac{dT_2}{dt} = -k[T_2(t) - T_1]$$

where $T_2(t)$ is the iron's temperature at time t, T_1 is the constant temperature of the water, and k is a proportionality constant to be determined from observable laboratory conditions.

If the water is held at 60°C and the iron is 100°C when $t = 0$, and if we find that after 60 seconds the immersed iron is down to 90°C, find the solution for $T_2(t)$ that predicts the temperature at any time t. How long will it take the iron to reach 65°C? When will the iron reach 60°C?

33. Although many aspects of cell growth need to be understood to eradicate cancer, it is thought that the termination of some inhibitor produced by the body can cause certain types of cancer. Since the growth of an organism does not proceed indefinitely, but eventually slows down and stops, let's assume there is a mutual inhibition of growth and multiplication by all growing and multiplying cells in the body. If the rate of growth and multiplication of cells is determined by a factor g, if each cell exerts upon every other cell an inhibiting effect

that is due to a factor j (some chemical substance), and if the body contains n cells, then the rate of multiplication of each cell is proportional to $ag - b(n-1)j$, where a and b are coefficients of proportionality. Hence

$$\frac{dn}{dt} = n(ag - b(n-1)j) = n[(ag + bj) - bjn]$$

Solve for $n(t)$ in terms of a, b, g, and j if $n(0) = n_0$. Note that the body will cease to grow when $dn/dt = 0$, that is, when

$$n = n_s = (ag + bj)/bj$$

Graph the growth curve from $t = 0$ to the time when growth stops.

34. Find the particular solution to the circuit equation

$$L\frac{di}{dt} + Ri = \cos t$$

for which $i(0) = R/(L^2 + R^2)$.

35. Molecular biologists are developing elaborate mathematical models of the steps taken by cells as they grow to produce their relatively permanent components, the macromolecules and nucleic acids. The kinetics of cell growth produces a differential equation governing cell volume V:

$$\frac{dV}{dt} = W + aM + \left(\frac{2c}{r_0}\right)(M - bV)$$

where:

b is a concentration of effective compounds outside the cell;

c is a permeation constant;

r_0 is radius of the cylindrical cell;

M is quantity of macromolecules per cell;

W is the constant pool of precursor;

a is a proportionality constant.

Show that the solution is

$$V = F + (V_0 - F)e^{-kt}$$

where
$$k = \frac{2bc}{r_0}$$

and
$$F = \frac{W}{k} + \frac{M}{k}\left(a + \frac{2c}{r_0}\right)$$

36. An equation of the form $dy/dx = p(x)y^2 + q(x)y + r(x)$ is called a **general Riccati equation**. There are a number of ways to attack an equation of this type. It may be transformed into a first-order linear differential equation in a new variable $v(x)$ defined by

$$v(x) = \frac{1}{y(x) - f(x)}$$

where $f(x)$ is one particular solution function found by any method. Under the assumption that we can obtain such a function $f(x)$, we write

$$y = f + \frac{1}{v}$$

and

$$y' = f'(x) - \frac{1}{v^2}v'(x)$$

Putting this into the Riccati equation produces

$$f' - \frac{1}{v^2}v' = p\left(f + \frac{1}{v}\right)^2 + q\left(f + \frac{1}{v}\right) + r$$

$$= (pf^2 + qf + r) + 2p\frac{f}{v} + \frac{p}{v^2} + \frac{q}{v}$$

From the original differential equation we have

$$\frac{df}{dx} = pf^2 + qf + r$$

Substitution gives us

$$(pf^2 + qf + r) - \frac{1}{v^2}v' = (pf^2 + qf + r) + 2p\frac{f}{v} + \frac{p}{v^2} + \frac{q}{v}$$

Canceling terms produces

$$-\frac{1}{v^2}v' = 2p\frac{f}{v} + \frac{p}{v^2} + \frac{q}{v}$$

or

$$-v' = 2pfv + qv + p$$

or

$$\frac{dv}{dx} = -(2pf + q)v - p$$

and finally

$$\frac{dv}{dx} + [2p(x)f(x) + q(x)]v = -p(x)$$

This is the desired first-order linear differential equation. Use this technique to solve the following Riccati equations when you are given that $f(x) = x$ is a solution (check it!).

(a) $\dfrac{dy}{dx} = -y^2 + 2xy - x^2 + 1;\ y(0) = 1$

(b) $\dfrac{dy}{dx} = 1 + y^2 - xy;\ y(0) = 3$

1.7 Geometry of Differential Equations

To help you get a better perspective of differential equations, this section explores some of their geometric properties.

Direction Fields

The process of finding the general solution to a first-order differential equation $y' = f(x, y)$ involves more than a simplistic statement such as "Find y as a function of x so that the equation is satisfied." The process of finding the solution is not always clear or free of ambiguities.

To better understand differential equations we should focus on the equation itself rather than on the solution. Geometrically, the given equation $y' = f(x, y)$ expresses that at a given point (x_0, y_0) where f is defined, the instantaneous rate of change of the function $y(x)$ with respect to movement in the x-direction is given by the number $f(x_0, y_0)$. Using this value we define a line segment (whose length is immaterial but generally taken to be very short), called a **lineal element**, of slope $f(x_0, y_0)$ through that point (see Figure 1.12). Since a lineal element exists for every point for

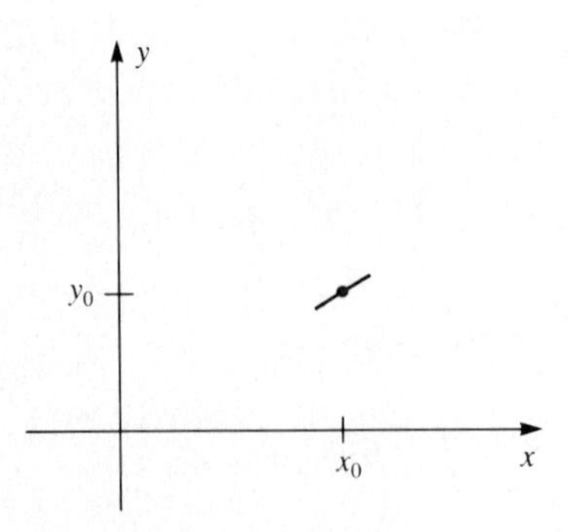

The lineal element at (x_0, y_0) having slope $f(x_0, y_0)$.

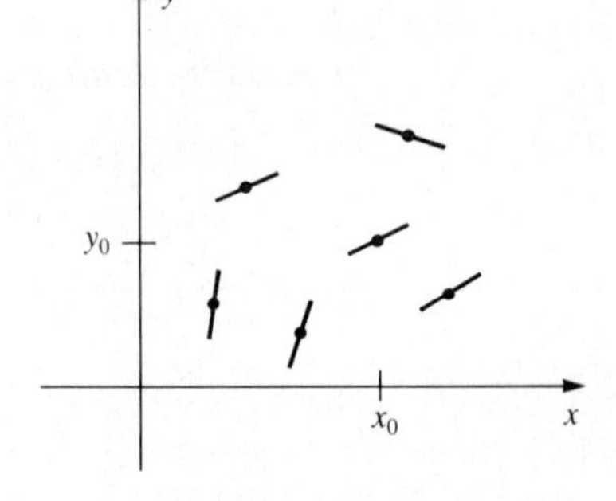

Additional lineal elements at (x_i, y_i) each having slope equal to $f(x_i, y_i)$ for $i = 1, 2, 3, \ldots$.

Figure 1.12

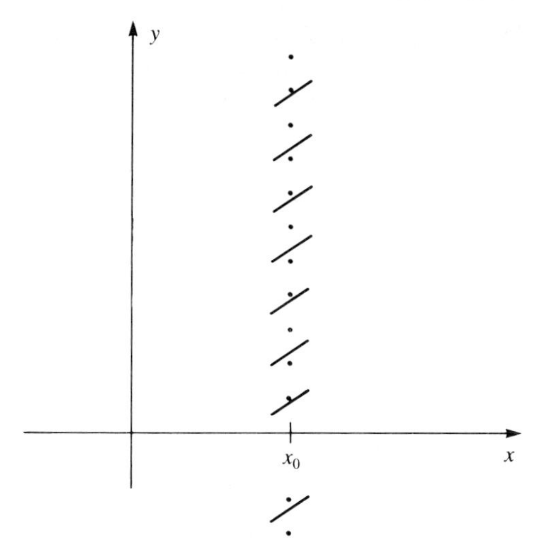

Figure 1.13 Lineal Elements at Points (x_0, y)

which the function is defined, we could plot an infinite collection of such small segments in the plane. This collection is called the **direction field**. Thus the differential equation $y' = f(x, y)$ defines a direction field. We can get a first look at a potential solution for the differential equation by examining this direction field.

First let's look at a simple example. We will approach the first-order differential equation $y' = 2x$, whose general solution is $y = x^2 + C$, from a purely geometric point of view. Since at any point in the plane the value of

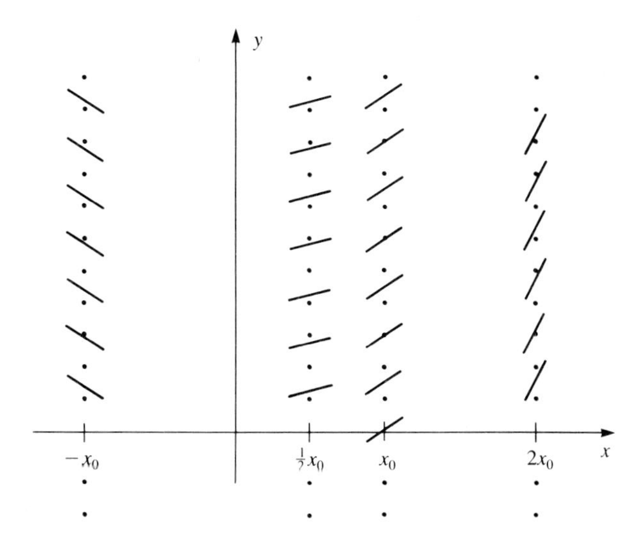

Figure 1.14 Lineal Elements at Points Related to x_0

$f(x, y) = 2x$ is exactly twice the x-coordinate, we find that the values of $y' = f(x, y) = 2x$ are slope values for lineal elements that get larger in magnitude as we move to the right or left. Figure 1.13 shows lineal elements at points (x_0, y), and Figure 1.14 shows additional lineal elements (note that slopes are twice the x-coordinate). These slope values are independent of the y-coordinate since $f(x, y) = 2x$. Now finishing out the direction field with a sufficient number of points (as in Figure 1.15) we get a better picture of the potential solution curves. When looking at this direction field, you will probably detect the parabolas $y = x^2 + C$ nested in the field with each vertex on the y-axis and the parabolas opening upward symmetrically.

Figure 1.16 shows a portion of the direction field for the equation $y' = 2y - x$, and Figure 1.17 shows the direction field whose domain is restricted and described by $y' = \sqrt{2 - x}$. When looked at geometrically, each first-order differential equation exhibits its unique definite pattern on the lineal elements.

Direction fields are also used in other disciplines. For example, if we observed the distribution of iron filings under the influence of a magnet, the filings would serve as the direction lines and their distribution would

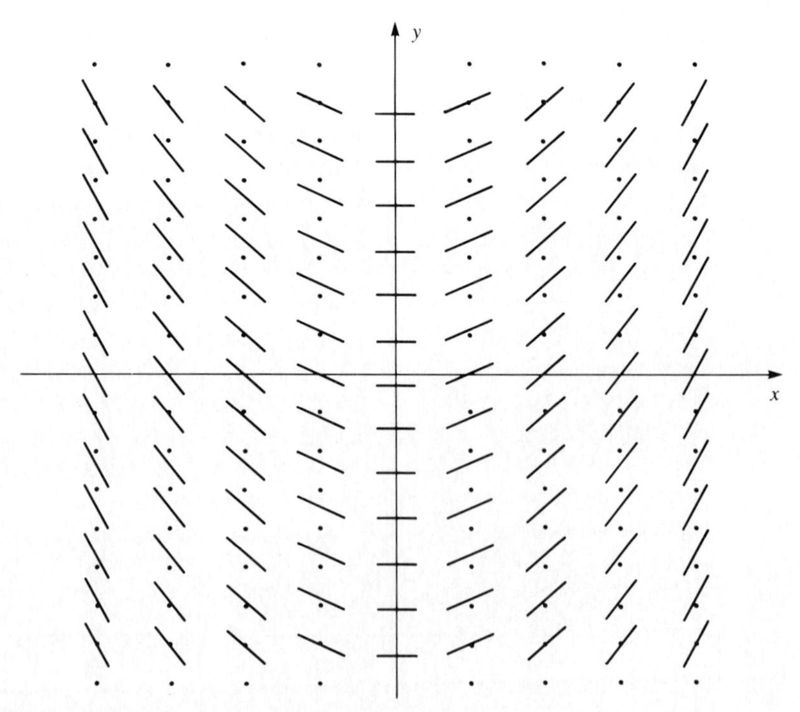

Figure 1.15 Direction Field for $y' = 2x$

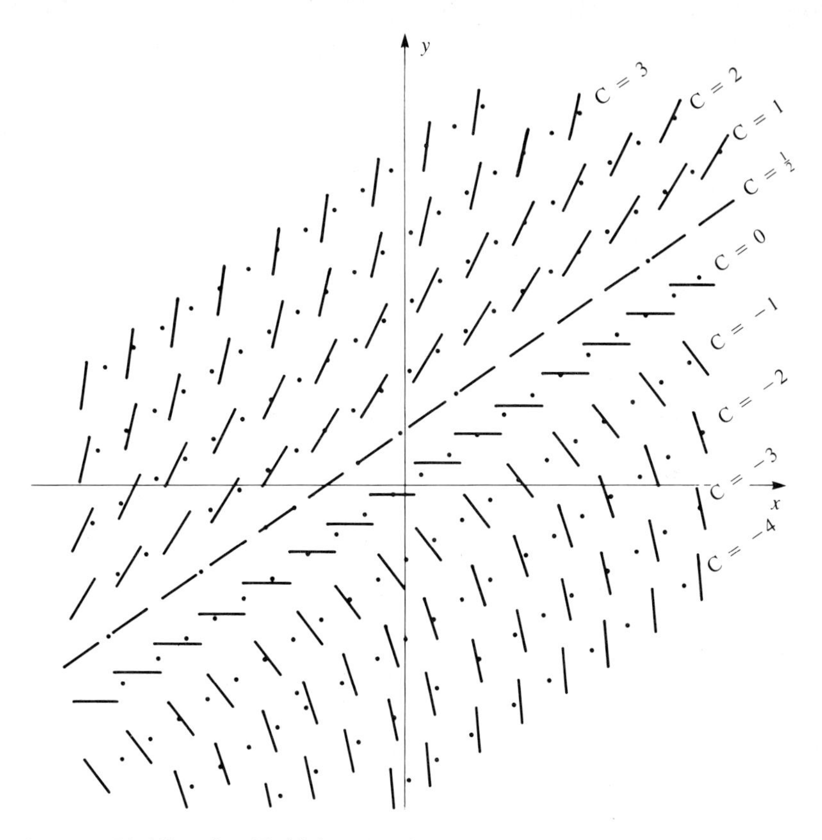

Figure 1.16 Direction Field for $y' = 2y - x$

suggest the curves of the magnetic field. A meteorologist also uses directed segments to indicate directions of currents and winds.

Another geometrical device is used to assist us in making the sketch of the direction field for $y' = f(x, y)$. The curves in the xy-plane defined by $f(x, y) = C$ are curves on which the value of y' is always the same, namely, the value of the constant. The curve $f(x, y) = C$ is called an **isocline** (equal inclination or equal slope for the lineal elements), and the value of y' is the same at every point of an isocline. Hence if we are able to draw an isocline, then at any of its points the lineal element has slope C and many lineal elements may simultaneously be drawn. By carefully choosing a selection of isoclines and on each of them drawing a collection of lineal elements we get a rather complete direction field. Isoclines were used in constructing the direction field for Figures 1.15, 1.16, and 1.17.

Once a reasonably complete direction field has been established in the xy-plane, we can construct a curve that is tangent at each of its points to the lineal element at that point. This curve is the graph of a function that is a

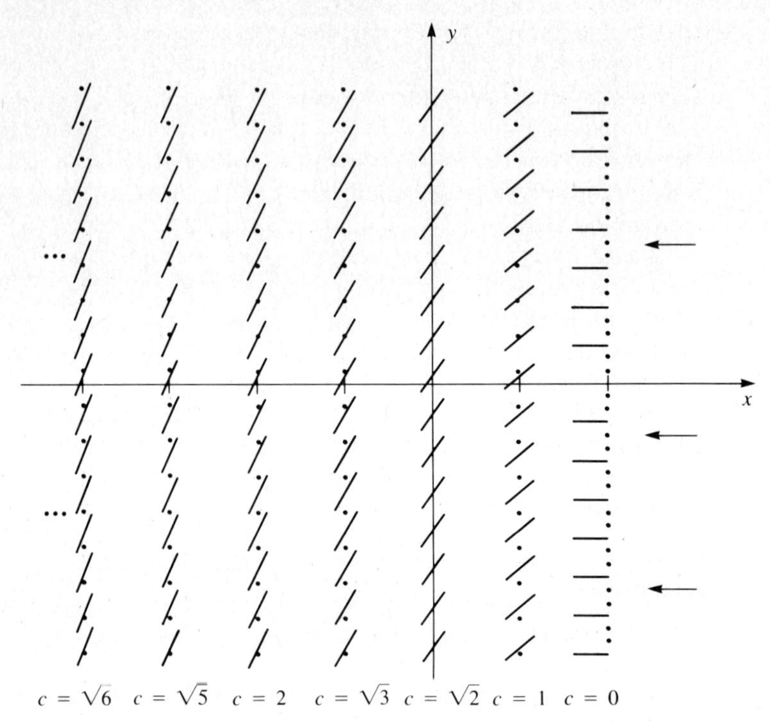

Figure 1.17 Direction Field for $y' = \sqrt{2 - x}$

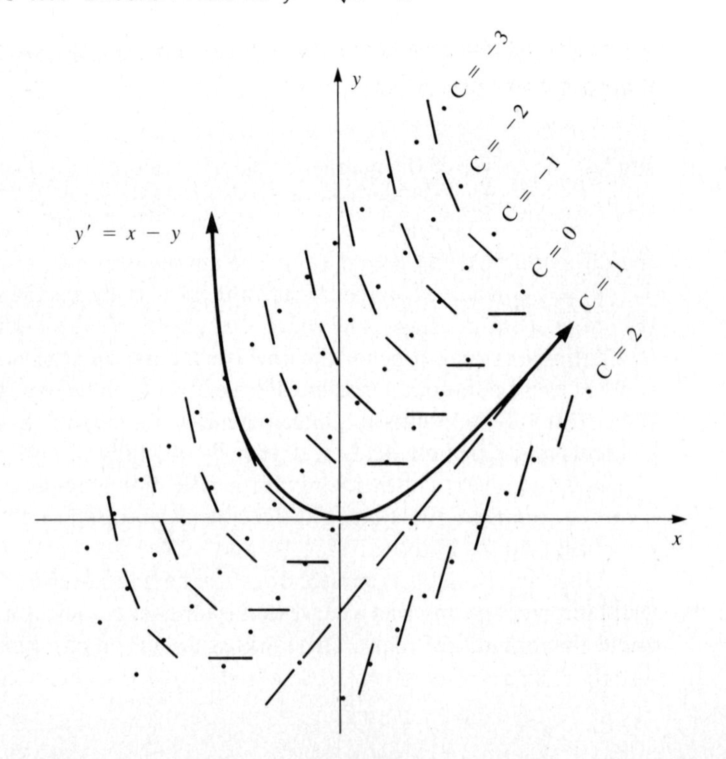

Figure 1.18

solution to the differential equation that produced the direction field. Conversely, if we have a function that is a solution to a differential equation, then it satisfies the differential equation so its slope y' is equal at every point to the value of $f(x, y)$ there. This means that it is tangent to the elements of the direction field. A geometrical approach to solving a differential equation would be to thread together the segments of the direction field into smooth differential curves. As we move from point to point in the field we would always be moving tangent to the solution curve and a smooth threading process would produce an actual solution curve.

For the direction field of $y' = x - y$ shown in Figure 1.18, one solution curve, which is drawn, passes through the origin. Note again the isoclines $x - y = C$ or $y = x - C$.

One approach to solving an arbitrary differential equation $y' = f(x, y)$ is to first draw the direction field to get an idea of the nature of the solution and then to solve it analytically, if possible, and draw in the solution curves to verify the correctness of both methods.

Example 45

Figure 1.19 illustrates this approach for solving several differential equations.

Differential Equation of a Family

Let $G(x, y)$ be a function whose domain is a set of points in the xy-plane. Then for every real number C in the range of G, the graph of $G(x, y) = C$ will be a curve in the plane. The entire collection of such graphs is called a **family of curves.**

Thus far we have begun with a differential equation $y' = f(x, y)$ and have worked toward getting a one-parameter family of curves as its solution family. Working now in a direction opposite to this approach, we will begin with a family of curves and find the differential equation that is satisfied by its members. The process is easy. It can be phrased as follows:

> Differentiate both sides of $G(x, y) = C$ with respect to x and solve for y'.

Example 46

Find the differential equation of the family $y = Ce^{2x}$. Rewriting the equation as $ye^{-2x} = C$, we differentiate with respect to x by using the

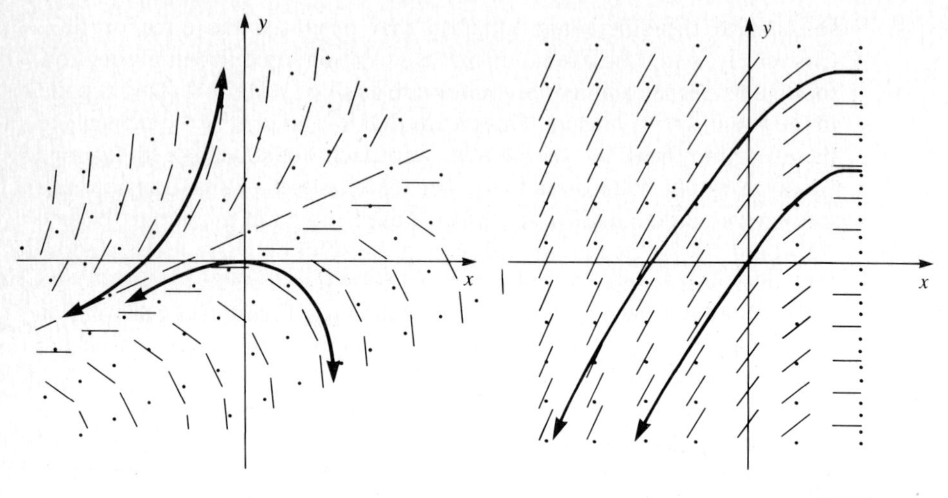

Direction Field for $y' = 2y - x$
with Solution Curves

Direction Field for $y' = \sqrt{2 - x}$
with Solution Curves

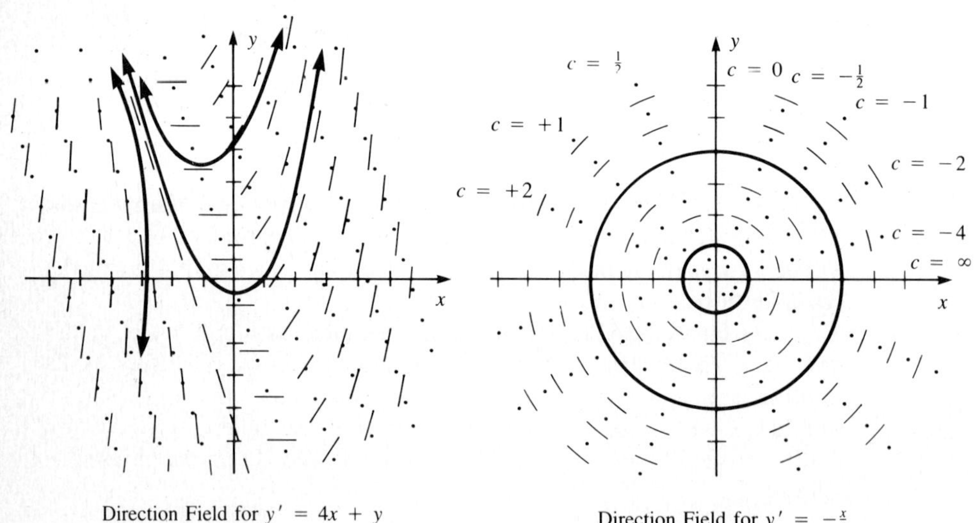

Direction Field for $y' = 4x + y$
with Solution Curves

Direction Field for $y' = -\frac{x}{y}$
with Solution Curves

Figure 1.19 Direction Fields

product rule and obtain

$$-2ye^{-2x} + e^{-2x}y' = 0$$

Dividing by the nonzero function e^{-2x} we obtain $y' - 2y = 0$. This is the differential equation whose solution family is $y = Ce^{2x}$.

Orthogonal Trajectories

As an application of this geometrical theory, we consider the problem of determining the orthogonal trajectories of a given family.

Definition: Let $F(x, y) = C$ be a given one-parameter family of curves in the xy-plane. Any curve that intersects each curve in this family at a right angle is called an **orthogonal trajectory** of the given family.

The family of such curves is called the **family of orthogonal trajectories** of the original family. Historically the problem of finding the orthogonal trajectories for a family was proposed by Bernoulli and Leibniz to Sir Isaac Newton near the end of his life. Newton's attempt at a solution was weak, but the subsequent development of calculus based on the work of these pioneers provides for an easy process.

As an example, the family of equilateral hyperbolas represented by $x^2 - y^2 = C$, shown in Figure 1.20, has as its orthogonal trajectories the family $xy = C$, also shown in Figure 1.20. To show that this relationship is correct, we will describe the procedure for finding orthogonal trajectories.

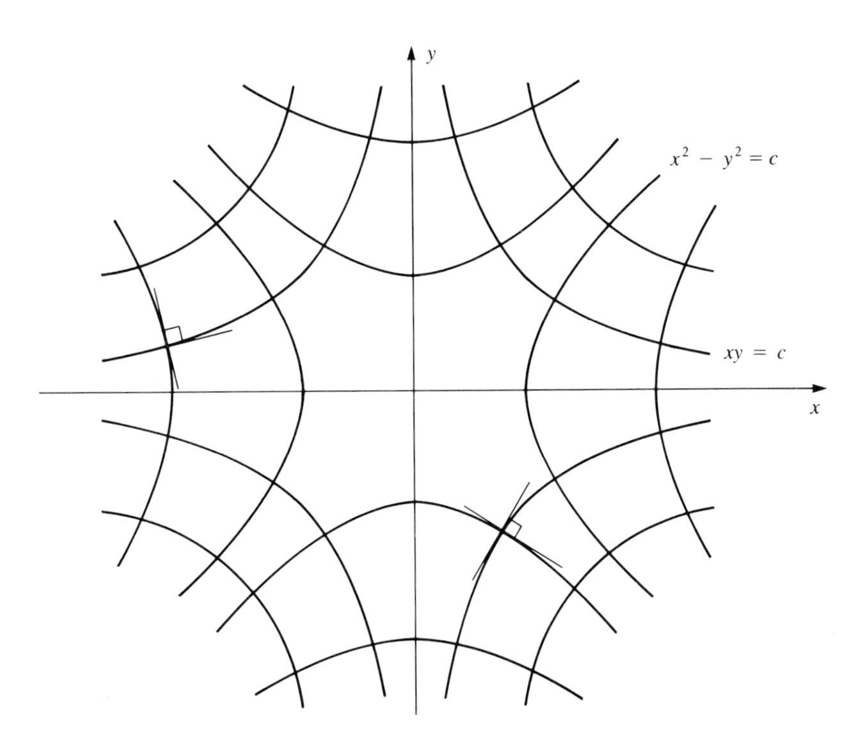

Figure 1.20

Starting with the original family we first use the steps described earlier to find the associated differential equation:

$$x^2 - y^2 = C$$

$$2x - 2yy' = 0$$

$$y' = \frac{x}{y}$$

This is the differential equation whose one-parameter family of solution functions is the original family. In its geometric interpretation the right hand side is the function describing the slope of each solution curve at a point. The orthogonal trajectories are perpendicular to the hyperbolas, so their slopes are the negative reciprocals of the slopes of the hyperbolas. For this example the differential equation for the orthogonal trajectories is $y' = -y/x$. Separation of variables gives $dy/y = -dx/x$, and integration gives $\ln|y| = -\ln|x| + \ln C$ or $y = C/x$. This, of course, is the family $xy = C$ we predicted in Figure 1.20.

Summarizing this, we see that to find the orthogonal trajectories for $F(x, y) = C$ we first obtain the differential equation of this family: $y' = f(x, y)$. We now replace $f(x, y)$ by its negative reciprocal, obtaining $y' = -1/f(x, y)$, which is the differential equation of the orthogonal trajectories. We can then solve this equation by any available method to obtain the trajectories.

Following are several examples of physical interpretations of orthogonal trajectories:

1. In meteorology, the isobars are curves connecting points of equal barometric pressure. Their orthogonal trajectories are wind direction lines from high- to low-pressure spots.
2. In electrostatics, the lines of potential are orthogonal trajectories to the field lines of force.
3. In fluid flow, the streamlines along which the fluid flows are orthogonal trajectories to lines of equal potential.

Example 47

Find the orthogonal trajectories of $y = Cx^2$.
Rewrite the equation as $yx^{-2} = C$ and differentiate $-2yx^{-3} + x^{-2}y' = 0$. This becomes $y' = 2y/x$. The negative reciprocal of the right side is now used to produce the new differential equation $y' = -x/2y$, which is solved to obtain $x^2 + 2y^2 = k^2$. This family of ellipses is sketched in Figure 1.21 with its original family of parabolas.

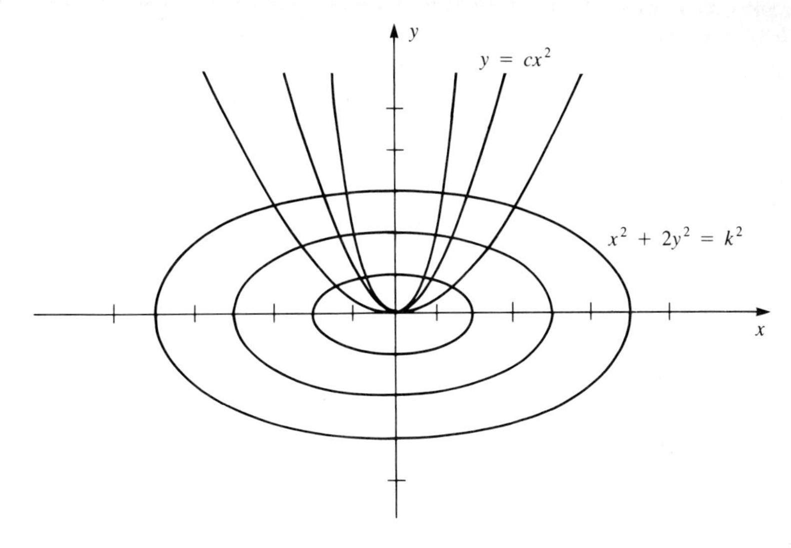

Figure 1.21

Example 48

The concept of orthogonal trajectories is used in oceanographic studies of wave trains. When a wave train approaches a beach, the lines of maximum energy are orthogonal to the wave contours. This relationship, illustrated in Figure 1.22, is important in the construction of offshore anchoring buoys for large ships, placement of offshore oil-drilling platforms for maximum stability, etc.

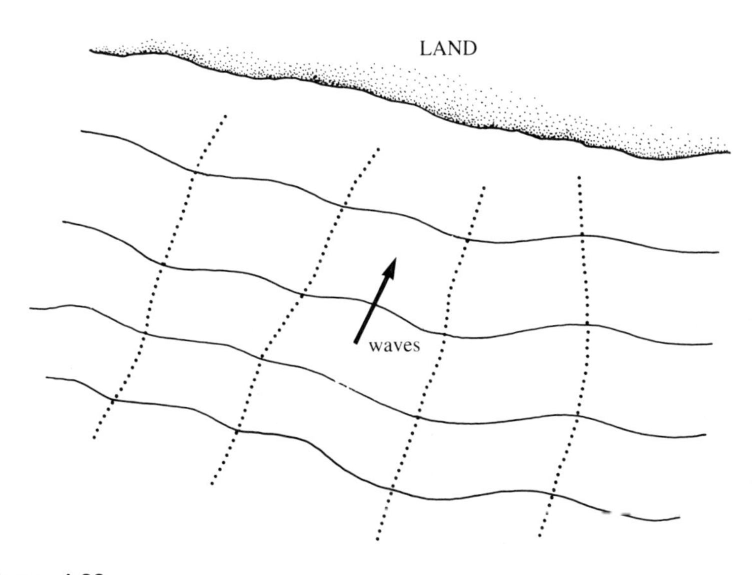

Figure 1.22

Exercises 1.7

Sketch the direction fields for each differential equation in Exercises 1–7. Also describe the solution curves generated.

1. $y' = 2x + 2$
2. $y' = xy$
3. $y' = 2x - y$
4. $y' = 1 - (x^2 + y^2)$
5. $yy' + x = 0$
6. $y' = \dfrac{4x + 3y}{3x + y}$
7. $y' = x + y + \dfrac{1}{4}$

Find the orthogonal trajectories for the families in Exercises 8–19.

8. $y^2 = x^2 + c$
9. $y = cx^5$
10. $xy = cx - 1$
11. $y = ce^{-x^2}$
12. $y^2 = cx^3$
13. $y = \dfrac{x}{cx + 1}$
14. $x^2y = k$
15. $e^x \cos y = k$
16. $\sin y = ke^{x^2}$
17. A family of straight lines through the origin.
18. A family of circles with variable radii, with centers on the y-axis, and which pass through the origin.
19. A family of parabolas with vertices at the origin and foci on the x-axis.
20. Show that $y^2 = 4p(x + p)$ is self-orthogonal.
21. Show that $x^2/a^2 + y^2/(a^2 - c^2) = 1$, where $(\pm c, 0)$ are the foci, is self-orthogonal.

Exercises 22–25 require the determination of curves from geometric properties.

22. At any point of a curve, let the slope of the curve equal the reciprocal of the x-coordinate. Find the equation of the family of curves having this property.
23. Find the equation of the family of curves having the following

property: At each point the line joining the point to the origin is perpendicular to the tangent line at the point.

24. Find the equation of the family of curves having the following property: At every point of the curve the projection onto the x-axis of that portion of the normal line from the curve to the x-axis has length L.
25. Find the equation of the family of curves for which the length of any arc equals the area above the x-axis and under the arc.

1.8 Equations Reducible to First-Order Equations

Certain second-order differential equations are actually disguised first-order equations and may be solved by successive applications of the present theory. If either the dependent variable or the independent variable is absent from the second-order equation, a substitution will accomplish the reduction of order. The given equation will then have one of the two forms: $F(x, y', y'') = 0$ if the dependent variable y is missing or $F(y, y', y'') = 0$ if the independent variable x is missing. We introduce the notation

$$p = \frac{dy}{dx}$$

Then

$$\frac{d^2y}{dx^2} = \frac{dp}{dx} = \frac{dp}{dy}\cdot\frac{dy}{dx} = p\frac{dp}{dy}$$

and thus we can substitute for the second derivative either of the expressions

$$y'' = \frac{dp}{dx} \quad \text{when } y \text{ is missing}$$

or

$$y'' = p\cdot\frac{dp}{dy} \quad \text{when } x \text{ is missing.}$$

Example 49

Consider the differential equation $y'' - 2y' = 0$. This second-order equation does not involve the dependent variable represented by the letter y. Therefore, we use the substitution $p = y'$, for which $p' = y''$, and reduce the equation to $p' - 2p = 0$, whose solution is easily found by separation of variables to be $p = c_1 e^{2x}$. We then set $y' = p = c_1 e^{2x}$ and solve for y by integration, obtaining $y = (c_1/2)e^{2x} + c_2$.

Example 50

Solve by reduction of order

$$\frac{d^2y}{dx^2} = x\left(\frac{dy}{dx}\right)^3$$

Since y does not appear, we set $y' = p$, $y'' = p'$ and obtain

$$\frac{dp}{dx} = xp^3$$

which is solved by separation of variables to $p = \pm 1/\sqrt{c_1^2 - x^2}$. Since $p = dy/dx$ we integrate this equation to obtain the solution

$$y = \text{Arcsin}\,\frac{x}{c_1} + c_2$$

When the independent variable x is not specifically present in the differential equation, the helpful substitution is $y' = p$ for which $y'' - p(dp/dy)$.

Example 51

Consider $(y + 1)y'' = (y')^2 + y'$. Since x is absent we set $y' = p$, $y'' = p(dp/dy)$ and obtain

$$(y + 1)p\frac{dp}{dy} = p^2 + p$$

or

$$(y + 1)\,dp = (p + 1)\,dy$$

Integration after separation produces

$$p + 1 = c_1(y + 1)$$

Then

$$\frac{dy}{dx} = p = c_1(y + 1) - 1$$

which produces

$$\ln|c_1(y + 1) - 1)| = c_1(x + c_2)$$

which is the general solution in implicit form.

Example 52

Consider $y'' = -y^{-3}$, a differential equation in which x is missing. Substitution changes this to $p(dp/dy) = -y^{-3}$. Separating the variables now gives $p\,dp = -y^{-3}\,dy$ which integrates to

$$\frac{p^2}{2} = \frac{1}{2y^2} + \frac{c_1}{2}$$

Setting $p = y'$ now gives

$$(y')^2 = \frac{1 + c_1 y^2}{y^2}$$

or

$$y' = \frac{\pm\sqrt{1 + c_1 y^2}}{y}$$

Separating again gives

$$\frac{y\,dy}{\pm\sqrt{1 + c_1 y^2}} = dx$$

This integrates to

$$\pm\int (1 + c_1 y^2)^{-1/2} y\,dy = x + c_2$$

$$\pm\frac{1}{c_1}(1 + c_1 y^2)^{1/2} = x + c_2$$

or

$$(1 + c_1 y^2)^{1/2} = \pm c_1(x + c_2)$$

or

$$1 + c_1 y^2 = c_1^2(x + c_2)^2$$

which is the general solution. This appears to be a two-parameter family of hyperbolas.

Example 53

In the study of astrodynamics* the differential equation

$$y'' = [1 + (y')^2]\cdot f(x, y, y')$$

is encountered. This may be integrated by reducing it to first order if the function $f(x, y, y')$ is linear in y' and is an exact total differential. If we choose a simple function such as $f(x, y, y') = y'$, we obtain the differential equation $y'' = y'[1 + (y')^2]$. Since both x and y are missing, we choose the easier substitution $p = y'$ and $p' = y''$ to obtain $p' = p(1 + p^2)$. This separates to

$$\frac{dp}{p(1 + p^2)} = dx \Rightarrow \frac{dp}{p} - \frac{p\,dp}{1 + p^2} = dx$$

---

* A. Ghaffari, "On the Integrability Pieces of the Equation of Motion for a Satellite in an Axially Symmetric Gravitational Field," *Celestial Mechanics*, 4 (1971), pp. 49–53.

To solve this for p as a function of x is difficult, so we back up and try the more difficult substitution $p = y'$ and $p(dp/dy) = y''$ to get $p(dp/dy) = p(1 + p^2)$. This separates to

$$\frac{dp}{1 + p^2} = dy \Rightarrow \text{Arctan } p = y + c$$

or

$$p = \tan(y + c)$$

$$\frac{dy}{dx} = \tan(y + c)$$

$$\frac{dy}{\tan(y + c)} = dx$$

$$\ln|\sin(y + c)| = x + k$$

$$\sin(y + c) = Ke^x$$

$$y + c = \text{Arcsin }(Ke^x)$$

$$\therefore\ y = \text{Arcsin }(Ke^x) - c$$

Exercises 1.8

In Exercises 1–10 find the general solution of each differential equation.

1. $y'y'' = 1$
2. $y'' = [1 + (y')^2]^{1/2}$
3. $y'' + 4(y')^2 = 0$
4. $xy'' - 3y' = x^4$
5. $(1 - x^2)y'' + 2xy' = 0$
6. $y'' = e^x(y')^2$
7. $xy'' = x^2 + y'$
8. $x^3y'' + 5x^2y' + 3 = 0$
9. $y^2y'' + (y')^3 = 0$
10. $yy'' = (y')^2$

In Exercises 11–16 find the particular solution indicated.

11. $x[1 - (y')^2] + y'(1 - x^2)y'' = 0;\ y(-2) = 0,\ y'(-2) = \dfrac{2}{\sqrt{3}}$
12. $(x - 1)(y'')^2 = y' - 1;\ y(2) = 1,\ y'(2) = 2,\ y'' > 0$
13. $y'' - 2(y')^2 = 2y';\ y(0) = 1,\ y'(0) = -1$
14. $yy'' - (y')^2 - y^2y' = 0;\ y(0) = y'(0) = 1$
15. $(x^2 + 1)y'' + 2xy' = 0$ } Computer integration may be needed.
16. $y'' + c^{-y}(y')^2 = 0$ }

17. The radius of curvature at a point (x, y) on a curve that has a horizontal tangent at the point $(3, 1)$ is $2y^{3/2}$. Find the equation of the curve.
18. Consider a 10-foot chain weighing 2 lb/ft that is being held down on a smooth table with two feet hanging over the frictionless edge, which is 15 feet from the floor. When the chain is released, its subsequent motion is analyzed as follows: On the vertical section where x is its length at time t, the forces are gravity $w = 2x$ and tension T at the top. Newton's law $F = ma$ gives $2x - T = \frac{2x}{32} \cdot \frac{d^2x}{dt^2}$. On the horizontal portion the only force is tension T. Thus $T = \frac{2(10 - x)}{32} \cdot \frac{d^2x}{dt^2}$. Adding these we obtain $2x = \frac{20}{32} \cdot \frac{d^2x}{dt^2}$. Solve this equation with the initial conditions implied in the problem to find:
    (a) how long it will take for the chain to slide off the table.
    (b) what its velocity will be at the moment it no longer touches the table.
    (c) when it will first touch the floor.
19. The Los Angeles area often suffers from temperature inversions in which a layer of cool air traps a layer of warm air beneath it and prevents the warm air from rising. The pollutants trapped in such an inversion create a massive smog dome containing—among other things—hydrogen sulfide H_2S and sulfur dioxide SO_2. To determine the concentration of these, $H(t)$ and $S(t)$, in order to predict the intensity of the irritation, we use the model

$$\frac{dH}{dt} = -\alpha H + \gamma$$

$$\frac{dS}{dt} = \alpha H - \beta S + \delta$$

where:

α is the conversion rate of H_2S into SO_2;

β is the conversion rate of SO_2 into sulfate;

γ is the production rate of H_2S;

δ is the production rate of SO_2.

Solve by using $\alpha = .35$, $\beta = .64$, $\gamma = .04$, $\delta = .05$, $H(0) = .08$, $S(0) = .02$. Estimate the levels of concentration that could be reached under a prolonged entrapment.
20. In the study of the mathematical relationships between the kinetics of glucose and insulin in blood plasma we let g = plasma glucose

concentration and h = plasma IRI (immunoreactive insulin) concentration and obtain

$$\frac{dh}{dt} = -c_1 h + c_2 g$$

$$\frac{dg}{dt} = -c_3 g - c_4 h$$

where the rate constants have the following values: $c_1 = .0457$; $c_2 = .0248$; $c_3 = .0630$; and $c_4 = .0030$. Solve for $h(t)$ and $g(t)$ by differentiating one equation and inserting the other.

21. In designing the punch arm on a large machine, the path that the tip of the arm will follow is of great importance. Suppose that in a given coordinate system, it is required that the tip pass through the point (3, 1) horizontally. If we specify the radius of curvature of the path to be $2y^{3/2}$, what path is traced out?
22. In an attempt to cast the human eye-positioning mechanism into a mathematical model, Cook and Stark* derived the following equation for an isometric contraction of the muscle:

$$T - \left(\frac{1.25\,T}{b + x'}\right)x' = K(x - L)$$

where T is muscle tension controlled by the nerve signal; L is the length of the elastic component; $b = 1500$ degrees/second; and $K = 6$ gm/degree. Solve for $x(t)$ in terms of T, L, and t. (*Hint:* Substitute for combinations of constants; for example, let $T/4 - 6L = 6V$ and let $250T + 1500L = u$. Then find $x(t)$ in terms of u, V, and t and then reverse substitute.)

1.9 Linear Motion of a Body

Many of the important applications of first-order equations are found in the field of elementary mechanics. We will consider here the motion of a solid body whose center of mass moves in a straight line. This motion follows the basic law of mechanics:

Newton's second law: The mass of the body times the acceleration of the center of mass is proportional to the force acting on the body.

---

* G. Cook and L. Stark, "Derivation of a Model for the Human Eye-Positioning Mechanism," *Bulletin of Mathematical Biophysics*, 29 (1967), pp. 153–74.

In mathematical language this is stated as

$$F = kma$$

If the mass of the body is not constant this may be written as

$$F = k\frac{d}{dt}(mv)$$

where the quantity mv is called the linear momentum. Newton's law would then be stated thus: The rate of change of momentum is proportional to the applied force.

Two systems of units are commonly used. In the centimeter-gram-second system (c-g-s), distance is in centimeters; mass in grams; time in seconds; and force in dynes. In the British system the basic units are feet for length, slugs for mass, seconds for time, and pounds for force. Both of these systems produce a proportionality constant $k = 1$.

If we now consider a body that is falling freely near the surface of the earth in the absence of air resistance, the only force on it is the gravitational force (exerted by the earth), which is equal to the weight w of the body. If g designates the acceleration due to gravity, then Newton's law $F = ma$ now takes the form $w = mg$ or $m = w/g$. This is a relation we will frequently use.

Let us now consider a body in motion along a straight line. A reference point 0, the origin, a distance unit, and a positive direction are selected. We will let the coordinate x designate the displacement from 0. Then the instantaneous velocity of the body is the time rate of change of x, $v = dx/dt$, and the instantaneous acceleration of the body is the time rate of change of v:

$$a = \frac{dv}{dt} = \frac{d^2x}{dt^2}$$

Newton's second law may now take three forms:

$$F = m\frac{dv}{dt}$$

$$F = m\frac{d^2x}{dt^2} \tag{18}$$

or

$$F = mv\frac{dv}{dx}$$

The correct form to use in a particular problem depends on the form of the resultant force F and what quantities are to be determined. For example, consider an object thrown directly upward from the surface of the earth with initial velocity v_0. If x is directed positively upward and we neglect air resistance then (18) becomes

$$m\frac{d^2x}{dt^2} = -mg$$

and the initial conditions are $x(0) = 0$, $dx/dt(0) = v_0$. After cancelling the m an easy integration of this differential equation provides

$$\frac{dx}{dt} = -gt + c_1$$

However, since $dx/dt(0) = v_0$, we have $v_0 = c_1$. Then $dx/dt = -gt + v_0$. This is a velocity determining equation.

Integrating again, we have

$$x(t) = -\frac{1}{2}gt^2 + v_0 t + c_2 \tag{19}$$

Since $x(0) = 0$, then $0 = c_2$ and thus $x(t) = -\frac{1}{2}gt^2 + v_0 t$ is the position determining equation.

Note that the velocity is positive until time $t = v_0/g$, when the object reaches its highest point, and thereafter v is negative as the body descends toward the surface. The maximum height from (19) is then

$$x_{\text{max}} = -\frac{1}{2}g\frac{v_0^2}{g^2} + v_0\frac{v_0}{g} = \frac{v_0^2}{2g}$$

This analysis produces the equations for vertical motion in a straight line with no forces except gravity acting on the body.

Now if a force other than the accelerative gravitational force is present we have a different problem. If a body falls through air toward the earth, it experiences the retarding force of air resistance. The amount of resistance depends on the body's velocity as well as its shape and the nature of the surrounding medium. No general law applies to this type of problem.

In some cases the damping or resistive force is proportional to the velocity. Since this force is opposite in direction to the gravitational force mg, we will designate it as

$$\text{force of damping} = F_d = -c(\text{velocity}) = -cv$$

where c is the positive proportionality constant. Then

$$F = m\frac{dv}{dt} = mg - cv \tag{20}$$

or

$$\frac{dv}{dt} + \frac{c}{m}v = g$$

This is a linear first-order equation whose integrating factor is

$$e^{\int c/m\,dt} = e^{ct/m}$$

The solution is then obtained by

$$ve^{ct/m} = \int ge^{ct/m}\,dt = \frac{ma}{c}e^{ct/m} + C_1$$

or

$$v(t) = \frac{mg}{c} + C_1 e^{-ct/m}$$

We consider the object as it is released from rest, so that $v(0) = 0$, and hence we have $C_1 = -mg/c$. Thus the solution is

$$v(t) = \frac{mg}{c}(1 - e^{-ct/m})$$

It is worth observing that as $t \to \infty$, the exponential term goes to zero and hence velocity approaches the limiting value $v_\infty = mg/c$. This implies that as the velocity of a falling object increases, the damping force increases in magnitude until it balances the gravitational force. The acceleration then tends to zero.

We get an equivalent analysis if we use the zero acceleration notion in equation (20). This equation $m(dv/dt) = mg - cv$ then has zero on the left, so that $mg = cv_\infty$ or $v_\infty = mg/c$. The value of c is called the **coefficient of resistance** of the medium; and if the resistance diminishes, the limiting velocity increases. This property would be readily observed if we simultaneously released from a high altitude a metal cube and a metal needle-nosed model rocket.

In contrast to the case in which the retarding force is proportional to the velocity, there are problems in which the air resistance is proportional to the square (or some other power) of the velocity. For example, consider the following case:

A female sky diver whose total weight (including clothes and parachute) is 192 pounds jumps from an air plane traveling parallel to the earth's surface. She experiences a vertical fall from rest. Before the parachute opens, the air resistance is equal to $(1/2)v$; when the chute opens 6 seconds later, the air resistance becomes $\frac{3}{4}v^2$. We will now find the velocity of the diver before the parachute opens and then the velocity after it opens.

For convenience of calculations we will place the origin at the jump point and let x be directed positively downward. We need to break the problem into two distinct parts. Figure 1.23 illustrates part one, the free fall; and the expression of Newton's second law is:

$$F_1 + F_2 = ma = \frac{w}{g}\cdot a = \frac{w}{g}\cdot\frac{dv}{dt}$$

$$192 - \frac{1}{2}v = \frac{192}{32}\frac{dv}{dt}$$

$$= 6\frac{dv}{dt}$$

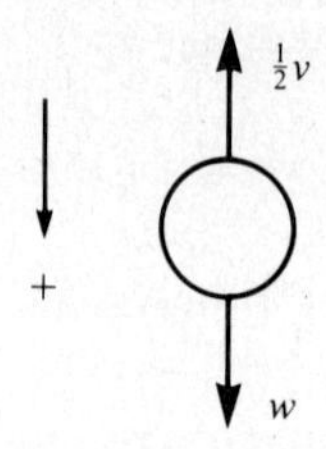

Figure 1.23

The given initial condition is $v(0) = 0$. In solving this differential equation we can use separation of variables to obtain

$$\frac{dv}{192 - (v/2)} = \frac{1}{6}\,dt$$

or

$$\frac{dv}{v - 384} = -\frac{1}{12}\,dt$$

Integrating this we get

$$\ln|v - 384| = -\frac{1}{12}t + c_0$$

or

$$v = 384 + ce^{-(1/12)t}$$

Since $v(0) = 0$, $c = -384$, so

$$v(t) = 384(1 - e^{-(1/12)t}) \tag{21}$$

which is valid in the time interval $0 \le t \le 6$. At the time the chute opens at 6 seconds

$$v(6) = v_6 = 384(1 - e^{-1/2}) \simeq 151.1 \text{ ft/sec}$$

For the second part of the problem, the resistive force F_2 has changed to $\frac{3}{4}v^2$, so the differential equation now is

$$192 - \frac{3}{4}v^2 = 6\frac{dv}{dt}$$

with $v(6) = v_6$ as previously obtained. If we now simplify and separate variables we get

$$\frac{dv}{v^2 - 256} = -\frac{1}{8}\,dt$$

Integrating yields

$$\ln\left|\frac{v - 16}{v + 16}\right| = -32\left(\frac{1}{8}t\right) + K = -4t + K$$

Therefore

$$e^{\ln|(v-16)/(v+16)|} = e^{-4t+K} = e^K \cdot e^{-4t}$$

and

$$\left|\frac{v-16}{v+16}\right| = e^K \cdot e^{-4t}$$

or

$$\frac{v-16}{v+16} = Ce^{-4t}$$

which simplifies to

$$v(t) = \frac{16(Ce^{-4t}+1)}{(1-Ce^{-4t})}$$

Using $v = v_6$ at $t = 6$ we get

$$v_6 = \frac{16(Ce^{-24}+1)}{1-Ce^{-24}}$$

which produces

$$C = e^{24}\left(\frac{135.1}{167.1}\right) = .8085\,e^{24}$$

Hence the valid velocity equation for time $t \geq 6$ is

$$v(t) = \frac{16(.8085\,e^{24-4t}+1)}{(1-.8085\,e^{24-4t})}$$

For large t, the limiting velocity is $v_\infty = 16$.

Note that by observing equation (21) for part one of the fall, the limiting velocity would be 384 feet per second if the chute failed to open. This would cause an abrupt stop when the sky diver reached the earth! In part two the limiting velocity is tolerable.

Let us now consider a new situation as an additional example of motion with a body of constant mass. A particle of mass m starts from a resting position at the origin and moves in a straight line. A force retarding the motion is a linear function of the velocity but is not dependent on displacement or time. Assuming that the retarding force is $F = a - bv$ with $a, b > 0$, we will find the equation of motion. If we use this value of F with Newton's second law and write the acceleration as dv/dt, we obtain

$$m\frac{dv}{dt} = a - bv$$

If we separate the variables and integrate, we arrive at

$$\frac{dv}{a-bv} = \frac{1}{m}\,dt$$

$$-\frac{1}{b}\ln|a-bv| = \frac{t}{m} + C$$

With a little algebra this reduces to

$$v = \frac{1}{b}[a - e^{-bc} \cdot e^{-bt/m}]$$

Since the given initial conditions are $y(0) = v(0) = 0$, we get

$$0 = a - e^{-bc}$$

so that

$$v(t) = \frac{a}{b}[1 - e^{-bt/m}]$$

As t grows large the terminal velocity is $v_{\max} = a/b$. If we now separate the variables in this velocity equation, we get

$$v = \frac{dy}{dt} = \frac{a}{b}[1 - e^{-bt/m}]$$

$$dy = \frac{a}{b}[1 - e^{-bt/m}]\, dt$$

so

$$y = \frac{a}{b}t + \frac{am}{b^2}e^{-bt/m} + K$$

Again, since $y(0) = 0$ we find $K = -am/b^2$, so

$$y = \frac{at}{b} + \frac{am}{b^2}(e^{-bt/m} - 1)$$

This equation describes the displacement of the body from the origin at any time $t \geq 0$.

Exercises 1.9

1. A ball is dropped from a height of 16 feet. When it hits the ground it bounces back up with a speed that is 5/8 of its speed of impact. How high does it go after the first bounce? After the second bounce?
2. A ball is dropped from a tall building. Two seconds later another ball is dropped from the same place. Do the two balls remain the same distance apart? Find a formula to express the separation distance as a function of the time after the second ball is released.
3. A ship weighing 48,000 tons starts from rest under a force of constant propeller thrust of 200,000 pounds. If the water resistance against the movement is 10,000v where v is velocity in ft/sec, find the general expression for the velocity as well as a value for terminal velocity.

4. An Olympic rowing crew starts a race from rest. The boat, including the crew, weighs 640 pounds. If the collective force exerted by the oars in the direction of motion is constantly 16 pounds and the water resistance is equal to twice the speed, find the speed 20 seconds after the race begins.
5. A barge is being towed at 9 ft/sec when the towline breaks. The barge continues in a straight line but slows down due to water resistance at a rate proportional to the square root of its velocity. Ninety seconds after the break, the barge is observed to be moving at 5 ft/sec. How far does the barge travel from the time of the break until it comes to rest?
6. A balloon is ascending vertically at the rate of 16 ft/sec. When the balloon is 880 feet above the ground a stone is released from it. Fix an origin in space at the point where the stone is released and find the equation of motion of the stone. What is the maximum height reached by the stone?
7. A rocket containing a lunar probe is launched from the surface of the earth with initial velocity v_0. Assume that the only force acting on the rocket is its weight. At x miles from the center of the earth, whose radius is $R = 4000$ miles, the weight of the rocket is mgR^2/x^2, where m is the mass and $g = \frac{32}{5280}$ mi/sec^2. By Newton's second law, $F = ma$, we obtain

$$\frac{-mgR^2}{x^2} = ma = m\frac{dv}{dt}, \text{ with } v(4000) = v_0$$

But

$$\frac{dv}{dt} = \frac{dv}{dx}\cdot\frac{dx}{dt} = v\frac{dv}{dx}$$

so that

$$\frac{-mgR^2}{x^2} = mv\frac{dv}{dt}$$

which is separable.

(a) Solve this problem to show that $v^2 = v_0^2 - 2gR + 2gR^2/x$.

(b) For the rocket to escape the earth's gravitational pull, v should be nonzero until x approaches infinity. Show that the initial velocity should be about 7 miles per second to accomplish this.

1.10 Mathematical Modeling

To give you a deeper understanding of the uses of differential equations, we present in Sections 1.10 and 1.11 a few detailed cases that demonstrate how differential equations are used in the real world.

Spread of an Infection through a Population

The first example results from notes on lectures delivered by Dr. Maynard Thompson (professor of mathematics at the University of Indiana and consultant to the Center for Disease Control in Atlanta, Georgia) at the Michigan Section, MAA Short Course on Problems in Mathematical Biology, Hope College, Holland, Michigan, July 1978.

In creating a model to study the spread of an infection through a population we will first need a number of definitions.

Let S be the number of persons susceptible (called the "susceptibles") to contracting a disease in a population of constant size N (births = deaths) and consider S to be a continuous time dependent variable for $t > 0$. The constant-size assumption is good for short-term diseases (e.g., bacterial infections) in Western countries. Let I be the number of persons in the population who are infected and spreading the infection (called the "infectives") and let $s(t)$ and $i(t)$ be the fractions of the population in the classes S and I at time t. Thus

$$s(t) + i(t) = 1$$

Assume that the population is homogeneous and uniformly mixing, that is, there is a contact rate λ, which is the number of individuals to whom the infection is passed per infective per day. Then the quantity λNsi represents the number of contacts per day involving susceptibles and infectives.

Consider the simple epidemic model first developed by Sir Ronald Ross (a British physician) in 1911 in which members of S move into I as time progresses. This is portrayed symbolically as the model $S \to I$ and represents a biologically short or nonexistant incubation and a long period of illness compared to the time horizon. In this model we may therefore conclude that the time rate of change of the quantity Ns, the number of susceptibles, is proportional to the number of contacts made per day between the susceptibles and the infectives and that the number of susceptibles decreases with time as they move into class I. We then obtain

$$\frac{d(Ns)}{dt} = -\lambda Nsi$$

where the proportionality constant is taken as 1 by adjusting units and $\lambda > 0$. By canceling the constant N we get

$$\frac{ds}{dt} = -\lambda si$$

and since $s + i = 1$ we may obtain a differential equation for either $s(t)$ or $i(t)$. Hence $ds/dt = -\lambda s(1 - s)$ with $s(0) = s_0$ where $s_0 \neq 1$. This may be solved by separation of variables

$$\frac{ds}{s(1-s)} = -\lambda\, dt$$

$$\int\left(\frac{1}{s} + \frac{1}{1-s}\right) ds = -\lambda \int dt = -\lambda t + C$$

$$\ln|s| - \ln|1-s| = -\lambda t + C$$

$$\ln\left|\frac{s}{1-s}\right| = -\lambda t + C$$

$$\left|\frac{s}{1-s}\right| = e^{-\lambda t + C} = e^{-\lambda t} \cdot e^{C} = C_1 e^{-\lambda t}$$

Then

$$\frac{s}{1-s} = Ke^{-\lambda t} \tag{22}$$

where K absorbs the absolute value sign. Solving this algebraically for s gives

$$s = \frac{Ke^{-\lambda t}}{1 + Ke^{-\lambda t}}$$

Using equation (22) and the initial value $s = s_0$ when $t = 0$, we easily obtain

$$K = \frac{s_0}{1 - s_0}$$

Thus

$$s(t) = \frac{s_0 e^{-\lambda t}}{1 - s_0 + s_0 e^{-\lambda t}}$$

A graph of $s(t)$ versus t shows the behavior of that fraction of the population which is in class S (see Figure 1.24).

If we had eliminated s instead, the initial value problem would have read

$$\frac{d(1-i)N}{dt} = -\lambda N(1-i)i,\ i(0) = i_0 \text{ where } i_0 \neq 1$$

or

$$\frac{di}{dt} = \lambda i(1-i),\ i(0) = i_0$$

This is similar to the earlier problem and is solved by a similar technique to

$$i(t) = \frac{i_0 e^{\lambda t}}{1 - i_0 + i_0 e^{\lambda t}}$$

The graph of $i(t)$ versus t shows the expected increase in the size of $i(t)$ as

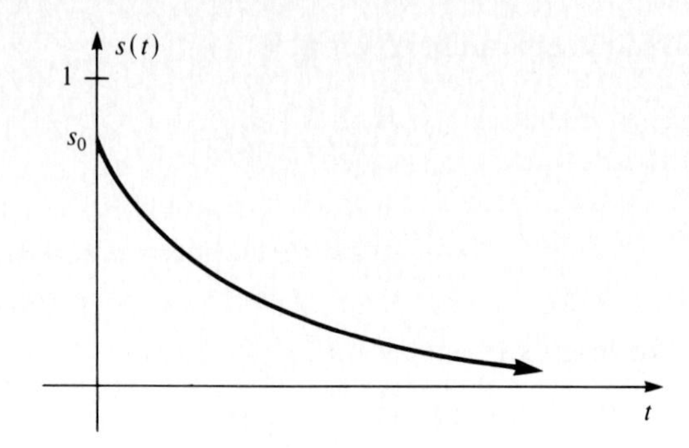

Figure 1.24

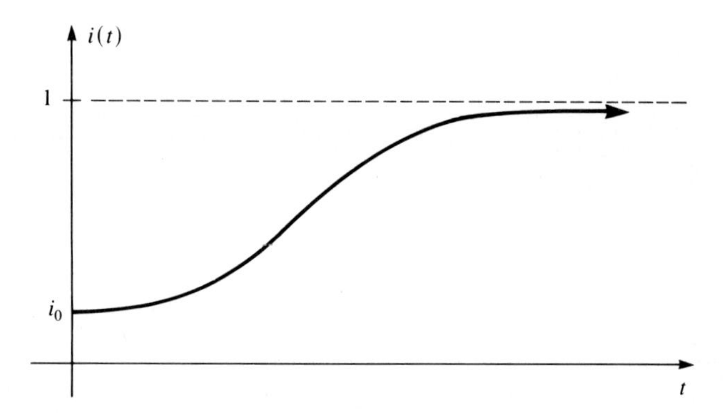

Figure 1.25

time progresses and as the susceptibles are transferred into class I (see Figure 1.25).

A slightly more complicated model—but one which is solved in a similar way—includes vital dynamics (i.e., births and deaths are accounted for in the population) and also allows for recovery from the illness. In this model, which is represented symbolically as $S \to I \to S$, the susceptibles move to the infective class, then recover and move back to susceptible class. We will assume that no immunity is acquired by the infectives as they recover and pass back into the susceptible class.

We will also assume all births are into class S and that deaths are uniform over the whole population (i.e., deaths are proportional to the number of individuals in various subpopulations). Let δ represent the death rate that is independent of class, constant in time, and not associated with the disease in question. Let γ represent the rate of removal (from class I) that is constant and due to the disease only. It gives a measure of how rapidly individuals recover and are removed from class I. Thus the total removal rate is $\delta + \gamma$. Again the constant size of the population is N.

The differential equation again analyzes the rate of change in Ns, but this model requires four terms on the right side:

$$\frac{d(Ns)}{dt} = -\lambda Nsi + \delta N + \gamma Ni - \delta Ns$$

The origin of the four terms is as follows:

1. The first represents the loss of persons from S to I that is due to their acquiring the infection by contact.
2. The second represents an influx to class S that is due to births. (Note that δ = death rate = birth rate is used.)
3. The third represents an influx to class S that is due to recoveries by persons in I. (Note that γ = removal rate is used.)
4. The fourth represents deletions from class S that are due to deaths.

We may cancel the constant N. Then $s(t) + i(t) = 1$ provides a differential equation in one variable only. Hence

$$\frac{ds}{dt} = -\lambda s(1 - s) + \delta + \gamma(1 - s) - \delta s; \quad s(0) = s_0 \neq 1$$

Or we could have

$$\begin{aligned}\frac{di}{dt} &= \lambda(1 - i)i - \delta - \gamma i + \delta(1 - i) \\ &= \lambda(1 - i)i - \delta - \gamma i + \delta - \delta i \\ &= i(\lambda - \delta - \gamma - \lambda i); \quad i(0) = i_0 \neq 1\end{aligned}$$

This is again solvable by separation of variables and partial-fraction decomposition. Let $\lambda - \delta - \gamma \equiv k$. Then

$$\frac{di}{dt} = i(k - \lambda i) \tag{23}$$

with $i(0) = i_0$. We can rewrite this as

$$\frac{di}{i(k - \lambda i)} = dt$$

Assuming first that $k \neq 0$, we have

$$\left(\frac{1/k}{i} + \frac{\lambda/k}{k - \lambda i}\right) di = dt$$

we can integrate to obtain

$$\frac{1}{k}\int\frac{di}{i}+\frac{1}{k}\int\frac{\lambda}{k-\lambda i}\,di=\int dt$$

$$\frac{1}{k}\ln|i|-\frac{1}{k}\ln|k-\lambda i|=t+C$$

$$\frac{1}{k}\ln\left|\frac{i}{k-\lambda i}\right|=t+C$$

$$\ln\left|\frac{i}{k-\lambda i}\right|=kt+C_1$$

Then

$$\left|\frac{i}{k-\lambda i}\right|=e^{kt}\cdot e^{C_1}=C_2e^{kt}$$

or

$$\frac{i}{k-\lambda i}=C_3e^{kt}$$

Using the condition that when $t=0$, $i=i_0$, we find

$$\frac{i_0}{k-\lambda i_0}=C_3$$

Before inserting this value we continue the algebraic reorganization:

$$i=C_3e^{kt}(k-\lambda i)=C_3ke^{kt}-C_3\lambda e^{kt}i$$

Solving for i gives

$$i=\frac{C_3ke^{kt}}{1+C_3\lambda e^{kt}}=\frac{\left(\dfrac{i_0}{k-\lambda i_0}\right)ke^{kt}}{1+\left(\dfrac{i_0}{k-\lambda i_0}\right)\lambda e^{kt}}$$

or

$$i(t)=\frac{ki_0e^{kt}}{k-\lambda i_0+\lambda i_0e^{kt}}\qquad\text{for } k\neq 0.$$

If $k=0$, we need to go back to equation (23):

$$\frac{di}{dt}=i(k-\lambda i)$$

with $i(0)=i_0$. When we let $k=0$, we have

$$\frac{di}{dt}=i(-\lambda i)=-\lambda i^2$$

Separating the variables gives

$$\frac{di}{i^2} = -\lambda\, dt$$

and we integrate to obtain

$$-\frac{1}{i} = -\lambda t + C$$

Since $i = i_0$ when $t = 0$, we find

$$-\frac{1}{i_0} = -\lambda \cdot 0 + C$$

or

$$C = -\frac{1}{i_0}$$

Thus

$$-\frac{1}{i} = -\lambda t - \frac{1}{i_0} = \frac{-(\lambda i_0 t + 1)}{i_0}$$

Therefore

$$i = \frac{i_0}{\lambda i_0 t + 1}$$

As we now analyze this solution we arrive at a threshold theorem:

If $k \leq 0$, $i(t) \to 0$ as $t \to \infty$.

If $k > 0$, $i(t) \to \dfrac{k}{\lambda} = \dfrac{\lambda - \delta - \gamma}{\lambda} = 1 - \left(\dfrac{\delta + \gamma}{\lambda}\right)$ as $t \to \infty$.

However, $1/(\delta + \gamma)$ is the average lifetime of an individual, and λ is the number of contacts per individual per unit of time and thus $\lambda/(\delta + \gamma)$ is the number of contacts per infective individual during the time in the infective class. This quantity is called the infectious contact rate and is denoted by σ. Hence for $k > 0$, $i(t) \to 1 - 1/\sigma$ as $t \to \infty$.

The threshold value σ enters this problem because of the vital dynamics that were included. The ultimate size of the class of infectives is controlled by the value of σ. Also the condition $k \leq 0$ means that $\lambda \leq \delta + \gamma$ or

$$\sigma = \frac{\lambda}{\delta + \gamma} \leq 1$$

which is interpreted as meaning that each infective contacts less than one susceptible. Under these conditions the disease will, obviously, die out and $i(t) \to 0$.

1.11 A Model for the Growth of Bacterial Populations

In certain bacterial colonies, toxic substances produced by the bacteria become a limiting factor to the further growth of the bacteria. The number of organisms will increase at a rate proportional to the number present at any time and decrease at a rate (per organism) proportional to the concentration of the toxic substance. The following differential equation describes the relation between the number of organisms in the culture and the concentration of the toxic metabolic product:

$$\frac{dn}{dt} = kn(1 - ac)$$

where: n is the number of organisms.
c is the concentration of toxic products.
k and a are positive constants.
If the toxic substance is formed at a constant rate r per organism we have $dc/dt = rn$. The two variables here are $n(t)$ and $c(t)$, and we have two first-order differential equations in the general form

$$\frac{dn}{dt} = f_1(n, c, t)$$

$$\frac{dc}{dt} = f_2(n, c, t)$$

If function f_1 did not contain variable c, that differential equation would be solvable for $n(t)$ immediately. Similarly if f_2 did not contain variable n, then the second differential equation would be solvable for $c(t)$ immediately. However, when this is not the case in either function, we say that the pair of equations is coupled together. In the present case we have a coupled pair with a nonlinear term. This type of problem is generally difficult; however, if we convert the second equation to

$$c = r\int_0^t n(t)\,dt$$

we can write

$$\frac{dn}{dt} = kn\left(1 - ar\int_0^t n(t)\,dt\right)$$

This integro-differential equation can be solved explicitly for $n(t)$ by a technique described by A. T. Reid*: Divide both sides by n and substitute

* A. T. Reid, "Note on the Growth of Bacterial Populations," *Bulletin of Mathematical Biophysics*, 14 (1952), pp. 313–16.

$y = \ln n$ or $n = e^y$. This gives

$$\frac{dy}{dt} = k - \alpha k \int_0^t e^{y(t)}\, dt$$

where $\alpha = ar$.

To eliminate the integral, differentiate with respect to t and obtain

$$\frac{d^2y}{dt^2} = -\alpha k e^y$$

Multiply by $2(dy/dt)$ to get $2y'y'' = -2\alpha k e^y y'$. This is integrable to $(y')^2 = -2\alpha k e^y + c$. (Check the derivative of each side.) However, when $t = 0$, we have $n = n_0$ and $dy/dt = k$. Thus $c = k^2 + 2\alpha k n_0$, and

$$\frac{dy}{dt} = \pm\sqrt{k^2 + 2\alpha k(n_0 - e^y)}$$

Reverse substitution now gives an equation in terms of n:

$$\frac{1}{n}\frac{dn}{dt} = \pm\sqrt{k^2 + 2\alpha k(n_0 - n)}$$

Separating the variables and integrating over the range where n' is positive, we get

$$\int dt = \int_{n_0}^{n} \frac{dn}{n\sqrt{k^2 + 2\alpha k(n_0 - n)}}$$

or

$$t = \frac{1}{A}\ln\left[\frac{(\sqrt{A^2 - 2\alpha k n} - A)(k + A)}{(\sqrt{A^2 - 2\alpha k n} + A)(k - A)}\right]$$

where $A = \sqrt{k^2 + 2\alpha k n_0}$

If we set $B = (k - A)/(k + A)$, we can solve for $n(t)$:

$$n(t) = \frac{Ce^{At}}{(1 - Be^{At})^2}$$

where $C = -2A^2B/\alpha k$. In this case, A, B, and C are constants.

We may now analyze the behavior of $n(t)$ by investigating the first and second derivatives and looking for maximum values and concavity. If we differentiate the solution $n(t)$ we get

$$\frac{dn}{dt} = ACe^{At}(1 - Be^{At})^{-2} + 2ABCe^{2At}(1 - Be^{At})^{-3}$$

If we set this equal to zero and solve for t, we get

$$t_c = \frac{1}{A}\ln\left(\frac{A + k}{A - k}\right)$$

as the critical point.

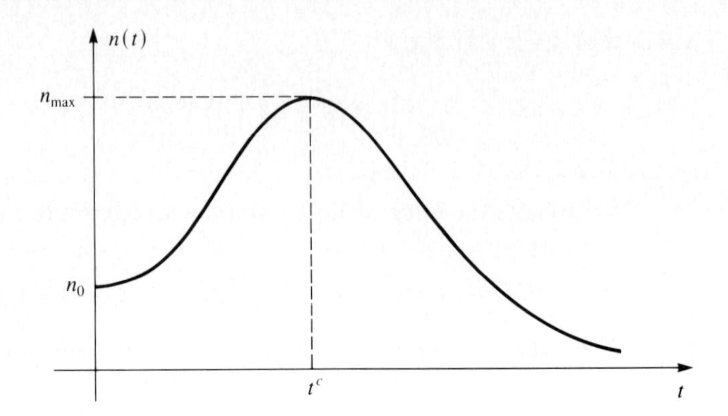

Figure 1.26

When we differentiate again and insert t_c for t, we find the second derivative is negative. Thus $n(t)$ has its maximum value at t_c; if the value of t_c is put into the solution for $n(t)$, we find that the maximum number of organisms is $n_{max} = n_0 + k/2\alpha$. The maximum number is a direct function of the initial number and parameters of the species involved.

Beyond the maximum point, we find dn/dt is negative but an analogous computation will show that the function $n(t)$ is symmetric about the line $t = t_c$. Hence the complete growth curve is as illustrated in Figure 1.26.

Miscellaneous Exercises for Chapter 1

In Exercises 1–50 solve the equations by one or more of the methods demonstrated in Chapter 1.

1. $y\,dx + (x + y)\,dy = 0$
2. $\dfrac{ds}{dt} = s^2t^2;\ s(0) = 1$
3. $y' - 2xy = xy^2$
4. $\dfrac{dv}{dt} = kvt$
5. $uv' = u^2v^2 + 2v$
6. $y(x^2 - y^2)\,dx - 2x^3\,dy = 0$
7. $xy' = y + x^2 \ln x;\ y(2) = 0$
8. $xy\,dy - 2(y + 3)\,dx = 0$
9. $(x - 2y)\,dx + (2x - y)\,dy = 0$
10. $\dfrac{dr}{d\theta} = r \tan\theta$

11. $y' = 2(3x - y)$
12. $[(2x)\tan y - 3y^2]\,dx + [x^2\sec^2 y - 6xy + 2]\,dy = 0$
13. $y' - y\sec x = 0$
14. $y(2 - 3xy)\,dx = x\,dy$
15. $y^2\,dx - x(2x + 3y)\,dy = 0$
16. $x\,dx + ye^{-x^2}\,dy = 0$
17. $(x^6 + y)\,dx - x\,dy = 0$
18. $(2xy - y)\,dx + (x^2 + x)\,dy = 0$
19. $e^{2x-y}\,dx + e^{x+y}\,dy = 0$
20. $xyy' - (y^2 - 1) = 0;\ y(2) = 0$
21. $v\dfrac{dv}{dx} = g;\ v(x_0) = v_0$
22. $y(2x^3 - x^2y + y^3)\,dx - x(2x^3 + y^3)\,dy = 0$
23. $3y' - 12y + 3x = 1$
24. $y^2\cos\sqrt{x}\,dx - 2\sqrt{x}e^{1/y}\,dy = 0$
25. $[2x\sin(x^2) + 3\cos y]\,dx - 3x\sin y\,dy = 0$
26. $(\cos x)y' + \sin x = 1 - y$
27. $2xy + (y^2 - 3x^2)y' = 0$
28. $(y - 2x)e^{y/x}\,dx + (2y - xe^{y/x})\,dy = 0$
29. $r\sec\theta\,dr - 2k^2\sin\theta\,d\theta = 0$
30. $L\dfrac{di}{dt} + Ri = E;\ i(0) = 0$
31. $\tan x\dfrac{dy}{dx} - 2y = C$
32. $2x^3y' + (y^3 \quad x^2y) = 0$
33. $(x^3 - x)y' - (3x^2 - 1)y = x^5 - 2x^3 + x;\ y(1) = 0$
34. $y' - \dfrac{1}{2}xy = \dfrac{1}{2}x$
35. $xy' + y = y^2\ln x;\ y(1) = 2$
36. $(x + y)\,dx + (3y + x)\,dy = 0;\ y(0) = 4$
37. $y' + \dfrac{y}{x\ln x} = \dfrac{1}{x};\ x > 0$
38. $xy' - \left(y + \cos\dfrac{1}{x}\right) = 0$
39. $(x + 1)y' - 3y = (x + 1)^5$
40. $2xy' - 2x^{5/2} - x^{3/2} - y$
41. $\dfrac{x}{y}y' = \dfrac{2x - y^2}{3y^2 - x}$

42. $\cos(y') = 1 - y$
43. $e^y(y' + 1) = 1$
44. $x^2 y' + (x^2 + 2x)y = 1$
45. $x \dfrac{dy}{dx} - y = \sqrt{x^2 + y^2}$
46. $y'' + \sin x = 0$
47. $y' - 2y = 4x$; $y(0) = 1$
48. $y' = \dfrac{\cos x}{3y^2 + e^y}$; $y(0) = 0$
49. $(x^2 + 1)y' - 2xy - x^2 - 1 = 0$; $y(1) = \pi$
50. $y' = e^x(1 - y^2)^{1/2}$; $y(0) = \dfrac{1}{2}$
51. In a study on visual adaptation by the cells of the retina to a flash of light, a simple first order differential equation was derived. The theory is that microscopic holes are punched in the retinal membrane by the flash at a rate proportional to the rate of absorption of light and that the holes close up at a rate proportional to their number. If p is the permeability; I, the illumination intensity; S, the sensitivity of the nerve ending; and α and β, constants; then $dp/dt = \alpha SI - \beta p$. Solve for $p(t)$ if $p(0) = p_0$.
52. In cancer research, a vitally important phenomenon is the relation between the time required for an irradiated cell to reach metaphase and the amount and rate of the radiation dose. The principal mechanism producing metaphase delay is the failure of chromosome coiling in prophase that is due to radiation induced cross-linking of chromosomal structural protein fibers.

    If: ξ represents the amount of coiling agent per chain length;

    g is a constant whose value depends on the elastic property of the protein fibers;

    L is the length of the chromosome;

    σ is the number of contractile sites per chain length;

    t is the time after cell birth (cleavage);

    then

    $$\frac{d}{dt}\left(\frac{L}{g}\right) = -k\sigma\xi\frac{L}{g}, \quad k \text{ is constant}$$

    Solve this equation for $L(t)$ with $L(0) = L_0$. (Note that this implies that each contractile fiber acts like a stretched spring.)
53. Medical research on the way that organic acids dissolve tooth enamel produced an interesting system of differential equations.

Dr. S. O. Zimmerman* showed that if we let $C = [Ca^{++}]$ and $P = [HPO_4^{--}]$, then the governing equations are

$$\frac{dC}{dt} = 10(K - kCP), \text{ with } C = P = 0 \text{ at time } t = 0$$

and $$\frac{dP}{dt} = 6(K - kCP), \text{ with } K \text{ and } k \text{ reaction-rate constants}$$

In analyzing these, we first see that $6(dC/dt) = 10(dP/dt)$ or

$$\frac{d}{dt}(6C - 10P) = 0$$

so that $(6C - 10P) =$ constant. At time $t = 0$, $C = P = 0$, so the constant is zero and $C = 5/3\ P$. Thus if we can find P we will also have C immediately.

Using $C = 5/3\ P$ in the dP/dt equation, solve for $P(t)$ to obtain

$$P(t) = \sqrt{\frac{3K}{5k}}\,\frac{1 - e^{-\sqrt{240Kk}t}}{1 + e^{-\sqrt{240Kk}t}}$$

54. One of the governing differential equations for the flow of a viscous fluid, such as oil, in a long cylindrical pipe is $x(dy/dx) = c^2x^2 - y$. Solve this equation by one of the methods in Chapter 1.
55. The manufacture of coloring base for printing inks involves a complex chain of chemical reactions. During the cooling process, a combination of solvents and resins at temperature T_1 flows continuously into a sealed mixing tank of volume V, which is already full of the mixture, and simultaneously an amount equal to the input is extracted at the bottom of the tank. The mixture in the tank is continuously stirred. Let $I(t)$ denote the rate of inflow of the mixture per cubic centimeter. The temperature $T_2(t)$ of the entire mixture must be carefully monitored, because some heat is carried off through the outlet pipe. Let $h(t)$ be the amount of excess heat present in one cubic centimeter of solution at any time t; then $H(t) = V \cdot h(t)$ is the total excess heat in the tank. At any time t, $T_2(t) - T_1(t)$ represents the excess temperature of the tank over the incoming liquid. Thus $h(t) = T_2(t) - T_1(t)$ gives the number of excess calories stored in one cubic centimeter, and $H(t) = [T_2(t) - T_1(t)] \cdot V$ is therefore the total excess heat in the tank. Since the tank is losing heat, we have $-dH/dt$ as the rate loss of calories as well as $[T_2(t) - T_1(t)]I(t)$.

$$\therefore \quad \frac{dH}{dt} = -[T_2(t) - T_1(t)]I(t)$$

---

* S. O. Zimmerman, "A Mathematical Theory of Enamel Solubility and the Onset of Dental Caries: I. The Kinetics of Dissolution of Powdered Enamel in Acid Buffer," *Bulletin of Mathematical Biophysics*, 28 (1966), pp. 417–32.

or
$$V\frac{dh}{dt} = -h(t)I(t)$$

$$\frac{dh}{dt} = -\frac{I(t)}{V} \cdot h(t)$$

Since the volume V is fixed, and $I(t)$ is a constant flow, we need only to have the value of h at time $t = 0$ to have a well-posed initial value problem. Assume that $T_2(0) = 100°C$, and $T_1(0) = 50°C$. Solve this differential equation first for $H(t)$ and then for $T_2(t)$, which was our goal.

56. The form of the body of a snake during locomotion gives rise to the following differential equation:

$$\frac{1}{L}\frac{dL}{dt} = \frac{1}{3\eta} \cdot a$$

where η is the density of the snake and a is a constant. Solve for $L(t)$ to see the way a section of the snake's body changes as it propels itself forward by muscle action.

57. Four grams of sulfur are placed in a solution of 75 cubic centimeters of benzol, which when saturated will hold 8 grams of sulfur. If in 20 minutes, 1.5 grams are in the solution, how many grams will be in the solution in 60 minutes?

Let x denote the grams of sulfur not yet dissolved. Then $4 - x$ is the amount dissolved, and $(4 - x)/75$ is the concentration. The saturated solution of benzol will have a sulfur concentration of $8/75$. Then

$$\frac{dx}{dt} = kx\left(\frac{8}{75} - \frac{4 - x}{75}\right)$$

Solve this equation for $x(t)$ and then answer the above question.

58. If the population of an ant colony doubles in 50 days, show that in 79 days the population will triple.

59. In a will, a man left a million dollars to a specific trust fund. However, the will provided that the money be invested and held for 200 years before being distributed. The will was contested by the U.S. Government on the grounds that the monetary wealth of the nation could be concentrated there. If the money earned an average 9% interest, approximately what would the value be at the end of 200 years?

60. A boat is being towed at the rate of 15 ft/sec. At the instant the tow line is cast off, the man in the boat begins to row in the direction of the motion with a force of 20 lbs. If the man and boat together weigh 320 lbs and the water resistance is equal to 5/4 times the velocity v, find the speed of the rowboat after 20 seconds.

61. In a recent study of the rhesus monkey, psychologists verified the relation between stimulus s and response r as:

$$\frac{dr}{ds} = k\frac{r^n}{s}$$

where k is a positive constant and n is a positive integer. Solve this equation for $n = 0, 1, 2$, and then for general n.

62. One year ago Professor A. G. Enius at University X discovered a new method of teaching calculus that could greatly benefit mathematics departments at 999 other institutions of higher leaning. During the past year, 10 of these schools adopted his method. Assume that the rate at which the number of adopters $N(t)$ increases is proportional to the number that have already adopted the technique and also to $1000 - N(t)$ (i.e., the number of the schools remaining in the dark). Thus

$$\frac{dN}{dt} = kN(1000 - N), \quad k > 0$$

(a) Solve the differential equation by using $N(0) = 10$ and a k value of .009.

(b) How many schools will adopt the method in the first 8 years following the discovery?

63. (a) A belt is wrapped around a drum of radius R, and tension in the belt in both directions, labeled T_1 and T_2 in Figure 1.27, causes the belt to be on the verge of slipping. Let θ be the angle of contact of the belt with the drum. If we let μ be the coefficient of friction between the belt and the drum surface, then the differential equation that governs the computation of the belt tension at an arbitrary contact point is

$$\frac{dT}{d\theta} = \mu T, \quad T(0) = T_1$$

Solve this differential equation for $T(\theta)$.

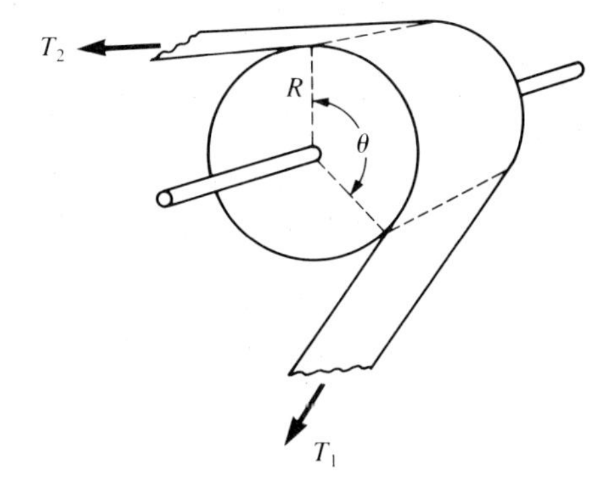

Figure 1.27

(b) Using the knowledge acquired in part (a), calculate the maximum tension in a cable wrapped around a cylinder when the coefficient of friction is $3/10$, the initial tension for impending slippage is 50 lbs, and the angle of contact is $\pi/3$. Now assume that the angle of contact is 2π, and calculate the maximum tension.

(c) A mooring line for a ship is wrapped around a post four times. The coefficient of friction between the rope and post is .25. If the ship is pulling on the rope with a force of 9000 lbs, what force must be exerted on the free end of the rope to avoid slippage of the rope?

64. In a radioactive decay process, the number of atoms that disintegrate per unit of time dN/dt is proportional to the number present N. Thus $dN/dt = kN$. Assume that the half-life of carbon 14 is 5730 years, so that a charcoal sample from wood that died 5730 years ago would produce half as many counts per minute as a tree cut down today. If s is an unknown number of years whose value we wish to find, we can combine $N(0)$ and $N(s)$ and use the given differential equation to find s. Show the steps to obtain $s = 1/k \ln N(s)/N(0)$ and the steps to obtain $k = -\ln 2/T$ where T is the half-life. Then apply these steps to the following problem: A piece of charcoal shows a count of 4.029 per minute, while living wood of the same type shows 6.68 counts per minute. How old is the wood in the charcoal sample?

65. Giant Company has an equity capital of 10^6 dollars at time t_0. Assume that the equation which represents Giant's rate of change of equity capital is:

$$\frac{dC}{dt} = (1 - N)rC + S$$

where: C = equity capital
N = dividend-payout ratio
r = rate of return
s = rate of new stock financing

Solve this equation for C as a function of time.

Chapter 2

Introduction to Numerical Methods

The Swiss mathematician Leonhard Euler (1707–1783) played an important role in the early development of techniques for finding solutions to differential equations. He was one of the first persons responsible for taking some newly developed concepts of calculus and extending them to differential equations. Euler (pronounced oiler) studied, understood, and published many of today's concepts of first-order equations and did a considerable amount of work on linear equations of higher order. He also developed some ingenious techniques. Euler provided insight into numerical solutions and introduced some elementary methods that form the basis for our study.

Since Euler's time sophisticated numerical solution processes have evolved. Many of these techniques constitute the subject matter of numerical analysis, an important advanced course that is taught at most colleges and universities, and this textbook introduces some of these processes.

Although the closed-form analytical techniques have been used for many years to solve first-order and many second-order equations, the differential equations that result from applications—particularly those in engineering and the natural sciences—generally cannot be solved by these analytical techniques. The advent of high-speed electronic computers led to an emphasis in scientific work to solve the resulting differential equations by a numerical method in which a finite set of points is generated. This set of points is an approximation to the actual solution function, and we will refer to this set as the **numerical solution** to the problem. This chapter looks at two numerical techniques, both attributed to Euler.

2.1 Preliminaries

Consider the general first-order initial value problem

$$\begin{Bmatrix} y' = f(x, y) \\ y(x_0) = y_0 \end{Bmatrix} \tag{1}$$

An integration approach to solving the differential equation in (1) is to first rewrite it in the equivalent form $dy = f(x, y)\,dx$. Let t be the variable of integration (i.e., replace x temporarily by t), and integrate both sides of the differential equation from the initial point x_0 to some arbitrary point x in its domain interval I.

$$\int_{x_0}^{x} dy = \int_{x_0}^{x} f(t, y(t))\,dt \Rightarrow y(x) - y(x_0) = \int_{x_0}^{x} f(t, y(t))\,dt \tag{2}$$

Using the initial value $y(x_0) = y_0$, equation (2) becomes

$$y(x) = y_0 + \int_{x_0}^{x} f(t, y(t))\,dt \tag{3}$$

This equation will supply a solution if the integration can be performed. If $f(x, y)$ does not depend on y explicitly, the equation can be integrated to the analytical solution.

Example 1

Consider

$$\begin{Bmatrix} y' = 3x^2 + 1 \\ y(2) = 5 \end{Bmatrix}$$

Note that $x_0 = 2$, $y_0 = 5$, $f(x, y) = 3x^2 + 1$. If we integrate both sides, the form of equation (3) becomes

$$y(x) = 5 + \int_{2}^{x} (3t^2 + 1)\,dt = 5 + \left[t^3 + t \Big|_2^x \right] = 5 + [x^3 + x - 8 - 2]$$

or

$$y(x) = x^3 + x - 5$$

This is easily verified as the unique solution to the initial value problem. This is a problem of the type we considered in Section 1.2.

If the function $f(x, y)$ given in the initial value problem *does* involve the unknown y, we will be unable to integrate the equation. Example 2 illustrates this type of problem.

Example 2

Consider

$$\begin{Bmatrix} y' = 2x^2 + xy^3 \\ y(1) = 5 \end{Bmatrix}$$

In terms of equation (3), which is equivalent to this initial value problem, we obtain

$$y(x) = 5 + \int_1^x [2t^2 + t \cdot y^3(t)]\, dt$$

Since $y(t)$ is unknown (it is what we are trying to find), this initial value problem cannot be solved analytically by this approach. When this case arises we may need a new approach.

2.2 Euler's Method

The most common numerical technique is to replace the function $f(x, y)$ by some approximating function on the interval $[x_0, x]$. We then use the expression in (3) to obtain a function $Y(x)$ that approximates the solution function $y(x)$. The ingenious Euler first proposed the following simple approach to find the approximate solution, a method commonly referred to as **Euler's method**.

We begin at the only point on the solution curve that we know, the initial point (x_0, y_0) and attempt to extend our solution to the right to the point $x_1 > x_0$. Since we will use the differential equation in the form (3) to provide the approximation to $y(x_1)$, we must pick an approximation to $f(x, y(x)) = f(t, y(t))$ on the interval $[x_0, x_1]$. Euler chose the simplest of all approximations, the constant $f(x, y(x)) = f(x_0, y_0)$. Using this constant function in equation (3), we may pull that part out of the integrand and get, on the interval $[x_0, x_1]$,

$$\begin{aligned} y(x_1) &\simeq y_0 + \int_{x_0}^{x_1} f(x_0, y_0)\, dt = y_0 + f(x_0, y_0) \int_{x_0}^{x_1} dt \\ &= y_0 + f(x_0, y_0)(x_1 - x_0) \end{aligned}$$

We use the notation Y_1 as the approximation to $y(x_1)$. Subsequently we use capital letters to denote approximate values and lowercase letters to denote precise values. If we let h designate the distance from x_0 to x_1 (called the **step size**), we can write

$$\begin{aligned} Y_1 &= y_0 + (x_1 - x_0) f(x_0, y_0) \\ &= y_0 + h \cdot f(x_0, y_0) \end{aligned} \tag{4}$$

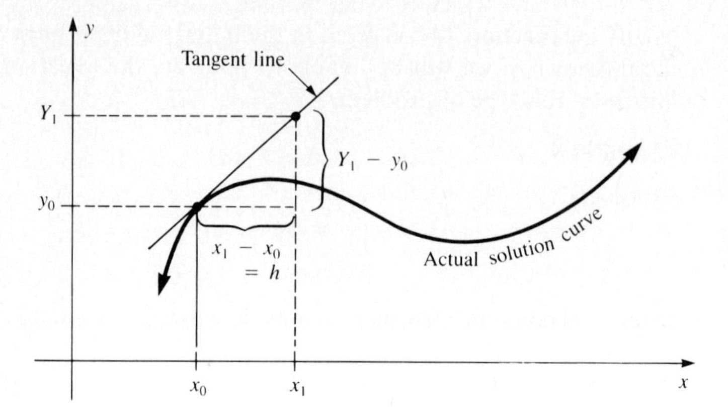

Figure 2.1

Notice that all the values we need to calculate on the right side of equation (4) are given in the initial value problem. We have now accomplished the first step in Euler's method. This simple Euler process has an elementary geometric interpretation, as illustrated in Figure 2.1.

At the initial point (x_0, y_0), the differential equation

$$y'\Big|_{(x_0, y_0)} = f(x_0, y_0)$$

gives a value of $y'(x_0)$, which is the slope of the tangent line to the solution curve at $x = x_0$. In Chapter 1 this was the slope of the lineal element at the point (x_0, y_0). Figure 2.1 shows this line drawn at (x_0, y_0) with slope $f(x_0, y_0)$. Above point x_1 we locate (x_1, Y_1) on this tangent line. Since its slope is the ratio of the vertical increase to the horizontal increase, we may write $f(x_0, y_0) = (Y_1 - y_0)/(x_1 - x_0)$. This is algebraically rearranged to $Y_1 - y_0 = (x_1 - x_0)f(x_0, y_0)$ or

$$Y_1 = y_0 + (x_1 - x_0)f(x_0, y_0) = y_0 + h \cdot f(x_0, y_0)$$

We now have moved from our initial point (x_0, y_0) to the next solution point (x_1, Y_1). The line segment joining the two points is our approximation to the solution curve.

Example 3 is a very simple initial value problem that demonstrates the simplicity of Euler's method and allows us to check its accuracy.

Example 3

Use Euler's method to obtain two points in the solution to:

$$\begin{Bmatrix} y' = y - x \\ y(0) = 2 \end{Bmatrix}$$

The given starting point (x_0, y_0) is $(0, 2)$ and the given function is $f(x, y) = y - x$. Hence $f(x_0, y_0) = f(0, 2) = 2$. Using a step size 0.1 to the

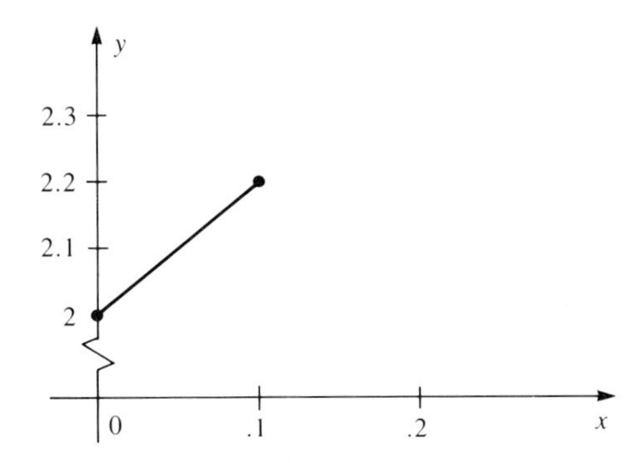

Figure 2.2

right, we obtain

$$Y_1 = y_0 + (x_1 - x_0)f(x_0, y_0)$$
$$= 2 + (0.1)(2) = 2.2$$

Now our solution set has two points: (0, 2) and (0.1, 2.2), as shown in Figure 2.2.

If we wish to extend the approximate solution farther to the right, for example, from point (x_1, Y_1) to (x_2, Y_2) where $x_2 > x_1$, we use (x_1, Y_1) as our base point and extend linearly by following an identical procedure to get

$$Y_2 = Y_1 + (x_2 - x_1)f(x_1, Y_1)$$

If we select points x_0, x_1, and x_2 equidistantly apart on the x-axis, the values $x_1 - x_0$ and $x_2 - x_1$ are each designated by h and we may write

$$Y_2 = Y_1 + h \cdot f(x_1, Y_1)$$

Since this process can be repeated many times, we obtain as a general expression of the simple Euler method:

$$\left\{\begin{array}{l} Y_{n+1} = Y_n + (x_{n+1} - x_n)f(x_n, Y_n);\ n = 0, 1, 2, \ldots \\ \text{where } x_1, x_2, x_3, x_4, \ldots \text{ are prechosen values and } Y_0 = y_0. \end{array}\right\} \quad (5)$$

Thus the simple Euler approach is to move from point to point on the numerical solution curve as we repeatedly move in the direction of the tangent line to the solution curve. We then calculate the new solution point at Y_{n+1} by using the value at Y_n and the functional value $f(x, y)$ at $x = x_n$, $y = Y_n$. The set of points thus obtained:

$$(x_0, Y_0), (x_1, Y_1), (x_2, Y_2), \ldots, (x_n, Y_n), \ldots$$

constitutes our approximate solution; when these points are joined together, they produce a polygonal curve that is the numerically obtained solution.

Example 4

Continue with Example 3 and obtain the next solution point. We begin at (0.1, 2.2) as our point (x_1, Y_1) and take $h = x_2 - x_1 = 0.1$. Now $f(x_1, Y_1) = Y_1 - x_1 = (2.2 - 0.1) = 2.1$ and

$$\begin{aligned} Y_2 &= Y_1 + h \cdot f(x_1, Y_1) \\ &= 2.2 + (0.1)(2.1) \\ &= 2.20 + .21 \\ &= 2.41 \end{aligned}$$

Thus the next solution point is (0.2, 2.41), as illustrated in Figure 2.3. We will continue with this problem in subsequent pages.

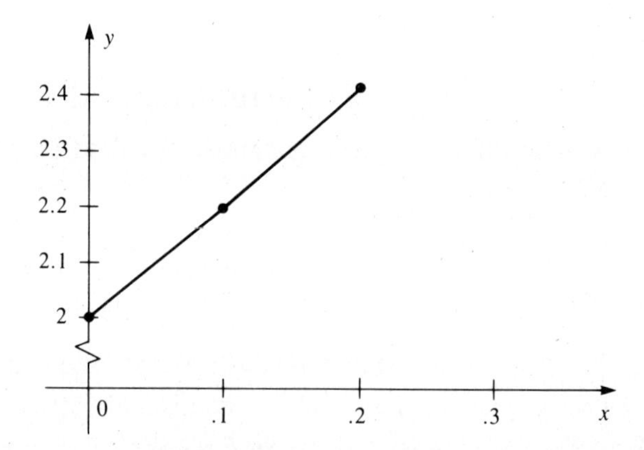

Figure 2.3

An analysis of what we now have gives us a simple geometric interpretation to Euler's method. Since $f(x_0, y_0) = y'(x_0)$ (this comes directly from the given problem), and Y_1 is the linear approximation to $y(x_1)$, then (x_1, Y_1) lies on the tangent line to the actual solution curve at (x_0, y_0). Similarly, to move to point (x_2, Y_2) we follow parallel to the tangent line to the solution curve at (x_1, Y_1). The concept of the direction field from Chapter 1 underlies this approach. We are moving horizontally in steps of size h and proceeding in the direction of the lineal element at each newly found point. As shown in Figure 2.4, the appropriate solution curve advances from point to point when we use this tangent method.

An alternative way to derive Euler's equations for the initial value problem

$$\begin{Bmatrix} y' = f(x, y) \\ y(x_0) = y_0 \end{Bmatrix}$$

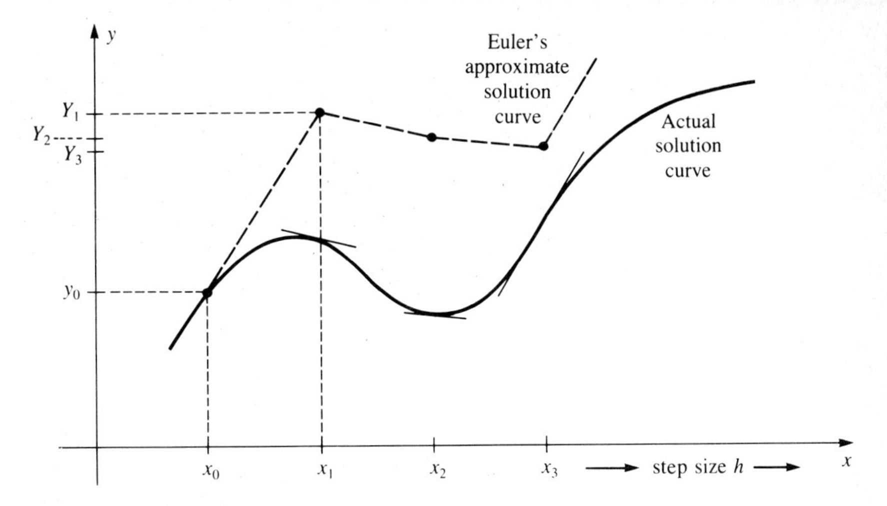

Figure 2.4

is to consider the standard Taylor series expansion formula from elementary calculus. Given a function $y(x)$ that is differentiable to all orders and a point x_0 in its domain, the Taylor series expansion for the function is:

$$y(x) = y(x_0) + y'(x_0)(x - x_0) + \frac{y''(x_0)}{2!}(x - x_0)^2 + \frac{y'''(x_0)}{3!}(x - x_0)^3 + \cdots$$

If we pick adjacent points, $x_0 = X_n$ and $x = X_{n+1}$, so that $x - x_0 = h$, we obtain

$$Y(X_{n+1}) \equiv Y_{n+1} = Y(X_n) + f(X_n, Y(X_n))\,(X_{n+1} - X_n) + \frac{y''(X_n)}{2!}(h)^2 + \cdots$$

By retaining the first two terms of the expansion we again get Euler's formula:

$$Y_{n+1} = Y_n + h \cdot f(x_n, Y_n)$$

Compare this with equation (5).

This technique is probably the oldest known numerical algorithm for solving elementary differential equations. The method has merit in its simplicity but weakness in its accuracy, as the following examples demonstrate.

Example 5

Continue with the computation of solution points for the initial value problem of Example 3.

$$y' = y - x; \; y(0) = 2$$

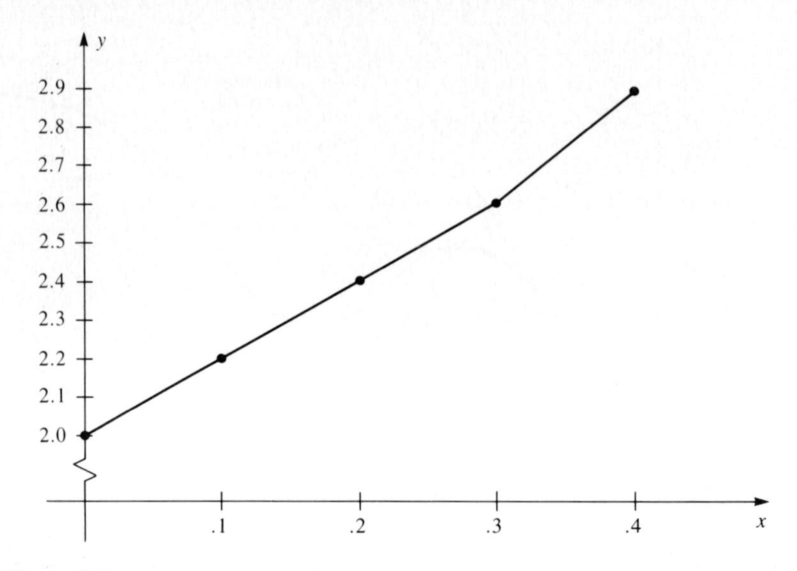

Figure 2.5

We have the points (0, 2), (0.1, 2.2), and (0.2, 2.41), and we will again let the step size $h = x_3 - x_2 = 0.1$. Then

$$\begin{aligned} Y_3 &= Y_2 + h \cdot f(x_2, Y_2) \\ &= 2.41 + (0.1)(2.41 - 0.2) \\ &= 2.41 + (0.1)(2.21) \\ &= 2.41 + 0.221 \\ &= 2.631 \end{aligned}$$

Thus we add the point (0.3, 2.631). Now we let $h = x_4 - x_3 = 0.1$ and compute

$$\begin{aligned} Y_4 &= Y_3 + h \cdot f(x_3, Y_3) \\ &= 2.631 + (0.1)(2.631 - 0.300) \\ &= 2.631 + (0.1)(2.331) \\ &= 2.631 + 0.2331 \\ &= 2.8641 \end{aligned}$$

and add the point (0.4, 2.8641) to the solution set (see Figure 2.5).

This process could continue indefinitely; but if we need a large collection of solution points, hand computations become time consuming. We see that this problem can be solved analytically if we write it as

$$y' - y = -x; \; y(0) = 2$$

and solve it as a first-order linear differential equation by the method we studied in Section 1.6. The unique solution is $y = e^x + 1 + x$. The actual

Table 2.1

Solution to $y' = y - x$; $y(0) = 2$ on $[0, 1]$ using $h = 0.1$

x	Y (Euler)	y (actual)
0.0	2	2
0.1	2.2	2.2052
0.2	2.41	2.4214
0.3	2.631	2.6499
0.4	2.8641	2.8918
0.5	3.1105	3.1487
0.6	3.3716	3.4221
0.7	3.6487	3.7138
0.8	3.9436	4.0255
0.9	4.2580	4.3596
1.0	4.5937	4.7183

values of y for any x can now be determined by simple substitution into this function. Nevertheless, we will continue with the numerical approach, but now we will use numbers provided by a computer instead of hand computations.

Simple computer programs written in the BASIC language can be used for printing out the Euler's method solution points as well as the actual analytically determined values. These programs are contained in Appendix B under the names EULER and YACTUL. Details of their construction will be discussed later in this chapter.

As expected, the first five entries in the Y(Euler) column of Table 2.1. are the values we found with hand computations. The accuracy of the approximations is *clearly weak*, especially as we move to larger values of x.

Example 6

Consider the initial value problem

$$\begin{Bmatrix} y' = 2x^2 + y \\ y(1) = 2 \end{Bmatrix}$$

This also is solvable analytically as a first-order linear equation, and we obtain $y(x) = 12e^{x-1} - 4 - 4x - 2x^2$. We will use this solution to check the accuracy of the Euler approach. We will form an approximate numerical solution on $[1, 2]$ by using a constant step size of 0.1. The first hand computation, which uses $f(x, y) = 2x^2 + y, x_0 = 1$, and $y_0 = 2$, yields

$$\begin{aligned} Y_1 &= y_0 + h \cdot f(x_0, y_0) \\ &- 2 + (0.1)[2(1)^2 + 2] \\ &= 2 + (0.1)(4) \\ &= 2.4 \end{aligned}$$

The second new point is obtained from

$$\begin{aligned} Y_2 &= Y_1 + h \cdot f(x_1, Y_1) \\ &= 2.4 + (0.1)[2(1.1)^2 + 2.4] \\ &= 2.4 + (0.1)(4.82) \\ &= 2.4 + 0.482 \\ &= 2.882 \end{aligned}$$

We now have solution points (1, 2), (1.1, 2.4), (1.2, 2.882). The computer can be used to generate the remaining approximate solution points as well as the actual solution values, as shown in Table 2.2. As in Example 5, the accuracy of the Euler result is poor, especially as the value of x increases. However, you should start at point $x = 1.2$ and do a couple more hand calculations to give you a feel for Euler's method.

Table 2.2

Solution to $y' = 2x^2 + y$; $y(1) = 2$ on [1, 2] using $h = 0.1$

x	Y (Euler)	y (actual)
1.0	2	2
1.1	2.4	2.4421
1.2	2.882	2.9768
1.3	3.4582	3.6183
1.4	4.1420	4.3819
1.5	4.9482	5.2847
1.6	5.8930	6.3454
1.7	6.9944	7.5850
1.8	8.2718	9.0265
1.9	9.7470	10.6952
2.0	11.4437	12.6194

Example 7

Consider the following initial value problem (which arises in hydraulics studies) whose analytical solution is hard to obtain.

$$\begin{Bmatrix} y' = 2 + y^{1.5} \\ y(0) = 0.7 \end{Bmatrix}$$

We wish to see the approximate solution function on the interval [0, 1.4] and will step in sizes $h = 0.2$. The first hand computation would use $x_0 = 0$, $y_0 = 0.7$, and $f(x, y) = 2 + y^{1.5}$. Then

$$\begin{aligned} Y_1 &= y_0 + h \cdot f(x_0, y_0) \\ &= 0.7 + (0.2)[2 + (0.7)^{1.5}] \\ &= 0.7 + (0.2)(2 + 0.585662) \\ &= 0.7 + (0.2)(2.585662) \\ &= 1.21713 \end{aligned}$$

Table 2.3

Solution to $y' = 2 + y^{1.5}$; $y(0) = 0.7$ on [0, 1.4] using $h = 0.2$

x	Y (Euler)	y (actual)
0.0	0.7	0.7
0.2	1.21713	1.2974
0.4	1.88569	2.1410
0.6	2.80358	3.4568
0.8	4.14243	5.7787
1.0	6.22865	10.6513
1.2	9.73766	24.2712
1.4	16.2150	95.7523

These are tedious computations so we will use the computer programs EULER and YACTUL for the remaining results, as shown in Table 2.3. You can see immediately that the Euler results are extremely inaccurate. This is a complicated problem which obviously needs further analysis.

One weakness, as observed before, is that the accuracy of the approximate result deteriorates as the integration continues. It is unfortunate that this deterioration is very common in solving differential equations numerically, especially in the case when the solution curve bends or oscillates rapidly. A smaller step size is one way to help adjust the situation, as illustrated in Example 8.

Example 8

Use smaller step sizes with Euler's method to obtain the results to the initial value problem of Example 6:

$$\begin{Bmatrix} y' = 2x^2 + y \\ y(1) = 2 \end{Bmatrix}$$

The second through sixth columns of Table 2.4 give the Euler results. By comparing the values in the actual solution to the best Euler result, we see

Table 2.4

x	Y ($h = 0.1$)	Y ($h = 0.05$)	Y ($h = 0.01$)	Y ($h = 0.005$)	Y ($h = 0.001$)	y (actual)
1.0	2	2	2	2	2	2
1.1	2.4	2.4203	2.4376	2.4398	2.4416	2.4421
1.2	2.882	2.9276	2.9667	2.9717	2.9758	2.9768
1.3	3.4582	3.5352	3.6011	3.6097	3.6166	3.6183
1.4	4.1420	4.2572	4.3561	4.390	4.3793	4.3819
1.5	4.9482	5.1096	5.2485	5.2665	5.2811	5.2847
1.6	5.8930	6.1099	6.2967	6.3210	6.3401	6.3454
1.7	6.9944	7.2772	7.5213	7.5530	7.5787	7.5850
1.8	8.2718	8.6328	8.9449	8.9855	9.0184	9.0265
1.9	9.7470	10.2001	10.5926	10.6437	10.6851	10.6952
2.0	11.4437	12.0049	12.4919	12.5553	12.6067	12.6194

that the Euler column with the smallest value of h gives three significant figures of accuracy; that is, for $x = 2$, the integers 1, 2, and 6 for Y are significant figures. (We ignore the decimal point when counting significant figures.)

We now are obtaining reasonably good results at the expense of a significantly large number of evaluations. Table 2.4 lists only 10 computations in each Y column, although many more computations were necessary for $h < .1$. For example, in the case of $h = .005$, the Euler method formula was used 200 times to obtain 200 new points in the solution set; in the case of $h = .001$, the computer made 1000 evaluations. Table 2.4 shows only an equally spaced subset.

Example 9

Use smaller step sizes with Euler's method to find the solution points for the initial value problem of Example 7:

$$\begin{Bmatrix} y' = 2 + y^{1.5} \\ y(0) = 0.7 \end{Bmatrix}$$

Table 2.5

x	$Y\,(h = 0.2)$	$Y\,(h = 0.05)$	$Y\,(h = 0.01)$	$Y\,(h = 0.001)$	y (actual)
0.0	0.7	0.7	0.7	0.7	0.7
0.2	1.2171	1.2734	1.2924	1.2969	1.2974
0.4	1.8857	2.0608	2.1237	2.1392	2.1410
0.6	2.8036	3.2378	3.4084	3.4518	3.4568
0.8	4.1424	5.1770	5.6400	5.7643	5.7787
1.0	6.2287	8.7946	10.1927	10.6028	10.6513
1.2	9.7377	16.8302	22.1751	24.0399	24.2712
1.4	16.2150	40.3896	74.5032	93.0666	95.7523

As we see from Table 2.5, in the case $h = .001$, with 1400 steps across the interval $[0, 1.4]$, the accuracy is better. The error is approximately 2.7 units or about 3%. However, this result is still unsatisfactory.

Rather than continuing to decrease h, which would drastically increase the number of computations, we will turn to a more sophisticated technique, which is described in the following section.

2.3 Improved Euler Method

The simple Euler method used a constant $f(x_0, y_0)$ to replace the variable $f(x, y)$ in the integrand of equation (3). This section shows how to improve Euler's method by improving the approximated choice for $f(x, y)$ on the interval $[x_0, x_1]$. This improved technique uses as an approximation to $f(x, y)$ the average of two values.

In the process of determining these values we first need the number obtained for y at x_1 by the simple Euler method. We will use the formula—see equation (4)—for the simple Euler computation and label this value $\bar{y}_1$:

$$\bar{y}_1 = y_0 + (x_1 - x_0)f(x_0, y_0)$$

Then using the value pair $(x_1, \bar{y}_1)$ we calculate at this point the function $f(x_1, \bar{y}_1)$. This calculated number is the slope of the tangent line to the solution curve at the place where $x = x_1$, but it is obtained from known values. We also use the previously calculated $f(x_0, y_0)$, which is the slope of the tangent line to the solution curve $x = x_0$. We now take the average of these two slope values to represent the integrand function $f(x, y)$ on $[x_0, x_1]$:

$$f(x, y) \simeq \frac{1}{2}[f(x_0, y_0) + f(x_1, \bar{y}_1)]$$

Note that this average value uses only information that is known when at the starting point (x_0, y_0). Putting this function into (3) and performing the simple mathematics give

$$y(x) = y_0 + \int_{x_0}^{x} f(t, y(t))\,dt$$

$$y(x_1) \simeq y_0 + \int_{x_0}^{x_1} \frac{1}{2}[f(x_0, y_0) + f(x_1, \bar{y}_1)]\,dt$$

$$\simeq y_0 + \frac{1}{2}[f(x_0, y_0) + f(x_1, \bar{y}_1)]\int_{x_0}^{x_1} dt$$

$$\simeq y_0 + \frac{1}{2}[f(x_0, y_0) + f(x_1, \bar{y}_1)](x_1 - x_0)$$

or

$$Y_1 = y_0 + \frac{1}{2}(x_1 - x_0)[f(x_0, y_0) + f(x_1, \bar{y}_1)]$$

where Y_1 represents the improved Euler approximate to $y(x_1)$.

This can, as before, be extended point by point to the right by extending our computations to (x_2, Y_2), (x_3, Y_3), ... using the successive pairs, as in equation (6).

$$\left\{\begin{aligned} \bar{y}_{n+1} &= Y_n + (x_{n+1} - x_n)f(x_n, Y_n) \\ Y_{n+1} &= Y_n + \frac{1}{2}(x_{n+1} - x_n)[f(x_n, Y_n) + f(x_{n+1}, \bar{y}_{n+1})] \\ n &= 0, 1, 2, 3, \ldots \end{aligned}\right\} \qquad (6)$$

If the x values are chosen to be equidistant on the interval, we again replace the step size $x_{n+1} - x_n$ by h.

This technique, called the **improved Euler method**, is the first of a class of predictor-corrector techniques that will be discussed in more detail later. To clarify this terminology, we will demonstrate this technique in Example 10. We will first compute a *predicted* value $\bar{y}_{n+1}$ by Euler's method; then using this value, we will compute the *corrected* Y_{n+1} as the approximation to $y(x_{n+1})$.

Example 10

Use the improved Euler's method on the initial value problem of Example 6:

$$y' = 2x^2 + y;\ y(1) = 2$$

Let the common step size $h = 0.1$. To get an appreciation of the content of this method, we use the equations in (6) for the hand computations:

$$n = 0 \begin{cases} \bar{y}_1 = 2 + 0.1\,[2\cdot(1)^2 + 2] = 2 + (0.1)(4) = 2.4 \\ Y_1 = 2 + 0.1\left\{\dfrac{(4) + [2\cdot(1.1)^2 + 2.4]}{2}\right\} = 2 + 0.1\left(\dfrac{4 + 4.82}{2}\right) \\ \quad = 2.4410 \end{cases}$$

$$n = 1 \begin{cases} \bar{y}_2 = 2.441 + 0.1\,[2\cdot(1.1)^2 + 2.441] = 2.441 + 0.4861 = 2.9271 \\ Y_2 = 2.441 + 0.1\left\{\dfrac{4.861 + [2\cdot(1.2)^2 + 2.9271]}{2}\right\} \\ \quad = 2.441 + 0.5334 = 2.9744 \end{cases}$$

We now use the output from the computer program IMEUL (listed in Appendix B) to complete the table with step size 0.1 (see Table 2.6). The large step size of $h = 0.1$ and the improved Euler method produces a greater degree of accuracy than the small step size $h = 0.005$ and the simple Euler method. The distinct decrease in computer time and effort needed with the improved method more than compensates for the few extra function evaluations required. Therefore, a method that improves on the simple Euler approach is usually the most feasible.

Table 2.6

x	Y (Improved Euler)	y (actual)
1.0	2	2
1.1	2.441	2.4421
1.2	2.9744	2.9768
1.3	3.6141	3.6183
1.4	4.3755	4.3819
1.5	5.2755	5.2847
1.6	6.3330	6.3454
1.7	7.5685	7.5850
1.8	9.0051	9.0265
1.9	10.6681	10.6952
2.0	12.5853	12.6194

Figure 2.6 shows pictorially the significant improvement in the approximate solution when we use the improved Euler method rather than Euler's method for the following type of problem:
Given: $x_0, y_0, f(x, y)$

To find: (x_1, Y_1) use
$$\begin{cases} \bar{y}_1 = y_0 + h \cdot f(x_0, y_0) \\ Y_1 = y_0 + h\left[\dfrac{f(x_0, y_0) + f(x_1, \bar{y}_1)}{2}\right] \end{cases}$$

(x_2, Y_2) use
$$\begin{cases} \bar{y}_2 = Y_1 + h \cdot f(x_1, Y_1) \\ Y_2 = Y_1 + h\left[\dfrac{f(x_1, Y_1) + f(x_2, \bar{y}_2)}{2}\right] \end{cases}$$

(x_3, Y_3) use
$$\begin{cases} \bar{y}_3 = Y_2 + h \cdot f(x_2, Y_2) \\ Y_3 = Y_2 + h\left[\dfrac{f(x_2, Y_2) + f(x_3, \bar{y}_3)}{2}\right] \end{cases}$$

$\vdots$

The averaging of the two values at each point in this improved process helps to keep the approximate solution closer to the actual solution. The original differential equation $y' = f(x, y)$ produces the various functional values that are used; and these values, which actually are values of y', are slopes. The extension to the next point is an averaging at the present point of two slopes, one from the Euler predicted next value and one from the current value.

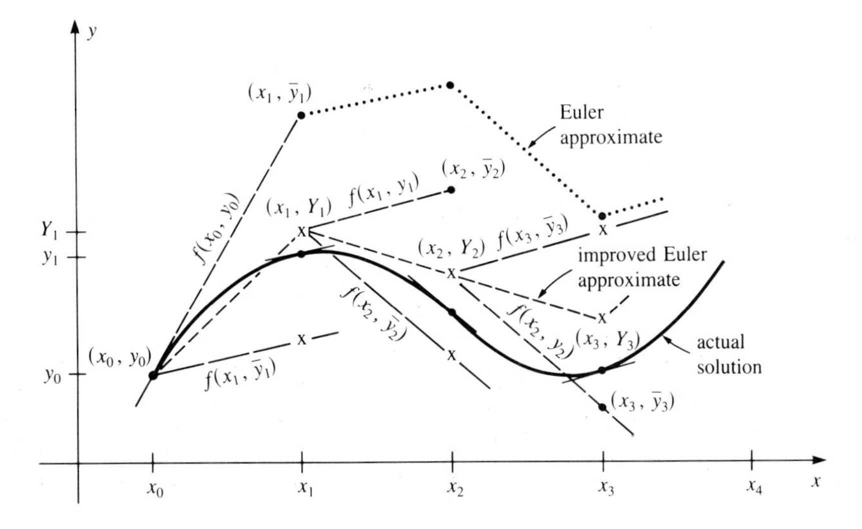

Figure 2.6

Example 11

Use the initial value problem of Example 7 to compare the Euler method with the improved Euler method.

We have

$$\begin{Bmatrix} y' = 2 + y^{1.5} \\ y(0) = .7 \end{Bmatrix}$$

and we solve on $[0, 1.4]$ in step sizes $h = .2$. We will go directly to the computer for our computations, as shown in Table 2.7. Immediately we can see the significant improvement at step size $h = 0.2$ of the improved Euler method over the Euler method. However, in this example the accuracy is still weak.

Table 2.7

x	Y (Euler)	Y (Improved Euler)	y (actual)
0.0	0.7	0.7	0.7
0.2	1.21713	1.2928	1.2974
0.4	1.88569	2.1199	2.1410
0.6	2.80358	3.3842	3.4568
0.8	4.14243	5.5347	5.7787
1.0	6.22865	9.7320	10.6513
1.2	9.73766	19.6907	24.2712
1.4	16.2150	51.853	95.7523

For further comparison, Examples 12 and 13 shorten the step size for Examples 10 and 11, respectively.

Example 12

We have $x_0 = 1$, $y_0 = 2$, and $f(x, y) = 2x^2 + y$. Table 2.8 shows results from the improved Euler method. From Table 2.8 we can also see that $h = 0.1$ produces two significant figures of accuracy; $h = 0.01$, five significant figures; and $h = 0.002$, almost exact results.

Table 2.8

x	$Y(h = 0.1)$	$Y(h = 0.01)$	$Y(h = 0.002)$	y (actual)
1.0	2	2	2	2
1.1	2.441	2.44204	2.44205	2.44205
1.2	2.9744	2.97681	2.97683	2.97683
1.3	3.6141	3.61826	3.61830	3.61831
1.4	4.3755	4.38183	4.38189	4.38190
1.5	5.2755	5.28456	5.28464	5.28466
1.6	6.3330	6.34529	6.34541	6.34543
1.7	7.5685	7.58485	7.58501	7.58504
1.8	9.0051	9.02626	9.02645	9.02650
1.9	10.6681	10.69490	10.6952	10.6952
2.0	12.5853	12.6190	12.6193	12.6194

Example 13

We have $x_0 = 0, y_0 = 0.7$, and $f(x, y) = 2 + y^{1.5}$. Table 2.9 gives results from the improved Euler method. Table 2.9 also shows that $h = 0.1$ produces two significant figures of accuracy; $h = 0.002$, three figures; and $h = 0.0005$, four.

In the type of problem in Example 13 it is significantly more difficult to obtain the degree of accuracy one might desire. This difficulty points to the need for even more detailed analysis of the differential equation and the derivation of more accurate techniques at large step sizes. These are presented in Chapter 5.

Table 2.9

x	$Y\ (h = 0.1)$	$Y\ (h = 0.01)$	$Y\ (h = 0.002)$	$Y\ (h = 0.0005)$	y (actual)
0.0	0.7	0.7	0.7	0.7	0.7
0.2	1.2962	1.2974	1.2974	1.2974	1.2974
0.4	2.1351	2.1409	2.1410	2.1410	2.1410
0.6	3.4359	3.4566	3.4568	3.4568	3.4568
0.8	5.7048	5.7778	5.7786	5.7787	5.7787
1.0	10.3511	10.6476	10.6511	10.6513	10.6513
1.2	22.5577	24.2473	24.2703	24.2712	24.2712
1.4	73.5745	95.3049	95.7338	95.7515	95.7523

Because of round-off errors, which occur when many machine computations are done, it is not always safe to say, "The smaller the step size, the better the results." There are cases for which a decrease in step size actually harms the numerical solution. How do we know when the result is of sufficient accuracy? This question is answered by the following rule of thumb:

> By observing the size of the interval $[A, B]$ over which the numerical integration is to be done, choose a beginning step size that uses a small number of steps, (e.g., 10). After obtaining this solution, run the program again with a new step size equal to half the original. Compare these results, particularly at the largest value of x. If they agree to the desired number of figures, accept the result. If not, reduce the step size again and integrate. Repeat the comparison of the results. Continue the process until you have attained the required accuracy.

The following flow chart (see Figure 2.7) and sample program for the improved Euler method provide a model and are used to integrate the differential equation $y' = f(x, y)$ over the x-interval $[A, B]$ by using $y(A) = y_0$ and taking N steps. Notice that the values of A, B, y_0, and N are

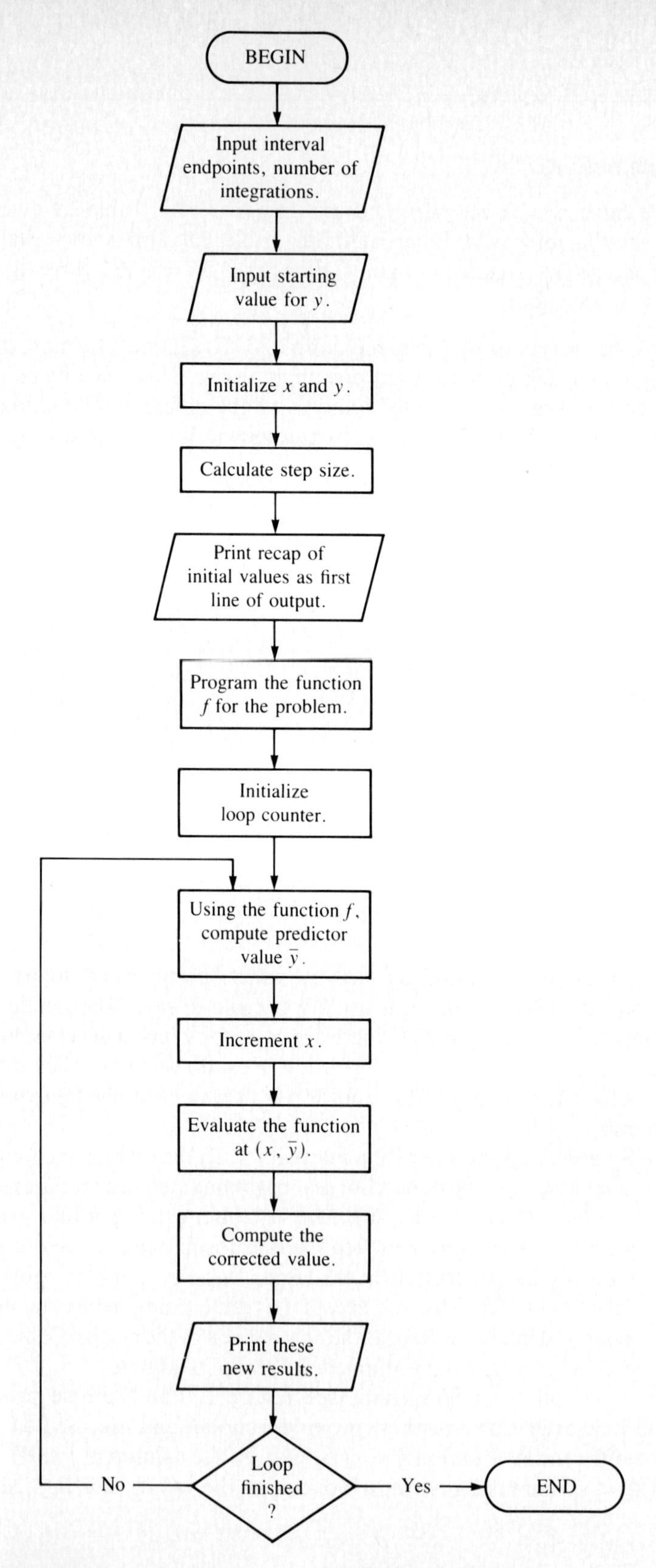
BEGIN
Input interval endpoints, number of integrations.
Input starting value for y.
Initialize x and y.
Calculate step size.
Print recap of initial values as first line of output.
Program the function f for the problem.
Initialize loop counter.
Using the function f, compute predictor value ȳ.
Increment x.
Evaluate the function at (x, ȳ).
Compute the corrected value.
Print these new results.
Loop finished ?
No
Yes
END

Figure 2.7

considered input data, so they can easily be changed. Observe that the successive steps of the program follow very closely the equations derived and mimic the hand-computation steps. Also the program can be easily modified to solve other first-order differential equations since only the evaluations of $F0$ and $F1$ require changing.

Sample Program in Basic for Solving $y' = f(x, y)$ with Improved Euler

Program Line	Explanation
100 INPUT $A, B, Y0, N$	Puts into the computer the interval and the initial values.
110 LET $X = A$	Initializes x at the left endpoint.
120 LET $Y = Y0$	Initializes y at the given value.
130 LET $H = (B - A)/N$	Sets the step size for integration.
140 PRINT X, Y, H	Provides the first line of solution, a recap of initial values.
150 FOR $K = 1$ TO N	Begins the loop for integration.
160 LET $F0 = \left\{\begin{matrix}\text{Put here} \\ \text{general} \\ \text{expression} \\ \text{for } f(x, y).\end{matrix}\right\}$	Inputs the given function.
170 LET $Y1 = Y + H * F0$	Computes the predictor.
180 LET $X = A + H*K$	Moves x to its next value.
190 LET $F1 = \left\{\begin{matrix}\text{Put here} \\ \text{general} \\ \text{expression} \\ \text{for } f(x, Y_1).\end{matrix}\right\}$	Evaluates the function at the predicted value.
200 LET $Y = Y + .5*H*(F0 + F1)$	Computes the corrected value for Y.
210 PRINT X, Y	Prints results of this computation.
220 NEXT K	Recycles back to beginning of loop.
230 END	Terminates program after last loop run.

Chapter 2 refers to three computer programs: EULER, IMEUL, and YACTUL. The program EULER, which does a numerical computation of the solution to the first-order initial value problem

$$\begin{Bmatrix} y' = f(x, y) \\ y(x_0) = y_0 \end{Bmatrix}$$

follows the steps of Euler's method almost exactly. The input to the program consists of values for the beginning point LEFT_X_ENDPNT and the ending point RIGHT_X_ENDPNT of the integration interval, the starting value of Y, called Y0, the number of integrations NUM_INTEGRATIONS to be performed, and the print step size parameter called PRINT_STEP_SIZE.

Note that the step size h is computed from h = ((RIGHT_X_ENDPNT) − (LEFT_X_ENDPNT))/(NUM_INTEGRATIONS), and PRINT_STEP_SIZE tells the computer how often, out of the NUM_INTEGRATIONS times, the value is to be printed. We will always obtain the starting values as output followed by (NUM_INTEGRATIONS)/(PRINT_STEP_SIZE) additional values ending at x = RIGHT_X_ENDPNT. The function $f(x, y)$ from the differential equation is inserted into the program at line 800.

The improved Euler method is programmed as IMEUL, as previously outlined. Details are similar to EULER. The program YACTUL provides the exact values of any function we input and prints as many of these as desired. The actual solution function is inserted at line 800. The user inputs the endpoints of the interval and the number of evaluations. The machine prints the desired values.

Problems

1. Consider
$$\begin{Bmatrix} 3xy' + y = 0 \\ y(1) = 2 \end{Bmatrix}$$

   (a) Solve analytically by separating the variables. Use the program YACTUL (listed in Appendix B), which evaluates an actual y solution function, to display the results on $[1, 2]$ in 10 steps.
   (b) Using the program EULER (listed in Appendix B) for the Euler method, numerically integrate from $x = 1$ to $x = 2$ using $h = 0.1$ and then $h = 0.05$.
   (c) Obtain four significant figures of accuracy by reducing the step size h appropriately.
   (d) Repeat the computations using the improved Euler method with the program IMEUL (listed in Appendix B). Get four significant figures in your result.

2. Consider
$$\begin{Bmatrix} y' = y^2 + 4 \\ y(0) = 0 \end{Bmatrix}$$

   (a) Solve analytically and use YACTUL to obtain the exact values.

(b) Using Euler, solve on $0 \leq x \leq 0.5$ with $h = 0.1$.
(c) Repeat using improved Euler. Make a table to compare results with those of (a) and (b).
(d) Solve the problem to five-figure accuracy by using the improved Euler method in program IMEUL.

3. Consider the separable differential equation

$$\begin{Bmatrix} \dfrac{dy}{dx} = y \sin x \\ y(0) = e^{-1} \end{Bmatrix}$$

(a) Use the EULER method with $h = 0.01$ to find $Y(\frac{\pi}{2})$, $Y(\pi)$, $Y(\frac{3\pi}{2})$, $Y(2\pi)$.
(b) Repeat (a) using IMEUL and $h = 0.01$.
(c) Find the analytical solution and use YACTUL to obtain the exact values.
(d) Determine the percent error made with the two numerical methods.

4. Consider $$\begin{Bmatrix} y' = 1 + y^{1/2} \\ y(0) = 0 \end{Bmatrix}$$

This may be solved analytically, but the solution is difficult to evaluate over an interval. Check this out in the following ways:
(a) Use EULER and $h = 0.1$, $h = 0.05$, and $h = 0.01$ to solve on $0 \leq x \leq 1$.
(b) Repeat (a), but use program IMEUL for the improved method.
(c) Repeat (a) with IMEUL and obtain four figures of accuracy.

5. Consider $$\begin{Bmatrix} y' = \cos(x + y) \\ y(0) = 0 \end{Bmatrix}$$

This equation cannot be solved analytically by methods of Chapter 1.
(a) Use EULER and $h = 0.1, h = 0.05, h = 0.01$ to solve on $0 \leq x \leq 2$.
(b) Repeat (a), but use IMEUL for the improved Euler method.
(c) Solve the problem with several values of h until you achieve three-figure accuracy for both methods. Put the results in table form and make some observations.

6. Consider the nonlinear problem

$$\begin{Bmatrix} y' = y - 2xy^{-1} \\ y(0) = 1 \end{Bmatrix}$$

The exact solution is $y = \sqrt{2x + 1}$. Compare this with the solution you obtain on $[0, 2]$ by using the following methods:
(a) The Euler method with $h = 0.1$ and $h = 0.01$.
(b) The improved Euler method with $h = 0.1$ and $h = 0.01$.

7. Given the initial value problem

$$\begin{cases} \dfrac{dy}{dx} = y^2 + x^3 \\ y(0) = 0 \end{cases}$$

use both EULER and IMEUL to solve on the interval [0, 1]. Obtain four significant figures of accuracy for each. Use a table to show your findings.

8. Solve the initial value problem

$$\begin{cases} \dfrac{dy}{dx} = (x^2 + y^2)^{-1} \\ y(0) = 0.6 \end{cases}$$

on the interval [0, 0.2] by using both EULER and IMEUL until four figures of accuracy are obtained.

9. Obtain a numerical solution to

$$\begin{cases} y' = 3.42 + 1.67\, y^{1.3} \\ y(0) - 0 \end{cases}$$

on some interval $[0, B]$ of your choosing and obtain two significant figures of accuracy.

Applications Problems

A–1. **Population—basic.** The growth of a certain colony of bees is being studied, and the population growth can be approximated by the exponential-curve initial value problem

$$\begin{cases} \dfrac{dP}{dt} = N \cdot P(t) \\ P(0) = P_0 \end{cases}$$

where N is a constant of proportionality related to the birth and death rates. The solution to this initial value problem is the exponential curve determined by $P(t) = P_0 e^{Nt}$. To allow for more facets of the population problem, the Verhulst differential equation $dP/dt = \mathrm{N} \cdot P(t) + I(t) - E(t)$ has been developed with I and E as functions of t that describe the rates of immigration and emigration. For the given colony of bees, experimental data suggest $N = 1.38$, $I = 0.58t$, and $E = 2.87t^4$. Find the population of the bee colony after 5 days if $P_0 = 15$, printing the population at intervals of 0.2. Graph the results.

A–2. **Population growth**. Suppose you have been assigned a task regarding U.S. population growth. (The same methods could be used for populations of spiders, elephants, viruses, etc.) If $y(x)$ represents the number of members at a point x in time, the simplest mathematical model for growth or decay of populations is to assume that the rate of change of y, or $y'(x)$, is proportional to the size of the population y (i.e., $y' = ky$). If the proportionality constant k is positive, the population experiences growth (see Figure 2.8a); if k is negative, then there is loss of population or decay (see Figure 2.8b).

Generally speaking, $k = b - d$ where b and d are the birth and death rates of the population respectively, but many physical factors change this simple situation. Reasoning from a theoretical point of view and verifying their conclusions with experiments on various populations, researchers* have found an equation that models a variety of population situations quite accurately:

$$y'(x) = k_1 y(x) - k_2[y(x)]^2$$

where k_1 and k_2 are positive constants. The term $-k_2 y^2$ usually represents some type of growth inhibiting effect. This equation is an acceptable model for representing the U.S. population for the past few hundred years if $k_1 = .03134$ and $k_2 = .0000001589$ where $y(x)$ represents the population (in thousands) and x is in years.

(a) The census of 1800 showed that the population of the U.S. was about 5,308,000 ($y(0) = 5308$). Use the improved Euler method

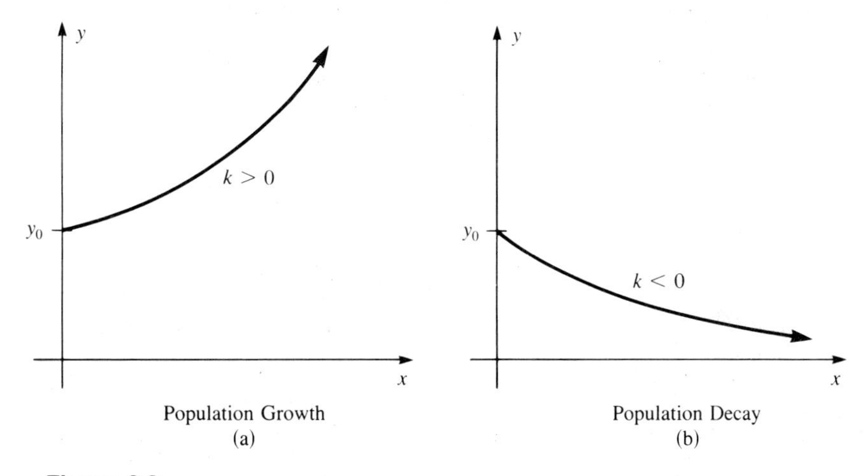

Figure 2.8

* A. J. Lotka, *Elements of Mathematical Biology* (New York: Dover, 1956).

Table 2.10

U.S. Population from 1800 to 1960

Year	Population (in thousands)
1800	5308
1820	9638
1840	17069
1860	31443
1880	50156
1900	75995
1920	105711
1940	131669
1960	179323

to compute an approximate solution to the problem for the years 1800 to 1900. If you go beyond 1900 the comparison deteriorates. This implies that a new model needs to be developed for the period 1900 on. Try to obtain two significant figures and compare your results with the actual census figures in Table 2.10.

(b) By solving $y' = k_1 y - k_2 y^2$, show that for positive k_1 and k_2 and an initial population size y_0 less than k_1/k_2, the population increases continuously and approaches the value k_1/k_2 as $x \to \infty$.

(c) Using the result of part (b), we will be able to adjust the constants k_1 and k_2 so that a population goal of fixed size will be attained. Set the value of k_2 to $(8.8)10^{-8}$ and find the value of k_1 so that the goal for U.S. population of 250 million is approached for large time value x.

(d) If we retain the model equation $y' = k_1 y - k_2 y^2$ and the new values of k_1 and k_2 from part (c), we may predict the population growth from 1970 to 2050. The official census in 1970 was 203,235,298. Run the numerical result to obtain two significant figures of accuracy.

A–3. **Thermodynamics**. Suppose a body has an initial temperature T_0 of 200°F and temperature T_1 of 150°F one hour later. Assuming the surrounding air is kept at a constant temperature T_A of 74°F, you wish to determine the relationship between temperature and time. The basic law of thermodynamics states that the rate at which a body changes temperature is proportional to the difference between its temperature and the temperature of the surrounding medium. This relationship, called Newton's law of cooling, implies

$T' = -k[T(t) - T_A]$, and the problem becomes

$$T' = -0.51(T - 74)$$

where $k = 0.51$ is obtained from observable laboratory conditions. Solve for $T(t)$ on $[0, 4]$ with 20 printouts using $h = 0.1$ and IMEUL. Compare with the actual solution $T(t) = 74 + 126e^{-.51t}$ obtained by solving the simple boundary value problem

$$T' = -k(T - 74),\ T(0) = 200,\ T(1) = 150$$

A–4. **Logistics.** A colony of cells in the early stages of growth tends to multiply exponentially, as suggested by the initial value problem

$$\left\{\begin{array}{l} \dfrac{dP}{dt} = N \cdot P(t) \\ \\ P(0) = P_0 \end{array}\right\}$$

whose solution is the exponential curve $P(t) = P_0 e^{Nt}$. This rapid growth results in crowding effects such that the proportionality constant N is better approximated by the term $N_0 - AP$. N_0 accounts for the increase in population due to division, and the parameter A represents the inhibition on growth due to crowding effects. The initial value problem becomes

$$\left\{\begin{array}{l} \dfrac{dP}{dt} = N_0 P - AP^2 \\ \\ P(0) = P_0 \end{array}\right\}$$

which is often referred to as the differential equation of logistics.

(a) For a particular colony of cells being studied, N_0 was found to be 1.25 and $A = 0.01$. While printing out the values for each half hour, find the population for the colony in 15 hours for an initial population value of 25; then repeat with an initial value of 75. Graph the results.

(b) Show that the analytical solution to the logistic differential equation is

$$P(t) = \frac{N_0 P_0}{AP_0 + [N_0 - AP_0]e^{-N_0 t}}$$

Run the solution on YACTUL for $P_0 = 25$ and $P_0 = 75$. Explain why P approaches $N_0 \cdot (1/A)$ independent of the starting population size.

A–5. **Irreversible chemical reactions.** In an irreversible chemical reaction, occurring at constant volume and temperature, a molecules of A combine with b molecules of B to give e molecules of E and f molecules of F, as indicated by the stoichiometric equation

$aA + bB \rightarrow eE + fF$. If, at any time t after the reaction starts, there are present N_A molecules of A and N_B molecules of B, then the rate law for the disappearance of A becomes

$$\frac{dN_A}{dt} = -k(N_A)^a(N_B)^b$$

where k is the velocity constant of the reaction. A similar equation for the rate of change of N_E is obtained by noting that for every a molecules of A disappearing in the reaction, e molecules of E will appear. Hence

$$\frac{dN_E}{dt} = -\frac{e}{a}\frac{dN_A}{dt} = k\frac{e}{a}(N_A)^a(N_B)^b \qquad (1)$$

(a) An example of this type of reaction is the combination of hydrochloric acid with sodium carbonate to produce sodium chloride, carbon dioxide, and water:

$$2HCl + Na_2CO_3 \rightarrow 2NaCl + CO_2 + H_2O$$

Assume that this reaction is started with 1000 molecules of HCl and 700 molecules of Na_2CO_3, the velocity constant is 24×10^{-8}, and the carbon dioxide bubbles out of the solution. Determine how long it will take to produce 600 molecules of sodium chloride. If we let $2x$ denote the number of molecules of NaCl present at time t, the rate at which NaCl changes is $2(dx/dt)$. Since in this reaction $a = 2$ and $b = 1$, the rate equation (1) produces

$$2\frac{dx}{dt} = (24 \times 10^{-8})(1000 - 2x)^2(700 - x)$$

with $x(0) = 0$. Therefore you must determine when the dependent variable $x = 300$. Use IMEUL with $h = 0.01$. Watch the naming of the variables.

(b) Other reactions would produce different differential equations. For example, the reaction of calcium carbide with cold water, giving calcium hydroxide (slaked lime) and acetylene gas may be represented by

$$CaC_2 + 2H_2O \rightarrow Ca(OH)_2 + C_2H_2$$

If the velocity constant is k, if we have n_1 molecules of CaC_2 and n_2 molecules of H_2O, and if the acetylene gas escapes, the differential equation for concentration of $Ca(OH)_2$ is

$$\frac{dx}{dt} = k(n_2 - 2x)^2(n_1 - x); \qquad x(0) = 0$$

Use the values $n_1 = 100$, $n_2 = 300$, and $k = 16 \times 10^{-7}$ to determine when $x = 150$.

A–6. **Psychological learning theory**. Psychologists have been able to describe memorization by a fairly simple mathematical model. If we refer to the amount of material memorized as attainment, then the attainment in time t will be denoted by $y(t)$. If the total amount of material to be memorized is T, called the maximum attainment, then the rate of change of attainment at time t is proportional to the amount of still-to-be-memorized material; that is, $y' = N(T - y)$, where N is a measure of natural learning ability. In the simplified version, N would be considered to be constant and the subject would have no material memorized at $t = 0$. However, if the learning process is lengthy, N becomes a function of time because of fatigue or forgetting. Since these factors slow the learning process, $N(t)$ is a positive monotone function.

(a) To account for fatigue, assume $N(t) = N_0/(1 + t)$ and $y(0) = 0$. With these in the differential equation, show that the actual solution is

$$y = T\left[1 - \left(\frac{1}{1 + t}\right)^{N_0}\right]$$

(b) Suppose a person has a natural-learning-ability factor of 0.7, which diminishes with fatigue, and begins to memorize a list of 100 words from a foreign language vocabulary. Using the program IMEUL, find the time t at which this person knows 50 words and 80 words. Use $h = 0.01$.

(c) To include the detrimental effects of forgetting, assume that the rate of forgetting is proportional to the amount already learned. When you include the forgetting factor in the differential equation, you have $y' = N(T - y) - Fy$, $y(0) = 0$, where $F > 0$ is constant. Using the values $N = 0.7$, $T = 100$, and $F = 0.2$, solve this initial value problem by numerical methods to find the time at which the attainment y is rising only very slowly.

A–7. **Modified logistics**. When studying the growth of a particular population, we assume that the rates of birth and death are proportional to existing population sizes. A typical differential equation would be $dp/dt = (b - d)p$. One of the popular variations of this equation that accounts for limitations imposed by the environment is

$$\frac{dp}{dt} = r\left(1 - \frac{p}{K}\right)p$$

where r is a constant related to the birth and death rates and K is the carrying capacity of the environment (i.e., the maximum population). Suppose the population to be studied is a particular species of fish in a specific lake and the differential equation for this situation is

$$\frac{dp}{dt} = 1.25\left(1 - \frac{p}{400}\right)p - c$$

where c is the rate at which fish are removed by fishing. The value of c determines whether the fish population approaches an equilibrium number or decreases to zero. The critical population level is related to the carrying capacity and the fishing rate by the equation

$$p_c = \frac{1}{2}K\left[1 - \left(1 - \frac{4c}{rK}\right)^{1/2}\right]$$

(a) When the value of c is less than $\frac{1}{4}rK$ and the initial population of the lake is greater than the critical population level, then the fish population approaches an equilibrium value equal to

$$\frac{1}{2}K\left[1 + \left(1 - \frac{4c}{rK}\right)^{1/2}\right]$$

However, if the starting population is less than P_c, the number of fish will steadily decrease to zero. Assume that during the month of May the value of c is 100 per day. Calculate the critical population level, P_c, and the anticipated equilibrium value. Solve the differential equation by using $P(0) = P_c + 0.5$ and find the number of fish after 25 days, printing out each day's number. Graph the results. Using $P(0) = P_c - 0.5$, determine the number of days until the species of fish is extinct.

(b) When the value of c is exactly $\frac{1}{4}rK$, the population reaches an equilibrium of $\frac{1}{2}K$ for any starting population greater than $\frac{1}{2}K$. Solve the equation by using $P(0) = 250$ and repeat using $P(0) = 300$; note the approach to the equilibrium value after 25 days.

(c) If $c > \frac{1}{4}rK$, the fish population will decrease to zero for any initial population size. During August the value of c is highest at 135. If the number of fish is 285 at the beginning of the month, how many fish remain after 20 days?

(d) What value of c would prevent extinction of the fish?

A–8. **Chemical mixtures.** A tank contains 10 gallons of water and 4 pounds of chemical C per gallon. To decrease the concentration of C, pure water is added to the container at the rate of 2 gallons per minute and the mixture is constantly stirred to keep the concentration consistent throughout the tank. The mixture is also tapped off at the rate of 2 gallons per minute: If $r(t)$ is the amount of chemical C in the container at time t, then $r' = (\text{rate in}) - (\text{rate out})$.

(a) Formulate the appropriate initial value problem and solve for $r(t)$ at any time t by analytical methods. Then use YACTUL to find $r(5)$.

(b) Solve numerically by using EULER and IMEUL on $[0, 5]$. Use $h = 0.05$ and print 20 values. Compare with actual values.

Note: A similar analysis would help to solve the following ecological problem. A pond is being polluted by the addition of a fixed amount of chemical each day. Springs under the pond add pure water at a constant rate and the polluted pond water spills off the dam at the same rate. A differential equation model will allow you to discuss the future of the lake and its pollution levels.

A–9. **Mixing**. The discharge valve on a 200-gallon tank that is full of water is opened at time $t = 0$ and 3 gallons per second flow out. At the same time 2 gallons per second of 1% chlorine mixture begin to enter the tank. Assume that the liquid is being stirred so that the concentration of chlorine is consistent throughout the tank. Your task is to determine the concentration of chlorine when the tank is half full.

It takes 100 seconds for this moment to occur, since we lose a gallon per second. If $y(t)$ is the amount of chlorine in the tank at time t, then the rate chlorine is entering is $\frac{2}{100}$ gal/sec and it is leaving at the rate $3[y/(200 - t)]$ gal/sec. Thus the differential equation is

$$\frac{dy}{dt} = \frac{2}{100} - 3 \cdot \frac{y}{200 - t}; \quad y(0) = 0$$

Analytical methods will yield the solution

$$y(t) = 2 - (0.01)t - 2[1 - 0.005t]^3$$

Thus $y(100) = 0.75\%$ concentration.

(a) Solve this problem by using the Euler technique on $[0, 100]$ with $h = 1$ but printing at intervals of 5.

(b) Repeat (a) but use improved Euler with $h = 1$.

(c) Compare results from (a) and (b) with YACTUL.

A–10. **Fluid flow**. A funnel has an outlet with a cross-sectional area of $\frac{1}{2}$ cm^2, and its cone has an outlet angle of 60° (see Figure 2.9). If the funnel is filled with water to height $h = 10$ cm and the valve in the outlet is opened at time $t = 0$, we use Torricelli's law $v = .6\sqrt{2gh}$ to derive the differential equation that determines the time at which

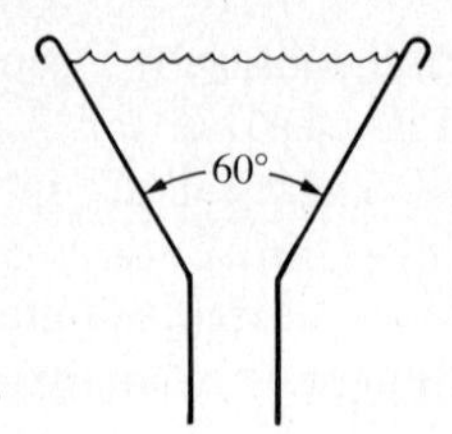

Figure 2.9

the funnel is empty. Since h is decreasing, the equation is

$$\frac{dh}{dt} = -\left(\frac{3}{2}\right)\frac{.6\sqrt{2gh}}{\pi h^2} = -\frac{.9\sqrt{2g}h^{-3/2}}{\pi}$$

Use the given conditions to find the time t for which $h = 0$.

A–11. **Population dynamics**. A company currently has 4000 employees, 20% of whom are women. Employees quit randomly at the rate of 60 per week, replacements are hired at 40 per week, and half of the newly hired employees are women. Therefore, at $t = 0$, $y(t)$, the number of women employed, is 800. Also the staff size at time t is $4000 - 20t$. Thus the change in $y(t)$ at time t is determined by the number being hired minus the number quitting or

$$\frac{dy}{dt} = \frac{1}{2}\cdot 40 - \left(\frac{60}{4000 - 20t}\right)y$$

Determine the size of the staff after 50 weeks and the percentage of women then employed. Use $y(0) = 800$ and program IMEUL to solve the problem on $[0, 50]$ with $h = 1$.

A–12. **Nutrition study**. V. S. Ivlev, a Russian biologist, has done a great deal of research in the flow of energy in the consumption of living organisms by fish and their usage of this consumption in their growing cycle. In a report* he presents a relation between the intensity of nutrition of fishes and the concentration of available food:

$$dr/dp = \alpha(R - r)$$

where r is the rate of consumption for a given concentration of food p, R is the maximal rate of consumption for infinite concentration, and α is a proportionality constant. The empirically determined value of R is $R = 15.5$ calories per hour. To find the

* V. S. Ivlev, "On the Utilization of Food by Planktophage Fishes," *Bulletin of Mathematical Biophysics*, 22 (1960), p. 371.

value of α, use the information that for $p = 2550$ cal/m^3, the value of r was 9.44 cal/hr.

(a) Obtain the analytical solution to the initial value problem for which $r = 0$ when $p = 0$. Use YACTUL to list the results for values of p from 0 to 5000.

(b) Use the EULER program to get a numerical result over this same interval. Reduce h until four significant figures of accuracy are obtained.

(c) Repeat part (b) using IMEUL.

A–13. **Meteorology**. In using large balloons to assist in weather forecasting, meteorologists often use the differential equation $dp/dh = -\rho$, which states that the rate of change of atmospheric pressure with height is the negative of the density (weight per unit volume) at that height. If we assume that the air expands without gaining or losing heat, then $p = k\rho^{7/5}$. Use sea level values of $p = 15$ lbs/in^2 and $\rho = 0.082$ lb/ft^3 to find pressure p as a function of height h.

(a) Find the analytical solution for p as a function of h. Then using YACTUL, record the values of p for sea level to 100,000 feet in 10 levels.

(b) Obtain the numerical solution by both EULER and IMEUL for the same values to three significant figures of accuracy.

A–14. **Electric circuits**. Consider an electric circuit containing both resistance R and capacitance C, whose values vary with time t. The differential equation for such a circuit is

$$y' + f(t)y = 0$$

where $y(t)$ is the charge and $f(t)$ is a function of resistance and capacitance given by $f(t) = [R(t)C(t)]^{-1}$. Assume the applied voltage $E = 0$. The charge on the capacitor at time $t = 0$ is $y(0) = C_0$. We wish to find the function that represents subsequent decay of the charge. Suppose at time $t = 0$ a unit charge is placed on the capacitor and that the resistance varies in the circuit while the capacitance is constant. Let the capacitance and resistance have values $C(t) = 1$ and $R(t) = (1 + b\cos t)^{-1}$, where b is a constant subject to $|b| < 1$. Note that the resistance oscillates about the value 1 in a periodic way.

(a) Set up the initial value problem and solve analytically. Analyze the solution by giving reasons for the following properties: (1) y is always positive; (2) $y \to 0$ as $t \to \infty$; (3) y is a monotonically decreasing function; and (4) The solution function is not periodic.

(b) Use IMEUL to solve the problem on $[0, 3]$ with $h = 0.01$ when $b = -\frac{1}{2}$, $b = 0$, and $b = \frac{1}{2}$. Graph each solution.

(c) Repeat (b) but use $f(t) = 1/[R(t)C(t)] = 1 - (\frac{1}{2})\sin t$, with $y(0) = 1$.

A–15. **Cooling**. In a foundry, it was observed that certain types of castings, when immersed in a special sand-liquid mixture, cooled in that surrounding medium according to the equation $du/dt = -.27(u - 60)^{5/4}$, with u in degrees Fahrenheit and t in minutes. Assume that the sand mixture has a constant temperature of 60°F and the proportionality coefficient in the equation is a function of the material in the casting. If the initial temperature of the casting is 320°F, how long will it take to cool the casting to 120°F? Use EULER and IMEUL with $h = 0.01$ to obtain the numerical result.

Chapter 3

Linear Differential Equations

3.1 Introduction to Linear Equations

The importance of linear differential equations in scientific applications cannot be overestimated. They occur in all fields of science and engineering. Chapter 3 is devoted to the second-order linear differential equation

$$a_0(x)y'' + a_1(x)y' + a_2(x)y = f(x)$$

as a special case of the most general second-order differential equation $F(x, y, y', y'') = 0$.

When a differential equation is encountered in scientific studies, one usually first attempts to solve it in closed form by analytical techniques. When this is not possible, one usually picks an appropriate numerical technique. With the first-order equations posed in Chapter 1, a reasonably large selection could be solved analytically, and Chapter 2 covered the simple numerical methods. In the present second-order case we are limited even more, because few of the nonlinear equations can be solved by *any* analytical method. Hence, Chapter 3 will not explore these nonlinear differential equations, but will leave to succeeding chapters the task of involving numerical methods in their solution.

Before we begin the present study, a few remarks are in order concerning solutions. As we begin working on the solution process for a given differential equation, one item of great interest is the existence of a solution. "Does this problem have a solution?" is a question of extreme importance. If this question can be answered in the affirmative, another question follows: "Is the solution that I have found the only solution to this problem?" This question also is of utmost importance.

These two questions are often answered simultaneously in a theorem called an **existence and uniqueness theorem**. When certain conditions are met by the parts of the problem, theorems of this kind guarantee that a solution exists and that it is the only solution to that problem.

For linear second-order equations the following existence and uniqueness theorem can be stated. Its proof may be found in advanced texts.

Theorem 3.1 (Existence and Uniqueness):

Let $a_0(x)$, $a_1(x)$, $a_2(x)$, and $f(x)$ all be continuous functions in a common interval I containing the point x_0 and let $a_0(x)$ be nonzero in I. Then the initial value problem

$$\begin{Bmatrix} a_0(x)y'' + a_1(x)y' + a_2(x)y = f(x) \\ y(x_0) = y_0,\ y'(x_0) = v_0 \end{Bmatrix} \tag{1}$$

possesses a unique solution that is twice continuously differentiable and defined throughout I.

If in equation (1) the function $f(x)$ is the zero function, we say that the equation is **homogeneous**. If not, we call it **nonhomogeneous**. If the coefficient functions $a_0(x)$, $a_1(x)$, and $a_2(x)$ are constant functions, the equation is said to have **constant coefficients**. Note that $y = 0$, the zero function, is always a solution of the homogeneous equation. It is called the **trivial** solution. As before, we are seeking nontrivial solutions.

Example 1

(a) In the initial value problem

$$\begin{Bmatrix} y'' - 5y' + 4y = 0 \\ y(1) = 3,\ y'(1) = -5 \end{Bmatrix}$$

the differential equation is second order, linear, and homogeneous, and it has constant coefficients.

(b) The differential equation $x^2y'' - 5xy' + 7y = 2e^{3x}$ is second order, linear, and nonhomogeneous, and it has variable coefficients.

3.2 Elementary Theory

Most application problems that arise from a study of motion in time and space produce differential equations that are second order (i.e., they involve distance, velocity, and acceleration). A vast number of both natural and man-made phenomena give rise to second-order equations.

Although we will focus most of our attention in this chapter toward second-order linear differential equations, much of the theory and many techniques apply equally well to higher-ordered equations. The results for these higher-ordered cases will follow easily as we extend the theory of the second-order equations to them in a natural way.

This section will list and explain many important basic properties of solutions to linear equations of any order. Proofs are done for the second-order case and are easily extendable. This will lay the theoretical groundwork for the chapter. Our discussion begins with the easier homogeneous equations, and Section 3.3 moves us to nonhomogeneous equations.

To assist in developing the theory, we will introduce some notation at this point. We introduce the differentiation operator D by using the notation Dy to replace y' for any differentiable function $y(x)$. This leads us to use

$$D^2y = D(Dy) = Dy' = y''$$

and

$$D^3y = D^2(Dy) = D(D^2y) = Dy'' = y'''$$

If we inserted this notation into a second-order linear differential equation such as (1), we would write $a_0(x)y'' + a_1(x)y' + a_2(x)y = f(x)$ as $a_0(x)D^2y + a_1(x)Dy + a_2(x)y = f(x)$. Now treating the operators as algebraic entities, we formally manipulate by factoring and obtain

$$[a_0(x)D^2 + a_1(x)D + a_2(x)]y = f(x) \tag{2}$$

We define the linear operator L to be the bracketed quantity:

$$L \equiv [a_0(x)D^2 + a_1(x)D + a_2(x)]$$

Then equation (1) written in revised notation as (2) may be simplified to

$$Ly = f(x)$$

In the above equations we have introduced only notation. Nothing has actually changed; only the form of the given differential equation is new. These operators are further demonstrated in Example 2.

Example 2

We can write $3xy'' - 4x^2y' + y = e^x$ as $3xD^2y - 4x^2Dy + y = e^x$ and then as $(3xD^2 - 4x^2D + 1)y = e^x$. Now with the operator defined by $L = (3xD^2 - 4x^2D + 1)$, the differential equation may be written $Ly = e^x$.

If in the differential equation, $f(x)$ is zero, we write $Ly = 0$. L can be thought of as an operator that acts on a twice differentiable function $y(x)$ to produce the function Ly.

The linearity of operator L is expressed in Theorem 3.2 and becomes an important property useful on many occasions in this chapter.

Theorem 3.2:

If $u_1(x)$ and $u_2(x)$ are twice differentiable functions and c_1 and c_2 are any constants, then $L(c_1u_1 + c_2u_2) = c_1Lu_1 + c_2Lu_2$.

Proof:

Using the definition of L, we have

$$\begin{aligned}
L(c_1u_1 + c_2u_2) &= [a_0(x)D^2 + a_1(x)D + a_2(x)](c_1u_1 + c_2u_2) \\
&= a_0(x)D^2(c_1u_1 + c_2u_2) + a_1(x)D(c_1u_1 + c_2u_2) \\
&\quad + a_2(x)(c_1u_1 + c_2u_2) \\
&= a_0(x)(c_1u_1'' + c_2u_2'') + a_1(x)(c_1u_1' + c_2u_2') \\
&\quad + a_2(x)(c_1u_1 + c_2u_2) \\
&= [a_0(x)c_1u_1'' + a_1(x)c_1u_1' + a_2(x)c_1u_1] \\
&\quad + [a_0(x)c_2u_2'' + a_1(x)c_2u_2' + a_2(x)c_2u_2] \\
&= [a_0(x)c_1D^2u_1 + a_1(x)c_1Du_1 + a_2(x)c_1u_1] \\
&\quad + [a_0(x)c_2D^2u_2 + a_1(x)c_2Du_2 + a_2(x)c_2u_2] \\
&= c_1[a_0(x)D^2 + a_1(x)D + a_2(x)]u_1 \\
&\quad + c_2[a_0(x)D^2 + a_1(x)D + a_2(x)]u_2 \\
&= c_1(L)u_1 + c_2(L)u_2 \\
&= c_1Lu_1 + c_2Lu_2 \qquad \blacksquare
\end{aligned}$$

This section also presents the theory that underlies the process of determining the general solution to a given linear differential equation. As a first step we state a few important theorems concerning solutions to homogeneous linear equations.

Theorem 3.3:

If $u_1(x)$ and $u_2(x)$ are solutions of the homogeneous linear equation $Ly = 0$, then any linear combination $c_1u_1 + c_2u_2$ is also a solution of $Ly = 0$.

Before looking at the proof for this theorem, consider Example 3.

Example 3

Easily verified solutions of $y'' + y = 0$ are $u_1(x) = \sin x$ and $u_2(x) = \cos x$. You should check that $5 \sin x - 3 \cos x$ is also a solution. In fact you should verify that $c_1 \sin x + c_2 \cos x$ is also a solution by differentiating and substituting. In this case the values for c_1 and c_2 may be arbitrarily picked.

Proof of Theorem 3.3:

Since u_1 is a solution we may substitute it for y in the equation and obtain the identity $Lu_1 = 0$. Since u_2 is a solution we similarly obtain $Lu_2 = 0$. But by the linearity property

$$L(c_1u_1 + c_2u_2) = c_1Lu_1 + c_2Lu_2 = c_1 \cdot 0 + c_2 \cdot 0 = 0$$

which implies that $c_1u_1 + c_2u_2$ is a solution to the equation $Ly = 0$.

A simple extension of the above argument could be now used to prove that if $u_1, u_2, u_3, \ldots u_n$ are solutions to the linear differential equation $Ly = 0$ which has order n, then $c_1u_1 + c_2u_2 + \cdots + c_nu_n$ is also a solution.

Note that the differential equation must be linear for Theorem 3.3 to apply.

Example 4

For the differential equation $y'' - 5y' + 4y = 0$ of Example 1, the functions $u_1 = e^x$ and $u_2 = e^{4x}$ are easily verified solutions. Thus the linear combination $6e^x + 13e^{4x}$ is a solution as well as $9e^x - 23e^{4x}$, and in general $c_1e^x + c_2e^{4x}$ is a way to write a two-parameter family of solutions.

One goal of this section is to build the theory for finding the general solution to a given differential equation. As we continue toward this goal we will need a basic concept from linear algebra.

Definition: Let f_1 and f_2 be two functions defined on an interval I. The two functions are said to be **linearly dependent** on the interval I if one of the functions is a constant multiple of the other, that is, if $f_1 = kf_2$. If f_1 and f_2 are not linearly dependent, they are said to be **linearly independent**.

Looking at this from another angle, we see that the pair of functions $\{f_1, f_2\}$ is **linearly dependent** on I if we can find a pair of constants c_1 and c_2, not both zero, such that $c_1f_1 + c_2f_2 = 0$ on I. Following up on this, we note that if $c_1 \neq 0$, we divide by c_1 and obtain $f_1 + (c_2/c_1)f_2 = 0$ or $f_1 = -(c_2/c_1)f_2 = kf_2$, which was our first definition for linear dependence. Linear independence may now be stated as follows:

The functions $\{f_1, f_2\}$ are linearly independent on interval I if the linear combination $c_1f_1 + c_2f_2 = 0$ implies that $c_1 = c_2 = 0$ for all x in I.

We may use either definition to determine the linear dependence or independence of the given functions. Examples 5–9 illustrate linear dependence and independence.

Example 5

The functions $\{x - 2, 6 - 3x\}$ are linearly dependent in every interval I since $f_2 = -3x + 6 = -3(x - 2) = -3f_1$. The functions $\{x, x^2\}$ are linearly independent in every interval I since x^2 clearly is not a constant multiple of x. One of these is linear; the other is not.

Example 6

The functions $\{3x, x^2 + 1\}$ are linearly independent in every interval I since $c_1 \cdot 3x + c_2(x^2 + 1) = 0$ implies that $c_2x^2 + 3c_1x + c_2 = 0$ and by equating coefficients of the polynomials on both sides we find that $c_2 = 0$, $3c_1 = 0$, $c_2 = 0$, or that $c_1 = c_2 = 0$ always.

Example 7

The functions $\{4x + 1, x - 3\}$ are linearly independent since

$$c_1(4x + 1) + c_2(x - 3) = 0$$

implies that $(4c_1 + c_2)x + (c_1 - 3c_2) = 0$, which provides

$$\left\{\begin{array}{l} 4c_1 + c_2 = 0 \\ c_1 - 3c_2 = 0 \end{array}\right\}$$

whose unique solution is $c_1 = c_2 = 0$.

Example 8

The functions $\{e^x, 5e^x\}$ are linearly dependent since $c_1e^x + c_2(5e^x) = 0$ is easily solved by picking $c_1 = -5$, $c_2 = 1$, which is a nonzero set. We also see that $f_1 = e^x = 1/5(5e^x) = (1/5)f_2$, so one function is a constant multiple of the other.

Example 9

The functions $\{e^{ax}, e^{bx}\}$ are linearly independent if $a \neq b$. To verify this, write $c_1e^{ax} + c_2e^{bx} = 0$. Since this is to be an identity that holds for all x, set $x = 0$ to obtain $c_1 + c_2 = 0$. Then set $x = 1$ to obtain $c_1e^a + c_2e^b = 0$. Using the first result to get $c_2 = -c_1$, the second becomes

$$c_1e^a - c_1e^b = c_1(e^a - e^b) = 0$$

But since $a \neq b$, $e^a \neq e^b$ so $c_1 = 0$, also giving $c_2 = 0$.

When only two functions are involved, the determination of linear dependence or independence is much easier than when three or more functions are being considered. Nevertheless, the definitions of linear dependence and linear independence extend to larger collections of functions in an analogous way.

Definition: A collection of functions $\{f_1, f_2, \ldots f_n\}$ is said to be **linearly dependent** on an interval I if there exists a set of constants $\{c_1, c_2, \ldots, c_n\}$, not all zero, such that $c_1f_1 + c_2f_2 + \cdots + c_nf_n = 0$ for all x in I. If no such set of constants exists, the functions are **linearly independent** on I. That is, the functions are linearly independent if $c_1f_1 + c_2f_2 + \cdots + c_nf_n = 0$ implies that $c_1 = c_2 = \cdots = c_n = 0$.

Another method for determining the linear dependence of a collection of functions involves the Wronskian. For a pair of differentiable functions

$\{f_1(x), f_2(x)\}$ the **Wronskian** (denoted by $W[f_1, f_2]$) is defined as the determinant:

$$W[f_1, f_2] \equiv \begin{vmatrix} f_1(x) & f_2(x) \\ f'_1(x) & f'_2(x) \end{vmatrix} = f_1 f'_2 - f'_1 f_2$$

Although we deal here with the Wronskian for only *pairs* of functions, the Wronskian for n functions would be the following determinant:

$$W(f_1, f_2, \ldots, f_n) = \begin{vmatrix} f_1 & f_2 & \cdots & f_n \\ f'_1 & f'_2 & \cdots & f'_n \\ \vdots & \vdots & & \\ f_1^{(n-1)} & f_2^{(n-1)} & \cdots & f_n^{(n-1)} \end{vmatrix}$$

Example 10

For the function pair $\{4x + 1, x - 3\}$, the Wronskian is

$$W[f_1, f_2] = \begin{vmatrix} 4x + 1 & x - 3 \\ 4 & 1 \end{vmatrix} = (4x + 1) - 4(x - 3) = 13$$

Example 11

For the function pair $\{e^x, 5e^x\}$, the Wronskian is

$$W[f_1, f_2] = \begin{vmatrix} e^x & 5e^x \\ e^x & 5e^x \end{vmatrix} = 5e^{2x} - 5e^{2x} = 0$$

Theorem 3.4:

If the Wronskian of two differentiable functions is not zero for at least one point in an interval I, then f_1 and f_2 are linearly independent over that interval. Conversely, if f_1 and f_2 are linearly independent solutions of $Ly = 0$ on I, then $W[f_1, f_2]$ is never zero on I.

Proof:

If f_1 and f_2 are not linearly independent on I, there exist constants c and $k = 1/c$ such that $f_1 = cf_2$ or $f_2 = kf_1$. Assuming the latter we find that $f'_2 = kf'_1$ and $W[f_1, f_2] = f_1 f'_2 - f'_1 f_2 = f_1(kf'_1) - f'_1(kf_1) = 0$. However, this contradicts the assumption that $W[f_1, f_2] \neq 0$. Hence the functions f_1 and f_2 must be linearly independent.

To prove the converse, assume that $W[f_1, f_2] = 0$ at $x_0 \in I$. Then consider the pair of equations

$$\begin{Bmatrix} c_1 f_1(x_0) + c_2 f_2(x_0) = 0 \\ c_1 f'_1(x_0) + c_2 f'_2(x_0) = 0 \end{Bmatrix} \quad (3)$$

where x_0 is a point in I. In this pair we consider c_1 and c_2 to be the unknowns whose values are to be found and consider $f_1(x_0)$, $f_2(x_0)$, $f'_1(x_0)$, and $f'_2(x_0)$ to be the constant coefficients of these unknowns. The theory from algebra we use here depends on the value of the determinant of these coefficients

$$\begin{vmatrix} f_1(x_0) & f_2(x_0) \\ f'_1(x_0) & f'_2(x_0) \end{vmatrix}$$

However, this is exactly $W[f_1, f_2]$.

A basic theorem from algebra states that a pair of linear equations that are homogeneous (zero values on the right side) and that have a zero coefficient determinant have a nonzero solution. Thus we are assured of a solution pair (c_1, c_2) that are not both zero.

Define $y(x) = c_1 f_1 + c_2 f_2$. Since f_1 and f_2 are hypothesized to be linearly independent solutions of $Ly = 0$ on I, then $y(x)$ is also a solution. But c_1 and c_2 solve the above pair (3), so

$$y(x_0) = c_1 f_1(x_0) + c_2 f_2(x_0) = 0$$

and then

$$y'(x_0) = c_1 f'_1(x_0) + c_2 f'_2(x) = 0$$

Thus $y(x)$ solves the initial value problem

$$\left\{ \begin{matrix} Ly = 0 \\ y(x_0) = 0,\ y'(x_0) = 0 \end{matrix} \right\}$$

However, this initial value problem also has the solution $\bar{y}(x) \equiv 0$. By Theorem 3.1 the solution of this initial value problem is unique, so $y(x) = \bar{y}(x) = 0$. Hence $y(x) = c_1 f_1 + c_2 f_2 = 0$ for all x in I, so f_1 and f_2 are linearly dependent, contrary to hypothesis. Our assumption then is incorrect and we have found that $W[f_1, f_2]$ is never zero on I. ■

Corollary:

If f_1 and f_2 are linearly dependent over an interval I, then the Wronskian $W[f_1, f_2] = 0$ on I.

This corollary is illustrated in Example 12.

Example 12

In Example 8, the pair $\{e^x, 5e^x\}$ were shown to be linearly dependent. Example 11 shows their Wronskian to be zero. In general let f_1 and f_2 be linearly dependent. Then $f_1 = kf_2$ for some nonzero k. The Wronskian now is

$$\begin{vmatrix} f_1 & f_2 \\ f'_1 & f'_2 \end{vmatrix} = \begin{vmatrix} kf_2 & f_2 \\ kf'_2 & f'_2 \end{vmatrix} = kf_2 f'_2 - kf_2 f'_2 = 0$$

Example 13

To see the importance of having f_1 and f_2 be solutions of $Ly = 0$ on interval I in the converse of Theorem 3.4, we examine the pair $f_1(x) = 1$ and

$f_2(x) = x^2$. This pair is linearly independent on $(-\infty, \infty)$ since f_1 is not a constant multiple of f_2, but

$$W[f_1, f_2] = \begin{vmatrix} 1 & x^2 \\ 0 & 2x \end{vmatrix} = 2x$$

takes on a zero value at $x = 0$, contrary to the conclusion of the theorem. We now observe that $f_1(x) = 1$ and $f_2(x) = x^2$ are not solutions of any second-order linear differential equation.

Example 14

Consider the differential equation $y'' + 4y = 0$. It is easy to verify that $u_1 = \sin 2x$ and $u_2 = \cos 2x$ are solutions. The Wronskian

$$W[u_1, u_2] = \begin{vmatrix} \sin 2x & \cos 2x \\ 2\cos 2x & -2\sin 2x \end{vmatrix} = -2$$

which is, of course, never zero. Hence the two solutions are linearly independent.

We are now in a position to prove the following important theorem.

Theorem 3.5:

If $u_1(x)$ and $u_2(x)$ are linearly independent solutions of $Ly = 0$, then $y(x) = c_1u_1(x) + c_2u_2(x)$ is the general solution of $Ly = 0$.

Proof:

At the point x_0 let the linear combination $y = c_1u_1 + c_2u_2$ have the constant value A and let $y'(x_0) = B$. Consider then

$$\left\{\begin{aligned} c_1u_1(x_0) + c_2u_2(x_0) &= A \\ c_1u_1'(x_0) + c_2u_2'(x_0) &= B \end{aligned}\right\}$$

as a pair of linear equations in the unknowns c_1 and c_2. The determinant of coefficients of the left side is

$$\begin{vmatrix} u_1 & u_2 \\ u_1' & u_2' \end{vmatrix} = W[u_1, u_2]$$

which is nonzero since the solutions are linearly independent. Thus, theory from algebra assures us that there is a unique pair (c_1, c_2) which solves this system. Using this pair, consider the function $v(x) = c_1u_1(x) + c_2u_2(x)$. It is clear that $v(x)$ also satisfies $Ly = 0$ since u_1 and u_2 are solutions and that $v(x)$ has initial values $v(x_0) = A$, $v'(x_0) = B$. By the uniqueness element of Theorem 3.1, only one function has these properties, so $v(x) \equiv y(x)$. Hence a proposed new solution $v(x)$ turns out to be just another member of the family $y(x)$ and thus $y(x) = c_1u_1(x) + c_2u_2(x)$ represents the general solution $Ly = 0$. ■

This theorem shows us that to find the general solution of an nth-order linear differential equation $Ly = 0$, we need only to find n linearly independent individual solutions and form the general linear combination of them. Theorem 3.6 is presented to show that this is always possible.

Theorem 3.6:

There exist two linearly independent solutions of

$$a_0(x)y'' + a_1(x)y' + a_2(x)y = 0$$

Proof:

Let $u_1(x)$ be the unique solution of $Ly = 0$ for which $u_1(x_0) = A \neq 0$ and $u_1'(x_0) = 0$. The existence of this u_1 is guaranteed by Theorem 3.1. Let $u_2(x)$ be the unique solution of $Ly = 0$ satisfying $u_2(x_0) = 0$, $u_2'(x_0) = B \neq 0$. Theorem 3.1 assures us that such a u_2 exists and is unique, but now $W[u_1, u_2]$ evaluated at x_0 is

$$\begin{vmatrix} u_1(x_0) & u_2(x_0) \\ u_1'(x_0) & u_2'(x_0) \end{vmatrix} = \begin{vmatrix} A & 0 \\ 0 & B \end{vmatrix} = AB \neq 0$$

so u_1 and u_2 are linearly independent solutions. ■

Example 15

The simple second-order linear differential equation $y'' - y = 0$ has solutions $u_1(x) = e^x$ and $u_2(x) = e^{-x}$. These are easily verified since $u_1''(x) = e^x$ and $u_2''(x) = e^{-x}$. Thus for both of these, $u'' - u = 0$. Since e^x and e^{-x} are linearly independent (see Example 9), the combination $c_1e^x + c_2e^{-x}$ is the general solution to this second-order differential equation. Note that it is a two-parameter family of functions.

Example 16

The general solution to the equation

$$4y'' + 5y = 0$$

is $y = c_1\cos(\frac{\sqrt{5}}{2})x + c_2\sin(\frac{\sqrt{5}}{2})x$. To show that this is true, we verify that $u_1(x) = \cos(\frac{\sqrt{5}}{2})x$ is a solution of $4y'' + 5y = 0$ and that $u_2(x) = \sin(\frac{\sqrt{5}}{2})x$ is also a solution of $4y'' + 5y = 0$. In the first case

$$u_1'(x) = \frac{\sqrt{5}}{2}\left(-\sin\frac{\sqrt{5}}{2}x\right)$$

$$u_1''(x) = -\left(\frac{\sqrt{5}}{2}\right)^2\cos\frac{\sqrt{5}}{2}x = -\frac{5}{4}\cos\frac{\sqrt{5}}{2}x$$

Thus $$4u_1'' + 5u_1 = 4\left(-\frac{5}{4}\cos\frac{\sqrt{5}}{2}x\right) + 5\left(\cos\frac{\sqrt{5}}{2}x\right) = 0$$

In the second case

$$u_2'(x) = \frac{\sqrt{5}}{2} \cos \frac{\sqrt{5}}{2} x$$

$$u_2''(x) = -\left(\frac{\sqrt{5}}{2}\right)^2 \sin \frac{\sqrt{5}}{2} x$$

Thus $$4u_2'' + 5u_2 = 4\left(-\frac{5}{4} \sin \frac{\sqrt{5}}{2} x\right) + 5 \sin \frac{\sqrt{5}}{2} x = 0$$

Also $u_1(x)$ and $u_2(x)$ are linearly independent since

$$W[u_1, u_2] = \begin{vmatrix} \cos \frac{\sqrt{5}}{2} x & \sin \frac{\sqrt{5}}{2} x \\ -\frac{\sqrt{5}}{2} \sin \frac{\sqrt{5}}{2} x & \frac{\sqrt{5}}{2} \cos \frac{\sqrt{5}}{2} x \end{vmatrix}$$

$$= \frac{\sqrt{5}}{2} \cos^2 \frac{\sqrt{5}}{2} x + \frac{\sqrt{5}}{2} \sin^2 \frac{\sqrt{5}}{2} x = \frac{\sqrt{5}}{2} \neq 0$$

Thus it is verified that the given y is the general solution.

Example 17

The second-order linear equation $xy'' + 2y' - xy = 0$, important in mathematical physics, has solutions

$$u_1(x) = \frac{\cosh x}{x}$$

and $$u_2(x) = \frac{\sinh x}{x}$$

To verify the linear independence of these solutions we form

$$W[u_1, u_2] = \begin{vmatrix} \dfrac{\cosh x}{x} & \dfrac{\sinh x}{x} \\ \dfrac{\sinh x}{x} - \dfrac{\cosh x}{x^2} & \dfrac{\cosh x}{x} - \dfrac{\sinh x}{x^2} \end{vmatrix}$$

$$= \frac{\cosh^2 x}{x^2} - \frac{\sinh^2 x}{x^2} = \frac{1}{x^2}$$

Note that any interval that does not include the point $x = 0$ is an interval of validity for the general solution

$$y(x) = c_1 \cdot \frac{\cosh x}{x} + c_2 \cdot \frac{\sinh x}{x}$$

Exercises 3.2

1. Verify that the functions $u_1(x) = e^{-4x}$ and $u_2(x) = e^{3x}$ are solutions to the differential equation $y'' + y' - 12y = 0$. Then form the general solution. From this solution family find the member for which $y(0) = 1$, $y'(0) = -2$.
2. Given the differential equation $Ly = 0$ where $L = D^2 + 6D + 9$, verify that $u_1(x) = e^{-3x}$ and $u_2(x) = xe^{-3x}$ are solutions. Show next that they are linearly independent by computing their Wronskian. Now form the general solution to $Ly = 0$.
3. Verify that $u_1(x) = x$ and $u_2(x) = xe^{2/x}$ are solutions of $x^3y'' + 2xy' - 2y = 0$ and that they are linearly independent. Now form the general solution.
4. Show that $u_1(x) = e^{-x}$, $u_2(x) = e^x$, and $u_3(x) = xe^x$ are linearly independent solutions of $y''' - y'' - y' + y = 0$. Now form the general solution.
5. Show that $u_1(x) = 1$, $u_2(x) = e^{3x}$, and $u_3(x) = e^{-3x}$ are linearly independent solutions of $y''' - 9y' = 0$. What is the general solution of this third-order differential equation?
6. Show that the functions $u_1 = x^3$, $u_2 = (\ln x)x^3$, and $u_3 = x^{-2}$ are solutions to $x^3y''' - x^2y'' - 6xy' + 18y = 0$. Use the 3×3 determinant $W[u_1, u_2, u_3]$ to show that these solutions are linearly independent. Form the general solution to this third-order differential equation.
7. Show that any set of functions that contains the zero function is a linearly dependent set.
8. Using the definitions, determine whether the functions given are linearly dependent or linearly independent on the stated interval.
   (a) $\{\sin x, \cos x\}$ on $[0, 2\pi]$
   (b) $\{x, x^2\}$ on $(0, \infty)$
   (c) $\{x + 1, x - 3, 2x + 5\}$ on $\mathbb{R}$
   (d) $\{e^x, xe^x, (x^2 + 2)e^x\}$ on $\mathbb{R}$
   (e) $\{x, |x|\}$ on $[1, 2]$
   (f) $\{x, |x|\}$ on $[-2, 2]$
9. Show that the listed function is *the general* solution to the given differential equation.
   (a) $y'' + 4y = 0$; $y = c_1 \cos 2x + c_2 \sin 2x$
   (b) $y''' - 3y'' + 2y' = 0$; $y = c_1 + c_2e^x + c_3e^{2x}$
   (c) $y''' + 6y'' + 12y' + 8y = 0$; $y = c_1e^{-2x} + c_2xe^{-2x} + c_3x^2e^{-2x}$
   (d) $y'' - 5y' + 4y = 3xe^{2x}$; $y = c_1e^{4x} + c_2e^x + \frac{3}{4}e^{2x} - \frac{3}{2}xe^{2x}$
   (e) $y'' - y = 5\cos 2x$; $y = c_1e^x + c_2e^{-x} - \cos 2x$
   (f) $x^2y'' - 3xy' + 3y = 0$; $y = c_1x^3 + c_2x$
   (g) $x^3y'' + 2xy' - 2y = 0$; $y = c_1x + c_2xe^{2/x}$
10. (a) If an attractive force is exerted on a particle by a source at the origin such that the force is proportional to the distance from the

origin, then simple harmonic motion on a line through the origin will result. The differential equation of motion is

$$F = ma$$

$$-kx = m\frac{d^2x}{dt^2}$$

or

$$\frac{d^2x}{dt^2} + \frac{k}{m}x = 0$$

Since k and m are positive constants we replace k/m by ω_0^2. Show that the general solution to this equation is $x(t) = c_1\cos\omega_0 t + c_2\sin\omega_0 t$.

(b) If we look at the motion as the projection on a diameter of a particle moving around a circle of radius R with angular velocity ω_0, then the above solution may be written $x(t) = R\sin(\omega_0 t + \delta)$, where δ is called the phase angle of x. Show that this is a unique solution if values would be given for the parameters.

11. Show that if the functions $u_1(x)$ and $u_2(x)$ are solutions of $y'' + 2y^2 = 0$, it does not necessarily follow that $c_1u_1 + c_2u_2$ is also a solution. Why?

3.3 The Nonhomogeneous Equation

The material of the preceding section for homogeneous equations sets the groundwork for characterizing the solutions of more complicated equations. Compare the following theorem with Theorem 3.5.

Theorem 3.7:

Let $u_1, u_2, u_3, \ldots, u_n$ be linearly independent solutions of $Ly = 0$ on some interval I and let u_p be any particular solution to the nonhomogeneous equation $Ly = f(x)$ on I. Then the general solution to the nonhomogeneous equation $Ly = f(x)$ is given by $y = u_h + u_p$, where $u_h = c_1u_1 + c_2u_2 + \cdots + c_nu_n$ is the general solution to the homogeneous equation.

Before attempting a proof of Theorem 3.7, consider Examples 18 and 19.

Example 18

Let $Ly = f$ be given by $y'' - 2y' + y = 4e^x$. Considering first $Ly = 0$, it can easily be verified that $u_1 = e^x$ and $u_2 = xe^x$ are solutions to this homogeneous equation. Since these are linearly independent, the general homogeneous solution, denoted by y_h, is $y_h = c_1e^x + c_2xe^x$. A particular solution to the nonhomogeneous equation can be verified to be $u_p = 2x^2e^x$.

Thus the general solution to $Ly = f$ is

$$y = y_h + y_p = c_1 e^x + c_2 x e^x + 2x^2 e^x$$

Example 19

To find the solution to the second-order linear nonhomogeneous differential equation with variable coefficients $x^2 y'' + 2xy' - 2y = x^2$, you should show that $u_1(x) = x$ and $u_2(x) = x^{-2}$ solve the homogeneous counterpart $x^2 y'' + 2xy' - 2y = 0$ and that they are linearly independent. Thus the general solution to the homogeneous equation is $y_h = c_1 u_1 + c_2 u_2 = c_1 x + c_2 x^{-2}$. Verify also that $\frac{1}{4}x^2$ is a particular solution to the nonhomogeneous equation. Therefore $y = y_h + y_p = c_1 x + c_2 x^{-2} + \frac{1}{4}x^2$ is the general solution of the differential equation posed in the problem.

Proof of Theorem 3.7:

Every function of the form $u_h + u_p$ is a solution of $Ly = f$ since

$$\begin{aligned} L(u_h + u_p) &= Lu_h + Lu_p = (c_1 Lu_1 + c_2 Lu_2 + \cdots + c_n Lu_n) + Lu_p \\ &= (\ 0 \ + \ 0 \ + \cdots + \ 0\ ) + f = f \end{aligned}$$

To show that every solution of $Ly = f$ is of the form $u_h + u_p$, we suppose initially that some other solution v exists. Then $Lv = f$, but since by hypothesis $Lu_p = f$, we have

$$L(v - u_p) = Lv - Lu_p = f - f = 0$$

so $v - u_p$ solves the homogeneous equation. However, the general solution to the homogeneous equation is a linear combination of the solutions u_1, $u_2, \ldots, u_n$. We then must have $v - u_p = c_1 u_1 + c_2 u_2 + \cdots + c_n u_n$, since all solutions of $Ly = 0$ are of this form. Thus

$$v = c_1 u_1 + c_2 u_2 + \cdots + c_n u_n + u_p = u_h + u_p$$

and we find v is really just another of our original family. Thus the entire collection of solutions is of the form $u_h + u_p$ and it then is the general solution. ■

Example 20

Given the differential equation $y'' + 9y = 3$, we can easily verify by differentiation and substitution that the solutions to the homogeneous equation $y'' + 9y = 0$ are $u_1(x) = \cos 3x$ and $u_2(x) = \sin 3x$. Thus $y_h = c_1 \cos 3x + c_2 \sin 3x$. One may also easily verify that a particular solution is $y_p = \frac{1}{3}$. This is simple, since $y_p' = y_p'' = 0$, so that $y_p'' + 9y_p = 0 + 9(\frac{1}{3}) = 3$. Hence the general solution to the given differential equation is

$$y = y_h + y_p = c_1 \cos 3x + c_2 \sin 3x + \frac{1}{3}$$

We now have a clear procedure for finding the general solution to the nonhomogeneous equation, provided we can by some method find one particular solution of the nonhomogeneous equation and can solve the associated homogeneous equation for its general solution.

Example 21

Consider the problem $y'' - 5y' + 4y = 8$. We saw in Example 4 that the associated homogeneous equation $y'' - 5y' + 4y = 0$ has the general solution $y_h = c_1e^x + c_2e^{4x}$. If we now carefully ponder the nonhomogeneous equation to find the one needed particular solution, it may occur after some deep thought that $y_p = 2$ does the trick. It *is* a particular solution and it was easy to find. Thus the general solution to this problem is the two-parameter family $y = y_h + y_p = c_1e^x + c_2e^{4x} + 2$.

In Example 21 a new technique was generated: We made an intelligent guess and found the solution. Surely not all second-order linear nonhomogeneous differential equations can be solved by intense meditation, but it is encouraging to know it works on occasion. In particular, if $f(x)$ in $Ly = f(x)$ is a constant, this simple process is successful. This assumes, of course, that the coefficient $a_2(x)$ that multiplies y in the differential equation is also a constant.

Thus we note that in

$$a_0y'' + a_1y' + a_2y = K; \quad K, \quad a_2 \text{ constant}$$

that

$$y_p = \frac{K}{a_2}$$

Example 22

Find the particular solution y_p for

$$2y'' - 17y' + 6y = 24$$

Since the function $f(x)$ is the constant 24 and the coefficient of y in the equation is the constant 6, we have $y_p = \frac{24}{6} = 4$. You should see that y_p' and y_p'' are both zero so that $y_p = 4$ does satisfy the entire differential equation.

Continuing with this newly found technique, we examine a slightly more complicated right side for $Ly = f(x)$. Consider the case $f(x) = Ce^{kx}$ where C and k are real constants.

In a problem such as $y'' - 5y' + 4y = 2e^{3x}$, we are attempting to find any particular function y_p that upon substitution will change the equation into an identity. The function y_p, its derivative y'_p, and its second derivative y''_p must all be inserted into the left side and then combined in an attempt to have the combination add up to $2e^{3x}$. Since we know from calculus that derivatives of exponential functions produce exponential functions, a logical choice is the exponential function itself. In fact, here we surely would expect to have y_p be some constant multiple of e^{3x}. Thus we are led to the natural choice $y_p = Ae^{3x}$, where A is a constant to be determined. By differentiating this trial function y_p twice and putting these into the equation, we obtain

$$9Ae^{3x} - 5(3Ae^{3x}) + 4(Ae^{3x}) = 2e^{3x}$$

Dividing out the nonzero exponential leaves $-2A = 2$ or $A = -1$. This provides the coefficient in our particular solution, which now is $y_p = -e^{3x}$. By Theorem 3.7 the general solution to the posed problem is

$$y(x) = c_1e^{4x} + c_2e^{x} - e^{3x}$$

A quick check will show that this does satisfy the original differential equation.

Example 23

Given that the solution to $y'' - 9y' + 20y = 0$ is $y_h = c_1e^{4x} + c_2e^{5x}$, find the general solution to

$$y'' - 9y' + 20y = 14e^{-2x}$$

We need to intelligently pick a trial function for y_p. Again the only logical choice is an exponential function of a particular form. Let $y_p = Ae^{-2x}$, where A is a constant to be determined. Then

$$y'_p = -2Ae^{-2x}$$

$$y''_p = 4Ae^{-2x}$$

$$\therefore \quad 4Ae^{-2x} - 9(-2Ae^{-2x}) + 20(Ae^{-2x}) = 14e^{-2x}$$

$$42Ae^{-2x} = 14e^{-2x}$$

Hence, $A = \frac{14}{42} = \frac{1}{3}$ and $y_p = \frac{1}{3}e^{-2x}$. We have determined the value of the previously undetermined coefficient. Thus the general solution is

$$y = y_h + y_p = c_1e^{4x} + c_2e^{5x} + \frac{1}{3}e^{-2x}$$

A third elementary case is the situation in which $f(x)$ in $Ly = f(x)$ is a simple polynomial.

Example 24

Consider $y'' - 5y' + 4y = 2x$. As in Example 21, the solution to the homogeneous counterpart $y'' - 5y' + 4y = 0$ is $y_h = c_1e^{x} + c_2e^{4x}$. The only logical choice for a trial y_p here is a polynomial. The degree of that polynomial must match the degree of the simple polynomial in $f(x)$ since the eventual solution y_p must be substituted for y in the differential equation and the two sides must match in degree. Thus we would pick as our trial solution

$$y_p = A_0 + A_1 x$$

where A_0 and A_1 are constants to be determined. Forcing this into the differential equation to help determine the two undetermined coefficients we would use

$$y'_p = A_1$$
$$y''_p = 0$$

The differential equation then becomes

$$0 - 5(A_1) + 4(A_0 + A_1 x) = 2x$$

or

$$(4A_0 - 5A_1) + 4A_1 x = 2x$$

Matching coefficients of like terms in the polynomials of the two sides gives

$$4A_0 - 5A_1 = 0$$
$$4A_1 = 2$$

Hence $A_1 = \frac{1}{2}$ and $A_0 = \frac{5}{8}$. We then have our particular solution

$$y_p = \frac{5}{8} + \frac{1}{2}x$$

and the general solution to the original differential equation is

$$y = y_h + y_p = c_1 e^x + c_2 e^{4x} + \frac{5}{8} + \frac{1}{2}x.$$

It should now be clear that our main concern in solving an nth-order linear nonhomogeneous differential equation will be focused on finding n linearly independent solutions to the nth-order homogeneous equation and only one particular solution to $Ly = f$.

Another theorem which will be of great value in solving linear equations is the following.

Theorem 3.8 (Principle of Superposition):

If u_1 is a solution to $Ly = F_1$ and u_2 is a solution to $Ly = F_2$, then $y = u_1 + u_2$ is a solution to $Ly = F_1 + F_2$.

Proof:

By hypothesis we have $Lu_1 = F_1$ and $Lu_2 = F_2$, so

$$L(u_1 + u_2) = Lu_1 + Lu_2 = F_1 + F_2. \quad \blacksquare$$

This theorem allows us to decompose a nonhomogeneous equation with a right side containing more than one additive term into a collection of equations each having a single term on the right side. We then solve each simpler problem and superimpose them by adding their respective solutions. This process is demonstrated in Examples 25–27.

Example 25

Consider $y'' - 9y = e^{2x} + 3$. Following the format of the principle of superposition, we rewrite this as a pair of differential equations $y'' - 9y = e^{2x}$ and $v'' - 9v = 3$. For the first one we can easily verify that e^{3x} and e^{-3x} each solve $y'' - 9y = 0$ and that they are linearly independent. We can also verify by the method of the earlier examples that $-\frac{1}{5}e^{2x}$ is a particular solution. Thus $y = c_1e^{3x} + c_2e^{-3x} - \frac{1}{5}e^{2x}$ is the general solution to this equation. The same homogeneous functions e^{3x} and e^{-3x} solve $v'' - 9v = 0$ and the v_p is $-\frac{1}{3}$ by the inspection process discussed earlier. Thus the general solution to this second-order equation is $v = c_1e^{3x} + c_2e^{-3x} - \frac{1}{3}$. The superimposed solution to the original problem now is

$$y = c_1e^{3x} + c_2e^{-3x} - \frac{1}{5}e^{2x} - \frac{1}{3}$$

Example 26

Use the principle of superposition to find the general solution to

$$y'' - 9y = -4e^{-x} + 5x - 18$$

From Example 25 we get

$$y_h = c_1e^{3x} + c_2e^{-3x}$$

On the right side of this differential equation we separate the function $f(x)$ into three parts:

$$f_1(x) = -4e^{-x}, \quad f_2(x) = 5x, \quad \text{and} \quad f_3(x) = -18$$

We need to find a particular solution for each of these parts and add their sum to the already found y_h. For the first part, we need to find the particular solution to $y'' - 9y = -4e^{-x}$, so we choose a trial $y_{p_1} = Ae^{-x}$. Then $y'_{p_1} = -Ae^{-x}$ and $y''_{p_1} = Ae^{-x}$, so upon substitution we have

$$Ae^{-x} - 9(Ae^{-x}) = -4e^{-x}$$

$$-8Ae^{-x} = -4e^{-x}$$

$$A = \frac{1}{2}$$

Thus

$$y_{p_1} = \frac{1}{2}e^{-x}$$

For the second part, we need to find the particular solution to $y'' - 9y = 5x$. Choose $y_{p_2} = A_0 + A_1 x$, for which $y'_{p_2} = A_1$ and $y''_{p_2} = 0$. Substituting, we have

$$0 - 9(A_0 + A_1 x) = 5x$$

Matching coefficients provides

$$-9A_0 = 0$$

$$-9A_1 = 5$$

Therefore $$y_{p_2} = 0 - \frac{5}{9}x$$

For part three, we have a constant term on the right. By inspection

$$y_{p_3} = \frac{-18}{-9} = 2$$

The general solution to the given differential equation is therefore

$$y = y_h + y_{p_1} + y_{p_2} + y_{p_3} = c_1 e^{3x} + c_2 e^{-3x} + \frac{1}{2}e^{-x} - \frac{5}{9}x + 2$$

Example 27

For the initial value problem

$$\begin{Bmatrix} 4y'' + y = 4 + 2x + \sin x \\ y(0) = 5, \ y'(0) = \frac{8}{3} \end{Bmatrix}$$

the solution y_h to the equation $4y'' + y = 0$ is

$$y_h = c_1 \cos\frac{1}{2}x + c_2 \sin\frac{1}{2}x$$

To get the particular solution to the given nonhomogeneous equation, we superimpose the particular solutions to $4y'' + y = 4$, $4y'' + y = 2x$, and $4y'' + y = \sin x$. The first and second are easily seen to be $y_{p_1} = 4$ and $y_{p_2} = 2x$. The third may be verified to be $y_{p_3} = -\frac{1}{3}\sin x$. Thus the general solution to the differential equation is

$$y(x) = c_1 \cos\frac{1}{2}x + c_2 \sin\frac{1}{2}x + 4 + 2x - \frac{1}{3}\sin x$$

Applying the initial condition $y(0) = 5$ produces $c_1 = 1$, and applying $y'(0) = 8/3$ produces $c_2 = 2$. Therefore the unique solution to the initial value problem is the function

$$y(x) = \cos\frac{1}{2}x + 2\sin\frac{1}{2}x + 4 + 2x - \frac{1}{3}\sin x$$

The graph of this function is shown in Figure 3.1.

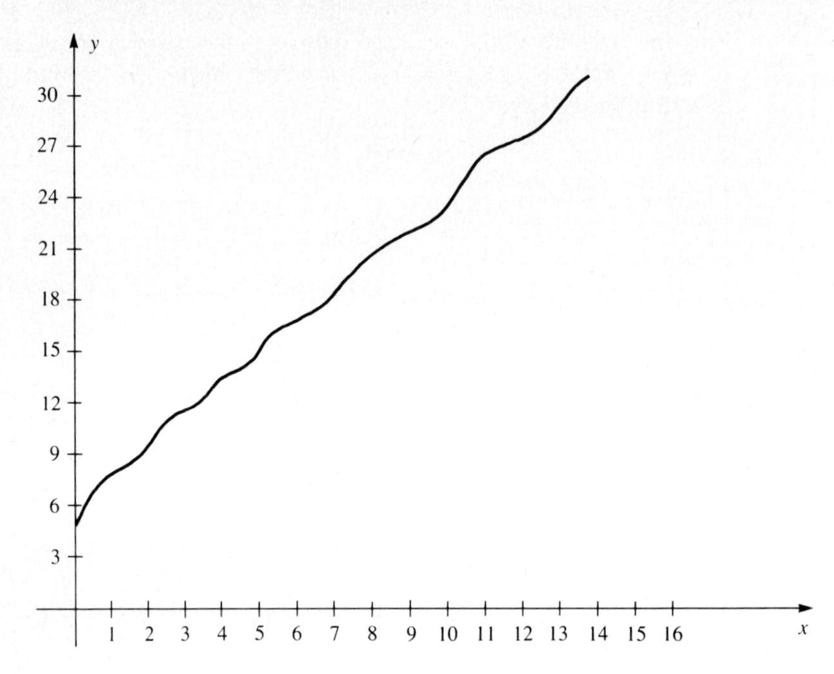

Figure 3.1

Exercises 3.3

1. Consider $Ly = f$ given by $y'' - 7y' + 6y = 12$. Verify that $u_1 = e^x$ and $u_2 = e^{6x}$ solve the equation $Ly = 0$. Now write the general solution to $Ly = f$.
2. For the differential equation $2y'' + y' - 6y = e^x + 6$, verify that $y = c_1e^{3x/2} + c_2e^{-2x}$ is the general solution to the homogeneous counterpart. Now find the general solution to this equation.
3. Solve $2y'' + y' - 6y = 5e^{3x} - x$.
4. Verify that $u_1 = e^{-2x}$ and $u_2 = e^{-x}$ are solutions of $y'' + 3y' + 2y = 0$. Then find the general solution of $y'' + 3y' + 2y = e^{2x}$.
5. Find the general solution of $y'' + 3y' + 2y = 6e^x + 4$.
6. Find the general solution of $y'' + 3y' + 2y = 2x^2 + 3x - 1$.
7. For the differential equation $2y'' - 11y + 5y = 4e^{3x} - 10$, verify that the homogeneous solution is $y_h = c_1e^{x/2} + c_2e^{5x}$. Now find the general solution.
8. Find the general solution to $2y'' - 11y' + 5y = e^x + 8x - 33$.
9. Find a particular solution to $y'' + y = x$. Verify it.
10. Find a particular solution to $y'' + y = 4x^2 + x - 6$.
11. Find a particular solution to $y'' - y' + y = 2e^x$. Verify it.
12. Find a particular solution to $y'' - y' + y = 6e^{9x} - 17x$.
13. Use the principle of superposition to find a particular solution to $(D^2 + 1)y = x^2 + 3x + 4$.

14. Use the principle of superposition to find a particular solution to $(D^2 - 4D + 3)y = e^{2x} + x + 6$. Treat the right side as two functions, e^{2x} and the polynomial $(x + 6)$.
15. Use Example 25 to help find the general solution to $y'' - 9y = 3x - 1$.
16. Use Example 27 to help find the general solution to $4y'' + y = e^{2x} - 4x + 1$.
17. Find the general solution to $y'' - 4y = 3x$.
18. Find the general solution to $y'' - 7y' + 6y = 3e^{4x} + e^{-x}$.
19. Find the general solution to $y'' + 3y' + 2y = 5e^x + (4x - 3)$.

3.4 Complex-Valued Solutions

To make the study of solutions to linear equations more systematic, we need to introduce the concept of a complex-valued solution. In calculus you were accustomed to studying real-valued functions of a real variable. In those functions, a real number from the domain was used to produce a real number in the range.

If $v(x)$ and $w(x)$ are real-valued functions of a real variable, then we define a **complex-valued function of a real variable** as $f(x) = v(x) + i \cdot w(x)$, where i is the imaginary unit with the property that $i^2 = -1$. Note that this is a function that assigns a complex number $v(x_0) + i \cdot w(x_0)$ to each real number x_0 in the domain of $f(x)$.

Example 28

Consider the function $f(x) = 3x + 2ix^2$. This is a complex-valued function of the real variable x, since for each real x in the domain of f, the function has as its output value a complex number. Using $x = 2$, we obtain $f(2) = 6 + 8i$. Using $x = -1$ produces $f(-1) = -3 + 2i$.

Example 29

Let $f(x) = \cos 3x + i \sin 3x$. This is also a complex-valued function of the real variable x. Using $x = \frac{\pi}{4}$, a real number from the domain of f, we obtain $f(\frac{\pi}{4}) = \cos \frac{3\pi}{4} + i \sin \frac{3\pi}{4} = -\frac{\sqrt{2}}{2} + i\frac{\sqrt{2}}{2}$, a complex number in the range of f. Using other real values for x that are in the domain of f will each produce a corresponding complex number.

The calculus of complex-valued functions follows almost all the rules of regular calculus. The derivative of a complex function $f(x) = v(x) + i \cdot w(x)$ is $f'(x) = v'(x) + i \cdot w'(x)$, provided both v' and w' exist. We simply differentiate independently the real functions $v(x)$ and $w(x)$. Higher derivatives are defined similarly.

We now need to define two concepts that are involved in obtaining complex-valued solutions.

Definition: The complex-valued function $f(x) = v(x) + i \cdot w(x)$ is **continuous** for those values of x for which both real-valued functions $v(x)$ and

$w(x)$ are continuous. The function f is **differentiable** at those values of x for which $v(x)$ and $w(x)$ are differentiable and the derivative of $f(x)$ is $f'(x) = v'(x) + i \cdot w'(x)$.

A special complex-valued function we will make great use of is the complex-exponential function written as $e^{(a+bi)x}$, where a and b are real numbers. We will also make a careful analysis of this function.

Since the exponent is a complex number, we will first write the function as $e^{(a+bi)x} = e^{ax+ibx} = e^{ax} \cdot e^{ibx}$. Since a and b are real numbers, the first part e^{ax} is a pure real function, familiar from elementary calculus, and the second part e^{ibx} is a pure imaginary function. It is this second part we now need to analyze. This imaginary part e^{ibx} may be written in a form that is much easier to work with if we use an identity, attributed to Euler, that breaks this function into a pair of real functions.

$$e^{ibx} = (\cos bx) + i(\sin bx) \quad \text{Euler's identity}$$

The result is still a complex-valued function of a real variable; but the real part, $\cos bx$, and the imaginary part, $\sin bx$, are real functions. An infinite-series proof of this identity is left as an exercise for the reader.

Example 30 gives two illustrations of the use of Euler's identity in analyzing complex-exponential functions.

Example 30

$$\begin{aligned} e^{(3+4i)x} &= e^{3x} \cdot e^{4ix} = e^{3x}(\cos 4x + i \sin 4x) \\ &= (e^{3x}\cos 4x) + i(e^{3x}\sin 4x) \end{aligned}$$

$$\begin{aligned} e^{(-2+5i)x} = e^{-2x+5ix} &= e^{-2x} \cdot e^{i5x} \\ &= e^{-2x}(\cos 5x + i \sin 5x) \\ &= (e^{-2x}\cos 5x) + i(e^{-2x}\sin 5x) \end{aligned}$$

Note that the complex-valued function is written as the sum of two real-valued functions called the real part and the imaginary part of the original function.

Example 31

A complex-exponential function of the form

$$f(x) = e^{\frac{2x}{3+i}}$$

must first be reorganized to

$$e^{\frac{2x}{3+i} \cdot \frac{3-i}{3-i}} = e^{\frac{(6-2i)x}{10}} = e^{\left(\frac{3}{5} - \frac{1}{5}i\right)x}$$

It then may be written

$$e^{\frac{2x}{3+i}} = e^{\frac{3x}{5}} \cdot e^{i\left(-\frac{x}{5}\right)} = e^{\frac{3x}{5}}\left[\cos\left(-\frac{x}{5}\right) + i\sin\left(-\frac{x}{5}\right)\right]$$

$$= e^{\frac{3x}{5}}\left(\cos\frac{x}{5} - i\sin\frac{x}{5}\right)$$

$$= \left(e^{\frac{3x}{5}}\cos\frac{x}{5}\right) + i\left(-e^{\frac{3x}{5}}\sin\frac{x}{5}\right)$$

In general notation we have $f(x) = v(x) + i \cdot w(x)$ where $v(x)$ and $w(x)$ are real functions.

Many times complex-exponential functions satisfy linear differential equations whose coefficients are real numbers. The following definition is illustrated in Examples 32 and 33.

Definition: The complex-valued function

$$f(x) = v(x) + i \cdot w(x) = e^{ax}\cos bx + i \cdot e^{ax}\sin bx$$

is a **solution** to an nth-order linear differential equation if f possesses n derivatives and the insertion of f into the equation converts it into an identity.

Example 32

The function $y = e^{2ix}$ satisfies $y'' + 4y = 0$ since $y' = 2ie^{2ix}$ and $y'' = 4i^2e^{2ix}$ produce $4i^2e^{2ix} + 4e^{2ix} = (-4 + 4)e^{2ix} = 0$. This complex-exponential function e^{2ix} could be written, using Euler's identity, as $y = (\cos 2x) + i(\sin 2x)$ and could be checked again as a valid solution.

Example 33

The function $y = e^{(1+i)x}$ satisfies the second-order differential equation

$$y'' - 2y' + 2y = 0$$

which is linear and has real constant coefficients. Checking this, we have $y' = (1 + i)e^{(1+i)x}$ and $y'' = (1 + i)^2e^{(1+i)x} = (2i)e^{(1+i)x}$. Substituting into the differential equation provides

$$(2i)e^{(1+i)x} - 2(1 + i)e^{(1+i)x} + 2e^{(1+i)x} = [2i - 2(1 + i) + 2]e^{(1+i)x} = 0$$

We have shown that $e^{(1+i)x}$ is a solution. Again we could write the solution in the form

$$e^{(1+i)x} = e^{x+ix} = e^x \cdot e^{ix} = e^x(\cos x + i\sin x)$$

$$= (e^x\cos x) + i(e^x\sin x)$$

To establish the connection between real and complex solutions of linear equations with real coefficients, we need the important finding expressed in Theorem 3.9.

Theorem 3.9:

If $f(x) = v(x) + i \cdot w(x)$ is a solution to the linear differential equation

$$a_0(x)y'' + a_1(x)y' + a_2(x)y = 0$$

where a_0, a_1, and a_2 are real functions, then the real and imaginary parts of $f(x)$, namely $v(x)$ and $w(x)$, are also solutions. The converse is also true.

Proof:

Write the differential equation in operator notation

$$[a_0(x)D^2 + a_1(x)D + a_2(x)]y = 0$$

Since the operator is linear, we denote the differential equation as $Ly = 0$. Since $f(x)$ is a given solution, we have $Lf = 0$; but then $Lf = L(v + iw) = Lv + iLw = 0$. This complex identity holds if and only if $Lv = 0$ and $Lw = 0$. Thus v and w are solutions of $Ly = 0$. ■

We must not lose sight of our goal: Find the general solution of the linear differential equation $a_0(x)y'' + a_1(x)y' + a_2(x)y = 0$ where the $a_i(x)$ are real-valued functions of a real variable. This solution should be expressed as a linear combination of *real* functions

$$y(x) = c_1u_1(x) + c_2u_2(x)$$

If we find a complex function $f(x)$ that is a solution to the given differential equation, then Theorem 3.9 provides us with two real solutions to that equation. We need then only to use the Wronskian or the definition to determine the linear independence of $v(x)$ and $w(x)$ and thus be able to write down the general solution. We actually save labor if we can find a single complex solution and apply Theorem 3.9.

Example 34

In Example 32 we saw that e^{2ix} satisfied $y'' + 4y = 0$. Theorem 3.9 tells us that the real and imaginary parts of this function are also solutions. To verify this we write

$$e^{2ix} = (\cos 2x) + i(\sin 2x)$$

and thus we must verify $v(x) = \cos 2x$ and $w(x) = \sin 2x$. We first calculate

$$v'(x) = -2\sin 2x$$

$$v''(x) = -4\cos 2x$$

So $$v'' + 4v = (-4\cos 2x) + 4(\cos 2x) = 0$$

Also $$w'(x) = 2\cos 2x$$

$$w''(x) = -4\sin 2x$$

so $$w'' + 4w = (-4\sin 2x) + 4(\sin 2x) = 0$$

These two functions $\{\cos 2x, \sin 2x\}$ are linearly independent since

$$W[\cos 2x, \sin 2x] = \begin{vmatrix} \cos 2x & \sin 2x \\ -2\sin 2x & 2\cos 2x \end{vmatrix} = 2\cos^2 2x + 2\sin^2 2x$$

$$= 2(\cos^2 2x + \sin^2 2x)$$

$$= 2$$

Using Theorem 3.5 we may now write the general solution to $y'' + 4y = 0$ as

$$y(x) = c_1 \cos 2x + c_2 \sin 2x$$

Example 35

Since in Example 33 we found that $e^{(1+i)x}$ solved $y'' - 2y' + 2y = 0$, we may conclude that the real and imaginary parts of $e^{(1+i)x}$, namely $(e^x \cos x)$ and $(e^x \sin x)$, also satisfy the same differential equation. You should verify that this is true. Also verify the linear independence of those functions by showing that their Wronskian has value e^{2x}, which is never zero. We see then that the general solution to the differential equation is

$$y(x) = c_1 e^x \cos x + c_2 e^x \sin x$$

Would you suspect that other complex functions solve the same differential equation? One other function, $e^{(1-i)x}$, has in its exponent the complex conjugate of $1 + i$. Breaking this function down into its real and imaginary parts produces two real-valued functions as before:

$$e^{(1-i)x} = e^{x-ix} = e^x \cdot e^{-ix}$$

$$= e^x[\cos(-x) + i\sin(-x)]$$

However, from trigonometry we know that for all x, $\cos(-x) = \cos x$ and $\sin(-x) = -\sin x$. Thus $e^{(1-i)x} = (e^x \cos x) + i(-e^x \sin x)$.

The real part for this function is the same as for the other complex solution $e^{(1+i)x}$. The imaginary part is the negative of the imaginary part of $e^{(1+i)x}$. It is then a constant multiple and hence linearly dependent with $(e^x \sin x)$. Of the four potential solutions obtained from $e^{(1-i)x}$ and $e^{(1-i)x}$, two are identical and two differ only by a negative sign. Thus we select the two remaining linear independent ones and form from them the general solution

$$y(x) = c_1(e^x \cos x) + c_2(e^x \sin x)$$

Even if we now create linear combinations of $e^{(1+i)x}$ and $e^{(1-i)x}$ in an attempt to determine other linearly independent solutions of $y'' - 2y' + 2y = 0$, we would arrive only at constant multiples of those already found. We could hardly expect new solutions since the general solution expresses all possible solutions to the original differential equation.

Example 36

Given the differential equation defined on $(0, \infty)$ with variable coefficients $x^2y'' + xy' + 4y = 0$, we may create the trial solution $y = x^c$ where c is a complex number. We would then have $y' = cx^{c-1}$ and $y'' = c(c-1)x^{c-2}$ (see Exercise 7). If we substitute these into the given equation, the differential equation becomes

$$x^2c(c-1)x^{c-2} + x\,cx^{c-1} + 4x^c = 0$$

or

$$[c(c-1) + c + 4]x^c = 0$$

Since we want $y = x^c$ to be a nontrivial solution, we then obtain

$$c(c-1) + c + 4 = 0$$

or

$$c^2 + 4 = 0$$

Thus x^c is a solution if $c = \pm 2i$. However,

$$x^{2i} = e^{2i\ln x} = e^{(2\ln x)i}$$

When we use Euler's identity, this becomes

$$[\cos(2\ln x)] + i[\sin(2\ln x)]$$

Hence $u_1(x) = \cos(2\ln x)$ and $u_2(x) = \sin(2\ln x)$ are two linearly independent real solutions. Again x^{-2i} produces no new linearly independent solutions. We then write the general solution to the given differential equation as

$$\begin{aligned} y(x) &= c_1u_1(x) + c_2u_2(x) \\ &= c_1\cos(2\ln x) + c_2\sin(2\ln x) \end{aligned}$$

You may want to verify that this is true.

Exercises 3.4

1. Write the following complex-valued functions in the form $f(x) = v(x) + i \cdot w(x)$:
   (a) e^{4ix} (b) $e^{(2-3i)x}$ (c) $e^{(-2+i)x}$
   (d) $e^{(-3-5i)x}$ (e) $e^{x/(1-i)}$ (f) $e^{ax/(i+2)}$
2. If $y = e^{-4ix}$ satisfies $y'' + 16y = 0$, find two linearly independent real functions that also satisfy this differential equation.
3. Show that $y = e^{5ix}$ solves $y'' + 25y = 0$. Now find two real solutions of this equation. Show that they are linearly independent.
4. Find two linearly independent real-function solutions of $y'' + 2y' + 5y = 0$ if a complex solution is $y = e^{(-1+2i)x}$. Use the real functions to form the general solution to the differential equation.
5. Show that the differential equation $2y'' + y' + y = 0$ has the complex function

$$y = e^{\left(-\frac{1}{4}+\frac{\sqrt{7}}{4}i\right)x}$$

as a solution. Which two real-valued functions also solve the differential equation? Are they linearly independent?

6. Given that t is a real number and n is an integer, show that $(e^{it})^n = e^{int}$ and hence that $(\cos t + i \sin t)^n = \cos nt + i \sin nt$. This result is known as De Moivre's theorem.
7. Let t be a positive real number and let $c = a + bi$ be a complex term. Then we define $t^c = e^{c \ln t}$. Show that

$$\frac{d}{dt}(t^c) = ct^{c-1}$$

8. Verify the following equations. (*Hint:* Use Euler's identity.)

(a) $\sin z = \dfrac{1}{2i}(e^{iz} - e^{-iz})$.

(b) $\cos z = \dfrac{1}{2}(e^{iz} + e^{-iz})$.

9. Find a Maclaurin series expansion for $\cos i$. Approximate its value.
10. Prove that the product of two conjugate complex numbers is a positive real number.
11. By writing the infinite series for e^z, show that $e^{z_1} \cdot e^{z_2} = e^{z_1 + z_2}$.
12. Use infinite series to prove Euler's identity by following this outline:
    (a) Write the Maclaurin series expansion for e^x.
    (b) Using a replacement for x, write the expansion for e^{ibx}.
    (c) Write the Maclaurin series expansion for $\cos x$ and $\sin x$.
    (d) With replacements, create the expressions for $\cos bx$ and $\sin bx$.
    (e) Put these together to show that $e^{ibx} = (\cos bx) + i(\sin bx)$.

3.5 Homogeneous Equations with Constant Coefficients

From our study of the theory in the previous sections, we know that a systematic procedure for solving a general linear differential equation of any order involves the determination of the linearly independent solutions to the homogeneous equation and the finding of a particular solution to the nonhomogeneous equation. The next few sections provide methods for these determinations.

We begin with the simplest case, the second-order linear homogeneous differential equation with constant coefficients, $Ly = a_0 y'' + a_1 y' + a_2 y = 0$. We assume $a_0 \neq 0$ (else we would have a first-order equation). After dividing by a_0 and renaming constants, we get $Ly = y'' + ay' + by = 0$. We will assume that a and b are real constants.

As a first step toward finding the solution(s) to this equation, we will look back at the corresponding first-order constant-coefficient differential equation $y' + ay = 0$, for which all the solutions were multiples of $y = e^{-ax}$. In this second-order case, $y'' + ay' + by = 0$, some form of exponential function would also be a reasonable trial solution and would utilize the property that the differentiation of an exponential function e^{rx} always yields a constant multiplied by e^{rx}.

Thus we let $y = e^{rx}$ and try to find the value(s) for the coefficient r. Then $Le^{rx} = r^2e^{rx} + are^{rx} + be^{rx} = 0$ or $(r^2 + ar + b)e^{rx} = 0$. Since e^{rx} is never zero, we obtain the following important result:

> e^{rx} is a solution to $Ly = y'' + ay' + by = 0$ if r is a solution of $P(r) = r^2 + ar + b = 0$.

This polynomial equation is called the **auxiliary polynomial equation** for $Ly = y'' + ay' + by = 0$. Another name often used is the **characteristic polynomial**. Notice that it is easily formulated by reverting to the operator form of L, namely $Ly = (D^2 + aD + b)y$. For this special case of constant-coefficient equations, call L a polynomial operator $P(D) = D^2 + aD + b$; then merely form the function $P(r)$ to get the auxiliary polynomial and set it equal to zero. Examples 37–39 illustrate this process.

Example 37

Consider $y'' - 3y' + 2y = 0$. In operator form this becomes

$$Ly = (D^2 - 3D + 2)y = 0 \text{ or } P(D)y = 0$$

Thus $P(r) = r^2 - 3r + 2 = 0$ is the auxiliary polynomial equation. In factored form $(r - 2)(r - 1) = 0$ gives roots $r = 2$ and $r = 1$. Thus e^{2x} and e^x are solutions to the differential equation.

Example 38

Consider $y'' + 2y' - 3y = 0$. This is a second-order homogeneous equation with constant coefficients. The auxiliary polynomial equation for this problem is easily formed, since we may write the differential equation in operator notation as $(D^2 + 2D - 3)y = 0$. Since $L = D^2 + 2D - 3$ is a polynomial operator, we write it as $P(D) = D^2 + 2D - 3$. From this we obtain $P(r) = r^2 + 2r - 3 = 0$. The roots of this equation are $r = 1$ and $r = -3$. Since we know that e^{rx} is a solution to $Ly = 0$ if r is a root of the auxiliary polynomial, we get the exponential functions $u_1(x) = e^x$ and $u_2(x) = e^{-3x}$ as solutions. These were shown to be linearly independent in Example 9.

Example 39

For the second-order homogeneous differential equation with constant coefficients

$$y'' - 2y' - 15y = 0$$

the auxiliary equation is

$$r^2 - 2r - 15 = 0$$

or

$$(r - 5)(r + 3) = 0$$

The roots are $r = 5$ and $r = -3$, and hence solutions are $u_1(x) = e^{5x}$ and $u_2(x) = e^{-3x}$. Since these are linearly independent, the general solution to the differential equation is $y(x) = c_1 e^{5x} + c_2 e^{-3x}$.

We have concluded that e^{rx} solves $y'' + ay' + by = 0$ if r is a root of $r^2 + ar + b = 0$. Using the quadratic formula, we see that these roots are

$$r = -\frac{a}{2} \pm \frac{\sqrt{a^2 - 4b}}{2}$$

Algebraic theory states that if the discriminant $\delta = a^2 - 4b$ is greater than zero, we obtain two distinct real roots; if $\delta = 0$ the roots are real and equal; and if $\delta < 0$ the roots are complex conjugate numbers. These cases must be treated separately, and the form of the solution functions for a given differential equation will depend on which of these three cases we obtain. Sections 3.6–3.8 deal with these three possibilities.

Exercises 3.5

For the differential equations in Exercises 1–6, formulate the auxiliary polynomial equation and then find the roots.

1. $y'' - 6y' + 5y = 0$
2. $2y'' + 7y' - 4y = 0$
3. $3y'' + 13y' - 10y = 0$
4. $y'' + 19y' + 84y = 0$
5. $y''' - 7y'' + 12y' = 0$
6. $y''' - y'' - 10y' - 8y = 0$

For the differential equations in 7–12, find the exponential solutions and test them for linear independence. Also form the general solution to the equation.

7. $y'' + 3y' - 10y = 0$
8. $2y'' + 9y' - 5y = 0$
9. $3y'' - 2y' - 8y = 0$
10. $2y'' + 7y' + 3y = 0$
11. $y'' + y' - 42y = 0$
12. $y''' - 4y'' + y' + 6y = 0$

3.6 Distinct Real Roots

When solving $y'' + ay' + by = 0$, a and b real constants, we must examine the roots of the auxiliary polynomial equation $r^2 + ar + b = 0$. If the discriminant $\delta = a^2 - 4b$ is positive, we will have distinct real roots. The theory behind this case has already been developed. If the roots, r_1 and r_2, of the auxiliary polynomial equation are real and distinct, then the solutions are $u_1(x) = e^{r_1 x}$ and $u_2(x) = e^{r_2 x}$. These were already shown to be linearly independent, so the general solution to $y'' + ay' + by = 0$ is $y = c_1 e^{r_1 x} + c_2 e^{r_2 x}$.

Example 40

Consider $y'' - 5y' + 4y = 0$. The auxiliary equation is $r^2 - 5r + 4 = 0$, and the roots are $r = 4$ and $r = 1$. Hence the general solution is

$$y(x) = c_1 e^{4x} + c_2 e^x$$

Example 41

Consider $3y'' + 16y' - 12y = 0$. This constant-coefficient equation has as its auxiliary polynomial equation $3r^2 + 16r - 12 = 0$. Factoring and finding the roots produce $r_1 = \frac{2}{3}$ and $r_2 = -6$. Thus the general solution is

$$y(x) = c_1 e^{2x/3} + c_2 e^{-6x}$$

3.7 Repeated Real Roots

If the discriminant $\delta = a^2 - 4b$ of the auxiliary polynomial equation equals zero, then its roots are real and equal. We do not have two distinct solutions but only one $u_1(x) = e^{rx}$. Furthermore, if we know that $u_1 = e^{rx}$ is a solution to $L(y) = y'' + ay' + by = 0$ where $r = -\frac{a}{2}$ and $a^2 - 4b = 0$, we can show that $u_2 = xe^{rx}$ is another solution:

We have $$u_2' = xre^{rx} + e^{rx} = (xr + 1)e^{rx}$$

and $$u_2'' = (xr + 1)(re^{rx}) + re^{rx} = (xr^2 + 2r)e^{rx}$$

Substituting these into the differential equation gives

$$\begin{aligned} L(xe^{rx}) &= (xr^2 + 2r)e^{rx} + a(xr + 1)e^{rx} + b(xe^{rx}) \\ &= (xr^2 + axr + bx)e^{rx} + (2r + a)e^{rx} \\ &= x(r^2 + ar + b)e^{rx} + (2r + a)e^{rx} \end{aligned}$$

Since the auxiliary polynomial equation for this differential equation is $r^2 + ar + b = 0$, then the first term above is zero.

Under the hypothesis of equal real roots with $\delta = a^2 - 4b = 0$, we have

$$r = -\frac{a}{2} \tag{4}$$

This is equivalent to $(2r + a) = 0$. Thus the second term above is also zero, so $L(xe^{rx}) = 0$, the differential equation is satisfied, and we have shown that for this case xe^{rx} solves the differential equation if e^{rx} does. Since neither of these is a constant multiple of the other, they are linearly independent.

Example 42

Consider $y'' - 4y' + 4y = 0$. The auxiliary equation is $r^2 - 4r + 4 = 0$, and its roots are $r_1 = 2$ and $r_2 = 2$. Hence the general solution is

$$y = c_1 e^{2x} + c_2 x e^{2x}$$

The method that generates this second solution is called the **variation of parameters**. It begins with the known solution $u_1 = c_1 e^{rx}$, changes the parameter c_1 to a variable $v_1(x)$ and attempts to find what form $v_1(x)$ should take to guarantee another solution. Since we need a second solution linearly independent from the given $u_1 = c_1 e^{rx}$, we cannot multiply by a constant, because that process would produce a linearly dependent solution. Thus we need to multiply by a function of x, namely $v_1(x)$.

We use a forcing process, that is, we let $u(x) = v_1(x)e^{rx}$ where r solves the auxiliary equation $r^2 + ar + b = 0$ and insert this function into the differential equation. Differentiating $u(x)$ we obtain

$$u'(x) = rv_1 e^{rx} + e^{rx}v_1' = e^{rx}(v_1' + rv_1)$$

and $\quad u''(x) = e^{rx}(v_1'' + rv_1') + re^{rx}(v_1' + rv_1) = e^{rx}(v_1'' + 2rv_1' + r^2 v_1)$

Substituting these results into the differential equation yields

$$e^{rx}(v_1'' + 2rv_1' + r^2 v_1) + e^{rx}(av_1' + arv_1) + e^{rx}(bv_1) = 0$$

Dividing out the nonzero term e^{rx} and rearranging, we get

$$v_1'' + (2r + a)v_1' + (r^2 + ar + b)v_1 = 0$$

Since under the assumption of this case $\delta = a^2 - 4b = 0$, equation (4) shows the root to be $r = -\frac{a}{2}$, so the coefficient of v_1' vanishes. Since r is a root of the auxiliary equation, the coefficient of v_1 vanishes and the equation becomes $v_1'' = 0$. This is easily integrated to $v_1 = c_1 + c_2 x$. Now since $u(x) = v_1 e^{rx}$ we obtain $u(x) = (c_1 + c_2 x)e^{rx} = c_1 e^{rx} + c_2 x e^{rx}$. This illustrates the second linearly independent solution xe^{rx} and provides the general solution to the differential equation in this case.

The Wronskian of this new pair of solutions is

$$W[f_1, f_2] = \begin{vmatrix} e^{rx} & xe^{rx} \\ re^{rx} & (rx+1)e^{rx} \end{vmatrix} = [(rx + 1) - rx]e^{2rx} = e^{2rx} \neq 0$$

Example 43

A problem in mechanical vibrations produces the equation $y'' - 6y' + 9y = 0$. Writing this equation as $(D^2 - 6D + 9)y = 0$, we form the auxiliary polynomial equation $r^2 - 6r + 9 = 0$, which factors into $(r - 3)^2 = 0$. We have a repeated root $r = 3$. The linearly independent solutions are e^{3x} and xe^{3x}, so the general solution is

$$y(x) = c_1 e^{3x} + c_2 x e^{3x}$$

The above process gives us the two linearly independent solutions to the second-order differential equation $y'' + ay' + by = 0$ when the roots of the auxiliary polynomial equation are equal. This concept is easily extended to higher-order equations if the higher-order equation is homogeneous and has constant coefficients. Theorem 3.10 provides for such an extension.

Theorem 3.10:

If an nth-order linear homogeneous differential equation with constant coefficients has an auxiliary polynomial equation with a root r of multiplicity m, then the linearly independent solutions corresponding to these roots are

$$e^{rx}, xe^{rx}, x^2 e^{rx}, \ldots, x^{m-1} e^{rx}$$

In addition, for each distinct real root r_i the linearly independent solution is $e^{r_i x}, r_i \neq r$.

Example 44 illustrates Theorem 3.10.

Example 44

The differential equation $y''' + 6y'' + 12y' + 8y = 0$ has an auxiliary equation $r^3 + 6r^2 + 12r + 8 = 0$ or $(r + 2)^3 = 0$. The root $r = -2$ is of multiplicity 3. Hence the general solution is

$$y(x) = c_1 e^{-2x} + c_2 x e^{-2x} + c_3 x^2 e^{-2x}$$

Notice in Example 45 how the general solution is affected when more than one root is repeated.

Example 45

If the roots of the auxiliary polynomial equation were the set $\{1, 1, 1, -3, -3, 2\}$, the general solution of the corresponding differential equation would be

$$y(x) = c_1 e^x + c_2 x e^x + c_3 x^2 e^x + c_4 e^{-3x} + c_5 x e^{-3x} + c_6 e^{2x}$$

Example 46

If the roots of the auxiliary polynomial were the set $\{2, 2, -3\}$, then the polynomial itself would be $(r - 2)(r - 2)(r + 3)$. The auxiliary polynomial equation would be $(r - 2)^2(r + 3) = 0$ or $r^3 - r^2 - 8r + 12 = 0$. Thus the differential equation that produces this polynomial is

$$y''' - y'' - 8y' + 12y = 0$$

Example 47

Suppose a particle moves vertically along the y-axis so that its acceleration plus its velocity has a value equal to six times its distance from the origin. It passes through the origin with velocity 2. Find a function which represents its distance at any time t.

Since acceleration is represented by y'' and velocity by y', we get $y'' + y' = 6y$. Thus we need to solve $y'' + y' - 6y = 0$, with $y(0) = 0$, $y'(0) = 2$. The auxiliary polynomial is $r^2 + r - 6 = 0$, with roots of -3 and 2. The general solution is $y(t) = c_1 e^{-3t} + c_2 e^{2t}$. The initial conditions of $y(0) = 0$, $y'(0) = 2$ produce $c_1 = -\frac{2}{5}$, $c_2 = \frac{2}{5}$. Thus $y(t) = -\frac{2}{5}e^{-3t} + \frac{2}{5}e^{2t}$. Figure 3.2 shows the graph of this equation.

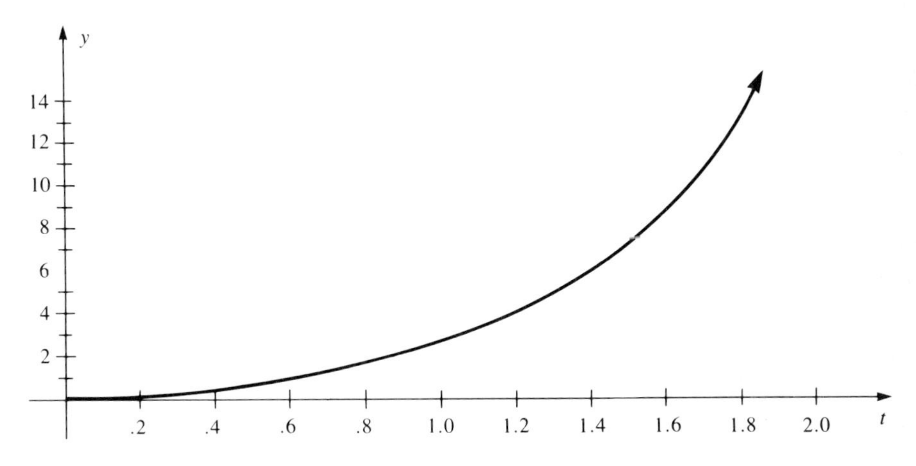

Figure 3.2

Example 48

In the process of diffusing gas into a liquid it was found that under certain conditions we can determine the concentration of gas in the liquid at time t if we know the concentration at two boundary points (e.g., 0 and A). The differential equation to be solved is

$$y'' = k\left(\frac{1}{D}\right)y$$

where $D > 0$ is the diffusion coefficient and k is a proportionality constant. The unique solution under the boundary conditions $y(0) = G$, $y(A) = 0$ is obtained as follows:

$$y'' - \frac{k}{D}y = 0 \Rightarrow r^2 - \frac{k}{D} = 0 \Rightarrow r = \pm\sqrt{\frac{k}{D}}$$

$$\therefore \quad y(x) = c_1 e^{\sqrt{k/D}\,x} + c_2 e^{-\sqrt{k/D}\,x}$$

if we further assume $k > 0$. (If not, we go to the case in Section 3.8.) Since $y(0) = G$ we get

$$G = c_1 e^0 + c_2 e^0 = c_1 + c_2$$

and since $y(A) = 0$ we get

$$0 = c_1 e^{\sqrt{k/D}A} + c_2 e^{-\sqrt{k/D}A}$$

or

$$c_1 = -c_2 e^{-2\sqrt{k/D}A}$$

Then by combining we get

$$G = -c_2 e^{-2\sqrt{k/D}A} + c_2$$

or

$$c_2 = \frac{G}{1 - e^{-2\sqrt{k/D}A}}$$

Then

$$c_1 = \frac{-Ge^{-2\sqrt{k/D}A}}{1 - e^{-2\sqrt{k/D}A}}$$

Thus the unique solution is

$$y(x) = \left(\frac{-Ge^{-2\sqrt{k/D}A}}{1 - e^{-2\sqrt{k/D}A}}\right)e^{\sqrt{k/D}x} + \left(\frac{G}{1 - e^{-2\sqrt{k/D}A}}\right)e^{-\sqrt{k/D}x}$$

Example 49

The differential equation

$$C\frac{d^2V}{dt^2} + \frac{1}{R}\frac{dV}{dt} + \frac{1}{L}V = 0$$

arises in the analysis of electrical circuits. Assume the capacitance C is 10^{-7}; the resistance R, 25; and the inductance L, $\frac{1}{3} \cdot 10^{-3}$. At time $t = 0$, the value of V is 15 and the value of dV/dt is $-27 \cdot 10^5$. Solve for $V(t)$.

Substituting the assigned values gives

$$10^{-7}V'' + \frac{1}{25}V' + 3 \cdot 10^3 V = 0; \; V(0) = 15, \; V'(0) = -27 \cdot 10^5$$

If we multiply by 10^7, the differential equation is rearranged to

$$V'' + 4 \cdot 10^5 V' + 3 \cdot 10^{10} V = 0$$

If we let $k = 10^5$, the auxiliary polynomial is $r^2 + 4kr + 3k^2 = 0$ and its roots are

$$r = \frac{-4k \pm \sqrt{16k^2 - 12k^2}}{2} = \frac{-4k \pm 2k}{2} = \begin{Bmatrix} -k \\ -3k \end{Bmatrix} = \begin{Bmatrix} -10^5 \\ -3 \cdot 10^5 \end{Bmatrix}$$

Thus

$$V = c_1 e^{-10^5 t} + c_2 e^{-3 \cdot 10^5 t}$$

Since $V(0) = 15$, we have $c_1 + c_2 = 15$; and since $V'(0) = -27 \cdot 10^5$, we have $-c_1 - 3c_2 = -27$. This pair of equations solves to $c_1 = 9$ and $c_2 = 6$, so that the unique solution is

$$V(t) = 9e^{-10^5 t} + 6e^{-3 \cdot 10^5 t}$$

Exercises 3.7

In Exercises 1–7, find the solution to each of the differential equations.

1. $y'' - 4y' + 3y = 0$
2. $y'' - 2y' + y = 0$
3. $y'' + 5y' + \frac{25}{4} y = 0;\ y(0) = 1,\ y'(0) = -2$
4. $4y'' + 4y' + y = 0$
5. $25y'' - 9y = 0$
6. $y''' - 6y'' + 9y' = 0$
7. $y^{(iv)} + 2y''' - y'' - 2y' = 0$
8. Which function has a third derivative that vanishes?
9. Find a curve having a slope 2 at the origin and satisfying the differential equation $y'' - 3y' = 0$.
10. Find a curve satisfying the differential equation $9y'' - 12y' + 4y = 0$ and having slope 2 when it passes through the point $(0, 1)$.
11. Find a curve satisfying the differential equation $4y'' + 20y' + 25y = 0$ and having slope $\frac{5}{2}$ when it passes through the point $(0, 2)$.
12. If the auxiliary polynomial equation has roots 2, 3, 3, and -1, what is the general solution?
13. In Exercise 12, what differential equation was needed to begin the solution process?
14. If the auxiliary polynomial equation for a given differential equation has roots -1, -1, -1, and -1, what is the general solution to the differential equation?
15. In a study* related to automobile driving, the researcher found that a driver's decision to make a correction in steering is correlated with his or her perception of nearness of the edge of the road and the rate at which he or she is approaching that edge. If x represents the distance of the center of the car from the center of the driving lane, then

$$\frac{d^3x}{dt^3} + k\frac{d^2x}{dt^2} + \gamma B\frac{dx}{dt} + Bx = 0; \quad k, \gamma, B \text{ positive constants}$$

If $k = .5$, $\gamma = 2$, and $B = .245$, solve for $x(t)$.

---

* N. Rashevsky, "Mathematical Biology of Automobile Driving," *Bulletin of Mathematical Biophysics*, 29 (1967), p. 181.

16. If a closed electric circuit contains a resistance R, an inductance L, and a capacitance C, then Kirchoff's second law states that the charge q on the capacitor satisfies

$$L\frac{d^2q}{dt^2} + R\frac{dq}{dt} + \frac{1}{C}q = 0$$

Assume the following conditions: $R = 4$ ohms, $L = .04$ henrys, $C = 0.002$ farads, the initial charge is 20 coulombs, and initially there is no current (so $q'(0) = 0$). Show that the expression for the charge at any time t is

$$q = 10e^{-50t}\sin(100t) + 20e^{-50t}\cos(100t)$$

3.8 Complex Conjugate Roots

We are studying the differential equation $y'' + ay' + by = 0$. When a and b are real constants, we examine the auxiliary polynomial equation $r^2 + ar + b = 0$. Its roots determine the form of the solution, and we separate the cases by inspecting the discriminant $\delta = a^2 - 4b$.

If $\delta < 0$, the roots of the auxiliary polynomial equation are a pair of complex conjugate numbers. If we call these $\alpha + i\beta$ and $\alpha - i\beta$, then the prior theory gives the complex function solutions

$$e^{(\alpha + i\beta)x} = e^{\alpha x}(\cos\beta x + i\sin\beta x)$$

and

$$e^{(\alpha - i\beta)x} = e^{\alpha x}(\cos\beta x - i\sin\beta x)$$

We saw in Section 3.4 that the real and imaginary parts of each of these provide solutions to the original equation. Of the four functions generated here, two functions $e^{\alpha x}\cos\beta x$ and $e^{\alpha x}\sin\beta x$ are linearly independent. Their Wronskian is

$$\begin{vmatrix} e^{\alpha x}\cos\beta x & e^{\alpha x}\sin\beta x \\ e^{\alpha x}(\alpha\cos\beta x - \beta\sin\beta x) & e^{\alpha x}(\beta\cos\beta x + \alpha\sin\beta x) \end{vmatrix} = \beta e^{2\alpha x}$$

Since the exponential function $e^{2\alpha x}$ is never zero, their Wronskian is zero only if $\beta = 0$; but we then would not have complex roots. Thus the Wronskian is never zero, and we obtain the two linearly independent solutions above. The general solution is

$$y(x) = c_1e^{\alpha x}\cos\beta x + c_2e^{\alpha x}\sin\beta x$$

Example 50

Consider the differential equation $y'' + 9y = 0$ with $y(0) = 2$, $y'(0) = -3$. The auxiliary polynomial equation is $r^2 + 9 = 0$, and its roots are $r = \pm 3i$. Then the complex exponential solutions are e^{3ix} and e^{-3ix}. Expanding

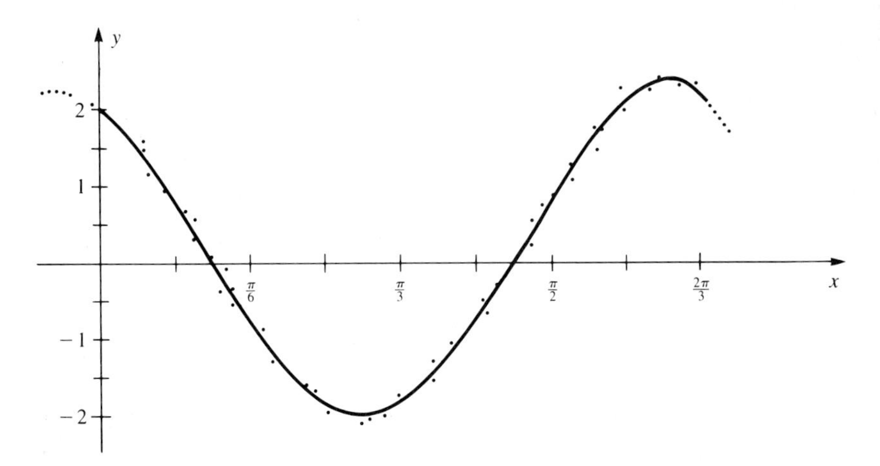

Figure 3.3

these, we have

$$e^{3ix} = \cos 3x + i \sin 3x$$

and
$$e^{-3ix} = \cos(-3x) + \sin(-3x)$$

Since the real and imaginary parts of a complex solution function are themselves solutions, we obtain the collection $\{\cos 3x, \sin 3x, \cos(-3x), \sin(-3x)\}$. Of these solutions, just two are linearly independent, so we choose the linearly independent pair $\{\cos 3x, \sin 3x\}$. The general solution then is $y(x) = C_1 \cos 3x + C_2 \sin 3x$. Now using the initial conditions, we get

$$2 = C_1 \cos 0 + C_2 \sin 0 = C_1 \Rightarrow C_1 = 2$$

and
$$-3 = -3C_1 \sin 0 + 3C_2 \cos 0 = 3C_2 \Rightarrow C_2 = -1$$

The unique solution now is

$$y(x) = 2 \cos 3x - \sin 3x$$

The graph of this equation is shown in Figure 3.3.

Example 51

Consider $y'' - 2y' + 2y = 0$. The roots of $r^2 - 2r + 2 = 0$ are $r = 1 + i$ and $r = 1 - i$, so the real and imaginary parts of $e^{(1+i)x}$ and $e^{(1-i)x}$ provide $e^x \cos x$ and $e^x \sin x$ as the linearly independent solutions; and the general solution is

$$y(x) = c_1 e^x \cos x + c_2 e^x \sin x$$

Example 52

Consider the initial value problem

$$y''(t) + y(t) = 0; \quad y(0) = A, \quad y'(0) = B$$

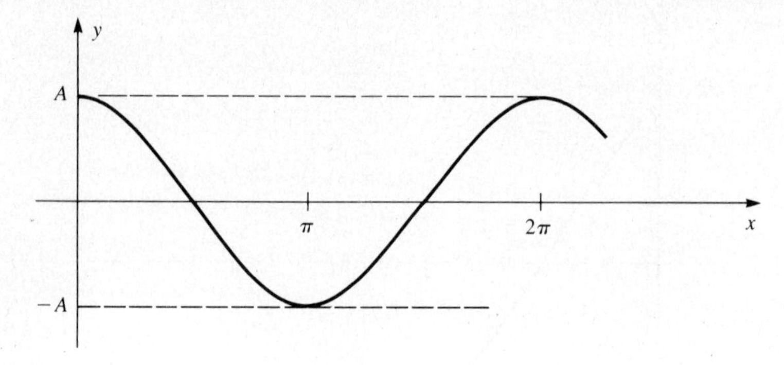

Figure 3.4

This is the equation of harmonic motion with initial displacement A and velocity B. Since the roots of the auxiliary equation $r^2 + 1 = 0$ are $\pm i$, we have e^{it} and e^{-it} as solutions; but the linearly independent real and imaginary parts are $\{\cos t, \sin t\}$. Thus the general solution to the differential equation is $y(t) = c_1\cos t + c_2\sin t$. Applying the initial conditions

$$A = y(0) = c_1\cos 0 + c_2\sin 0 = c_1$$

and

$$B = y'(0) = -c_1\sin 0 + c_2\cos 0 = c_2$$

we find the unique solution

$$y(t) = A\cos t + B\sin t$$

The special case of $B = 0$ gives an oscillation of amplitude A and a period of 2π, as shown in Figure 3.4.

We now have the techniques for solving any differential equation that is linear and homogeneous and that has constant real coefficients. We were able to generate these techniques because of the theorems from algebra that state that every polynomial with real coefficients is factorable into linear and irreducible quadratic factors. Our auxiliary polynomial equation then can be factored and all its roots determined. Therefore, we can create the general solution to any such equation.

Example 53 gives you additional experience with solutions involving complex conjugate roots.

Example 53

A particle moves linearly and passes through the origin with velocity 2. Its acceleration $a(t)$ is given by the difference between its velocity and its displacement. Thus

$$a(t) = v(t) - s(t)$$

or

$$\frac{d^2s}{dt^2} = \frac{ds}{dt} - s$$

or

$$s'' - s' + s = 0$$

The auxiliary polynomial equation is $r^2 - r + 1 = 0$, and the roots are $r = \frac{1}{2} \pm \frac{\sqrt{3}}{2} i$. Determining the real and imaginary parts of $e^{[(1/2)+(\sqrt{3}/2)i]t}$, we obtain the linearly independent solutions

$$s_1(t) = e^{t/2} \cos \frac{\sqrt{3}}{2} t$$

and

$$s_2(t) = e^{t/2} \sin \frac{\sqrt{3}}{2} t$$

Thus the general solution is

$$s(t) = C_1 e^{t/2} \cos \frac{\sqrt{3}}{2} t + C_2 e^{t/2} \sin \frac{\sqrt{3}}{2} t$$

Since the initial conditions give $s(0) = 0$ and $s'(0) = 2$, then $0 = C_1$ and $2 = \frac{1}{2} C_1 + \frac{\sqrt{3}}{2} C_2$ or $C_2 = \frac{4}{3}\sqrt{3}$. The equation of motion which gives the distance s from the origin at any time t is

$$s(t) = \frac{4}{3} \sqrt{3} e^{t/2} \sin \frac{\sqrt{3}}{2} t$$

Exercises 3.8

Find the general solutions to the differential equations in Exercises 1–3.

1. $y'' + 6y' + 9y = 0$
2. $y''' + y' = 0$
3. $y''' - y'' - y' + y = 0$
4. A particle moves so that its acceleration is always negative (i.e., it is decelerating) and proportional to its displacement from the origin. Find its equation of motion if it passes through the origin with velocity $\frac{3}{2}$.

Find the general solution to the differential equations in Exercises 5–20.

5. $y''' - 3y'' + 3y' - y = 0$
6. $2y'' - y' - 3y = 0$
7. $y'' + 2y' + 3y = 0$
8. $2y'' + y' - y = 0$
9. $y''' - 8y' + 3y = 6$
10. $y^{(iv)} - 6y'' + 5y = 20$
11. $y^{(iv)} - 6y'' + 9y = 0$
12. $y^{(iv)} - 6y'' + y = 0$
13. $(D^5 - 4D^3)y = 0$
14. $y'' + 2y' - 3y = 0$; $y(0) = 1$, $y'(0) = 2$

15. $y'' = 3y + y'; y(0) = \frac{2}{10}, y'(0) = \frac{17}{10}$

16. $y'' + 4y = 0; y\left(\frac{\pi}{8}\right) = A, y'\left(\frac{\pi}{8}\right) = B$

17. $y''' - y'' + 4y' - 4y = 0; y(0) = 1, y'(0) = 0, y''(0) = 2$

18. $\frac{d^2y}{dt^2} + 2y = 0$

19. $y'' + 4y' + 5y = 0; y(0) = 1, y'(0) = 0$
20. $2y'' + 4y' + 5y = 0; y(0) = 2, y'(0) = 3$
21. A particle moves so that its acceleration is equal to half of the sum of its velocity and its displacement. Find its equation of motion if it passes through the origin with velocity 3.
22. Find a curve having slope $\frac{1}{2}$ at the origin and satisfying the differential equation $y'' = y - 2y'$.
23. If a simple pendulum with a weight attached is released from rest, it will undergo oscillatory motion. The differential equation that results from applying Newton's second law is

$$\frac{d^2\theta}{dt^2} + \frac{g}{l}\sin\theta = 0$$

where $g = 32$ ft/sec^2. The fact that a small θ results in $\theta \simeq \sin\theta$ is often used to linearize this nonlinear equation. Using the linearized equation for a two-foot-long pendulum that is released from rest at position $\theta = \frac{\pi}{8}$ at time $t = 0$, find the angular velocity $\omega = d\theta/dt$ at the point at the bottom of the arc. At what time does the pendulum reach that point?
24. If a clock pendulum of length 38.4 inches produces a tick each time it crosses the bottom equilibrium position, how many ticks will it make in one hour?
25. If a floating body is depressed vertically from its equilibrium position by an amount y, the resulting motion is described by

$$\frac{d^2y}{dt^2} + \frac{\rho A}{m}y = 0$$

where ρ is liquid density, A is the body's cross-sectional area, and m is its mass. A cubical block of wood measuring 1.5 feet on a side and weighing 72 pounds is floating in water ($\rho = 62.5$). It is depressed 0.5 foot and released. Find the equation of motion and the amplitude of the motion. How far below the water line is the bottom of the block when freely floating?
26. When technicians test certain metals for use in a spacecraft, a tortional stiffness test is applied to a slender vertical rod of the test metal to

which a solid metal cylinder is attached. If r is the radius of the cylinder and k is a tortional stiffness constant, then

$$\frac{mr^2}{2}\frac{d^2\theta}{dt^2} + k\theta = 0$$

is the differential equation that is solved for the angle θ through which the cylinder is twisted about an axis through the vertical rod. If a newly created alloy has a k value of 1.5 and if the mass of the system is 300 grams and the cylinder's radius is 10 cm, find θ as a function of time.

27. In mining certain minerals, the cost rises linearly with cumulative production. In seeking the optimum process for mining this type of minerals, researchers* found the differential equation

$$\frac{d^2x}{dt^2} - r\frac{dx}{dt} - \frac{r\beta}{b}x = -\frac{rK}{b}$$

where r, β, b, and K are nonzero constants. Solve this equation.

28. In studying the relation between the national income and the burden of debt incurred by the citizens, E. D. Domar** concluded that the applicable differential equation is $d^2D/dt^2 - kD = 0$, where D is total public debt and the income is assumed to grow at a constant relative rate k $(0 < k < 1)$. Write the solution in exponential functions and then in hyperbolic functions. [*Hint:* $\sinh t = (e^t - e^{-t})/2$ and $\cosh t = (e^t + e^{-t})/2$.]

3.9 Nonhomogeneous Equations Revisited

In Section 3.3 we looked at the technique for solving a nonhomogeneous linear equation $Ly = f(x)$. This method consisted of determining linearly independent solutions to the associated homogeneous equation $Ly = 0$ and then, by any available method, finding one particular solution to the nonhomogeneous equation.

In that section we found that if the differential equation, represented in operator form as Ly, had constant coefficients and if the function $f(x)$ on the right side of the differential equation had a specific form, we were able to find the general solution to the given equation. Specifically, when $f(x)$ was a constant, a polynomial, or an exponential function of the form $e^{\alpha x}$ with α a real constant, we formed a trial solution and then carried our computations

---

* O. C. Herfindahl and A. V. Kneese, *Economic Theory for Natural Resources* (Columbus, Ohio: Charles E. Merrill, 1974).

** E. D. Domar, "The Burden of Debt and the National Income," *American Economic Review* (December 1944), p. 798.

to completion of the final general solution. The principle of superposition allowed us to extend those simple cases to any additive combination of the functions $f_1 = \text{constant}$, $f_2 = a_0 + a_1x + a_2x^2 + \cdots + a_nx^n$, and $f_3 = e^{\alpha x}$.

There is one case that we have not investigated—the case when $f(x)$ is a trigonometric function such as $f(x) = \sin \beta x$ or $f(x) = \cos \beta x$. Some of our previously developed theory will help us obtain a solution in this case.

Consider a linear differential equation in the form $Ly = F(x)$ where $F(x)$ is a complex function. Write the equation as $Ly = F = f(x) + i \cdot g(x)$ where f and g are real functions. Suppose that we were able by some method to find a complex-valued function $u(x) = v(x) + i \cdot w(x)$, v and w real functions, that solves this equation. We would then have $Lu(x) = F(x)$ or $L[v(x) + i \cdot w(x)] = f(x) + i \cdot g(x)$. By the linearity of L we get

$$Lv(x) + i \cdot Lw(x) = f(x) + i \cdot g(x)$$

and by the identification property of complex numbers this implies that $Lv(x) = f(x)$ and $Lw(x) = g(x)$. This result now tells us:

> The real part of the complex solution to $Ly = f(x) + i \cdot g(x)$ solves the equation $Ly = f(x)$ and the imaginary part of the complex solution to $Ly = f(x) + i \cdot g(x)$ solves the equation $Ly = g(x)$.

The bonus is that we now solve two real differential equations simultaneously when we solve just one complex differential equation.

Example 54

Consider the complex differential equation $Ly = F = f(x) + i \cdot g(x)$ given by

$$y'' - 3y' + 2y = e^{-2ix} = \cos 2x - i \sin 2x$$

Following the technique developed in Section 3.3 for $Ly = f(x)$ for exponential function $f(x)$, we propose a particular solution to this complex equation $y_P = Ae^{-2ix}$ where A is the undetermined coefficient whose value is to be determined. Taking derivatives of this proposed solution gives $y_P' = -2iAe^{-2ix}$ and $y_P'' = -4Ae^{-2ix}$. Substituting these into the differential equation produces $(-4A + 6iA + 2A)e^{-2ix} = e^{-2ix}$. Dividing out the exponential leaves $(-2 + 6i)A = 1$ or

$$A = \frac{1}{-2 + 6i} = \frac{-2 - 6i}{(-2 + 6i)(-2 - 6i)} = \frac{-1 - 3i}{20}$$

Putting this value of A into our trial solution y_P, we obtain the complex function that solves the original $Ly = F$:

$$y_P = \left(-\frac{1}{20} - \frac{3}{20}i\right)e^{-2ix}$$

We replace e^{-2ix} by its expansion from Euler's identity and multiply to get

$$y_P = \left(-\frac{1}{20} - \frac{3}{20}i\right)\left(\cos 2x - i\sin 2x\right)$$

Splitting this into its real and imaginary parts gives

$$\begin{aligned} y_P &= \left(-\frac{1}{20}\cos 2x - \frac{3}{20}\sin 2x\right) + i\left(-\frac{3}{20}\cos 2x + \frac{1}{20}\sin 2x\right) \\ &= v(x) + i\cdot w(x) \end{aligned}$$

We now collect our bonus: We find that the particular solution $v(x) = -\frac{1}{20}\cos 2x - \frac{3}{20}\sin 2x$ (which is the real part) solves $y'' - 3y' + 2y = \cos 2x$, and $w(x) = -\frac{3}{20}\cos 2x + \frac{1}{20}\sin 2x$ (which is the imaginary part) solves $y'' - 3y' + 2y = -\sin 2x$. These may be checked by direct substitution. If this solution is now added to the homogeneous solution $y_h = c_1e^{2x} + c_2e^x$, we will obtain the general solution to either equation. That is, the general solution to

$$y'' - 3y' + 2y = \cos 2x$$

is

$$\begin{aligned} y &= y_h + y_P = y_h + v(x) \\ &= c_1e^{2x} + c_2e^x - \frac{1}{20}\cos 2x - \frac{3}{20}\sin 2x \end{aligned}$$

and the general solution to

$$y'' - 3y' + 2y = -\sin 2x$$

is

$$\begin{aligned} y &= y_h + y_p + y_h + w(x) \\ &= c_1e^{2x} + c_2e^x - \frac{3}{20}\cos 2x + \frac{1}{20}\sin 2x \end{aligned}$$

Example 55

Suppose that in working with a certain mechanical system you needed to solve the differential equation $3y'' + 2y' - y = 10e^{ix}$. First write this as $3y'' + 2y' - y = 10(\cos x + i\sin x)$ and then as two equations (using the real and imaginary parts of the right side function):

$$3y'' + 2y' - y = 10\cos x \tag{5}$$

$$3y'' + 2y' - y = 10\sin x \tag{6}$$

Look for a complex-valued function as the particular solution, trying $y_P = Ae^{ix}$. Then $y_P' = iAe^{ix}$ and $y_P'' = -Ae^{ix}$, when inserted into the differential equation, give

$$-3Ae^{ix} + 2iAe^{ix} - Ae^{ix} = 10e^{ix}$$

or

$$-4A + 2iA = 10$$

Then $A = \dfrac{10}{-4+2i} = \dfrac{10}{-4+2i} \cdot \dfrac{-4-2i}{-4-2i} = \dfrac{-40-20i}{20} = -2 - i$

Thus $y_P = (-2 - i)e^{ix} = (-2 - i)(\cos x + i \sin x)$
$= (-2\cos x + \sin x) + i(-\cos x - 2\sin x)$

Using the auxiliary polynomial equation

$$3r^2 + 2r - 1 = 0$$

or

$$(3r - 1)(r + 1) = 0$$

we easily find the homogeneous solution to be $y_h = c_1 e^{x/3} + c_2 e^{-x}$. The solution to the original differential equation then is

$$y(x) = c_1 e^{x/3} + c_2 e^{-x} + (-2\cos x + \sin x) + i(-\cos x - 2\sin x)$$

As our bonus, we get the solution to equation (5) by using the real part of this y_P,

$$y(x) = c_1 e^{x/3} + c_2 e^{-x} - 2\cos x + \sin x$$

and the solution to equation (6) by using the imaginary part of this y_P,

$$y(x) = c_1 e^{x/3} + c_2 e^{-x} - \cos x - 2\sin x.$$

We can now outline a method for solving equations that are similar to equations (5) and (6).

Procedure: To solve a differential equation of the type

$$Ly = \cos \beta x \quad \text{or} \quad Ly = \sin \beta x$$

change temporarily to the problem $Ly = e^{i\beta x}$ and find the complex exponential solution. Break this solution down into its real and imaginary parts. The real part will solve $Ly = \cos \beta x$, and the imaginary part will solve $Ly = \sin \beta x$.

The process just described will become an integral part of the more complicated sections in which we introduce ways to solve $Ly = F$ for any given F. Example 56 illustrates the solution to a problem of the type given in equation (6).

Example 56

Consider

$$y'' - 3y' - 4y = 25 \sin 3x$$

Change the problem to $y'' - 3y' - 4y = 25e^{3ix}$ and look for a complex exponential solution. Use as the trial solution

$$y_P = Ae^{3ix}$$

Then

$$y'_P = 3iAe^{3ix}$$

and

$$y''_P = -9Ae^{3ix}$$

Substitute these into the complex differential equation to get

$$-9Ae^{3ix} - 3(3iAe^{3ix}) - 4(Ae^{3ix}) = 25e^{3ix}$$

Divide by e^{3ix} and collect terms to obtain

$$(-13 - 9i)A = 25$$

$$A = \frac{25}{-13 - 9i} = \frac{25}{-13 - 9i} \cdot \frac{-13 + 9i}{-13 + 9i}$$

$$= \frac{25(-13 + 9i)}{250} = -\frac{13}{10} + \frac{9}{10}i$$

The trial solution is now

$$y_P = \left(-\frac{13}{10} + \frac{9}{10}i\right)e^{3ix}$$

$$= \left(-\frac{13}{10} + \frac{9}{10}i\right)(\cos 3x + i \sin 3x)$$

$$= -\frac{13}{10}\cos 3x + \frac{9}{10}i\cos 3x - \frac{13}{10}i\sin 3x + \frac{9}{10}i^2\sin 3x$$

Separate out the real and imaginary parts to obtain

$$y_P = \left(-\frac{13}{10}\cos 3x - \frac{9}{10}\sin 3x\right) + i\left(\frac{9}{10}\cos 3x - \frac{13}{10}\sin 3x\right)$$

$$= v(x) + i \cdot w(x)$$

Since the original problem was

$$y'' - 3y' - 4y = 25 \sin 3x$$

and the function on the right is the imaginary part of $25e^{3ix}$, retain only the imaginary part of this particular solution, namely,

$$y_P = \frac{9}{10}\cos 3x - \frac{13}{10}\sin 3x$$

This is now added to the homogeneous solution y_h, easily obtained from the auxiliary polynomial equation

$$r^2 - 3r - 4 = (r - 4)(r + 1) = 0$$

to get

$$y = c_1e^{4x} + c_2e^{-x} + \frac{9}{10}\cos 3x - \frac{13}{10}\sin 3x$$

Substitute this solution in the original differential equation:

$$y' = 4c_1e^{4x} - c_2e^{-x} - \frac{27}{10}\sin 3x - \frac{39}{10}\cos 3x$$

$$y'' = 16c_1e^{4x} + c_2e^{-x} - \frac{81}{10}\cos 3x + \frac{117}{10}\sin 3x$$

so
$$\begin{aligned} y'' - 3y' - 4y = {} & 16c_1e^{4x} + c_2e^{-x} - \frac{81}{10}\cos 3x + \frac{117}{10}\sin 3x \\ & -12c_1e^{4x} + 3c_2e^{-x} + \frac{117}{10}\cos 3x + \frac{81}{10}\sin 3x \\ & -4c_1e^{4x} - 4c_2e^{-x} - \frac{36}{10}\cos 3x + \frac{52}{10}\sin 3x \\ = {} & 0 + 0 + 0 + \frac{250}{10}\sin 3x \\ = {} & 25\sin 3x \end{aligned}$$

and the solution is correct.

Our progress is recapped in outlines *A* and *B*.

A. We can solve any homogeneous linear equation with constant coefficients provided we find the roots of the auxiliary polynomial equation (A.P.E.).

Type of Root of A.P.E.	**Linearly Independent Solutions**
Real and Distinct	
$r_1 \neq r_2$	e^{r_1x} and e^{r_2x}
Real and Equal	
$r_1 = r_2$	e^{r_1x} and xe^{r_1x}
Complex Conjugate	
$\left.\begin{matrix} r_1 = \alpha + i\beta \\ r_2 = \alpha - i\beta \end{matrix}\right\} \beta \neq 0$	$e^{\alpha x}\cos \beta x \quad e^{\alpha x}\sin \beta x$

B. We can solve the nonhomogeneous linear equation with constant coefficients $Ly = F$ if the right side function F has certain special forms.

Form of $F(x)$	**Solution Form**
$F = \text{constant} = k$	$y_P = k/a_n$
$F = \text{real polynomial of degree } n$	$y_P = A_0 + A_1x + A_2x^2 + \cdots + A_nx^n$
$F = \text{real exponential} = ce^{kx}$	$y_P = Ae^{kx}$, if k is not a root of auxiliary polynomial.
$F = \begin{cases} \text{complex exponential } ce^{i\beta x} \\ \text{pure sine function } c\sin\beta x \\ \text{pure cosine function } c\cos\beta x \end{cases}$	$y_P = Ae^{i\beta x}$

Example 57 combines some of the techniques you have learned.

Example 57

Consider
$$y'' - 4y = 2x + 5\cos x$$

We first solve the associated homogeneous equation by using the auxiliary polynomial equation $r^2 - 4 = 0$. The roots are ± 2, so we get

$$y_h = c_1e^{2x} + c_2e^{-2x}$$

Using the principle of superposition, we will obtain two particular solutions, one for the polynomial $f_1(x) = 2x$ and another for the trigonometric $f_2(x) = 5\cos x$. If we let

$$y_{P_1} = A_0 + A_1x$$

then
$$y'_{P_1} = A_1$$

and
$$y''_{P_1} = 0$$

Putting these into $y'' - 4y = 2x$ to get the values of A_0 and A_1, we have

$$0 - 4(A_0 + A_1x) = 2x$$

Therefore $-4A_0 = 0$ and $-4A_1 = 2$ or $A_0 = 0$ and $A_1 = -\frac{1}{2}$. Then

$$y_{P_1} = -\frac{1}{2}x$$

Now we let $y_{P_2} = Ae^{ix}$, and we must remember to retain only the real part of this result since

$$f_2(x) = 5\cos x = \text{real part of } 5e^{ix}$$

Then
$$y'_{P_2} = Aie^{ix}$$
and
$$y''_{P_2} = Ai^2e^{ix} = -Ae^{ix}$$

We will substitute these into $y'' - 4y = 5e^{ix}$, producing

$$-Ae^{ix} - 4(Ae^{ix}) = 5e^{ix}$$

$$\therefore \quad -5A = 5$$

or

$$A = -1$$

$$\therefore \quad y_{P_2} = -1e^{ix} = -1(\cos x + i\sin x)$$

$$= -\cos x + i(-\sin x)$$

The real part of this, which we retain as part of our solution, is $-\cos x$. Therefore,

$$y = y_h + y_{P_1} + y_{P_2} = c_1e^{2x} + c_2e^{-2x} - \frac{1}{2}x - \cos x$$

Exercises 3.9

1. If you were asked to solve the following differential equations, what complex-valued function would you use to find the particular solution?
   (a) $y'' - 5y' + 4y = 3\cos 2x$ (b) $y'' - 2y' + y = 5\sin 4x$
   (c) $2y'' + 9y' - 5y = 4\cos 6x$
2. For the given complex function, find the real part and the imaginary part.
   (a) $y = (-3 + 5i)e^{-4ix}$ (b) $y = 12ie^{5ix}$
   (c) $y = (-2 + 5i)e^{-2ix}$ (d) $y = (-4 + 2i)e^{3x}$
   (e) $y = 5e^{-7ix}$ (f) $y = (-7 + 9i)e^{(-8i+2)x}$
3. Solve the complex equation $y'' - y' - 12y = e^{ix}$.
4. Use the results of Exercise 3 to solve $y'' - y' - 12y = \sin x$.
5. (a) Solve the differential equation $y'' + 2y' - 15y = 4\cos x$.
   (b) Solve the differential equation $y'' + 2y' - 15y = 4\sin x$.

In Exercises 6–19, solve the given differential equation.

6. $y'' + y' - 6y = 5\sin 4x$
7. $y'' + 5y' + 6y = 16\cos x$
8. $y'' - 9y = 9\sin 5x$
9. $2y'' + 7y' - 4y = \cos 2x$
10. $y'' - 4y' + 3y = 2(\cos x + \sin x)$

11. $3y'' - y' - 2y = x + 4\cos x$
12. $2y'' - 3y' - 2y = \sin x + \cos x$
13. $4y'' + 4y' + y = x + 3\sin 3x$
14. $y'' - y = 3 + 2x + 4\sin 2x$
15. $y'' + y = -\frac{1}{2}x + e^{2x} + 4\cos 2x$
16. $y'' + 4y' + 4y = 24 + 2\sin 3x + 2\cos 2x$
17. $9y'' + 12y' + 4y = \alpha + \beta e^{2x} + \gamma\cos 5x$ for any real α, β, γ.
18. $y'' - 5y' + 6y = x + \sin\left(x + \frac{\pi}{2}\right)$
19. $y'' - y' - 12y = -3 + \sin\left(x + \frac{\pi}{4}\right)$

3.10 The Method of Undetermined Coefficients

The summaries at the end of Section 3.9 gave explicit solution techniques for a variety of cases. Those were the simple cases, and the techniques were also simple. To cover the wealth of cases that remain, Section 3.10 proposes one sweeping theorem, which—if carefully studied and digested—will provide the procedures needed for many new types of problems.

The theorem involves the construction of a trial solution to be used for finding the needed particular solution y_P to the nonhomogeneous equation. This trial solution is then differentiated, and the solution and its derivatives are substituted back into the original differential equation. By analyzing this equation we determine the unknown coefficients in the trial solution. When this step has been accomplished, we are able to obtain our final particular solution.

At this point you should feel confident of your ability to find the general solution to a homogeneous linear equation with constant coefficients. Thus, a general method is now needed to find the particular solution required to finish the process of solving a nonhomogeneous linear equation with constant coefficients.

The following theorem is a comprehensive statement covering a large variety of problems. At first you may be perplexed because of its length and coverage; but the theorem is quite understandable, and a careful study of the examples that follow will help you to master it.

Theorem 3.11 (The Sweeping Theorem):

Consider the linear nth-order nonhomogeneous differential equation $Ly = f(x)$ *where* $f(x)$ *takes on the form:*

(a) *a polynomial;*

(b) *an exponential function;*

(c) *a sum, difference, or product of sine or cosine functions; or*
(d) *a sum, difference, or product of any of the above.*

A valid trial solution for a particular solution to $Ly = f(x)$ *takes the form:*

$$y_P = (A_0 + A_1x + A_2x^2 + \cdots + A_nx^n)x^j e^{cx}$$

where e^{cx} *matches the exponential in* f (c *is zero if no exponential exists and is complex if sines and/or cosines are present),* n *is the degree of the polynomial portion of* f *(if it exists), and* j *is the multiplicity of* c *as a root of the auxiliary polynomial equation for* L.

Listed below are some examples of problems that are in the category of Theorem 3.11 and that can therefore now be solved:

$$Ly = xe^{2x}$$

$$Ly = (3 + 2x)\sin 5x$$

$$Ly = 2e^{4x} - \cos 2x \sin 7x$$

$$Ly = 4x^2e^x \cos 3x$$

Some of the problems which *cannot* be solved under Theorem 3.11 include $Ly = \tan x$ (a quotient of trigonometric functions) and $Ly = \frac{1}{x}e^x$ (a quotient of an exponential and a polynomial). Later we will discuss ways of solving even these.

One of the more difficult parts to determine in obtaining the trial solution is the value of j. Example 58 illustrates the process.

Example 58

Let $Ly = y'' - 5y' + 6y = 2e^{3x}$. Then $P(r) = r^2 - 5r + 6 = 0$, and the roots are $r_1 = 2$ and $r_2 = 3$. The value of c, which is the coefficient of x in the exponential part of $f(x) = 2e^{3x}$, is 3. In the list of roots of $P(r) = 0$, the value 3 appears once. Hence, the value of j (the multiplicity of 3 as a root) is 1.

Now let's look at a problem that will produce a different value for j. Let $y'' - 6y' + 9y = 4e^{3x}$. Then $P(r) = r^2 - 6r + 9 = 0$, and the roots are $r_1 = r_2 = 3$. Now from $f(x) = 4e^{3x}$ we have $c = 3$ and it is a root of multiplicity 2, so j has value 2 in this example.

The value of j might also be zero, as illustrated in the following problem. If we let $y'' - 5y' + 4y = 5e^{3x}$, we find $j = 0$ since $c = 3$ is not the value of either of the roots $r_1 = 4$ or $r_2 = 1$.

We need to include the part x^j in the trial solution only when we find that the coefficient of x in the exponential term is in fact a root of $P(r) = 0$. The reason for putting this term in will be made clear shortly.

The remaining parts of the trial solution are more easily determined. The value n of the polynomial portion $A_0 + A_1x + A_2x^2 + \cdots + A_nx^n$ of the trial solution matches the degree of the polynomial part of the given $f(x)$.

The exponential portion e^{cx} is inserted to match whatever exponential part appears in $f(x)$. Two important aspects must be remembered:

1. When no real exponential part exists in f, we assign the value of zero to c and mentally insert e^{0x} into the trial solution.
2. When f contains $\cos \beta x$ or $\sin \beta x$, we insert $e^{i\beta x}$ and use the techniques developed in Section 3.9 to find the particular solution.

The best way to solidify these ideas is practice. In the following examples we will first formulate only the trial solution and then work out a few problems in detail.

Example 59

Consider $y'' - 2y' + y = 2e^{2x}$. Let's analyze the function f: no polynomial part, hence $n = 0$; no sine or cosine part, but an exponential function e^{2x}, hence c is 2. The roots of $P(r) = 0$ are $r_1 = r_2 = 1$, hence $j = 0$. Therefore

$$y_p = A_0 e^{2x}$$

Example 60

Consider $y'' - 2y' + y = 2e^x$. Analyzing f finds: no polynomial part, hence $n = 0$; no sine or cosine part, hence $c = 1$. The roots of $P(r) = 0$ are $r_1 = r_2 = 1$, hence $j = 2$ (since the value of c is in the list twice). Therefore

$$y_p = A_0 x^2 e^x$$

Example 61

Consider $y'' - 3y' + 2y = 2xe^x$. Analyzing f finds: polynomial part is $2x$, which is first degree, hence $n = 1$; no sine or cosine part, hence $c = 1$. The roots of $P(r) = 0$ are $r_1 = 2$ and $r_2 = 1$, hence $j = 1$ (since the value of c appears once in the collection of roots). Therefore

$$y_P = (A_0 + A_1 x)x^1 e^x = (A_0 x + A_1 x^2)e^x$$

Example 62

Consider $y'' - 3y' + 2y = 2e^x \cos 2x$. Analyzing f finds: no polynomial part, hence $n = 0$; cosine part implies that e^{2ix} must be inserted as well as e^x, hence $c = 1 + 2i$ (since the combination of e^x times e^{2ix} produces $e^{(1+2i)x}$). The roots of $P(r) = 0$ are $r_1 = 2$ and $r_2 = 1$, hence $j = 0$. Therefore

$$y_P = A_0 e^{(1+2i)x}$$

Example 63

Consider $y'' - 2y' + 5y = 3xe^x \sin 2x$. Analyzing f finds: polynomial part is first degree, hence $n = 1$; sine part implies that e^{2ix} must be inserted as well as e^x, hence $c = 1 + 2i$. The roots of $P(r) = r^2 - 2r + 5 = 0$ are $r_1 = 1 + 2i$ and $r_2 = 1 - 2i$, hence $j = 1$. Therefore

$$y_P = (A_0 + A_1 x)x^1 e^{(1+2i)x} = (A_0 x + A_1 x^2)e^{(1+2i)x}$$

We continue the illustrations but will now include more detail.

Example 64

Suppose an L-R-C electrical circuit in an optical scanning device requires you to solve the differential equation $y'' - 5y' + 4y = 3xe^{2x}$. The polynomial part of $f(x)$ is $3x$; the exponential part is e^{2x}; no trigonometric part exists. We must set $n = 1$ and $c = 2$. The roots of the auxiliary polynomial equation are 4 and 1, so c is not in this list, hence $j = 0$. Therefore

$$y_P = (A_0 + A_1 x)e^{2x}$$
$$y_P' = 2(A_0 + A_1 x)e^{2x} + A_1 e^{2x}$$
$$y_P'' = [(4A_0 + 4A_1) + 4A_1 x]e^{2x}$$

Thus the differential equation becomes

$$[(4A_0 + 4A_1) + (4A_1 x)]e^{2x} - 5(2A_0 + A_1 + 2A_1 x)e^{2x} + 4(A_0 + A_1 x)e^{2x} = 3xe^{2x}$$

$$(4A_1 - 10A_1 + 4A_1)x + 4A_0 + 4A_1 - 10A_0 - 5A_1 + 4A_0 = 3x$$

or $$(-2A_1)x + (-2A_0 - A_1) = 3x$$

By equating coefficients of like terms on each side, we find

$$-2A_1 = 3$$

and $$-2A_0 - A_1 = 0$$

Therefore $$A_0 = \frac{3}{4}$$

and $$A_1 = -\frac{3}{2}$$

The undetermined coefficients A_0 and A_1 are now determined. Therefore the final $y_P = (\frac{3}{4} - \frac{3}{2}x)e^{2x}$. The general solution to the original equation may now be formed:

$$y(x) = y_h + y_p = (C_1 e^{4x} + C_2 e^{x}) + \left(\frac{3}{4}e^{2x} - \frac{3}{2}xe^{2x}\right)$$

Example 65

Consider $y'' - y = 5\cos 2x$. An analysis of the function $f(x)$ dictates that $n = 0$ and $c = 2i$. The roots are $r_1 = 1$ and $r_2 = -1$, so $j = 0$. Thus the trial solution is

$$y_P = A_0 e^{2ix}$$

Now $$y_P' = 2iA_0 e^{2ix}$$

and $$y_P'' = -4A_0 e^{2ix}$$

so the transformed differential equation $y'' - y = 5e^{2ix}$ becomes

$$-4A_0e^{2ix} - A_0e^{2ix} = 5e^{2ix}$$

Therefore $-5A_0 = 5$ or $A_0 = -1$. Substituting this value, we get $y_P = -1e^{2ix}$. Since our original problem contained the real part of e^{2ix} on its right side, we use only the real part of this result and hence obtain $y_P = -1(\cos 2x)$. We then add the homogeneous solution $y_h = C_1e^x + C_2e^{-x}$ to obtain the general solution:

$$y = y_h + y_P = C_1e^x + C_2e^{-x} - \cos 2x$$

The next example indicates the justification for including the x^j term in the trial solution.

Example 66

Consider $y'' - 2y' + y = 4e^x$. The analysis of $f(x) = 4e^x$ clearly shows that $n = 0$ and $c = 1$. We also see that the roots of $P(r) = 0$ are $r_1 = r_2 = 1$. Suppose we are unaware of the need for $j = 2$ in this problem, use the values $n = 0$ and $c = 1$, and create the incorrect $y_P = A_0e^x$. Then $y_P' = A_0e^x$ and $y_P'' = A_0e^x$, and the differential equation becomes $A_0e^x - 2A_0e^x + A_0e^x = 4e^x$ or $0 = 4$.

The inconsistency arose because the trial y_P was a solution to the homogeneous equation and thus could not solve the given nonhomogeneous equation. Using A_0xe^x would not suffice, because it also is a solution to $Ly = 0$. (This is transparent when we write $y_h = C_1e^x + C_2xe^x$). We need a y_P function that is linearly independent from y_h and has the simplest, yet correct, form. Thus we are led to $y_P = A_0x^2e^x$. This is exactly the y_P that our analysis predicted; it has $j = 2$ as was previously indicated. With this corrected y_P we have $y_P' = A_0(x^2 + 2x)e^x$ and $y_P'' = A_0(x^2 + 4x + 2)e^x$. Now the differential equation becomes

$$A_0(x^2 + 4x + 2)e^x - 2A_0(x^2 + 2x)e^x + A_0x^2e^x = 4e^x$$

Cancelling the exponential and rearranging terms produces

$$(A_0 - 2A_0 + A_0)x^2 + (4A_0 - 4A_0)x + 2A_0 = 4$$

or

$$A_0 = 2$$

Therefore the final $y_P = 2x^2e^x$. The general solution to the original equation is now

$$y(x) = (C_1e^x + C_2xe^x) + 2x^2e^x$$

At this point one very important note is in order. The principle of superposition plays a major role in formulating the general solution. In the examples we have just studied, the function on the right consisted of a single term. If more than one term appears, we apply the principle and work separate problems for each of the terms. This procedure also allows us to sort out a more complicated combination of polynomials, exponentials, and trigonometric functions and write it as a sum of terms whose forms fit the criteria of the sweeping theorem. Listed below are some examples of nonhomogeneous linear equations whose right-side functions require such analysis.

1. $Ly = (2x^2 + 3)e^x$ should be written as $Ly_1 = 2x^2e^x$ and $Ly_2 = 3e^x$.
2. $Ly = 3\cos^2 2x$ should be written as $Ly = 3(\frac{1}{2} + \frac{1}{2}\cos 4x)$ and then as $Ly_1 = \frac{3}{2}$ and $Ly_2 = \frac{3}{2}\cos 4x$.
3. $Ly = e^{2x}\sin^3 x$ should be written as $Ly = e^{2x}\sin x(1 - \cos^2 x)$ and then as

$$Ly = e^{2x}\sin x - e^{2x}\sin x\left(\frac{1}{2} + \frac{1}{2}\cos 2x\right)$$

$$= e^{2x}\sin x - \frac{1}{2}e^{2x}\sin x - \frac{1}{2}e^{2x}(\sin x\cos 2x)$$

and then as

$$Ly_1 = \frac{e^{2x}\sin x}{2}$$

and

$$Ly_2 = -\frac{1}{2}e^{2x}(\sin x\cos 2x)$$

$$= -\frac{1}{2}e^{2x}\left(\frac{1}{2}\sin 3x - \frac{1}{2}\sin x\right)$$

4. $Ly = 4\sin(3x)\cos(2x)$ should be written as $Ly = 4(\frac{1}{2}\sin 5x + \frac{1}{2}\sin x)$ and then as $Ly_1 = 2\sin 5x$ and $Ly_2 = 2\sin x$.

The preceding illustrations use the following trigonometric identities, which may not be familiar to you.

$$\sin A\cos B = \frac{1}{2}\sin(A + B) + \frac{1}{2}\sin(A - B)$$

$$\sin A\sin B = \frac{1}{2}\cos(A - B) - \frac{1}{2}\cos(A + B)$$

$$\cos A\cos B = \frac{1}{2}\cos(A - B) + \frac{1}{2}\cos(A + B).$$

The great value of equations (7) is that they allow us to change a product into a sum.

Example 67 illustrates the use of one of these identities in solving a differential equation.

Example 67

Consider the second order nonhomogeneous equation

$$y'' - y' - 6y = 2\cos 3x \cos 2x$$

Applying the appropriate trigonometric identity from above, we change the right side to $\cos x + \cos 5x$. The principle of superposition is now applied as we work two separate problems:

$$y'' - y' - 6y = \cos x = Re(e^{ix}) \tag{8}$$

and

$$y'' - y' - 6y = \cos 5x = Re(e^{5ix}) \tag{9}$$

where Re(function) denotes the real part of the function.

The homogeneous solution y_h is easy to find since the auxiliary polynomial equation is $r^2 - r - 6 = (r-3)(r+2) = 0$. Therefore we have $y_h = c_1 e^{3x} + c_2 e^{-2x}$. Working on the particular solution to equation (8), we formulate our trial solution, which we will designate as y_{P_8},

$$y_{P_8} = A_0 e^{ix}$$

For this function

$$y'_{P_8} = iA_0 e^{ix}$$

and

$$y''_{P_8} = -1A_0 e^{ix}$$

so the equation (8) becomes $-A_0 e^{ix} - iA_0 e^{ix} - 6A_0 e^{ix} = e^{ix}$. This is solved for A_0 to find

$$A_0 = \frac{-1}{7+i} = \frac{-1(7-i)}{50}$$

Thus our trial solution is

$$y_{P_8} = \left(-\frac{7}{50} + \frac{1}{50}i\right)e^{ix} = \left(-\frac{7}{50} + \frac{1}{50}i\right)(\cos x + i\sin x)$$

$$= \left(-\frac{7}{50}\cos x - \frac{1}{50}\sin x\right) + i\left(\frac{1}{50}\cos x - \frac{7}{50}\sin x\right)$$

We use only the real part of this result (since equation (8) had the real part of e^{ix} as its right side). Therefore the final y_{P_8} is

$$y_{P_8} = -\frac{7}{50}\cos x - \frac{1}{50}\sin x$$

Working now on the particular solution to equation 9, we formulate our trial solution

$$y_{P_9} = B_0 e^{5ix}$$

For this function

$$y'_{P_9} = 5iB_0 e^{5ix}$$

and
$$y''_{P_9} = -25B_0e^{5ix}$$

so the equation (9) becomes

$$-25B_0e^{5ix} - 5iB_0e^{5ix} - 6B_0e^{5ix} = e^{5ix}$$

This is solved for B_0 to produce

$$B_0 = \frac{-1}{31+5i} = \frac{-1(31-5i)}{(31+5i)(31+5i)} = \frac{-31+5i}{31^2+25} = \frac{-31+5i}{986}$$

Thus our trial solution is

$$y_{P_9} = \left(\frac{-31}{986} + \frac{5}{986}i\right)e^{5ix} = \left(\frac{-31}{986} + \frac{5}{986}i\right)(\cos 5x + i\sin 5x)$$

$$= \left(-\frac{31}{986}\cos 5x - \frac{5}{986}\sin 5x\right) + i\left(\frac{5}{986}\cos 5x - \frac{31}{986}\sin 5x\right)$$

Again we use only the real part of this trial solution:

$$y_{P_9} = -\frac{31}{986}\cos 5x - \frac{5}{986}\sin 5x$$

The total problem now has as its solution the sum of the homogeneous solution and the two particular solutions:

$$y(x) = C_1e^{3x} + C_2e^{-2x} - \frac{7}{50}\cos x - \frac{1}{50}\sin x - \frac{31}{986}\cos 5x - \frac{5}{986}\sin 5x$$

If initial conditions had been posed, we would now use them to calculate the values for C_1 and C_2.

Example 68

In a problem of motion of a particle under an input forcing function $f(t)$, we get the differential equation

$$\frac{d^2x}{dt^2} + \omega_0^2 x = \frac{1}{m}f(t)$$

This is the nonhomogeneous counterpart to simple harmonic motion described earlier and is called forced undamped motion. If the input function is of the form $f(t) = mF\sin\omega t$, a comparison must be made of the natural frequency ω_0 and the forcing frequency ω.

To solve the differential equation above with the given $f(t)$, we will assume that $\omega \neq \omega_0$ and that ω_0/ω is a rational number. Then we can show that the solution provides

$$x_h = c_1\cos\omega_0 t + c_2\sin\omega_0 t$$

and
$$x_P = \frac{F}{\omega_0^2 - \omega^2}\sin\omega t$$

This is stable periodic motion since the solution is bounded. If $\omega = \omega_0$ we get unstable motion as

$$x_h = c_1 \cos \omega_0 t + c_2 \sin \omega_0 t$$

and
$$x_p = \frac{-F}{2\omega_0} t \cos \omega_0 t$$

and the x_p term gets unboundedly large as $t \to \infty$.

Example 69

If the motion of a system such as the one in Example 68 takes place in a medium where damping resistance occurs, the differential equation takes the form

$$\frac{d^2x}{dt^2} + 2\alpha \frac{dx}{dt} + \omega_0^2 x = \frac{1}{m} f(t)$$

where α is the coefficient of resistance of the system.

If the shock absorber on an automobile spring has an α value of 5 and the natural frequency of the spring is 4 with a forcing function due to a bumpy road $f(t) = 4m \sin 3t$, we obtain the motion differential equation

$$x'' + 10x' + 16x = 4 \sin 3t$$

Reasonable initial conditions are $x(0) = 1$ and $x'(0) = 0$. Solving the differential equation first, we find the auxiliary polynomial equation to be

$$(r + 8)(r + 2) = 0$$

so
$$x_h = C_1 e^{-8t} + C_2 e^{-2t}$$

Using the procedure of this section, we let

$$x_P = Ae^{3it}$$

Then
$$x_P' = 3iAe^{3it}$$

$$x_p'' = -9Ae^{3it}$$

so the transformed differential equation $x'' + 10x' + 16x = 4e^{3it}$ becomes

$$[-9A + 10(3iA) + 16A]e^{3it} = 4e^{3it}$$

or
$$(7 + 30i)A = 4$$

Hence
$$A = \frac{4}{7 + 30i} \cdot \frac{7 - 30i}{7 - 30i} = \frac{28 - 120i}{949}$$

The trial x_P becomes

$$x_P = \left(\frac{28}{949} - \frac{120}{949} i\right) e^{3it} = \left(\frac{28}{949} - \frac{120}{949} i\right)(\cos 3t + i \sin 3t)$$

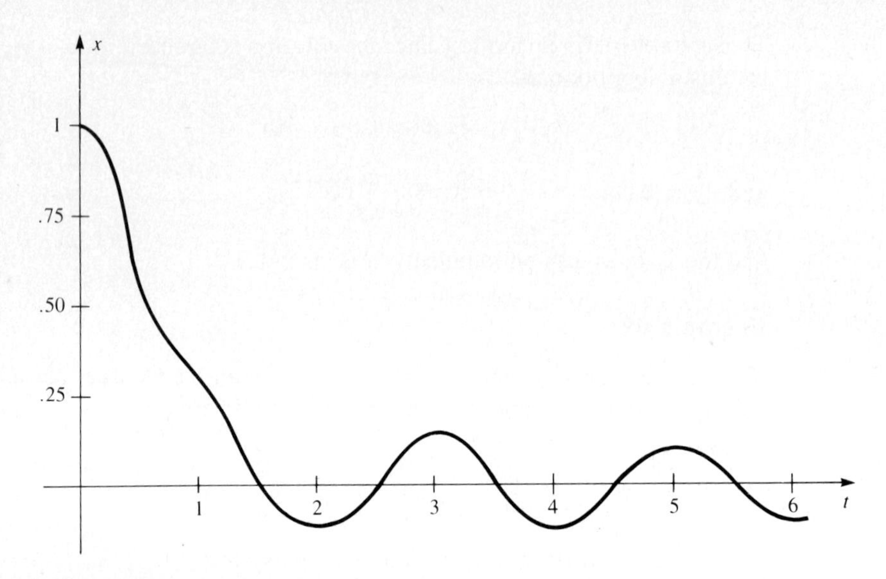

Figure 3.5

Since we are to retain only the imaginary part of this particular solution, we get

$$x_p = -\frac{120}{949}\cos 3t + \frac{28}{949}\sin 3t$$

The general solution to the original differential equation is

$$\begin{aligned} x(t) &= x_h + x_p \\ &= C_1 e^{-8t} + C_2 e^{-2t} - \frac{120}{949}\cos 3t + \frac{28}{949}\sin 3t \end{aligned}$$

Using the given initial conditions, we find that the values of C_1 and C_2 give

$$x(t) = -0.35e^{-8t} + 1.47e^{-2t} - 0.126\cos 3t + 0.029\sin 3t$$

Figure 3.5 shows the graph of this equation. Note that the homogeneous-solution part is decaying exponentially and the particular-solution part is bounded and oscillatory.

Exercises 3.10

Find the general solution to the differential equations in Exercises 1–23.

1. $y'' + y' - 6y = 2e^{4x}$
2. $y'' + 2y' + y = e^{-2x}$
3. $y'' - 3y' + 2y = x^2 - 2e^x$

4. $y'' - y' - 6y = x^3$
5. $y'' + 3y' + 5y = 2\sin 3x$
6. $y'' - y' - 6y = 3e^{-2x}$
7. $y'' - 3y' + 3y = xe^x$
8. $(D^2 - 2D - 1)y = 3\sin x$
9. $(D^2 + 2D)y = -4$
10. $y'' + 4y - 2xe^x = 0$
11. $y'' + 4y = \sin 2x \cos 3x$
12. $y'' + 4y = 6x^2 - 16x\cos 2x$
13. $y^{(iv)} - y = 2e^x$
14. $y'' = 4y + 12x^2e^{-2x}$
15. $y'' + y = 4x\sin x + \cos 2x$
16. $y''' - y'' = 5x^3 + e^x$
17. $y''' - 4y'' = x^2 + 8$
18. $y''' - 4y'' + y' - 4y = e^{4x}\sin x$
19. $y'' + y = 3x\sin x$; $y(0) = 2$, $y'(0) = 1$
20. $y'' + 5y' + 4y = 2e^x$; $y(0) = 2$, $y'(0) = 0$
21. $y'' - y' - 12y = 2x$; $y(0) = -1$, $y'(0) = 2$
22. $y'' - 2y' - 8y = 3e^{4x}$; $y(0) = 0$, $y'(0) = 3$
23. $y'' - 10y' + 25y = x\sin x$; $y(0) = 3$, $y'(0) = -2$
24. Determine x so that it will satisfy the differential equation

$$\frac{d^2x}{dt^2} - 3\frac{dx}{dt} + 2x = e^{3t}$$

and will vanish when $t = 0$ and when $t = \ln 2$.

25. Find a solution of the differential equation $y'' = 4y' - x$ if $y = \frac{1}{4}$ and $y' = 0$ when $x = 0$.
26. If the equation of motion of a mechanical system is $x'' + 2x = 8\sin t$, find the general solution. Is the motion stable or unstable?
27. A 6-pound body stretches a given spring 2 inches. If a forcing function $f(t) = \frac{3}{16}\sin(8\sqrt{6}\,t)$ is attached to the system, find the motion equation and determine whether the motion is stable or unstable. [*Hint:* Use $mx'' + kx = f(t)$ where k is the spring constant.]
28. The damped motion of a certain system has an α value of 2.5, and $\omega_0 = \sqrt{6}$ with $f(t) = m(e^{-t}\sin 2t)$. Solve for $x(t)$ and describe the motion when t (time) is large. Use $x(0) = 0$, $x'(0) = 1$.

3.11 Equidimensional Equations

In the preceding sections we learned techniques for solving linear differential equations—both homogeneous, where $Ly = 0$, and nonhomogeneous, where $Ly = f(x)$. In every case the unknown function $y(x)$ and all of its derivatives had constant coefficients, and the homogeneous solution was relatively easy to find.

If we consider the general linear equation $Ly = f(x)$ with variable coefficients, we have a much more difficult case; and only in special cases can we find the general solution by analytic methods. In Chapters 4 and 5 we will see how easy it is to get very accurate numerical solutions to these equations. However, our present knowledge is sufficient to generate the general solution for equidimensional (or Cauchy-Euler) equations, which contain—as we shall see in the following definition—variable coefficients.

An **equidimensional differential equation** is defined to be an equation of the form

$$b_0 x^n y^{(n)} + b_1 x^{n-1} y^{(n-1)} + \cdots + b_{n-2} x^2 y'' + b_{n-1} x y' + b_n y = f(x) \tag{10}$$

where b_i are constants. The distinguishing feature of this equation is that each term on the left side has as its coefficient a constant times a power of x equal to the order of the derivative.

The method that solves our dilemma with a problem of this type is a familiar procedure: we change the variables. In equation (10), y is the dependent variable and x is the independent variable. We will make the change of independent variable by letting $x = e^t$ or $t = \ln x$ where we will assume $x > 0$. As we will see, this transformation will change the equidimensional equation to a constant coefficient equation in y and t, which we can solve by previous methods. A reverse change of variables using $t = \ln x$ will give us the general solution to the original variable-coefficient equidimensional differential equation.

As before, both homogeneous and nonhomogeneous equations will be considered, but we will limit our attention to second-order equations for the present discussion. The higher-order case will be solved by an easy extension of this method.

The change-of-variable equation is $x = e^t$ or $t = \ln x$, and by the chain rule for differentiation

$$\frac{dy}{dx} = \frac{dy}{dt} \cdot \frac{dt}{dx} = \frac{dy}{dt} \cdot \frac{1}{x} \tag{11}$$

Also

$$\begin{aligned}
\frac{d^2y}{dx^2} &= \frac{d}{dx}\left(\frac{dy}{dx}\right) = \frac{d}{dx}\left(\frac{1}{x} \cdot \frac{dy}{dt}\right) = \frac{1}{x} \cdot \frac{d}{dx}\left(\frac{dy}{dt}\right) + \left(\frac{dy}{dt}\right)\frac{d}{dx}\left(\frac{1}{x}\right) \\
&= \frac{1}{x}\left(\frac{d}{dt} \cdot \frac{dt}{dx}\right)\frac{dy}{dt} + \left(\frac{dy}{dt}\right)\left(-\frac{1}{x^2}\right) \\
&= \frac{1}{x}\left(\frac{d}{dt} \cdot \frac{1}{x}\right)\frac{dy}{dt} - \frac{1}{x^2}\frac{dy}{dt} = \frac{1}{x^2}\frac{d^2y}{dt^2} - \frac{1}{x^2}\frac{dy}{dt} = \frac{1}{x^2}\left(\frac{d^2y}{dt^2} - \frac{dy}{dt}\right)
\end{aligned} \tag{12}$$

Hence from (11):

$$x\frac{dy}{dx} \equiv xy' = \frac{dy}{dt} \tag{13}$$

Using the operator notation we may also write this as

$$xy' = Dy(t)$$

Also from (12):

$$x^2\frac{d^2y}{dx^2} \equiv x^2y'' = \frac{d^2y}{dt^2} - \frac{dy}{dt} \tag{14}$$

This may be written

$$x^2y'' = (D^2 - D)y(t)$$

These formulas provide the means of removing the variable coefficient portion of the equidimensional equation. If we substitute these into the general second-order equidimensional equation

$$b_0x^2y'' + b_1xy' + b_2y = f(x)$$

we get
$$b_0\left(\frac{d^2y}{dt^2} - \frac{dy}{dt}\right) + b_1\left(\frac{dy}{dt}\right) + b_2y = f(e^t)$$

This may be reorganized into

$$b_0\frac{d^2y}{dt^2} + (b_1 - b_0)\frac{dy}{dt} + b_2y = f(e^t)$$

which is now solvable by earlier methods since it is a constant-coefficient equation. In this problem the independent variable t will vary over all the real numbers; but because $x = e^t$, the original independent variable x will vary over the interval $(0, \infty)$. This was the underlying reason for our original assumption of $x > 0$. If x were less than zero, we would substitute $x = -e^t$, and minor changes would occur in the equations. We will assume for our examples that $x > 0$.

Example 70

Consider $x^2y'' - 3xy' + 3y = 0$. Let $x = e^t$ or $t = \ln x$. Using equations (13) and (14), we change the differential equation into

$$\left(\frac{d^2y}{dt^2} - \frac{dy}{dt}\right) - 3\left(\frac{dy}{dt}\right) + 3y = 0$$

or
$$\frac{d^2y}{dt^2} - 4\frac{dy}{dt} + 3y = 0$$

The general solution to this transformed equation is easily seen to be

$$y(t) = C_1e^{3t} + C_2e^t$$

The reverse change of variables, using $t = \ln x$ and then the identity $e^{\ln u} = u$, gives

$$y(x) = C_1e^{3\ln x} + C_2e^{\ln x} = C_1e^{\ln x^3} + C_2e^{\ln x} = C_1x^3 + C_2x$$

This is easily verified as the correct general solution.

Example 71

Consider $x^2y'' + 2xy' - 2y = x^2$. Using the transformation $x = e^t$ we obtain

$$\left(\frac{d^2y}{dt^2} - \frac{dy}{dt}\right) + 2\left(\frac{dy}{dt}\right) - 2y = (e^t)^2 = e^{2t}$$

or

$$\frac{d^2y}{dt^2} + \frac{dy}{dt} - 2y = e^{2t}$$

This nonhomogeneous equation is now solved by the method of undetermined coefficients of the previous section. The auxiliary polynomial equation is $r^2 + r - 2 = 0$ whose roots are $r_1 = -2$ and $r_2 = 1$. The right-side function is e^{2t}; we have no polynomial part, so $n = 0$; and since the exponential part indicates that $c = 2$, we know that $j = 0$. Thus $y_P = Ae^{2t}$; $y'_P = 2Ae^{2t}$; and $y''_P = 4Ae^{2t}$. Therefore the new differential equation becomes

$$4Ae^{2t} + 2Ae^{2t} - 2(Ae^{2t}) = e^{2t}$$

Then $A = \frac{1}{4}$, and the particular solution $y_P = \frac{1}{4}e^{2t}$ is added to the homogeneous solution $y_h = C_1e^{-2t} + C_2e^t$ to provide the general solution. However, we must reverse substitute, so

$$y(x) = C_1e^{-2\ln x} + C_2e^{\ln x} + \frac{1}{4}e^{2\ln x} = C_1x^{-2} + C_2x + \frac{1}{4}x^2$$

Remember that the general solution to any differential equation may be verified by substituting it and its derivatives into the differential equation. Before substituting in this problem, we determine that

$$y'(x) = -2c_1x^{-3} + c_2 + \frac{1}{2}x$$

and

$$y''(x) = 6c_1x^{-4} + \frac{1}{2}$$

Then

$$x^2y'' + 2xy' - 2y = \left(6c_1x^{-2} + \frac{1}{2}x^2\right) + (-4c_1x^{-2} + 2c_2x + x^2)$$

$$+\left(-2c_1x^{-2} - 2c_2x - \frac{1}{2}x^2\right) = x^2$$

This verifies that the solution is correct.

In working with differential equations you will notice that the solution to an equation or an initial value problem can always be checked for accuracy. It is a rare branch of mathematics that allows this latitude, so take advantage of this luxury.

In order to apply the method of this chapter to higher-order equations, we carry the chain-rule derivative shown earlier to one higher order. This produces

$$\frac{d^3y}{dx^3} = \frac{1}{x^3}\left(\frac{d^3y}{dt^3} - 3\frac{d^2y}{dt^2} + 2\frac{dy}{dt}\right)$$

Using the operator notation, we find a shorter way of writing this identity:

$$x^3y''' = D^3y(t) - 3D^2y(t) + 2Dy(t) = [D(D-1)(D-2)]y(t)$$

The earlier identities provided

$$x^2y'' = [D(D-1)]y(t)$$

and

$$xy' = Dy(t)$$

The process of mathematical induction could now be employed to prove that

$$x^ny^{(n)} = \{D[D-1][D-2][D-3]\cdots[D-(n-1)]\}y(t)$$

Example 72

Consider $x^3y''' - x^2y'' - 6xy' + 18y = 0$. The variable change produces

$$[D(D-1)(D-2)]y(t) - [D(D-1)]y(t) - 6Dy(t) + 18y(t) = 0$$

or

$$[D^3 - 3D^2 + 2D - (D^2 - D) - 6D + 18]y(t) = 0$$

or

$$(D^3 - 4D^2 - 3D + 18)y(t) = 0$$

Factoring the operator into $(D-3)(D-3)(D+2)$ displays the auxiliary polynomial equation as $(r-3)^2(r+2) = 0$, and its roots are $r_1 = r_2 = 3$ and $r_3 = -2$. Thus the solution of the $y(t)$ equation is

$$y(t) = C_1e^{3t} + C_2te^{3t} + C_3e^{-2t}$$

The reverse change now gives the general solution

$$y(x) = C_1x^3 + C_2(\ln x)x^3 + C_3x^{-2}$$

If the equidimensional equation is nonhomogeneous, the transformation $x = e^t$ applied to the right-side function may produce a new function that cannot be handled by the theorem for undetermined coefficients studied in Section 3.10. This case will be covered by a new method in Section 3.12.

In anticipation of the new technique, we recognize the need to obtain two linearly independent solutions to the homogeneous second-order equation $a_0(x)y'' + a_1(x)y' + a_2(x)y = 0$. Our earlier work showed the ease of obtaining these for the constant-coefficient case, and the work just concluded helps in the equidimensional case.

One further situation deserves mention. If one nontrivial solution to the homogeneous equation can be found, the problem of finding another can be accomplished by a change of variables. In this situation we reduce the order of the equation (e.g., from second order to first order) and then solve the equation by methods of Chapter 1. This technique can be illustrated as follows:

Suppose the known nontrivial solution of $a_0(x)y'' + a_1(x)y' + a_2(x)y = 0$ is $u_1(x)$. This solution might have been obtained by a variety of methods. (As we observed in some earlier problems, a good guess is often helpful. Usually good guesses require experience.) In this method we introduce a new dependent variable $v(x)$ by letting $y(x) = v(x)\cdot u_1(x)$. Then $y'(x) = vu_1' + v'u_1$, and $y''(x) = vu_1'' + 2v'u_1' + v''u_1$. The differential equation then becomes

$$a_0(vu_1'' + 2v'u_1' + v''u_1) + a_1(vu_1' + v'u_1) + a_2vu_1 = 0$$

Rearranging this produces

$$a_0v''u_1 + (2a_0u_1' + a_1u_1)v' + (a_0u_1'' + a_1u_1' + a_2u_1)v = 0$$

Since u_1 is a known nontrivial solution of the homogeneous equation, the coefficient of v is zero. Therefore we get

$$a_0v''u_1 + (2a_0u_1' + a_1u_1)v' = 0$$

We now make another variable change $v' = z$ to reduce the order. This implies $v'' = z'$, so the differential equation becomes

$$a_0u_1z' + (2a_0u_1' + a_1u_1)z = 0$$

This is solved for z by separating the variables and integrating. When we find the new value of z we can let $z = v'$ and integrate again to obtain $v(x)$. Inserting this into $y(x) = v(x)u_1(x)$ produces the general solution to $Ly = 0$. The two parameters enter via the two integrations just performed. We can now easily identify the two linearly independent solutions to $Ly = 0$.

Example 73

Consider $x^3y'' + 2xy' - 2y = 0$. Bearing in mind that the coefficients are simple polynomials, examine this equation and call on your experience in differentiating to obtain a polynomial solution. You should find that the simplest polynomial $y = x$ is a solution. This is easy to verify since $y' = 1$ and $y'' = 0$. To obtain the general solution we follow the method just detailed and set $y = vx$, $y' = v + xv'$, and $y'' = 2v' + xv''$. Then the differential equation becomes

$$x^3(xv'' + 2v') + 2x(xv' + v) - 2vx = 0$$

or

$$x^4v'' + (2x^3 + 2x^2)v' = 0$$

This is missing the variable v so setting $z = v'$ and dividing by x^2 gives

$$x^2 z' + (2x + 2)z = 0$$

Separating the variables gives

$$\frac{dz}{z} = -2\left(\frac{x+1}{x^2}\right) dx = \left(-\frac{2}{x} - \frac{2}{x^2}\right) dx$$

or
$$\ln|z| = -2\ln|x| + \frac{2}{x} + \ln C_1 = \ln\frac{C_1}{x^2} + \frac{2}{x}$$

Hence
$$z = \frac{C_1}{x^2} e^{2/x}$$

Since $z = v'$, we have $v' = C_1 x^{-2} e^{2/x}$. Integrating this gives $v = C_2 e^{2/x} + C_3$. Thus the solution to the original problem is

$$y(x) = v(x) \cdot x$$

or
$$y(x) = C_3 x + C_2 x e^{2/x}$$

We were aware that x was a solution; now we find that $xe^{2/x}$ is another linearly independent solution.

Example 74

In designing a heat exchanger, the engineer may encounter the differential equation $(x + 2)y'' - 2(x + 1)y' + xy = 0$. This is a second-order linear differential equation with variable coefficients. The number of ways to solve this type of equation is small. To apply the present technique we must produce one solution to this equation. If we add the coefficients, we find that the sum is zero and we then know that $y = e^x$ is a solution. Using $y = y' = y''$ we obtain

$$\begin{aligned}(x + 2)e^x - 2(x + 1)e^x + xe^x &= [(x + 2) - 2(x + 1) + x]e^x \\ &= [x + 2 - 2x - 2 + x]e^x \\ &= 0 \cdot e^x \\ &= 0\end{aligned}$$

Knowing one solution we now change variables with $y = ve^x$. Then $y' = e^x(v + v')$ and $y'' = e^x(v'' + 2v' + v)$. The differential equation then becomes

$$e^x(x + 2)(v'' + 2v' + v) - e^x(2)(x + 1)(v' + v) + e^x(x)v = 0$$

or
$$(x + 2)v'' + 2v' = 0$$

Again v is a missing variable, so with the substitution $v' = z$, this becomes

$$\frac{dz}{z} = \frac{-2}{x + 2} dx$$

Integrating gives

$$z = \frac{c_1}{(x+2)^2}$$

or
$$v' = c_1(x+2)^{-2}$$

Then
$$v = c_2(x+2)^{-1} + c_3$$

So
$$y(x) = ve^x = c_3 e^x + c_2 \cdot \frac{e^x}{x+2}$$

Thus we observe that our second linearly independent solution of the original homogeneous equation is

$$\frac{e^x}{x+2}$$

Ordinarily this type of problem is hard to solve by any method. In this special case we obtained the solution. When you encounter this type of problem, try one or more of the above nonstandard methods. You will discover there is no loss if it does not work and a big win if it does!

Exercises 3.11

In Exercises 1–18 solve the equations by any available method.

1. $x^2y'' + xy' - 4y = 0$
2. $x^2y'' - 3xy' + 13y = 0$
3. $x^2y'' - 2xy' + 2y = x^2 + 2$
4. $x^2y'' - 6y = \ln x,\ x > 0$
5. $x^2y'' - 4xy' + 6y = 0$
6. $2x^2y'' + xy' - y = 3x$
7. $x^2y'' - 5xy' + 13y = 0$
8. $x^2y'' + 7xy' + 5y = 10 - \dfrac{4}{x}$
9. $x^2y'' - xy' - 3y = x^2 \ln x,\ x > 0$
10. $4x^2y'' + 4xy' - y = 0$
11. $y'' = \dfrac{2}{x^2}\, y,\ x > 0$
12. $x^3y''' + 6x^2y'' + 7xy' + y = 0$
13. $x^3y''' + 2x^2y'' - 10xy' - 8y = 0$
14. $xy''' - 2y'' = 0$
15. $3x^3y''' + 4x^2y'' - 10xy' + 10y = 4x^{-2}$
16. $xy'' - (x+2)y' + 2y = x^3 + x$
17. $y'' - 2xy' + 2y = 4$
18. $x^2y'' + 4xy' + 2y = 2\ln x,\ x > 0$

19. Astronomer Z. Kopal* obtained the following equation:

$$r\frac{d^2\psi}{dr^2} + 4\frac{d\psi}{dr} = 0, \; r > 0$$

Solve this equidimensional differential equation.

20. If we are given an equation of the type

$$a_0(k_1x + k_2)^2y'' + a_1(k_1x + k_2)y' + a_2y = 0$$

it can be changed into an equidimensional equation by the change of variable $u = k_1x + k_2$. Use this information to solve the following equations:
(a) $(3x + 1)^2y'' + (3x + 1)y' - 3y = 9x$
(b) $(2x - 3)^2y'' = 2x - y$
(c) $(2x - 1)^3y''' + (2x - 1)y' - 2y = 4x$

21. Find the solution of $x^2y'' - xy' = 3x - x^2$ if $y = 3$ and $y' = 2$ when $x = 1$.

22. Find the equation of a curve that satisfies the differential equation $4x^2y'' + 4xy' - y = 0$ and crosses the x-axis at an angle of 45° at $x = 1$.

23. Suppose a hollow spherical shell has an inner radius r_1 and outer radius r_2 and the temperature at the inner and outer surfaces are u_1 and u_2, respectively. Then the temperature u at a distance r from the center ($r_1 \le r \le r_2$) is given by the differential equation

$$r\frac{d^2u}{dr^2} + 2\frac{du}{dr} = 0$$

Solve this equation for $u(r)$ if $u = u_1$ when $r = r_1$ and $u = u_2$ when $r = r_2$.

24. A steam pipe has temperature u_1 at its inner surface $r = r_1$ and temperature u_2 at its outer surface $r = r_2$. The temperature at a radial distance r, ($r_1 < r < r_2$) is given by

$$r\frac{d^2u}{dr^2} + \frac{du}{dr} = 0$$

Solve this equation for $u(r)$.

In Exercises 25–26, solve the equations by any available method.

25. $xy'' - (x - 1)y' - y = 0$
26. $x^2y'' - 2xy' + 2y = x^3 \ln x, \; x > 0$
27. Solve $xy'' - (1 + 2x^2)y' = x^3e^{x^2}$, given that $u_1 = e^{x^2}$ is a solution to $Ly = 0$. Try to determine another solution by inspection.

---

* Z. Kopal, "Stress History of the Moon and the Terrestrial Planet," *Icarus*, 2 (1963).

28. Solve the following equation by any available method:
$(\sin 4x)y'' - 4(\cos^2 2x)y' = 2\sin^2 2x \cos 2x$
29. Solve $(2x + 1)(x + 1)y'' + 2xy' - 2y = (2x + 1)^2$, given that $u_1 = x$ is a solution to $Ly = 0$.
30. Solve $(\sin^2 x)y'' - (\sin 2x)y' + (2 - \sin^2 x)y = \sin^3 x$, given $u_1 = \sin x$ is a solution to $Ly = 0$.

In Exercises 31–32, find u_1 by inspection; then find u_2.

31. $(x^2 - 2x)y'' + (2 - x^2)y' - 2(1 - x)y = 5x^2(x - 2)^2 e^x$
33. $x^2y'' - x(x + 2)y' + (x + 2)y = x^3$

3.12 Variation of Parameters

This method* furnishes a technique for finding a particular solution to the general nonhomogeneous linear equation $Ly = f(x)$ provided that the general homogeneous solution to $Ly = 0$ can first be found. This method offers not only an alternative procedure to the method of undetermined coefficients, but it also provides methodology for cases that cannot be handled by the earlier sweeping theorem. For example, the rather simple looking equation

$$y'' + y = \tan x$$

is not solvable by the method of undetermined coefficients.

We will show that there exists a general method (called the **variation of parameters**) of obtaining the general solution of any linear second-order differential equation, whether it has constant or variable coefficients, provided we have the homogeneous solution portion y_h.

Consider $a_0(x)y'' + a_1(x)y' + a_2(x)y = f(x)$. Suppose that $u_1(x)$ and $u_2(x)$ are linearly independent solutions of the associated homogeneous equation. Then $y_h = c_1u_1(x) + c_2u_2(x)$. In our search for a satisfactory trial solution to the nonhomogeneous equation, it is reasonable that we should consider something at least related to the given y_h. We know that the particular solutions we seek must be linearly independent from the homogeneous solutions u_1 and u_2. This independence may be expressed by saying that y_P is not a constant multiple of either u_1 or u_2. If this is true then the y_P we seek must have the property that y_P/u_1 and y_P/u_2 are not constant. We should replace both c_1 and c_2 in y_h by functions of x and look for a particular solution of the form $y_P = c_1(x)u_1(x) + c_2(x)u_2(x)$. With this form y_P/u_1 is a function of x and y_P/u_2 is a function of x. Thus we verify the nonconstant-ratio property referred to above. We have replaced the

* The method was first formulated by Joseph L. Lagrange (1736–1813), a pioneer in the development of the theory of both ordinary and partial differential equations.

parameters c_1 and c_2 by the variable functions $c_1(x)$ and $c_2(x)$. Hence we get the name for this method, the variation of parameters.

When we propose this trial y_P, we realize that two functions must now be determined. We know u_1 and u_2 and we must determine $c_1(x)$ and $c_2(x)$. If this trial function y_P and its derivatives are put into the differential equation, we obtain one relation in the two unknown functions. Since we have two functions but only one relation, we are free to impose a second condition provided it does not violate the first one. This second relation will be obtained in the derivation process. We calculate the first derivative of the trial function:

$$y'_P = c_1 u'_1 + c_2 u'_2 + (c'_1 u_1 + c'_2 u_2) \tag{15}$$

To keep our computations as simple as possible we exercise the above-mentioned freedom and arbitrarily set the last two terms equal to zero. This provides the second needed condition that complements the differential equation. Since the functions $c_1(x)$ and $c_2(x)$ are being determined, we ask that they satisfy the new imposed condition $c'_1 u_1 + c'_2 u_2 = 0$. From equation (15) we now obtain

$$y''_P = c_1 u''_1 + c_2 u''_2 + c'_1 u'_1 + c'_2 u'_2$$

Hence the original differential equation becomes

$$a_0(c_1 u''_1 + c_2 u''_2 + c'_1 u'_1 + c'_2 u'_2) + a_1(c_1 u'_1 + c_2 u'_2) + (a_2)(c_1 u_1 + c_2 u_2) = f(x)$$

Reorganizing this provides

$$c_1(a_0 u''_1 + a_1 u'_1 + a_2 u_1) + c_2(a_0 u''_2 + a_1 u'_2 + a_2 u_2) + a_0(c'_1 u'_1 + c'_2 u'_2) = f(x)$$

Remembering that u_1 and u_2 are solutions of the homogeneous equation, we know the terms multiplied by c_1 and c_2 are zero. Thus

$$c'_1 u'_1 + c'_2 u'_2 = \frac{f(x)}{a_0(x)}$$

We now have obtained a pair of simultaneous equations

$$\begin{Bmatrix} c'_1 u_1 + c'_2 u_2 = 0 \\ c'_1 u'_1 + c'_2 u'_2 = \dfrac{f(x)}{a_0(x)} \end{Bmatrix} \tag{16}$$

We need to solve this set of equations for c'_1 and c'_2 by elimination and integrate those solutions to c_1 and c_2. The equations can be algebraically solved provided the determinant of the coefficients of c'_1 and c'_2 does not vanish. The determinant is

$$\begin{vmatrix} u_1 & u_2 \\ u'_1 & u'_2 \end{vmatrix} = W[u_1, u_2] \neq 0$$

Hence the solvability is immediate since the Wronskian of linearly independent functions is always nonzero.

There exists a variety of solution techniques from elementary algebra for solving two equations in two unknowns. Using simple elimination or using Cramer's rule, we obtain

$$c_1' = \frac{-f(x)u_2(x)}{a_0(x)W[u_1, u_2]} \Rightarrow c_1(x) = \int \frac{-f(x)u_2(x)}{a_0(x)W[u_1, u_2]}\,dx \tag{17}$$

and

$$c_2' = \frac{f(x)u_1(x)}{a_0(x)W[u_1, u_2]} \Rightarrow c_2(x) = \int \frac{f(x)u_1(x)}{a_0(x)W[u_1, u_2]}\,dx \tag{18}$$

In these we add no constant of integration since we are obtaining a particular solution. Using these newly found functions, we determine that the final solution to $Ly = f$ is

$$y_P(x) = c_1(x)u_1(x) + c_2(x)u_2(x)$$

Examples 75–78 will give you some experience in using the variation of parameters method.

Example 75

Consider $y'' + y = \cot x$. Since the auxiliary polynomial equation is $r^2 + 1 = 0$, it is easily determined that $u_1(x) = \sin x$ and $u_2(x) = \cos x$ and the homogeneous solution is

$$y_h = c_1 \sin x + c_2 \cos x$$

We propose the particular solution

$$y_P = c_1(x)\sin x + c_2(x)\cos x$$

Here the pair of simultaneous equations (16) is

$$c_1' \sin x + c_2' \cos x = 0$$

$$c_1' \cos x + c_2'(-\sin x) = \cot x$$

By algebraic elimination we find

$$c_1' = \cot x \cos x$$

$$c_2' = -\cot x \sin x$$

e $$c_1(x) = \int \cot x \cos x\,dx = \int \frac{\cos^2 x}{\sin x}\,dx = \int \frac{1 - \sin^2 x}{\sin x}\,dx$$

$$= \int \csc x\,dx - \int \sin x\,dx = \ln|\csc x - \cot x| + \cos x$$

and

$$c_2(x) = -\int \cot x \sin x\,dx = -\int \cos x\,dx = -\sin x$$

We now can write

$$\begin{aligned} y_P &= c_1(x)u_1(x) + c_2(x)u_2(x) \\ &= [(\sin x)\ln|\csc x - \cot x| + (\sin x)\cos x] + (\cos x)(-\sin x) \\ &= (\sin x)\ln|\csc x - \cot x| \end{aligned}$$

The general solution to the original problem is

$$y(x) = K_1 \sin x + K_2 \cos x + (\sin x)\ln|\csc x - \cot x|$$

Example 76

Consider

$$y'' - 3y' + 2y = \frac{e^x}{1 + e^x}$$

The easily obtained solution to the associated homogeneous equation is

$$y_h = c_1 u_1(x) + c_2 u_2(x) = c_1 e^x + c_2 e^{2x}$$

where c_1 and c_2 are constant. We propose the particular solution $y_P = c_1(x)e^x + c_2(x)e^{2x}$ and will use the integral forms (17) and (18) to obtain

$$c_1(x) = \int \frac{-\left(\dfrac{e^x}{1 + e^x}\right)e^{2x}}{1 \cdot W[e^x, e^{2x}]}\, dx$$

and

$$c_2(x) = \int \frac{\left(\dfrac{e^x}{1 + e^x}\right)e^{x}}{1 \cdot W[e^x, e^{2x}]}\, dx$$

We first find

$$W[e^x, e^{2x}] = \begin{vmatrix} e^x & e^{2x} \\ e^x & 2e^{2x} \end{vmatrix} = 2e^{2x}e^x - e^x e^{2x} = e^{3x}$$

Thus

$$c_1(x) = \int \frac{-\left(\dfrac{e^x}{1 + e^x}\right)e^{2x}}{e^{3x}}\, dx = -\int \frac{dx}{1 + e^x}$$

$$= -\int \frac{e^{-x}\, dx}{e^{-x}(1 + e^x)} = \int \frac{-e^{-x}}{e^{-x} + 1}\, dx = \ln|e^{-x} + 1|$$

and

$$c_2(x) = \frac{\left(\dfrac{e^x}{1 + e^x}\right)e^{x}}{e^{3x}}\, dx = \int \frac{dx}{e^x(1 + e^x)} = \int \frac{e^{-2x}\, dx}{e^{-2x} \cdot e^x(1 + e^x)}$$

$$= \int \frac{e^{-2x}\, dx}{e^{-x} + 1} = \int \left(e^{-x} - \frac{e^{-x}}{e^{-x} + 1}\right) dx$$

$$= -e^{-x} + \ln|e^{-x} + 1|$$

Using these values, we find that the particular solution is

$$y_P = \ln(1 + e^{-x})e^x + [\ln(1 + e^{-x}) - e^{-x}]e^{2x}$$

$$= \ln(1 + e^{-x})(e^x + e^{2x}) - e^x$$

Combining this with the given y_h we have the general solution

$$\begin{aligned} y = y_h + y_P &= c_1 e^x + c_2 e^{2x} + \ln(1 + e^{-x})(e^x + e^{2x}) - e^x \\ &= (c_1 - 1)e^x + c_2 e^{2x} + \ln(1 + e^{-x})(e^x + e^{2x}) \\ &= c_3 e^x + c_2 e^{2x} + \ln(1 + e^{-x})(e^x + e^{2x}) \end{aligned}$$

Example 77

Consider $y'' + 4y = \sec 2x$. Here $u_1(x) = \sin 2x$ and $u_2(x) = \cos 2x$. By the derived expressions

$$c_1(x) = \int \frac{-\sec 2x \cos 2x}{1 \cdot (-2)}\,dx = \frac{1}{2}\int dx = \frac{x}{2}$$

$$c_2(x) = \int \frac{\sec 2x \sin 2x}{1 \cdot (-2)}\,dx = \frac{1}{4}\int \frac{-2\sin 2x}{\cos 2x}\,dx = \frac{1}{4}\ln|\cos 2x|$$

$$\therefore \quad y_P = \frac{x}{2}\sin 2x + \frac{1}{4}(\cos 2x)\ln|\cos 2x|$$

and $$y(x) = c_1 \sin 2x + c_2 \cos 2x + \frac{x}{2}\sin 2x + \frac{1}{4}(\cos 2x)\ln|\cos 2x|$$

The extension of the previous theory to a higher-ordered equation is not difficult. For a third-order linear nonhomogeneous equation

$$a_0(x)y''' + a_1(x)y'' + a_2(x)y' + a_3(x)y = f(x)$$

we first need the three linearly independent functions $u_1(x)$, $u_2(x)$, and $u_3(x)$ that solve the associated homogeneous equation in order to obtain

$$y_h = c_1 u_1(x) + c_2 u_2(x) + c_3 u_3(x)$$

Then the variation of the parameters would produce

$$y_P(x) = c_1(x)u_1(x) + c_2(x)u_2(x) + c_3(x)u_3(x)$$

which would again be substituted into the original differential equation.

As this is condensed, two *condition equations* are produced in addition to the differential equation. This set is the third-order counterpart to equations (16):

$$\left\{\begin{aligned} c_1' u_1 + c_2' u_2 + c_3' u_3 &= 0 \\ c_1' u_1' + c_2' u_2' + c_3' u_3' &= 0 \\ c_1' u_1'' + c_2' u_2'' + c_3' u_3'' &= \frac{f(x)}{a_0(x)} \end{aligned}\right\} \tag{19}$$

Again the determinant of this set is $W[u_1, u_2, u_3]$, which is nonzero by hypothesis. This set may then be solved for c_1', c_2', and c_3' and integrated to c_1, c_2, and c_3. Using these newly found functions $c_1(x)$, $c_2(x)$, and $c_3(x)$, we arrive at the general solution:

$$y = y_h + y_P = [K_1u_1 + K_2u_2 + K_3u_3] + [c_1(x)u_1(x) + c_2(x)u_2(x) + c_3(x)u_3(x)]$$

Example 78

Consider $y''' + y' = \sec x$. The auxiliary polynomial equation for the associated homogeneous equation is $r^3 + r = 0$. This factors to $r(r^2 + 1) = 0$. Since the roots are $r = 0$ and $r = \pm i$, we obtain

$$y_h = c_1 \cdot (1) + c_2 \cdot \cos x + c_3 \cdot \sin x$$

and the Wronskian of this set $\{1, \cos x, \sin x\}$ is

$$W[u_1, u_2, u_3] = \begin{vmatrix} 1 & \cos x & \sin x \\ 0 & -\sin x & \cos x \\ 0 & -\cos x & -\sin x \end{vmatrix} = 1(\sin^2 x + \cos^2 x) = 1$$

The equations (19) for this problem are

$$c_1'(1) + c_2'(\cos x) + c_3'(\sin x) = 0$$

$$c_1'(0) + c_2'(-\sin x) + c_3'(\cos x) = 0$$

$$c_1'(0) + c_2'(-\cos x) + c_3'(-\sin x) = \sec x$$

Using the technique called Cramer's rule to solve this set, we get

$$c_1' = \frac{\begin{vmatrix} 0 & \cos x & \sin x \\ 0 & -\sin x & \cos x \\ \sec x & -\cos x & -\sin x \end{vmatrix}}{W[1, \cos x, \sin x]} = \frac{\sec x(\cos^2 x + \sin^2 x)}{1} = \sec x$$

$$c_2' = \frac{\begin{vmatrix} 1 & 0 & \sin x \\ 0 & 0 & \cos x \\ 0 & \sec x & \sin x \end{vmatrix}}{W[1, \cos x, \sin x]} = \frac{1(-\sec x \cos x)}{1} = -1$$

$$c_3' = \frac{\begin{vmatrix} 1 & \cos x & 0 \\ 0 & -\sin x & 0 \\ 0 & -\cos x & \sec x \end{vmatrix}}{W[1, \cos x, \sin x]} = \frac{1(-\sin x \sec x)}{1} = -\tan x$$

We now integrate each of these results

$$c_1(x) = \int \sec x \, dx = \ln|\sec x + \tan x|$$

$$c_2(x) = \int -1 \, dx - -x$$

$$c_3(x) = \int -\tan x \, dx = \ln|\cos x|$$

Thus the particular solution is

$$\begin{aligned} y_P &= c_1(x)u_1(x) + c_2(x)u_2(x) + c_3(x)u_3(x) \\ &= (\ln|\sec x + \tan x|)1 + (-x)(\cos x) + (\ln|\cos x|)(\sin x) \end{aligned}$$

and the general solution to the differential equation $y''' + y' = \sec x$ is

$$y = c_1 + c_2 \cos x + c_3 \sin x + \ln|\sec x + \tan x| - x \cos x + (\sin x)\ln|\cos x|$$

The preceding examples worked nicely with this method because the integration was not difficult, but experience shows that this is not usually the case. The integration can become difficult when only minor modifications are made in the equation. For instance, Example 75 would have been tough if there had been an extra xe^x on the right side. We would have encountered the integrals

$$\int xe^x \csc x \, dx$$

and

$$\int xe^x \sin x \, dx$$

These are not generally covered in elementary calculus texts. Example 77 would have been considerably more difficult if Arctan $3x$ had also appeared on the right side. Even linear differential equations can be very difficult to solve.

These examples illustrate the fact that when you work on applications problems in various disciplines, you should not expect the resulting differential equation to be in a standard, easy-to-solve form. If it is, you have studied the tools for solving it. If it is not, you will need to subject it to numerical treatment. The next two chapters will guide you through the techniques needed.

Exercises 3.12

Use the variation of parameters technique to solve the differential equations in Exercises 1–25.

1. $y'' + y = \csc x$
2. $y'' + y = \tan x$
3. $y'' + b^2 y = \sec bx$
4. $y'' + y = \cot x$
5. $y'' - 2y' + y = \dfrac{e^x}{(x-1)^2}$
6. $y'' + a^2 y = \sec ax$

7. $y'' + 9y = \sec 3x \csc 3x$
8. $y'' + 9y = \csc 3x$
9. $y'' + y' = \cosh x$
10. $y'' + 2y' + y = e^{-x} \ln|x|$
11. $xy'' + y' = x + 1$
12. $y'' - 2y' + 2y = e^{x}(\tan x + \cot x)$
13. $y'' - y = \dfrac{1}{(1 + e^{x})}$
14. $(D^2 + 1)y = 2 \sec^3 x$
15. $(D^2 + 2D - 8)y = (6x^{-1} - x^{-2})e^{2x}$
16. $(D^2 + 4)y = 8 \tan^2 2x$
17. $(D - 1)^3 y = 18x^{-4} e^{x} \ln|x|$
18. $4y'' - 8y' + 5y = e^{x} \tan^2(\frac{x}{2})$
19. $9y'' + y = \tan^2(\frac{x}{3})$
20. $y'' + y = \sec x \tan x$
21. $4y'' - 4y' + y = e^{x/2} \ln|x|$
22. $y'' + 2y' + 2y = e^{-x} \sec x$
23. $y'' + 4y = 4 \cot 2x$
24. $y''' + y' = \tan x$
25. $y''' + 4y' = \sec 2x$
26. In the theory of electric circuits that contain inductance L, resistance R, capacitance C, and electromotive force $E(t)$, Kirchhoff's second law still applies and the charge q on the capacitor satisfies

$$L \frac{d^2 q}{dt^2} + R \frac{dq}{dt} + \frac{1}{C} q = E(t)$$

Let $L = 1$ henry, $R = 5$ ohms, and $C = \frac{}{6}$ farad.
(a) If $E(t) = e^{kt}$, solve this equation by variation of parameters.
(b) If $E(t) = \sin \alpha t$, solve this equation by variation of parameters.
(c) If $E(t) = \cos \beta t$, solve this equation by variation of parameters.

27. If an object of mass m is falling near the surface of the earth, then Newton's second law of motion indicates that the position y satisfies

$$m \frac{d^2 y}{dt^2} = F$$

where F is the sum of the gravitational force and the air resistance force directed oppositely. If the force due to gravity is mg and the air resistance is proportional to the speed dy/dt with constant of proportionality $-k$, the above differential equation becomes

$$m \frac{d^2 y}{dt^2} = mg - k \frac{dy}{dt}$$

Find the general solution by any method.

3.13 Application: The Motion of a Spring

A law of physics, called Hooke's law, states that if a spring is stretched or compressed from its natural length L, the spring exerts a force $F(x)$ that tends to return it to its natural length. This force is proportional to the amount x of the stretch. If we designate the proportionality constant by k (called the spring constant), then Hooke's law is expressed as

$$F(x) = -kx$$

This constant is unique to each spring and is a function of the material from which it is made as well as its construction.

If a solid object, such as a mass of steel, is attached to the spring, the spring will stretch from its natural length by an amount E, called the **elongation**. If the weight of the spring itself is negligible compared to the weight of the steel, then in its stretched resting position there is a balancing of forces. By Hooke's law the spring is exerting a restoring force of kE while the weight of the steel has an oppositely directed gravitational force of mg. We then may write $kE = mg$.

Suppose we create a system by attaching a heavy object to a spring of length L. The object will stretch the spring and then come to rest. The length of the spring is now $L + E$. We set the origin at the point corresponding to the top of the now resting object with the positive direction downward. We now apply pressure to make the object stretch the spring more, and we thereby displace the object from its resting position. In doing this we create a set of forces, the restoring force of the spring and the oppositely directed force of gravity, which dictate the position x of the object as a function of time t (see Figure 3.6). The object will have velocity dx/dt and acceleration

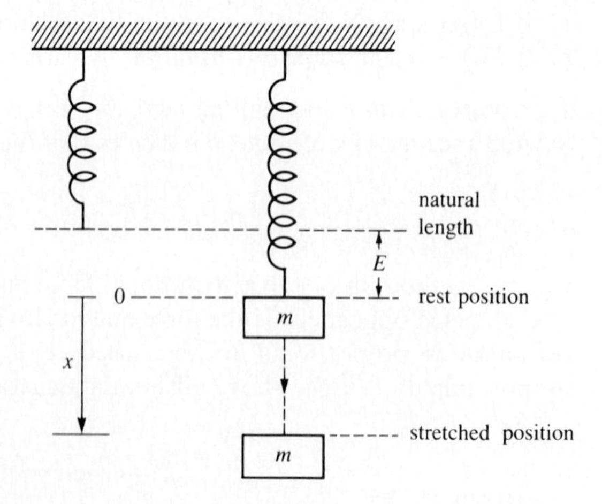

Figure 3.6

d^2x/dt^2. From Newton's second law of motion we get

$$m\frac{d^2x}{dt^2} = f(x)$$

where $f(x)$ is the directed sum of the forces on the object at position x.

If additional forces are present, such as the resistance of the medium in which the spring is located, the motion will be affected. Such a resistive force will produce **damped motion**. If no such force exists, the motion is called **undamped** and the spring undergoes free **vibration**.

Undamped Free Vibration

Suppose we are now in an undamped free-vibration mode and we displace the mass to a positive position x units from the origin. The spring is stretched to $x + E$, so Hooke's law gives a force of $-k(x + E)$. The gravitation force is still mg, so the total force is now $f(x) = mg - k(x + E)$. Since we have already determined that $mg = kE$, then

$$f(x) = mg - kx - kE = kE - kx - kE = -kx$$

Hence Newton's law gives

$$m\frac{d^2x}{dt^2} = -kx$$

or

$$\frac{d^2x}{dt^2} + \frac{k}{m}x = 0$$

This is the second-order linear differential equation for undamped free vibration. If the initial stretched position is to the value $x = x_0$ at time $t = 0$ and a velocity v_0 is given to the mass at the time $t = 0$, we will have initial conditions

$$x(0) = x_0 \quad \text{and} \quad x'(0) = v_0$$

Example 79

Suppose a 3-pound weight stretches a given spring 6 inches. We pull the weight four inches farther down and release it with no initial velocity. We derive a function that gives the position x of the weight at any subsequent time t. The initial value problem is

$$\frac{d^2x}{dt^2} + \frac{k}{m}x = 0; \quad x(0) = \frac{4}{12}, \quad x'(0) = 0$$

(Since we will be using $g = 32$ ft/sec^2, we need to convert the inches into feet. Therefore, we have divided by 12.) We first find k/m by using $kE = mg$.

$$k\left(\frac{6}{12}\right) = m \cdot 32$$

or
$$\frac{k}{m} = \frac{32 \cdot 12}{6} = 64$$

Hence
$$x'' + 64x = 0$$

The auxiliary polynomial equation is $r^2 + 64 = 0$, and its roots are $r = \pm 8i$, so the general solution is

$$x(t) = C_1 \cos 8t + C_2 \sin 8t$$

Since
$$x(0) = C_1 \cos 0 + C_2 \sin 0 = \frac{4}{12}$$

we have
$$C_1 = \frac{1}{3}$$

But
$$x'(t) = -8C_1 \sin 8t + 8C_2 \cos 8t$$

and
$$x'(0) = -\left(\frac{1}{3}\right) \sin 0 + 8C_2 \cos 0 = 0$$

so we have
$$C_2 = 0$$

The particular solution then is

$$x(t) = \frac{1}{3} \cos 8t$$

and represents the undamped free vibration. This function (which is sketched in Figure 3.7) would oscillate indefinitely, which in physical standards cannot happen. Air resistance or internal friction would eventually stop the motion. These forces need to be accounted for in further study.

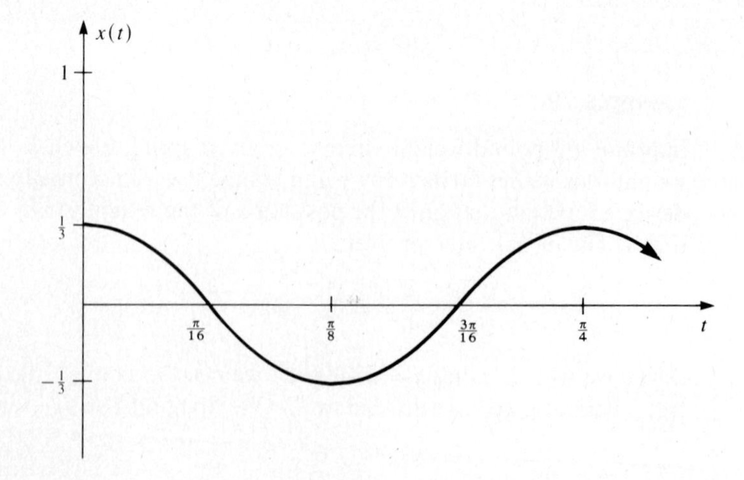

Figure 3.7

To solve the general undamped free-vibration problem, consider

$$\frac{d^2x}{dt^2} + \frac{k}{m}x = 0; \quad x(0) = x_0, \quad x'(0) = v_0$$

The general solution to the differential equation is

$$x(t) = C_1\cos\sqrt{\frac{k}{m}}\,t + C_2\sin\sqrt{\frac{k}{m}}\,t$$

This is often referred to as **simple harmonic motion**.

Using the initial conditions gives

$$x(0) = C_1\cos 0 + C_2\sin 0 = x_0$$

or

$$C_1 = x_0$$

Also

$$x'(t) = \left(-\sqrt{\frac{k}{m}}\right)(x_0)\sin\sqrt{\frac{k}{m}}\,t + \left(\sqrt{\frac{k}{m}}\right)C_2\cos\sqrt{\frac{k}{m}}\,t$$

$$\therefore \quad x'(0) = -\sqrt{\frac{k}{m}}\,x_0\sin 0 + \sqrt{\frac{k}{m}}\,C_2\cos 0 = v_0$$

or

$$C_2 = v_0\sqrt{\frac{m}{k}}$$

Thus the general solution is

$$x(t) = x_0\cos\sqrt{\frac{k}{m}}\,t + v_0\sqrt{\frac{m}{k}}\,\sin\sqrt{\frac{k}{m}}\,t$$

Now define an acute angle θ such that

$$\tan\theta = \frac{x_0}{v_0\sqrt{m/k}}$$

and call the hypotenuse value β (see Figure 3.8). Then

$$\cos\theta = \frac{v_0\sqrt{m/k}}{\beta}$$

and

$$\sin\theta = \frac{x_0}{\beta}$$

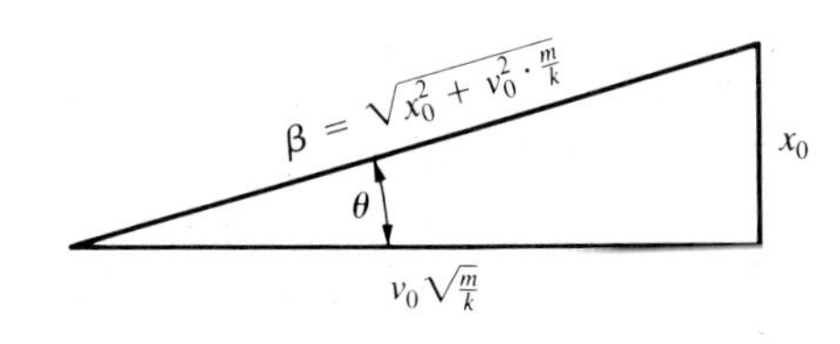

Figure 3.8

Now use the trigonometric identity

$$\sin(A + B) = \sin A \cos B + \cos A \sin B$$

where $A = \sqrt{k/m}\, t$ and $B = \theta$. Then

$$\sin\left(\sqrt{\frac{k}{m}}\, t + \theta\right) = \sin\sqrt{\frac{k}{m}}\, t \cos\theta + \cos\sqrt{\frac{k}{m}}\, t \sin\theta$$

$$= \left(\sin\sqrt{\frac{k}{m}}\, t\right)\left(\frac{v_0\sqrt{m/k}}{\beta}\right) + \left(\cos\sqrt{\frac{k}{m}}\, t\right)\left(\frac{x_0}{\beta}\right)$$

Rearrange this into

$$\beta\sin\left(\sqrt{\frac{k}{m}}\, t + \theta\right) = v_0\sqrt{\frac{m}{k}}\sin\sqrt{\frac{k}{m}}\, t + x_0\cos\sqrt{\frac{k}{m}}\, t$$

or

$$\beta\sin\left(\sqrt{\frac{k}{m}}\, t + \theta\right) = x_0\cos\sqrt{\frac{k}{m}}\, t + v_0\sqrt{\frac{m}{k}}\sin\sqrt{\frac{k}{m}}\, t$$

where the right side coincides exactly with the right side of the general solution to $x(t)$.

$$\therefore \quad x(t) = \beta\sin\left(\sqrt{\frac{k}{m}}\, t + \theta\right)$$

where

$$\beta = \sqrt{x_0^2 + v_0^2\frac{m}{k}}$$

and

$$\theta = \arctan\left(\frac{x_0}{v_0\sqrt{m/k}}\right)$$

By inspecting this equation we can see that the body oscillates with amplitude β and that the sinusoidal oscillation has period $2\pi\sqrt{m/k}$. We then define the frequency of the system as the number of complete oscillations per unit time and

$$f = \frac{\sqrt{k/m}}{2\pi}$$

Damped Motion

We now consider the case of damped motion in which we account for other forces. We will assume the object experiences damping forces proportional to the velocity dx/dt of the body. To $f(x)$ we must then add the term $-\gamma(dx/dt)$ where γ is a positive constant called the **damping constant**. (We assume γ fixed with time and position.) Newton's force equation now becomes

$$m\frac{d^2x}{dt^2} = -k(x + E) + mg - \gamma\frac{dx}{dt}$$

Again using $mg = kE$ we obtain

$$m\frac{d^2x}{dt^2} = -kx - \gamma\frac{dx}{dt}$$

or

$$\frac{d^2x}{dt^2} + \frac{\gamma}{m}\cdot\frac{dx}{dt} + \frac{k}{m}x = 0$$

Appropriate initial conditions to accompany this would be $x(0) = x_0$ and $x'(0) = v_0$. We again have a second-order linear homogeneous differential equation and well-posed initial conditions.

Example 80

Suppose that a weight attached to a spring extends the spring 8 inches and that the spring constant is 6. If the damping constant for this system is 2, the formula for the position of the weight at any time t is the solution to the differential equation

$$\frac{dx}{dt^2} + \frac{\gamma}{m}\cdot\frac{dx}{dt} + \frac{k}{m}x = 0$$

where $k = 6$, $E = \frac{8}{12}$, and $\gamma = 2$. We find m from $mg = kE$

$$m = \frac{kE}{g} = \frac{6(8/12)}{32} = \frac{1}{8}$$

Thus the differential equation is

$$\frac{d^2x}{dt^2} + \frac{2}{1/8}\cdot\frac{dx}{dt} + \frac{6}{1/8}x = 0$$

or

$$\frac{d^2x}{dt^2} + 16\frac{dx}{dt} + 48x = 0$$

The auxiliary polynomial equation is $r^2 + 16r + 48 = 0$ or $(r + 12)(r + 4) = 0$, and its roots are $r_1 = -12$ and $r_2 = -4$. Therefore the position equation is

$$x(t) = C_1e^{-12t} + C_2e^{-4t}$$

As $t \to \infty$, the value of $x(t) \to 0$; but the manner of oscillation is governed by the values of C_1 and C_2. If C_1 and C_2 have the same sign, $x(t)$ will never be zero and the system will never quite return to its initial position (see Figure 3.9). If C_1 and C_2 have opposite signs, then $x(t)$ will equal zero for at most one value of t and the object will return to its starting position once, overshoot it, and then slowly approach the resting position again (see Figure 3.10). An oscillation of this type is called **overdamped**.

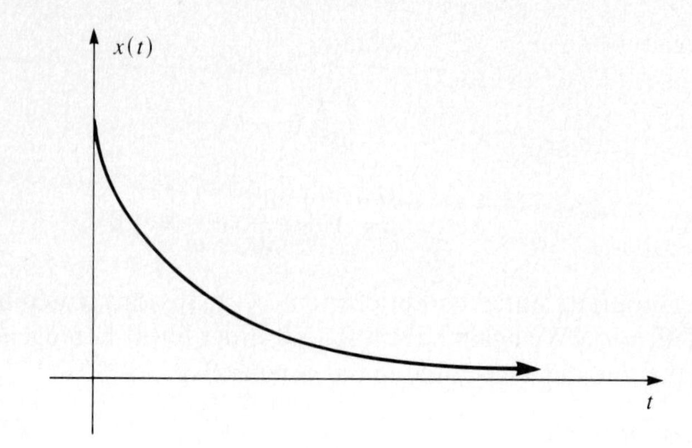

Figure 3.9

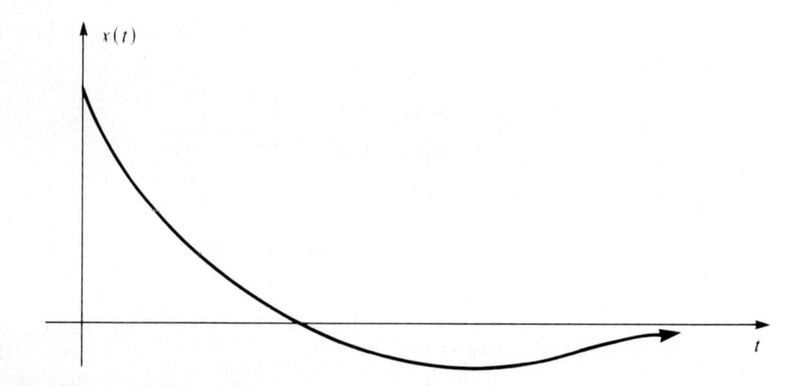

Figure 3.10

The solution to the general damped-vibration problem

$$\frac{d^2x}{dt^2} + \frac{\gamma}{m}\frac{dx}{dt} + \frac{k}{m}x = 0; \quad x(0) = x_0, \quad x'(0) = v_0$$

is obtained as follows: The solution to the differential equation is obtained by using the auxiliary polynomial equation

$$r^2 + \left(\frac{\gamma}{m}\right)r + \left(\frac{k}{m}\right) = 0$$

whose roots are

$$r = \frac{-\frac{\gamma}{m} \pm \sqrt{\left(\frac{\gamma}{m}\right)^2 - 4(1)\left(\frac{k}{m}\right)}}{2}$$

$$= -\frac{\gamma}{2m} \pm \frac{\sqrt{\gamma^2 - 4km}}{2m}$$

The nature of the roots depends on the value of the discriminant $\gamma^2 - 4km$. If it is positive, then we get two negative real roots and the solution to the differential equation is

$$x(t) = C_1 e^{\left(\frac{-\gamma + \sqrt{\gamma^2 - 4km}}{2m}\right)t} + C_2 e^{\left(\frac{-\gamma - \sqrt{\gamma^2 - 4km}}{2m}\right)t}$$

Since the coefficients of t are negative, the exponentials will become smaller, as t gets larger, regardless of the values of C_1 and C_2. If the discriminant is zero, the root is $-\gamma/2m$ and the solution is

$$x(t) = C_1 e^{(-\gamma/2m)t} + C_2 t e^{(-\gamma/2m)t}$$

Again as t gets large, the size of $x(t)$ becomes very small. For a positive value of C_1, we would have the curve illustrated in Figure 3.11. This type of oscillation occurs in a medium of high viscosity, such as oil. For the case of a negative discriminant, we have $\gamma^2 - 4km < 0$ and the general solution is

$$x(t) = C_1 e^{(-\gamma/2m)t}\left[\cos \frac{\sqrt{4km - \gamma^2}}{2m} t\right] + C_2 e^{(-\gamma/2m)t}\left[\sin \frac{\sqrt{4km - \gamma^2}}{2m} t\right]$$

In each of these terms, the exponential factor is the damping part while the trigonometric factor is the oscillating part. For a given set of initial conditions in which $C_1 > 0$ we would have damped oscillations of the type shown in Figure 3.12.

Example 81

Suppose we have a spring with $k = 2$ and attach a 4-pound weight to it. In two different substances into which the system is immersed we find: (a) $\gamma = \frac{1}{2}$ and (b) $\gamma = 1$. At time $t = 0$ we stretch the spring one unit and

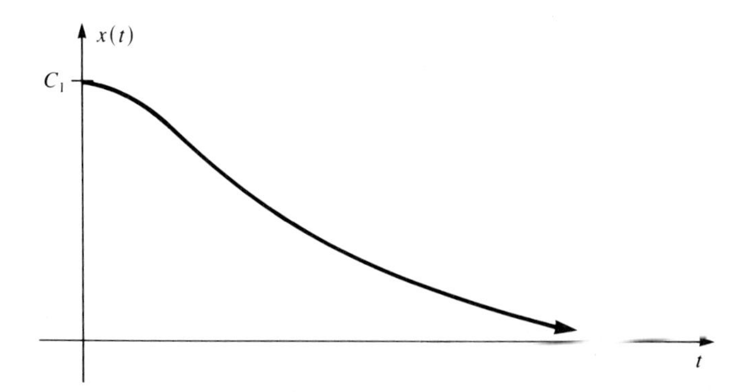

Figure 3.11

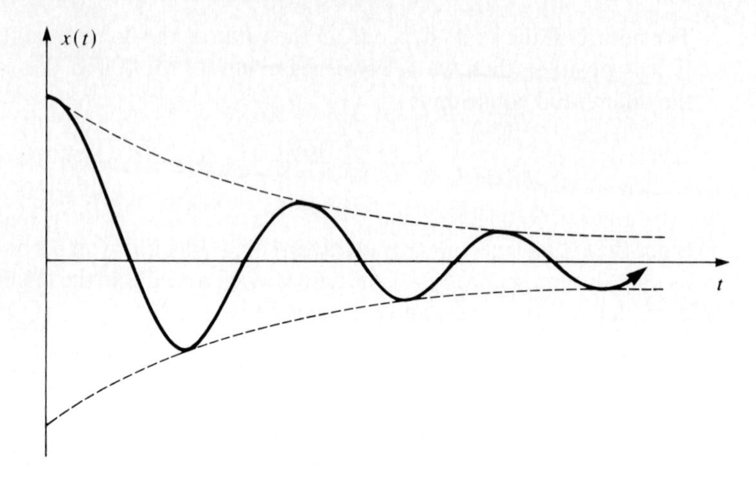

Figure 3.12

release it with velocity 4. Find the motion formula for each case. In case (a) the constants are $k = 2$, $m = \frac{1}{8}$, and $\gamma = \frac{1}{2}$, so we have

$$\frac{d^2x}{dt^2} + 4\frac{dx}{dt} + 16x = 0; = x(0) = 1, \quad x'(0) = 4$$

From $r^2 + 4r + 16 = 0$ we get roots

$$r = \frac{-4 \pm \sqrt{16 - 4(1)(16)}}{2} = \frac{-4 \pm 4\sqrt{3}\,i}{2} = -2 \pm 2\sqrt{3}\,i$$

So the general solution to the differential equation is

$$x(t) = C_1 e^{-2t}\cos 2\sqrt{3}\,t + C_2 e^{-2t}\sin 2\sqrt{3}\,t$$

When correctly substituted, the initial conditions provide $C_1 = 1$ and $C_2 = \sqrt{3}$. Thus

$$x(t) = e^{-2t}(\cos 2\sqrt{3}\,t + \sqrt{3}\sin 2\sqrt{3}\,t)$$

This type of oscillation, in which the object passes through its rest position a number of times, is referred to as **underdamped**. The graph of the equation is illustrated in Figure 3.13.

In case (b) the constants are $k = 2$, $m = \frac{4}{32} = \frac{1}{8}$, and $\gamma = 1$, so we have

$$\frac{d^2x}{dt^2} + \frac{1}{(1/8)} \cdot \frac{dx}{dt} + \frac{2}{(1/8)}x = 0$$

$$\frac{d^2x}{dt^2} + 8\frac{dx}{dt} + 16x = 0$$

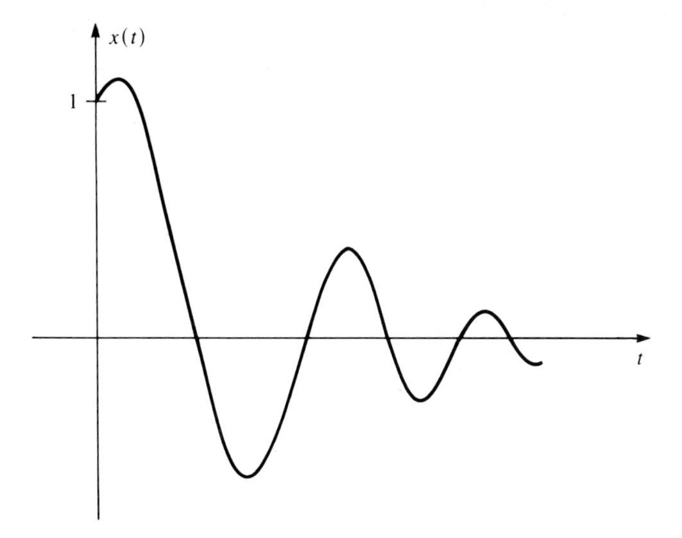

Figure 3.13

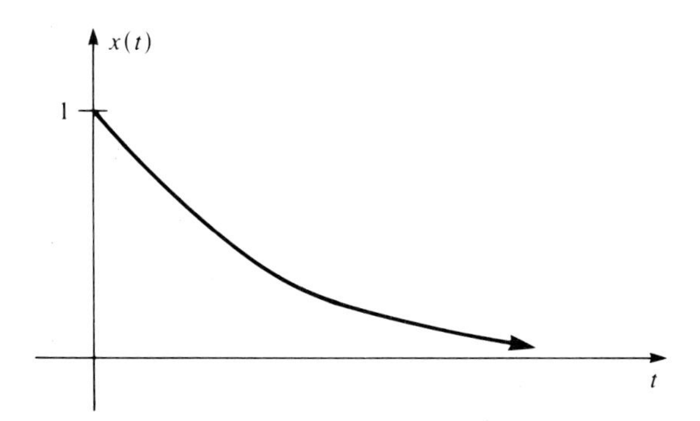

Figure 3.14

The auxiliary polynomial equation is $r^2 + 8r + 16 = 0$, and its roots are $r_1 = r_2 = -4$. Therefore

$$x(t) = C_1 e^{-4t} + C_2 t e^{-4t}$$

The initial conditions $x(0) = 1$ and $x'(0) = 4$ give $C_1 = 1$ and $C_2 = 1$ so the unique solution is

$$x(t) = e^{-4t}(1 + t)$$

This type of oscillation is referred to as **critically damped**. The curve of the equation is shown in Figure 3.14.

Equation solutions for the four types of oscillation are summarized in Table 3.1.

Table 3.1.

Summary of Solutions to Differential Equations for Oscillations

Type of Oscillation	Constant Values*	General Solution Type	Reference Example
Undamped free vibration	$\gamma = 0$ (vacuum)	$x(t) = c_1 \cos \sqrt{\frac{k}{m}}\, t + c_2 \sin \sqrt{\frac{k}{m}}\, t$	79
Overdamped vibration	$\gamma^2 - 4km > 0$	$x(t) = c_1 e^{-r_1 t} + c_2 e^{-r_2 t}$	80
Underdamped vibration	$\gamma^2 - 4km < 0$	$x(t) = c_1 e^{-\alpha t} \cos \beta t + c_2 e^{-\alpha t} \sin \beta t$	81(a)
Critically damped vibration	$\gamma^2 - 4km = 0$	$x(t) = e^{-rt}(c_1 + c_2 t)$	81(b)

* k = spring constant; m = mass of suspended object; γ = damping constant.

Miscellaneous Exercises for Chapter 3

In Exercises 1–50 solve each of the equations by one or more methods presented in this chapter.

1. $y'' + y = 2e^x$
2. $x^2 y'' + xy' - y = 0$
3. $(D^5 + 2D^3 + D)y = 0$
4. $D^2(D - 1)^2 y = 0$
5. $2u'' - 5u' + 2u = 0$
6. $x^2 z'' + 4xz' + 2z = 0$
7. $y'' - 2y' + 5y = 1$
8. $(D^2 - 1)y = \sin^2 x$
9. $(D^2 + 2D + 1)y = e^x + e^{-x}$
10. $x^2(x + 1)y'' - 2y = 0$; given a solution $y_1 = 1 + \frac{1}{x}$
11. $(D^2 + 2)y = x \cos x$
12. $y^{(v)} - 10y''' + 9y' = 0$
13. $xy^{(iv)} = 1$
14. $(D^3 + 1)y = 0$
15. $(D^2 + 4)(D + 1)y = \cos 2x$
16. $y'' - 3y' + 2y = 2^x$
17. $(D^2 + 1)y = 3 \sin \dfrac{x}{2}$
18. $y'' + 4xy' + (4x^2 + 2)y = 0$; given a solution $y_1 = e^{ax^2}$
19. $y'' + y = 0$; $y(0) = 0$, $y(\pi) = 0$

20. $y''' - 7y'' + 14y' - 8y = 0$
21. $y'' + 3y' + 2y = x \sin x$
22. $D(D + 1)y = \sinh x$
23. $y'' + 4y = \cos x \cos 3x$
24. $yy'' = 2x(y')^2;\ y(2) = 2,\ y'(2) = \frac{1}{2}$
25. $(D^2 + 1)y = x \cos x + e^x$
26. $x^3 y'' + xy' - y = 0$
27. $y^4 - y^3 y'' = 1$
28. $(D^3 + D + 2)y = e^{-x} \sin 2x$
29. $y^{(iv)} - y = 0;\ y(0) = y'(0) = y''(0) = 0,\ y'''(0) = 1$
30. $x^3 y''' - 3x^2 y'' + 6xy' - 6y = 0$
31. $y'' + 4y = \sec^2 2x$
32. $y'' + 4y' + 13y = 5 \sin 2x;\ y(0) = 2,\ y'(0) = -1$
33. $x^2 y'' - 3xy' + 13y = 0$
34. $y'' + 4y' + 4y = 0;\ y(0) = 2;\ y'(0) = 5$
35. $9y'' + 6y' + 5y = 5;\ y(0) = 4,\ y'(0) = 0$
36. $y'' - 5y' + 6y = 3 + \frac{6}{5}x$
37. $(x^2 + 1)y'' - 2xy' + 2y = 0$
38. $y'' = \left(\frac{1}{y}\right)(y')^2$
39. $y'' + 4y' + 5y = xe^x$
40. $(D^4 - 1)y = 0$
41. $(D - 2)^3(D - 1)y = 6(x^2 + 2x)e^{2x}$
42. $(D^2 + D - 6)y = 1 - 6x;\ y(0) = 3,\ y'(0) = -7$
43. $(D + 2)^2 y = 4e^{2x} + x$
44. $(D - 1)^3 y = 6e^x$
45. $(2 + 4D - D^2)y = 4x^3$
46. $(D^2 + a^2)y = \sec^2 ax$
47. $(xD^2 - D)y = 4/x$; given that x^2 solves $Ly = 0$
48. $(D^3 - D^2)y = 4 \cos x$
49. $(D^3 + a^3)y = 0$
50. $y^{(iv)} - 15y'' + 10y' + 24y = 0$
51. To physiologists concerned with blood circulation, the problem of velocity of waves propagated through elastic tubes presents an important mathematical study. If u is the radial displacement of the artery wall, then the equation determined by Mirsky* is

$$\frac{d^2u}{dr^2} + \frac{1}{r}\frac{du}{dr} - m^2 \frac{u}{r^2} = 0$$

---

* I. Mirsky, "Pulse Velocities in an Orthotropic Elastic Tube," *Bulletin of Mathematical Biophysics*, 29(1967), pp. 311–18.

where m is a constant derived from stress-and-strain components. Solve this equation for $u(r)$.

52. In building design it is sometimes useful to use supporting columns that are special geometrical designs. In studying the buckling of columns of varying cross sections, we obtain the following differential equation:

$$x^n \frac{d^2y}{dx^2} + k^2 y = 0; \quad k > 0, n \text{ a positive integer}$$

(a) If $n = 1$, the column is rectangular with one dimension constant.
(b) If $n = 2$, we get another important case.
(c) If $n = 4$, the column is a truncated pyramid or cone.
Solve the differential equation in each of these cases.

53. A particle executes simple harmonic motion according to

$$\frac{d^2x}{dt^2} + \omega_0^2 x = 0$$

where $\omega_0 > 0$ and x is positive. At the end of each $\frac{3}{4}$ seconds, it passes through the equilibrium or rest point with a velocity of ± 8 ft/sec.
(a) Find the equation describing the motion.
(b) Find the period, frequency, and amplitude of the motion.

54. A helical spring with negligible internal friction stretches 4 inches when a 5-pound body is attached to it. After it comes to rest, it is stretched an additional 2 inches and released. Find the equation of its motion, period, frequency, and amplitude.

55. A spherical body of radius R is in equilibrium when half of it is submerged in a liquid. If the body is pushed down and then released, it will execute harmonic motion. If y is the distance displaced from equilibrium, then the differential equation of motion is

$$\frac{d^2y}{dt^2} = -\frac{g}{2}\left[3 \cdot \frac{y}{R} - \left(\frac{y}{R}\right)^3\right], \text{ with } y(0) = \sqrt{6}R,\ y'(0) = 0$$

Solve for the function $y(t)$ with g and R as unspecified parameters.

56. A particle is said to execute damped harmonic motion if its equation of motion satisfies

$$m\frac{d^2y}{dt^2} + 2mr\frac{dy}{dt} + m\omega_0^2 y = 0$$

where the coefficient $2mr > 0$ is called the coefficient of resistance of the system, and ω_0 is the natural or undamped frequency. (Note that $m\omega_0^2 = k$, the spring coefficient.)

If $r^2 > \omega_0^2$, we get nonoscillatory motion, because the damping force (r) overpowers the restoring force (ω_0).

Verify the preceding statement for the following case: A helical spring is stretched to 24 inches by a 2-pound weight. It is brought to rest, immersed in a heavy fluid whose coefficient of resistance is $\frac{1}{2}$, given an additional pull of 6 inches, and released. Find the equation of motion and draw a rough graph.

57. A particle is moving on a horizontal line in accordance with the law

$$\frac{d^2x}{dt^2} + 6\frac{dx}{dt} - 16x = 0$$

At $t = 0$, it is at $x = 2$ and moving to the left with velocity 10 ft/sec. Will the particle change direction eventually? When and how often?

58. At time $t = 0$, a solid cylinder is placed at the top of an inclined plane and released from rest. If it rolls without slipping and if a frictional force acts in opposition to its motion, the differential equation of motion is

$$\frac{3}{2} \cdot \frac{d^2s}{dt^2} = g \sin \alpha$$

where s is the distance displaced. Find $s(t)$ if $\alpha = 30°$.

Chapter 4

Numerical Methods for Second-Order Equations

4.1 Introduction

In the early part of Chapter 1, the general definition of a second-order equation was an equation of the form $y'' = f(x, y, y')$ where f is an arbitrary combination of those three variables. If we consider all of the functional combinations that could be created on the right side and think back to the analytical techniques for solving second-order differential equations in Chapter 3, we will conclude that only a small portion of the possible second-order differential equations are solvable by analytical methods.

Very few nonlinear equations are solvable. Of the linear ones, the constant coefficient equations and a small scattering of others are solvable in closed form. Hence Chapter 4 is very important in the study of differential equations, for here we will study methods of obtaining an *approximate* solution to *any* second-order differential equation, linear or nonlinear. We will repeat the techniques of Chapter 2 and extend them to the general second-order differential equation.

Our first task in solving numerically a second-order initial value problem

$$\left.\begin{cases} y'' = f(x, y, y') \\ y(x_0) = y_0,\ y'(x_0) = v_0 \end{cases}\right\} \tag{1}$$

is to break it down by a variable-renaming process into a pair of first-order equations, each with its initial condition. If we write $y' = v$ and consequently $y'' = v'$, then equation (1) may be written as the pair

$$\left.\begin{cases} y' = v \\ v' = f(x, y, y) \end{cases}\right\}$$

The variable renaming also produces the initial conditions $y(x_0) = y_0$ and $v(x_0) = v_0$. Note here that if the function $f(x, y, v)$ does, in fact, specifically involve the variable y, then the pair of first-order differential equations is coupled together in that the y' equation contains v and the v' equation contains y. Under these conditions we will need to solve the equations simultaneously.

Examples 1 and 2 illustrate how this process can be used to prepare an equation for a numerical solution.

Example 1

Consider $(x^2 + 4)y'' - (2x - 9)y' + 6x^3y = 3e^{2x} - 7$. This differential equation is linear, but it is not solvable by any of the elementary analytical techniques that you have studied. However, by following the above process we can prepare it for a numerical solution. Let $y' = v$, so that $y'' = v'$. Then we obtain the pair

$$\left.\begin{cases} y' = v \\ v' = \left(\dfrac{2x - 9}{x^2 + 4}\right)v - \left(\dfrac{6x^3}{x^2 + 4}\right)y + \dfrac{3e^{2x} - 7}{x^2 + 4} = f(x, y, v) \end{cases}\right\}$$

Example 2

Consider

$$\left.\begin{cases} y'' + 5y^3y' - 2\cos y = \tan 4x \\ y(1) = 4,\ y'(1) = -3 \end{cases}\right\}$$

This differential equation is nonlinear, and it is not solvable by analytical methods. We can, however, write

$$y'' = -5y^3y' + 2\cos y + \tan 4x = f(x, y, y')$$

If we let $y' = v$ and $y'' = v'$, the initial conditions become $y(1) = 4$ and $v(1) = -3$. Thus we have the pair

$$\left.\begin{cases} y' = v;\ y(1) = 4 \\ v' = -5y^3v + 2\cos y + \tan 4x = f(x, y, v);\ v(1) = -3 \end{cases}\right\}$$

4.2 Euler's Method for Second-Order Equations

Euler's method for finding approximate solutions to second-order equations closely parallels the method we studied in Chapter 2. The unique solution function we are attempting to approximate passes through the point (x_0, y_0) with the slope v_0. The point (x_0, y_0) is the first point in the solution set, and we will again attempt to find the corresponding values of y at the values $x_1 = x_0 + h$, $x_2 = x_0 + 2h$, $x_3 = x_0 + 3h, \ldots$, where h is the predetermined step size. The approximate solution will be denoted by the set of points (x_k, Y_k), $k = 0, 1, 2, \ldots$, where Y_k is the approximation to $y(x_k)$. We will simultaneously obtain the set (x_k, V_k), $k = 0, 1, 2, \ldots$, where V_k is the approximation to $v(x_k) = y'(x_k)$. This will give us a record of the slope at each solution point. We follow the method of Chapter 2 and obtain the Euler approximates.

We will integrate the two coupled first-order equations one at a time. The first equation $y' = v$ is written in differential form as $dy = v\,dx$ and integrated from x_0 to an arbitrary point x to the right of x_0.

$$\int_{x_0}^{x} dy = \int_{x_0}^{x} v\,dx$$

This may be evaluated on the left to obtain

$$y(x) - y(x_0) = \int_{x_0}^{x} v\,dx$$

or

$$y(x) = y_0 + \int_{x_0}^{x} v\,dx$$

At the specific value $x = x_1$ this gives

$$y(x_1) = y_0 + \int_{x_0}^{x_1} v\,dx \tag{2}$$

Using the Euler approximation, we replace $v(x)$ by the given constant v_0 and have

$$Y_1 = y_0 + v_0 \int_{x_0}^{x_1} dx = y_0 + v_0(x_1 - x_0)$$

$$= y_0 + hv_0$$

To keep the notation consistent, we use $y_0 = Y_0$ and $v_0 = V_0$ and obtain

$$Y_1 = Y_0 + hV_0$$

We now must find V_1, for it is used in the computation of Y_2. Using the equation $v' = f(x, y, v)$, which in differential form is $dv = f(x, y, v)\,dx$, we integrate from x_0 to x_1 and obtain

$$\int_{x_0}^{x_1} dv = \int_{x_0}^{x_1} f(x, y, v)\,dx$$

or
$$v(x_1) - v(x_0) = \int_{x_0}^{x_1} f\,dx$$

But $v(x_0) = v_0$, so

$$v(x_1) \equiv y'(x_1) = V_0 + \int_{x_0}^{x_1} f\,dx \tag{3}$$

Again we use Euler's approach and approximate $f(x, y, v)$ by the constant function $f(x_0, y_0, v_0) = f(x_0, Y_0, V_0)$. Thus

$$\begin{aligned} V_1 = v(x_1) &= V_0 + (x_1 - x_0)f(x_0, y_0, v_0) \\ &= V_0 + hf(x_0, Y_0, V_0) \end{aligned}$$

Further extensions to points to the right of x_1 (e.g., $x_2, x_3, \ldots$) are carried out in the same manner. Extending the above concepts, we see that the general step becomes

$$\left.\begin{cases} Y_{k+1} = Y_k + hV_k \\ V_{k+1} = V_k + h \cdot f(x_k, Y_k, V_k) \end{cases}\right\} \quad k = 0, 1, 2, 3, \ldots \tag{4}$$

The geometrical interpretation of this scheme is shown in Figure 4.1. Notice how the new value for Y_{k+1} follows on a line whose slope is determined by the value of V_k and how the new value for V_{k+1} follows on a line whose slope is determined by a value of y'', the rate of change of the slope.

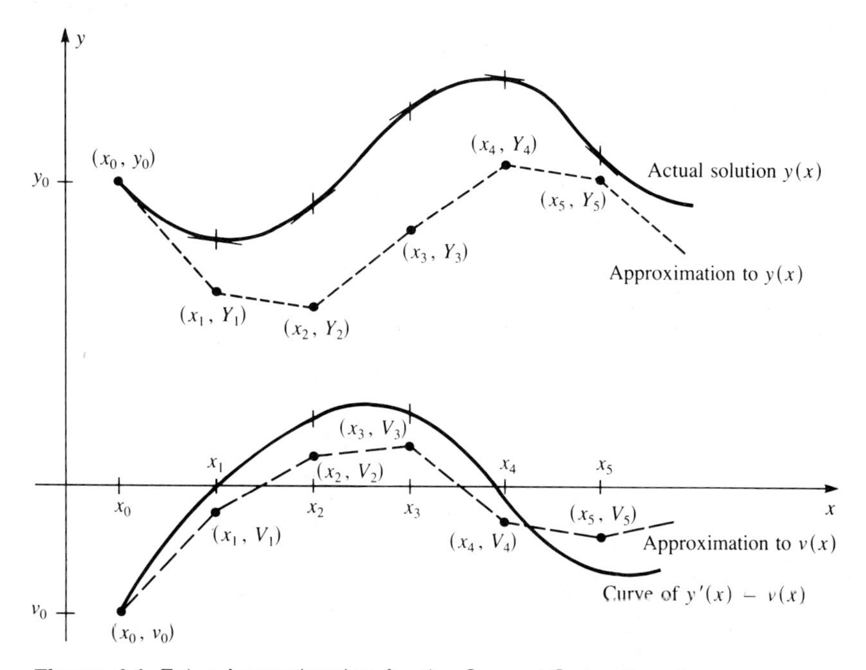

Figure 4.1 Euler Approximation for the Second-Order Equation

Example 3

For the initial value problem

$$\begin{Bmatrix} y'' - y' + 2y = 0 \\ y(0) = 1, \, y'(0) = 1 \end{Bmatrix}$$

we will apply the Euler formulas to obtain points in the approximate solution and use a step size $h = 0.1$. First we make the replacement $y' = v$ and obtain the pair

$$\begin{Bmatrix} y' = v \\ v' = -2y + v \end{Bmatrix}$$

with initial conditions $y(0) = 1$, $v(0) = 1$.
Since $x_0 = 0$, $Y_0 = 1$, $V_0 = 1$, $h = 0.1$, and $f(x, Y, V) = -2Y + V$, we apply the formulas of equations (4):

$$k = 0 \begin{Bmatrix} Y_1 = Y_0 + hV_0 = 1 + (0.1)(1) = 1.1 \\ V_1 = V_0 + h \cdot f(x_0, Y_0, V_0) \\ = 1 + (0.1)[-2(1) + 1] = 1 + (0.1)(-1) = 0.90 \end{Bmatrix}$$

$$k = 1 \begin{Bmatrix} Y_2 = Y_1 + hV_1 = (1.1) + (0.1)(0.90) = 1.19 \\ V_2 = V_1 + h \cdot f(x_1, Y_1, V_1) \\ = 0.90 + (0.1)[-2(1.1) + 0.90] = 0.90 + (0.1)(-1.3) = 0.77 \end{Bmatrix}$$

$$k = 2 \begin{Bmatrix} Y_3 = Y_2 + hV_2 = 1.19 + (0.1)(0.77) = 1.267 \\ V_3 = V_2 + h \cdot f(x_2, Y_2, V_2) \\ = 0.77 + (0.1)[-2(1.19) + 0.77] = 0.77 + (0.1)(-1.61) = 0.609 \end{Bmatrix}$$

$$k = 3 \begin{Bmatrix} Y_4 = Y_3 + hV_3 = 1.267 + (0.1)(0.609) = 1.3279 \\ V_4 = V_3 + h \cdot f(x_3, Y_3, V_3) \\ = (0.609) + (0.1)[-2(1.267) + (0.609)] = 0.4165 \end{Bmatrix}$$

$$\vdots$$

These approximations and the actual solution points are listed in Table 4.1.

Table 4.1

Solution to $y'' - y' + 2y = 0$; $y(0) = 1$, $y'(0) = 1$

x	Y	V	y (actual)	v (actual)
0.0	1.0	1.0	1.0	1.0
0.1	1.1	0.90	1.09450	0.884855
0.2	1.19	0.77	1.17595	0.739018
0.3	1.267	0.609	1.24127	0.562326
0.4	1.3279	0.4165	1.28740	0.355241

The values in the last two columns were obtained from the actual solution

$$y = e^{(1/2)x} \cos \frac{\sqrt{7}}{2} x + \frac{1}{\sqrt{7}} e^{(1/2)x} \sin \frac{\sqrt{7}}{2} x$$

$$y' = \frac{-3}{\sqrt{7}} e^{(1/2)x} \sin \frac{\sqrt{7}}{2} x + e^{(1/2)x} \cos \frac{\sqrt{7}}{2} x$$

This comparison allows us to assess the potential accuracy of the Euler method.

Example 4

We will now apply the Euler method on the interval [0, 1] to the nonlinear problem

$$y'' - 2y^2y' - \frac{1}{5}y = -12x; \qquad y(0) = 1, \qquad y'(0) = \frac{1}{2}$$

Replacing y' by v we obtain the equivalent system

$$\begin{Bmatrix} y' = v;\ y(0) = 1 \\ v' = 2y^2v + 0.2y - 12x;\ v(0) = \dfrac{1}{2} \end{Bmatrix}$$

Checking the first steps by hand computations and using step size $h = 0.1$ and the given $x_0 = 0$, $x_1 = 0.1$, $Y_0 = 1$, $V_0 = \frac{1}{2}$, $f(x, y, v) = 2y^2v + 0.2y - 12x$, and equations (4), we get

$$k = 0 \begin{Bmatrix} Y_1 = Y_0 + hV_0 = 1 + (0.1)(0.5) = 1.05 \\ V_1 = V_0 + h \cdot f(x_0, Y_0, V_0) \\ \quad = (0.5) + (0.1)[2(1^2)(0.5) + (0.2)(1) - 12(0)] \\ \quad = 0.5 + (0.1)(1.2) = 0.62 \end{Bmatrix}$$

$$k = 1 \begin{Bmatrix} Y_2 = Y_1 + hV_1 = 1.05 + (0.1)(0.62) = 1.112 \\ V_2 = V_1 + h \cdot f(x_1, Y_1, V_1) \\ \quad = (0.62) + (0.1)[2(1.05)^2(0.62) + (0.2)(1.05) - 12(0.1)] \\ \quad = (0.62) + (0.1)(.3771) = 0.65771 \end{Bmatrix}$$

$$\vdots$$

The completion of this problem would be tedious, and we would normally let the computer finish it.

The only advantage of the Euler scheme for second-order equations lies in its simplicity, thus providing an easy introduction to numerical schemes. Its tendency to great inaccuracy, as illustrated in Example 3, rules it out as a practical method of operation. Therefore, we advance to an improved scheme following the developments of Chapter 2, and will use equations (4) as a first approximation.

4.3 Improved Euler Scheme

In attempting to find the values of $Y(x)$ close to the actual values $y(x)$, let's assume that we have computed the values Y_k and V_k and that we want to extend our approximate solution one more step to the right to find Y_{k+1}, the approximation to $y(x_{k+1})$, and to find V_{k+1}, the approximation to $v(x_{k+1}) = y'(x_{k+1})$.

This improved Euler method will, as in Chapter 2, use the average of two values, one at the current point and one in the direction field ahead. The integration of the two first-order differential equations again produces, from (2) and (3),

$$\begin{Bmatrix} Y_{k+1} = Y_k + \displaystyle\int_{x_k}^{x_{k+1}} v\,dx \\ V_{k+1} = V_k + \displaystyle\int_{x_k}^{x_{k+1}} f\,dx \end{Bmatrix} \tag{5}$$

To integrate these equations, we must furnish replacements for v and f. The earlier derived Euler values provide our first approximations, called the **predictors**,

$$\begin{Bmatrix} \bar{y}_{k+1} = Y_k + hV_k \\ \bar{v}_{k+1} = V_k + h \cdot f(x_k, Y_k, V_k) \end{Bmatrix} \tag{6}$$

Then using these values from the direction field ahead, we calculate

$$f \simeq f(x_{k+1}, \bar{y}_{k+1}, \bar{v}_{k+1})$$

Thus our averaged values will be

$$v \simeq \frac{V_k + \bar{v}_{k+1}}{2}$$

$$f \simeq \frac{f(x_k, Y_k, V_k) + f(x_{k+1}, \bar{y}_{k+1}, \bar{v}_{k+1})}{2}$$

Inserting these into (5) we get the improved Euler **corrector** equations:

$$\begin{Bmatrix} Y_{k+1} = Y_k + \left(\dfrac{V_k + \bar{v}_{k+1}}{2}\right)(x_{k+1} - x_k) \\ \quad = Y_k + \dfrac{h}{2}(V_k + \bar{v}_{k+1}) \qquad k = 0, 1, 2, 3, \ldots \\ V_{k+1} = V_k + \left(\dfrac{f(x_k, Y_k, V_k) + f(x_{k+1}, \bar{y}_{k+1}, \bar{v}_{k+1})}{2}\right)(x_{k+1} - x_k) \\ \quad = V_k + \dfrac{h}{2}[f(x_k, Y_k, V_k) + f(x_{k+1}, \bar{y}_{k+1}, \bar{v}_{k+1})] \end{Bmatrix} \tag{7}$$

$$k = 0, 1, 2, 3, \ldots$$

The pair (6) is called the predictor set because it estimates the future values of Y and V, and the pair (7) is called the corrector set because it calculates the new averaged values of Y and V.

Example 5

Apply the improved Euler method to the simple initial value problem of Example 3, using step size $h = 0.1$.

$$\left\{\begin{matrix} y'' - y' + 2y = 0 \\ y(0) = 1,\ y'(0) = 1 \end{matrix}\right\}$$

We have $x_0 = 0$, $Y_0 = 1$, $V_0 = 1$, $h = 0.1$, and $f(x, Y, V) = -2Y + V$. We apply formulas (6) and (7):

$$k = 0 \left\{\begin{aligned} \bar{y}_1 &= Y_0 + hV_0 = 1.1 \\ \bar{v}_1 &= V_0 + h \cdot f(x_0, Y_0, V_0) = 0.9 \\ Y_1 &= Y_0 + \frac{h}{2}[V_0 + \bar{v}_1] = 1 + \frac{(0.1)}{2}(1 + 0.9) = 1.095 \\ V_1 &= V_0 + \frac{h}{2}[f(x_0, Y_0, V_0) + f(x_1, \bar{y}_1, \bar{v}_1)] \\ &= 1 + \frac{(0.1)}{2}[(-2)(1) + 1 + (-2)(1.1) + 0.9] \\ &= 1 + (0.05)[-2.3] \\ &= 1 - 0.115 = 0.885 \end{aligned}\right\}$$

$$k = 1 \left\{\begin{aligned} \bar{y}_2 &= Y_1 + hV_1 = 1.095 + (0.1)(0.885) = 1.1835 \\ \bar{v}_2 &= V_1 + h \cdot f(x_1, Y_1, V_1) = 0.885 + (0.1)[-2(1.095) + 0.885] \\ &\qquad = 0.8850 - 0.1305 \\ &\qquad = 0.7545 \\ Y_2 &= Y_1 + \frac{h}{2}[V_1 + \bar{v}_2] = 1.095 + \frac{(0.1)}{2}(0.885 + 0.7545) \\ &\qquad = 1.176975 \\ V_2 &= V_1 + \frac{h}{2}[f(x_1, Y_1, V_1) + f(x_2, \bar{y}_2, \bar{v}_2)] \\ &= 0.885 + \frac{(0.1)}{2}[(-2)(1.095) + 0.885 \\ &\quad + (-2)(1.1835) + 0.7545] \\ &= 0.739125 \end{aligned}\right\}$$

If you refer back to Table 4.1, you will immediately see that the results of the improved Euler method are more accurate than those obtained with simple Euler in Example 3. Results obtained from the computer by using a program (listed in Appendix B) called IMEUL2, which incorporates the improved Euler equations (6) and (7), are shown in Table 4.2. This table also shows actual solutions and IMEUL2 solutions for y when $h = 0.01$. Notice that the $h = 0.01$ column gives five significant digits.

Table 4.2

Solution to $y'' - y' + 2y = 0$; $y(0) = 1$, $y'(0) = 1$

	$h = 0.1$		$h = 0.01$		
x	y	v	y	y (actual)	v (actual)
0.0	1.0	1.0	1.0	1.0	1.0
0.1	1.095	0.885	1.0945	1.09450	0.884855
0.2	1.176975	0.739125	1.17596	1.17595	0.739018
0.3	1.24281	0.562177	1.24129	1.24127	0.562326
0.4	1.28941	0.354593	1.28742	1.28740	0.355241
0.5	1.31375	0.117503	1.31137	1.31134	0.118918
0.6	1.31295	−0.147223	1.31029	1.31027	−0.144757
0.7	1.28436	−0.436929	1.28159	1.28157	−0.433123
0.8	1.22564	−0.748154	1.22296	1.22294	−0.742721
0.9	1.13483	−1.07611	1.13248	1.13246	−1.06928
1.0	1.01044	−1.41721	1.00869	1.00868	−1.40774

Example 6

Apply the improved Euler scheme on the interval $[1, 2]$ to the following initial value problem:

$$\left\{\begin{array}{l} 2xy'' - 3y' + 5x^2y = 1 - x \\ y(1) = -0.5,\ y'(1) = 0.2 \end{array}\right\}$$

Because of the variable coefficients, this differential equation does not lend itself to an analytical solution. Our solutions must also stay away from $x = 0$. First replace y' by v and obtain the equivalent system

$$\left\{\begin{array}{l} y' = v;\ y(1) = -0.5 \\ v' = (1 - x - 5x^2y + 3v)/2x;\ v(1) = 0.2 \end{array}\right\}$$

To establish the idea on how the improved Euler scheme works, let's take one step by hand. Picking $h = 0.1$ and using $x_0 = 1$, $y_0 = -0.5$, $v_0 = 0.2$, and $f(x, y, v) = (1 - x - 5x^2y + 3v)/2x$ in equations (6), we get

$$\bar{y}_1 = Y_0 + hV_0 = -0.5 + (0.1)(0.2) = -0.48$$

$$\bar{v}_1 = V_0 + h \cdot f(x_0, Y_0, V_0)$$

$$= (0.2) + (0.1)[1 - 1 - 5(1)^2(-0.5) + 3(0.2)]/2 = 0.355$$

Then using these values in (7), we obtain

$$Y_1 = Y_0 + \frac{h}{2}[V_0 + \bar{v}_1] = -0.5 + \frac{(0.1)}{2}[0.2 + 0.355] = -0.47225$$

$$V_1 = V_0 + \frac{h}{2}[f(x_0, Y_0, V_0) + f(x_1, \bar{y}_1, \bar{v}_1)]$$

$$= (0.2) + \frac{(0.1)}{2}[1.55 + 1.75864] = 0.365432$$

Having now found Y_1 and V_1, we reapply (6) and (7) with $k = 1$ and compute Y_2 and V_2, then Y_3 and V_3, etc., obtaining the approximate solution points: $(1, -0.5)$, $(1.1\ -0.47225)$, $(1.2, Y_2)$, $(1.3, Y_3), \ldots, (2, Y_{10})$. Instead of completing these manually, we can turn to the computer program IMEUL2 and produce the results in Table 4.3.

Table 4.3

Solution to $2xy'' - 3y' + 5x^2y = 1 - x$; $y(1) = -0.5$, $y'(1) = 0.2$ (Improved Euler Method)

x	Y (h = 0.1)	Y (h = 0.02)	Y (h = 0.01)
1.0	−0.5	−0.5	−0.5
1.1	−0.47225	−0.471906	−0.471895
1.2	−0.426949	−0.426358	−0.426341
1.3	−0.362738	−0.36208	−0.362062
1.4	−0.27926	−0.278758	−0.278749
1.5	−0.177242	−0.177287	−0.177296
1.6	−0.590392E-01	−0.059959	−0.599978E-01
1.7	0.715626E-01	0.694093E-01	0.693305E-01
1.8	0.209145	0.205456	0.20533
1.9	0.34678	0.341377	0.341199
2.0	0.476278	0.469169	0.468942

Nowhere in the derivation of the Euler or the improved Euler equations did we assume that f was linear. We need only to program the computer to evaluate the given function f in (4) or in (6) and (7). Thus, numerical methods are no more difficult or time consuming for nonlinear equations than for linear ones. This feature expands greatly the number and types of problems we can work.

Example 7

The initial value problem

$$\begin{Bmatrix} y'' - 2y^3y' = x^2 \\ y(0) = 0,\ y'(0) = 1 \end{Bmatrix}$$

in which the differential equation is nonlinear is handled with ease by using the improved Euler method. Again we let $y' - v$, so $y'' = v'$. We then get

$$\begin{Bmatrix} y' = v;\ y(0) = 0 \\ v' = x^2 + 2y^3v;\ v(0) = 1 \end{Bmatrix}$$

Table 4.4

Solution to $y'' - 2y^3y' = x^2$; $y(0) = 0$, $y'(0) = 1$

x	$Y\ (h = 0.1)$	$Y\ (h = 0.02)$	$Y\ (h = 0.01)$
0.0	0.0	0.0	0.0
0.1	0.1	0.100009	0.100009
0.2	0.20012	0.200164	0.200165
0.3	0.300801	0.300915	0.300918
0.4	0.402931	0.403167	0.403174
0.5	0.508009	0.508443	0.508457
0.6	0.618384	0.619149	0.619174
0.7	0.737712	0.739072	0.739118
0.8	0.871906	0.874454	0.874543
0.9	1.03137	1.03666	1.03685
1.0	1.23707	1.25015	1.25068

Using the computer for our computations, we obtain the values in Table 4.4. The computer program IMEUL2 (improved Euler method for a second-order equation) listed in the Appendix B will allow the numerical integration of any second-order initial value problem. At line 800 of the program, the differential equation is inserted in BASIC. To do this step, solve the differential equation for y'' and code the function $f(x, y, y')$.

Example 8

Given

$$\left\{\begin{matrix} y'' - 4x^2y' + 3y^2 = \sin x \\ y(-1) = 2,\ y'(-1) = 7 \end{matrix}\right\}$$

we would write

$$y'' = f(x, y, y') = \sin x - 3y^2 + 4x^2y'$$

With $y' = v$, this becomes

$$v' = f(x, y, v) = \sin x - 3y^2 + 4x^2v$$

and the initial conditions are $y(-1) = 2$ and $v(-1) = 7$. Then line 800 becomes

```
800 DEF FNF(X, Y, V) = SIN(X) - 3 * Y * Y + 4 * X * X * V
```

and the response to the questions for input might be $-1, 1, 2, 7, 20, 2$. This would provide a step size of $\frac{1-(-1)}{20} = \frac{1}{10}$ and would print every second computed value. On the other hand, the input set $-1, 1, 2, 7, 100, 5$ would provide a step size of $\frac{1-(-1)}{100} = \frac{1}{50}$ and would print every fifth computed value.

If we were given a third-order differential equation with appropriate initial conditions, we would begin its analysis by breaking it into three first-order equations. For example, the equation $y''' = f(x, y, y', y'')$ would become $y' = v, v' = w, w' = f(x, y, v, w)$; and the initial conditions $y(x_0) = y_0$, $y'(x_0) = v_0$, and $y''(x_0) = w_0$ would become $y(x_0) = y_0$, $v(x_0) = v_0$, and $w(x_0) = w_0$, respectively. A computer program IMEUL3 incorporates all of these into a technique for finding the numerical solution of an arbitrary third-order initial value problem. Again at line 800 of IMEUL3 we program the function $f(x, y, v, w)$. Inspect the listing of this program in Appendix B to see how it includes all of the above.

For summarization and comparison, we will now list the two methods for attacking second-order differential equations:

Given:
$$\begin{cases} \text{The function } f(x, y, v) \\ \text{The beginning points } (x_0, y_0) \text{ and } (x_0, v_0) \\ \text{The } x\text{-values } x_{i+1} = x_i + h \end{cases}$$

Euler's method:

$$\begin{cases} Y_1 = y_0 + h(v_0) \\ V_1 = v_0 + h[f(x_0, y_0, v_0)] \end{cases}$$

$$\begin{cases} Y_2 = Y_1 + h(V_1) \\ V_2 = V_1 + h[f(x_1, Y_1, V_1)] \end{cases}$$

$$\vdots$$

Improved Euler method:

$$\begin{cases} \left.\begin{aligned} \bar{y}_1 &= Y_0 + hV_0 \\ \bar{v}_1 &= V_0 + h[f(x_0, Y_0, V_0)] \end{aligned}\right\} & \text{predictors} \\ \left.\begin{aligned} Y_1 &= Y_0 + \frac{h}{2}(V_0 + \bar{v}_1) \\ V_1 &= V_0 + \frac{h}{2}[f(x_0, Y_0, V_0) + f(x_1, \bar{y}_1, \bar{v}_1)] \end{aligned}\right\} & \text{correctors} \end{cases}$$

$$\begin{cases} \left.\begin{aligned} \bar{y}_2 &= Y_1 + hV_1 \\ \bar{v}_2 &= V_1 + h[f(x_1, Y_1, V_1)] \end{aligned}\right\} & \text{predictors} \\ \left.\begin{aligned} Y_2 &= Y_1 + \frac{h}{2}(V_1 + \bar{v}_2) \\ V_2 &= V_1 + \frac{h}{2}[f(x_1, Y_1, V_1) + f(x_2, \bar{y}_2, \bar{v}_2)] \end{aligned}\right\} & \text{correctors} \end{cases}$$

$$\vdots$$

To facilitate the programming of the improved Euler technique for second-order equations a flow chart is given in Figure 4.2.

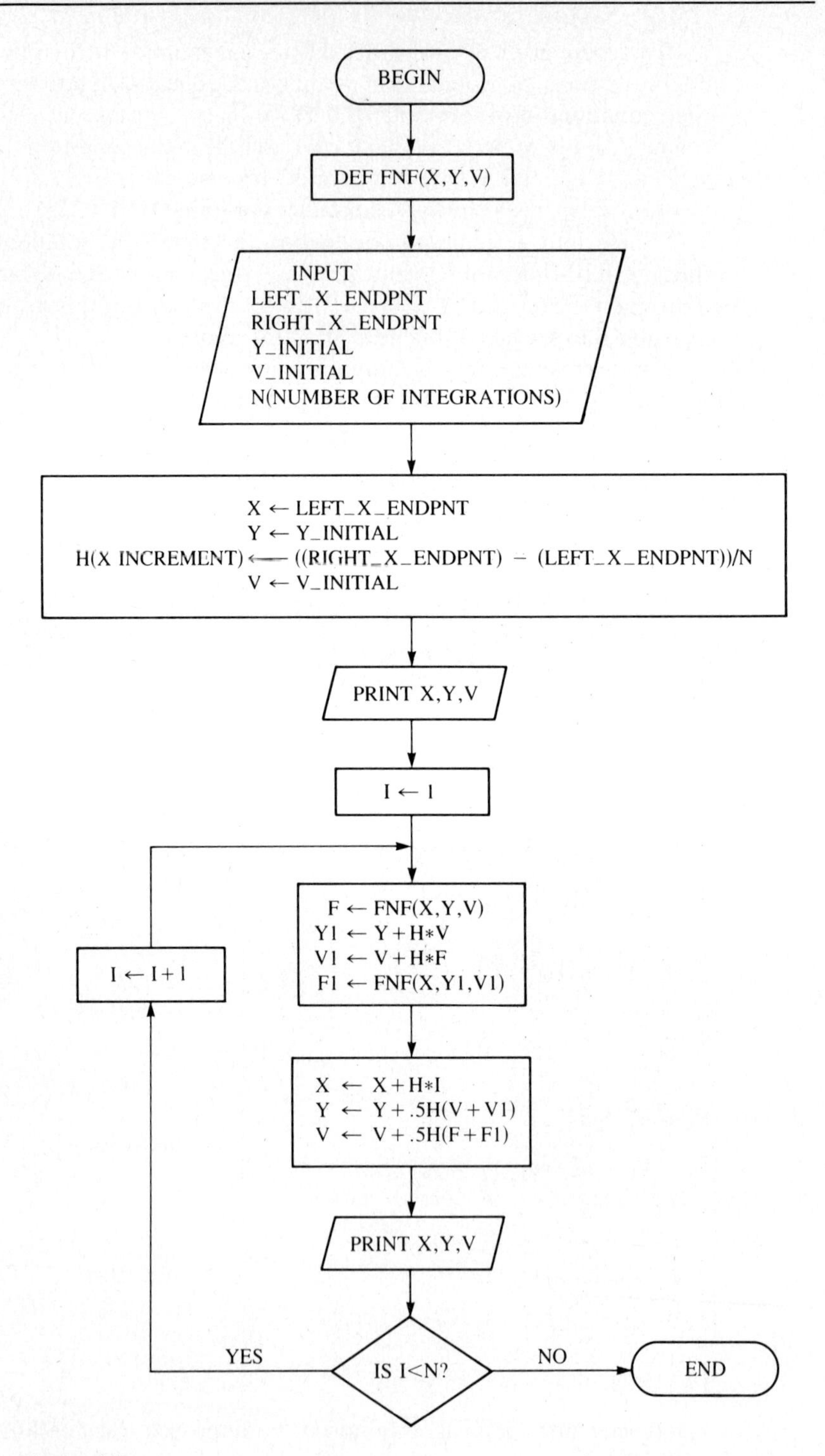
BEGIN
DEF FNF(X,Y,V)
INPUT
LEFT_X_ENDPNT
RIGHT_X_ENDPNT
Y_INITIAL
V_INITIAL
N(NUMBER OF INTEGRATIONS)
X ← LEFT_X_ENDPNT
Y ← Y_INITIAL
H(X INCREMENT) ← ((RIGHT_X_ENDPNT) − (LEFT_X_ENDPNT))/N
V ← V_INITIAL
PRINT X,Y,V
I ← 1
F ← FNF(X,Y,V)
Y1 ← Y + H*V
V1 ← V + H*F
F1 ← FNF(X,Y1,V1)
I ← I + 1
X ← X + H*I
Y ← Y + .5H(V + V1)
V ← V + .5H(F + F1)
PRINT X,Y,V
YES
IS I<N?
NO
END

Figure 4.2

Problems

1. Consider $y'' - 2y - 15 = 0$; $y(0) = -3$, $y'(0) = 0$.
   (a) Solve this initial value problem analytically. Put the solution function into YACTUL on $[0, 1]$.
   (b) With the formulas and by hand calculations, using step size $h = 0.1$, find three new points in the solution set by using Euler's method. Compare with the result from part (a).
   (c) Using the improved Euler scheme, repeat (b) and create a comparison table.
   (d) Solve the problem on $[0, 1]$ by using the computer with the improved method, first with step size $h = 0.1$ and then $h = 0.05$, or until your results are accurate to three significant figures.

2. Consider $y'' + x^2y' - 3xy = x$; $y(0) = 0$, $y'(0) = 1$.
   (a) Carefully following the formulas and using $h = 0.1$, calculate two points in the solution set with Euler's method and hand computation.
   (b) Using the improved Euler scheme, repeat part (a).
   (c) Solve the problem on the interval $[0, 1]$ with the improved Euler method and obtain the result to three-figure accuracy.

3. Given:

$$(x^2 + 1)y'' - (2x + 3)y' + 5x^2y = e^x + 2; \quad y(0) = 3, \quad y'(0) = -\tfrac{1}{2}$$

   Solve by using the improved Euler method. Using a computer, run on $[0, 1]$ and note the behavior near 1. Also observe that we may step negatively by creating the solution on $[0, -1]$.

4. Solve the following equation on the interval $[1, 2]$ and plot the graph:

$$y'' + (2 + \sqrt{x})y = 0; \quad y(1) = 3, \quad y'(1) = 1$$

   Begin with a large step size and then shorten it to obtain four significant figures of accuracy.

5. Solve the following equation on the interval $[0, 2]$ and plot the graph:

$$y^2y'' + (2 - 3x)y' - y = 4x; \quad y(0) = \frac{1}{2}, \quad y'(0) = 0$$

   Obtain four significant figures of accuracy.

6. Consider $y''' + y'' + y' + y = 0$; $y(0) = 1$, $y'(0) = 1$, $y''(0) = 1$.
   (a) Solve analytically.
   (b) Use IMEUL3 to obtain a numerical solution with $h = 0.1$ and then $h = 0.01$. Compare to the actual solution on the interval $[0, 7]$.

7. Use IMEUL3 to solve the following equation on the interval $[0, 2]$:

$$y''' + 2xy'' - x^2y + 4, \quad y(0) = 1; \quad y'(0) = 1, \quad y''(0) = 0$$

   Obtain three-figure accuracy.

8. If an axial load is applied to a prismatic bar and the buckling effect is studied, we obtain

$$\frac{d^3y}{dx^3} + k^2(l - x)\frac{dy}{dx} = 0$$

with $k,\ l > 0$. Using $k = 2.3$, $l = 1.6$, $y(0) = 0.4$, $y'(0) = -0.3$, and $y''(0) = 1.63$, solve for $y(x)$ on $[0, 4]$.

Applications Problems

A–1. **Curves of pursuit**. Suppose that a hawk at point $(a, 0)$ spots a sparrow at the origin flying along the positive vertical axis and proceeding at speed s. The hawk immediately sets out in pursuit of his dinner by flying directly toward the sparrow at a constant speed h (see Figure 4.3). To determine the hawk's path we consider the following: The hawk begins at time $t = 0$; after t seconds (or minutes or hours) the sparrow is at the point $P = (0, st)$, because the distance equals the rate multiplied by the time, and the hawk is at $Q = (x, y)$. Since the line joining P to Q is tangent to the hawk's path, we find that its slope is expressed by

$$y' = \frac{y - st}{x - 0}$$

or

$$xy' - y = -st$$

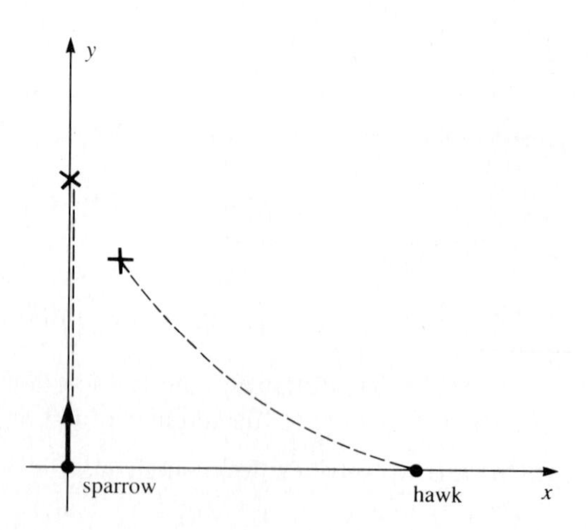

Figure 4.3

The length of the hawk's path can be found by using the arc length formula from calculus

$$-\int_a^x \sqrt{1 + (y')^2}\, dx = (h)(t)$$

Solving both equations for t and equating the results provide

$$\frac{y - xy'}{s} = \frac{-1}{h} \int_a^x \sqrt{1 + (y')^2}\, dx$$

Differentiation provides

$$xy'' = \frac{s}{h}\sqrt{1 + (y')^2}$$

or

$$y'' = \left(\frac{s}{h}\right)\frac{\sqrt{1 + (y')^2}}{x}$$

Natural initial conditions are $y(a) = 0$ and $y'(a) = 0$.

(a) If $s = 16$ mph, $h = 20$ mph, and $a = \frac{1}{8}$ mile, use the program IMEUL2 to find the spot on the y-axis at which the interception is made. Use a small enough step size to get some accuracy.

(b) Try the problem again with $s = 20$, $h = 20$, and $a = \frac{1}{8}$.

A–2. **Oscillation**. Consider a weight attached to two springs and disturbed slightly from the position of rest (see Figure 4.4). The initial value problem for the small oscillations of the mass is

$$y'' = -y^3; \quad y(0) = \frac{2}{10}, \quad y'(0) = 0$$

The solution is a periodic oscillation of period about 40. First use IMEUL2 to run the solution on a quarter period [0, 10] in steps of 0.5 to see the form of the solution. Now use $h = 0.1$ with IMEUL2

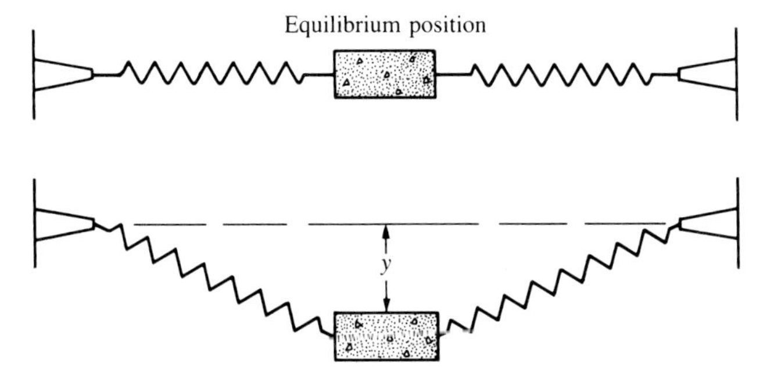

Figure 4.4

on $[0, 10]$. Then with IMEUL2, run on $[0, 80]$ using $h = 0.1$ but with 40 printouts. Sketch the graph and comment on it.

A–3. **Fluid flow**. When one formulates the general dynamic equations to represent the flow of a fluid, it is unfortunate that they are exceedingly difficult to solve. One of the main mathematical difficulties lies in the fact that the equations are nonlinear. Suppose an object is immersed in a fluid and the lines of flow around the object are desired. The problem that is formulated is a problem in boundary-layer theory. In particular, whenever a fluid of low viscosity flows past a solid surface, the transition from the velocity of the surface to that of the stream is accomplished in a narrow layer near the surface. The German mathematician Prandtl first used approximation techniques to develop boundary-layer theory in 1904. In 1908 Blasius first formulated a solvable problem for the steady two-dimensional motion along a flat plate placed edgeways to the stream. With the origin at the forward edge, the general dynamic equations reduce to a third-order nonlinear differential equation

$$y''' = -yy'', \quad y(0) = y'(0) = 0, \quad y''(0) = 1$$

Solve on $[0, 2]$ by using the improved Euler algorithm for third-order equations. For checking your answer, the solution by power-series methods is $y(2) = 1.78809$ and the solution by an advanced numerical method is 1.7880597.

A–4. **Nonlinear mechanics**. Suppose we have a pendulum bob of point mass M on a rigid rod of length L and negligible mass swinging from a friction-free pivot. If we assume there is no air resistance, Newton's laws are sufficient to describe the motion. We use $F = -Mg \sin \theta$, and since force = (mass) (acceleration) we obtain

$$-Mg \sin \theta = M \frac{d}{dt}\left(L \frac{d\theta}{dt}\right)$$

or

$$\frac{d^2\theta}{dt^2} + \frac{g}{L} \sin \theta = 0$$

where θ is displacement from rest position (see Figure 4.5). This nonlinear equation is difficult to handle analytically. The simplifying approach is to assume θ is small so that $\sin \theta \approx \theta$. This gives the equation $\theta''(t) + (g/L)\theta = 0$, which is linear.

(a) If $g = 32$, $L = 2$, and $\theta(0) = \frac{1}{4}$ radian, solve the linear problem analytically. Assume the pendulum is released with no initial velocity.

(b) Solve the nonlinear problem numerically on the time interval $[0, \frac{\pi}{4}]$ obtaining three-figure accuracy. Compare this solution with the solution from part (a).

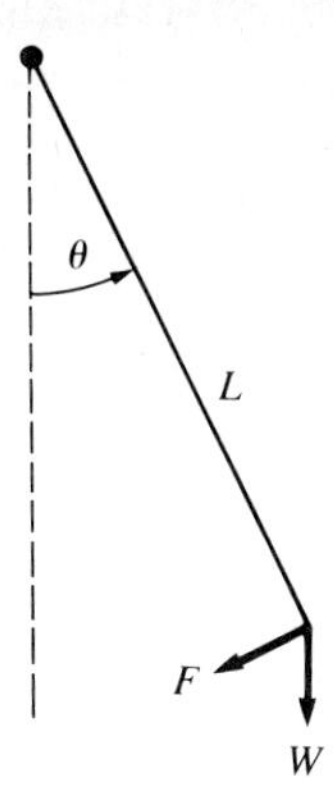

Figure 4.5

(c) Change $\theta(0)$ to 1 radian and repeat parts (a) and (b).
(d) Leaving $\theta(0) = \frac{1}{4}$ radian, change L to 4 and repeat parts (a) and (b). Compare your results with previous answers.

A–5. **Gun installation.** When a shell is fired from a large artillery gun, the barrel recoils on a well-lubricated guide and its rapid motion is braked by a battery of heavy springs, which first stop the recoil and then push the barrel back into firing position. The return motion is also braked by a hydraulic cylinder, called a dashpot, consisting of a closed cylinder within which slides a piston. The differential equations that describe the motion of the barrel, if we assume the firing is horizontal and friction in the guide is negligible, are

$$\frac{d^2x}{dt^2} + \frac{k}{m}x = 0; \quad x(0) = x_0, \quad x'(0) = v_0 \qquad \text{(recoil)}$$

$$\frac{d^2x}{dt^2} + \frac{\mu}{m} \cdot \frac{dx}{dt} + \frac{k}{m}x = 0; \quad x(t_1) = x_1, \quad x'(t_1) = 0 \text{ (return)}$$

where x_1 is the maximum displacement from rest, occurring at time t_1. In these equations m is the mass of the moving system, k is the spring constant of the braking system, and μ is the dashpot coefficient. Consistent constants for the return-motion differential equation are $m = 1$, $k = 16$, $x_1 = 15$, and $\mu = 10$. Let $t_1 = 0$.

(a) Solve the return equation analytically to obtain the actual solution $x(t) = 20e^{-2t} - 5e^{-8t}$. Use YACTUL and observe the smooth return of the barrel on the interval $[0, 8]$.
(b) Solve by using IMEUL2 and compare results.
(c) Change μ to 0.8 (remember that μ measures the coefficient of proportionality between dashpot friction and the speed of the piston). Solve now by using IMEUL2 on $[0, 5]$. Use at least

50 printouts and observe the damped oscillatory motion with period 1.58 seconds. Sketch the graph and note the periods.

A–6. **Earth dams.** In constructing the dikes to hold out the North Sea and reclaim the lowlands, the Dutch created earth dams of the type shown in Figure 4.6. The water that collects on the low side is continually pumped back up to the high side by windmills. The sizing of the earth dam involves solving the seepage problem to minimize the passage of water. This problem falls into the class called free-boundary problems, and one particular application involves the following equation:

$$y'' + 2f(x)y + y^{n/3} = 0; \quad y(0) = 0, \quad y'(0) = 0.75$$

(a) Using $f(x) = \frac{3}{2}\sin 2x$ and $n = 4$, solve for $y(x)$ on $[0, 1.5]$.
(b) Using $f(x) = 4.6\sin 3x$ and $n = 5$, solve for $y(x)$ on $[0, 1]$.

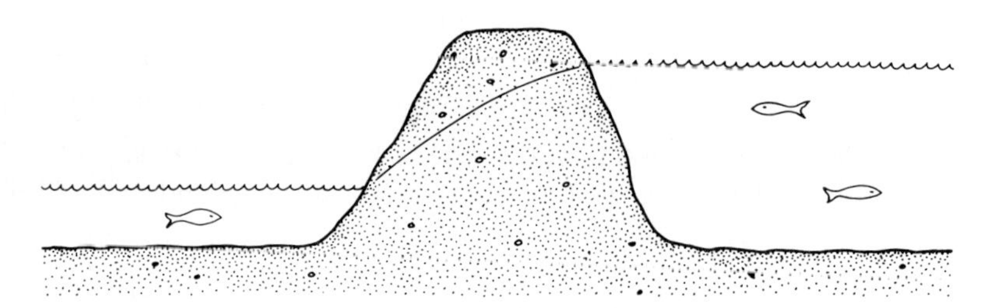

Figure 4.6

A–7. **Stellar structure.** An equation of the type

$$\frac{d^2y}{dx^2} + \frac{2}{x}\cdot\frac{dy}{dx} + f(y) = 0$$

where $f(y)$ is some given function of y, was first studied by Emden in examining the thermal behavior of spherical clouds of gas acting on gravitational equilibrium and subject to the laws of thermodynamics. The case where $f(y) = y^n$, $n > 1$, was treated by Chandrasekhar in his study of stellar structure. The natural initial conditions are $y(0) = 1$ and $y'(0) = 0$, and the only case that can be solved exactly is $n = 5$. In the other cases $n > 1$, solutions can be found only by numerical integration. Use IMEUL2 to solve this equation on $[0.1, 10]$ with $h = 0.1$ for the case $n = 2, 3, 4, 5$. You may shift the initial conditions to take effect at $x = 0.1$. Compare the analytical solution for $n = 5$,

$$y_5 = \left(1 + \frac{1}{3}x^2\right)^{-1/2}$$

with the approximate solution.

A–8. **Vibrations and resonance**. The simplest system available for the study of vibratory motion consists of a weight W supported by a vertically suspended spring whose spring constant is k. Hooke's law and Newton's law produce

$$-kx = \frac{W}{g} \cdot \frac{d^2x}{dt^2}$$

where positive x is downward. It was found experimentally that a 10-pound weight stretches a certain spring 6 inches. The weight is now pulled 4 inches below equilibrium position and released from rest.

(a) Set up the initial value problem and solve it numerically on $[0, \frac{\pi}{2}]$. Sketch the graph.

(b) Since the results of part (a) are unrealistic for actual applications, we must include a damping term that results from friction, air resistance, etc. Usually this force is proportional to the velocity of the weight, so the differential equation becomes

$$-\beta \frac{dx}{dt} - kx = \frac{W}{g} \cdot \frac{d^2x}{dt^2}$$

If for the above spring the damping force is 1.5 times the instantaneous velocity, set up the initial value problem and solve it numerically on $[0, \frac{\pi}{2}]$. Sketch the graph.

(c) Increase the value of β to 4 and solve on $[0, 1]$. This is the case of overdamping in which all oscillation is suppressed.

(d) When oscillation is suppressed only to the degree that any decrease in damping allows oscillations, we have "critical damping." Repeat part (b) with $\beta = 3$. Is this the value for critical damping?

(e) The previous cases have considered only the restoring force of the spring and damping forces of the environment. Suppose now that additional external forces, which are time dependent, act on the spring. These include moving the support or applying a push to the weight at prescribed times. The differential equation now becomes

$$-kx - \beta \frac{dx}{dt} + F(t) = \frac{W}{g} \cdot \frac{d^2x}{dt^2}$$

Assume that $\beta = 1.5$, and let $F(t) = 18 \cos 4t$. Solve the equation on $[0, 4]$ and sketch the graph.

(f) Let the external force now be $3 \cos 8t$, with no damping term included. Solve on $[0, 2\pi]$ and sketch the graph. This depicts the phenomenon of resonance that occurs when the frequency of the applied external force is equal to the natural frequency of the undamped system. When damping is included, resonance occurs when the external force frequency is slightly less than the natural frequency.

(g) Suppose now a new situation for this spring, which produces the differential equation

$$\frac{d^2x}{dt^2} + 100x = 36 \cos 8t; \quad x(0) = x'(0) = 0$$

Solve this equation on [0, 6.3] using 63 printouts, and sketch the graph. This exhibits the phenomenon of amplitude modulation or beats used in music, optics, and electricity.

A–9. **Bending of a beam.** A cantilever beam fixed at the left end and carrying a uniform load of intensity q is shown in Figure 4.7. To obtain the deflection curve for this beam, we use the basic differential equation for uniform loading that utilizes the relationship between curvature and bending moment M:

$$K = -\frac{M}{EI}$$

where I is the moment of inertia and E is the constant modulus of elasticity. Since

$$K = \frac{1}{\rho} = \frac{d^2y}{dx^2}$$

where y is the deflection from the original position, we obtain

$$\frac{d^2y}{dx^2} = -\frac{M}{EI}$$

Now $$M = \frac{9}{2}(L - x)^2$$

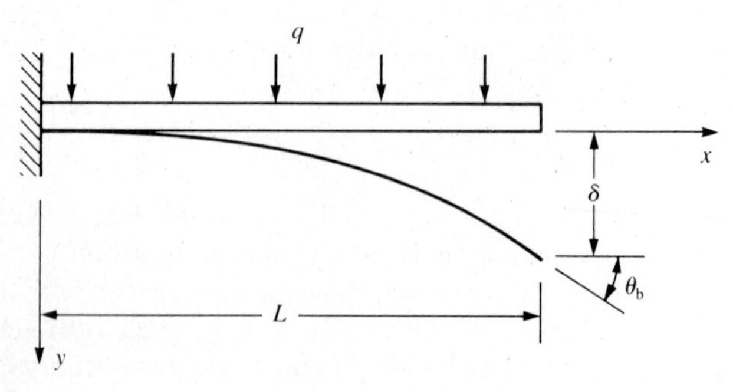

Figure 4.7

where q is the load and L is the length of the beam. Therefore

$$(EI)y'' = \frac{q}{2}(L - x)^2; \quad y'(0) = 0, \quad y(0) = 0$$

To analyze the bending we will use $\theta_b = y'(L)$ and $\delta = y(L)$. Assume that $L = 20$ feet, $q = 1600$ lb/ft, $E = 1.2 \times 10^6$ lb/ft^2, and $I = 2$ (units ft^4). Solve this problem on $[0, 20]$. Compare with the analytical result obtained by integrating. Show that $\theta_b = qL^3/6EI$ and $\delta = qL^4/8EI$. How big is the actual deflection at the free end?

A–10. **Nonlinear mechanics**. In the study of nonlinear mechanics, an equation that arises is van der Pol's equation

$$y'' + \varepsilon(y^2 - 1)y' + y = 0$$

with initial conditions $y(0) = 1$ and $y'(0) = 1$. The parameter ε equals 1 for the elementary case. Use IMEUL2 to solve this equation on $[0, 8]$ with $h = 0.1$, but print at every $\frac{1}{2}$ unit. Now do a hand plot in the yv-plane for the 17 points generated. This is called a phase-plane plot.

Chapter 5

Higher-Order Methods

We have seen that the simple (perhaps crude?) Euler method of solving an initial value problem produces a viable solution but that dramatic improvements in accuracy are attained when the improved Euler technique is used. Chapter 5 discloses some more accurate schemes for numerically integrating a differential equation with appropriate initial conditions.

To gain a good understanding of the techniques, we will concentrate initially on first-order equations. Once the fairly complicated algorithm has been established for the first-order equation, it is easily extended to second-order and then higher-ordered equations. This extension will be studied in Chapter 10.

The investigations by Runge (1856–1927) around 1894 and a follow-up by Kutta (1867–1944) a few years later produced a family of algorithms commonly referred to as Runge-Kutta schemes. These are often called **self-starting** methods.

Independent work by Adams, Bashforth, Moulton, Milne, and others provided alternative techniques known as **predictor-corrector** methods. They often are more efficient on the machine, but generally they are more difficult to use. Frequently a combination of self-starting and predictor-corrector methods is employed in practical situations. The derivations of the methods mentioned here are beyond the scope of this book. They may be found in many advanced numerical analysis texts such as Lambert,* Milne,** and Ralston and Rabinowitz.† However, we will list the techniques, set up the computer routines, and examine a few examples.

5.1 Runge-Kutta Formulas

The methods developed by Runge and Kutta are a particular set of self-starting numerical methods; that is, sufficient initial information is contained in the problem to start the integration. The simple Euler method is a first-order Runge-Kutta method, and improved Euler is a Runge-Kutta method of order two. However, greatly increased efficiency is attained if further refinements are made to these early methods.

In the Euler method we based our replacement value in the integrand only on the value of the function at the last computed point. The improvement made in the second-order method was to look ahead in the direction field for an additional value, which was then averaged with the current value to create the replacement. Our third-order method will combine two slope values in the field ahead with the current value, while the fourth-order method will calculate three slope values in the direction field ahead of the point for combination with current information and a subsequent replacement.

We look first at the actual equations used to move from the current point (x_n, Y_n) in the numerical approximate solution to the next point (x_{n+1}, Y_{n+1}). The formula for the third-order Runge-Kutta (or Runge-Kutta III) method is:

$$Y_{n+1} = Y_n + \frac{1}{6}(k_1 + 4k_2 + k_3) \qquad n = 0, 1, 2, \ldots \tag{1}$$

where

$$k_1 = h \cdot f(x_n, Y_n)$$

$$k_2 = h \cdot f\left(x_n + \frac{1}{2}h,\ Y_n + \frac{1}{2}k_1\right)$$

$$k_3 = h \cdot f(x_n + h,\ Y_n - k_1 + 2k_2)$$

The three points at which function computations are made are those with x-coordinates $x_n, x_n + \frac{1}{2}h$, and $x_n + h$, where $x_n + h = x_{n+1}$. In calculating a numerical solution to an initial value problem using the Runge-Kutta III

---

* J. D. Lambert, *Computational Methods in Ordinary Differential Equations* (New York: John Wiley & Sons, 1973).

** W. E. Milne, *Numerical Solution of Differential Equations* (New York: John Wiley & Sons, 1953).

† A. Ralston and P. Rabinowitz, *A First Course in Numerical Analysis* (New York: McGraw Hill, 1978).

method, one first computes the various k's and then determines Y_{n+1}. Since each k depends on x_n and Y_n, the set of three k values must be newly computed each time a new n is encountered.

The formula for the fourth-order Runge-Kutta method is:

$$Y_{n+1} = Y_n + \frac{1}{6}(k_1 + 2k_2 + 2k_3 + k_4) \tag{2}$$

where
$$k_1 = h \cdot f(x_n, Y_n)$$
$$k_2 = h \cdot f\left(x_n + \frac{1}{2}h,\ Y_n + \frac{1}{2}k_1\right)$$
$$k_3 = h \cdot f\left(x_n + \frac{1}{2}h,\ Y_n + \frac{1}{2}k_2\right)$$
$$k_4 = h \cdot f(x_n + h,\ Y_n + k_3)$$

The points at which function computations are made are those with x-coordinates $x_n, x_n + \frac{1}{2}h$ (twice), and $x_n + h$. When using the Runge-Kutta IV method, we need to recalculate the entire set of four k values each time we move forward to a new point. Runge-Kutta IV requires slightly more machine work than do previous methods, but its increased accuracy compensates adequately for that.

A general study of these methods indicates that a Runge-Kutta scheme of order p requires exactly p function evaluations if $p \leq 4$, and more than p evaluations if $p > 4$. Thus the Runge-Kutta IV method usually gives optimal results.

The pictorial interpretation of this method is more complicated than those presented earlier. The concept of the direction field of a differential equation is crucial here. If this concept is not clear to you, review it in Section 1.7 before reading further.

5.2 Geometric Interpretation of Runge-Kutta Methods

Because of their simplicity, the Euler methods, cannot adequately account for rapid bends in a solution curve. The Runge-Kutta IV scheme uses as much information as can be gained from the differential equation itself in the interval (x_i, x_{i+1}) to get a better estimate for the next solution point (x_{i+1}, Y_{i+1}).

In this method we estimate the true chord slope m, providing the path to the next point, by a weighted average of slopes in the direction field at four specifically chosen points. Beginning at our initial point P_0: (x_0, Y_0) the first slope estimate is k_1, the slope of the tangent line at current point P_0 (see Figure 5.1). This value $k_1 = h \cdot f(x_0, y_0)$ is the Euler slope value studied previously.

We continue by locating the point $\bar{P}_1$: $(x_0 + \frac{1}{2}h, \bar{y}_1)$ halfway along the P_0 tangent line. As indicated in Figure 5.2, the coordinate $\bar{y}_1$ is the average

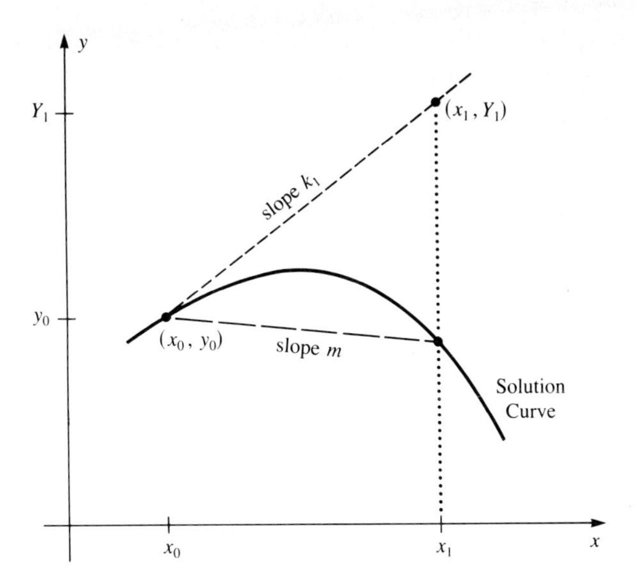

Figure 5.1

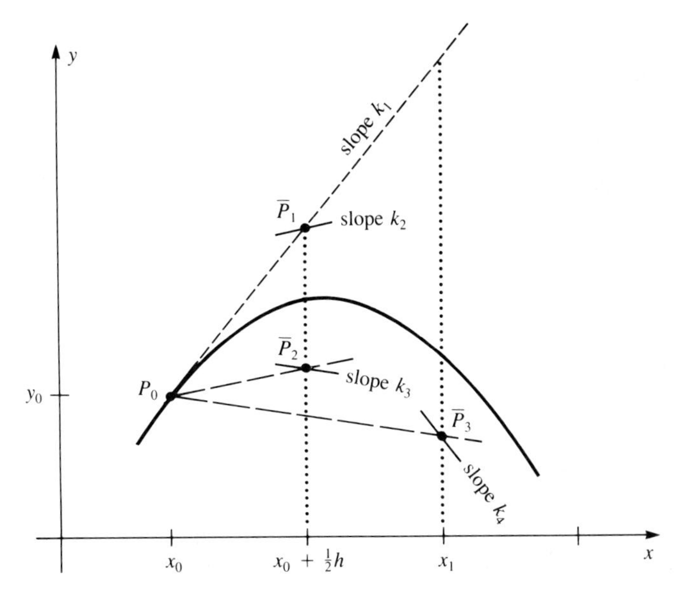

Figure 5.2

of Y_1 and y_0. Since the value of Y_1 is the Euler estimate $Y_1 = y_0 + h \cdot f(x_0, y_0)$, we get

$$\bar{y}_1 = \frac{(Y_1 - y_0)}{2} = \frac{1}{2} h \cdot f(x_0, y_0)$$

or

$$\bar{y}_1 = \frac{1}{2} k_1$$

We now calculate at the point $\bar{P}_1$ the slope of the direction field by using the differential equation

$$y'|_{\bar{P}_1} = f\left(x_0 + \frac{1}{2}h, \bar{y}_1\right)$$

and call it k_2. This is a first estimate for the slope of the solution curve at the actual midpoint $x_0 + \frac{1}{2}h$.

Now shoot a chord from P_0 with slope k_2 to locate another point $\bar{P}_2{:}(x_0 + \frac{1}{2}h, \bar{y}_2)$ halfway between x_0 and x_1. The slope of the direction field there is $k_3 = f(x_0 + \frac{1}{2}h, \bar{y}_2)$. Since $\bar{P}_2$ is on a line from (x_0, y_0) to $(x_0 + \frac{1}{2}h, \bar{y}_2)$ having slope k_2, we find

$$\frac{\bar{y}_2 - y_0}{\frac{1}{2}h} = k_2$$

or

$$\bar{y}_2 = y_0 + h\left(\frac{1}{2}k_2\right)$$

The value of k_3 is a second estimate for the slope of the solution curve at the midpoint $x_0 + \frac{1}{2}h$.

We again shoot a chord from P_0 with slope k_3 to locate $\bar{P}_3{:}(x_0 + h, \bar{y}_3)$ at the right end of the interval. We calculate the slope of the direction field there as $k_4 = f(x_1, \bar{y}_3)$. Since the line from (x_0, y_0) to $(x_1\ \bar{y}_3)$ has slope k_3,

$$\frac{\bar{Y}_3 - Y_0}{x_1 - x_0} = k_3$$

or

$$\bar{y}_3 = y_0 + h(k_3)$$

The computation of k_4 gives a fourth estimate for the slope value of the solution curve, this one calculated at the value x_1.

Now estimate the true chord slope m to be a weighted combination of these:

$$m = \frac{1}{6}(k_1 + 2k_2 + 2k_3 + k_4)$$

The y-coordinate for the new solution point P_1 is

$$Y_1 = Y_0 + m$$

which is exactly the Runge-Kutta IV formula.

Iteration of the process just described produces the successive points on the approximate solution curve. Examples 1 and 2 will give you some experience with the Runge-Kutta methods.

Example 1

Find $Y(1)$ for the following equation by using the third-order Runge-Kutta with $h = 0.1$.

$$y' = y + x; \quad y(0) = 1$$

Here $f(x, y) = y + x$. Using equation (1) we need to compute with $n = 0, 1, \ldots, 9$. We begin the computation:

$$n = 0 \begin{cases} x_0 = 0,\ y_0 = 1 \\ k_1 = h \cdot f(x_0, y_0) = (0.1)(1 + 0) = 0.1 \\ k_2 = h \cdot f\left(x_0 + \frac{1}{2}h,\ y_0 + \frac{1}{2}k_1\right) = h \cdot f\left(0 + \frac{1}{2}(0.1),\ 1 + \frac{1}{2}(0.1)\right) \\ \quad = h \cdot f(0.05, 1.05) = (0.1)(1.05 + 0.05) = 0.11 \\ k_3 = h \cdot f(x_0 + h,\ y_0 - k_1 + 2k_2) = h \cdot f(0.1,\ 1 - 0.1 + 0.22) \\ \quad = h \cdot f(0.1,\ 1.12) = (0.1)(1.12 + 0.1) = 0.122 \\ Y_1 = y_0 + \frac{1}{6}(k_1 + 4k_2 + k_3) = 1 + \frac{1}{6}[0.1 + 4(0.11) + 0.122] \\ \quad = 1 + \frac{1}{6}(0.662) = 1.11033 \end{cases}$$

$$n = 1 \begin{cases} x_1 = 0.1,\ Y_1 = 1.11033 \\ k_1 = h \cdot f(x_1, Y_1) = h \cdot f(0.1, 1.11033) = (0.1)(1.21033) = 0.121033 \\ k_2 = h \cdot f\left(x_1 + \frac{1}{2}h,\ Y_1 + \frac{1}{2}k_1\right) \\ \quad = h \cdot f(0.15, 1.17085) = (0.1)(1.32085) = 0.132085 \\ k_3 = h \cdot f(x_1 + h,\ Y_1 - k_1 + 2k_2) \\ \quad = h \cdot f(0.2, 1.25347) = (0.1)(1.45347) = 0.145347 \\ Y_2 = Y_1 + \frac{1}{6}(k_1 + 4k_2 + k_3) \\ \quad = 1.11033 + \frac{1}{6}(0.121033 + 0.52834 + 0.145347) = 1.24279 \end{cases}$$

$$n = 2 \begin{cases} x_2 = 0.2,\ Y_2 = 1.24279 \\ k_1 = h \cdot f(x_2, Y_2) = h \cdot f(0.2, 1.24279) = (0.1)(1.44279) = 0.144279 \\ k_2 = h \cdot f\left(x_2 + \frac{1}{2}h,\ Y_2 + \frac{1}{2}k_1\right) = (0.1)(1.56493) = 0.156493 \\ k_3 = h \cdot f(x_2 + h,\ Y_2 - k_1 + 2k_2) = (0.1)(1.7115) = 0.17115 \\ Y_3 = Y_2 + \frac{1}{6}(k_1 + 4k_2 + k_3) \\ \quad = 1.24279 + \frac{1}{6}(0.144279 + 0.625972 + 0.17115) = 1.39969 \end{cases}$$

Continuing in this manner through $n = 9$ would finally produce $y(1) = Y_{10} = 3.43636$.

Example 2

Find $Y(1)$ for the following equation by using the fourth-order Runge-Kutta method with $h = 0.1$.

$$\{y' = y + x; \quad y(0) = 1\}$$

Again $f(x, y) = y + x$, and we now use (2) with $n = 0, 1, 2, \ldots, 9$.

$$n = 0 \begin{cases} x_0 = 0,\ y_0 = 1 \\ k_1 = h \cdot f(x_0, y_0) = 0.1 \\ k_2 = h \cdot f\left(x_0 + \frac{1}{2}h,\ y_0 + \frac{1}{2}k_1\right) = h \cdot f(0.05,\ 1.05) = 0.11 \\ k_3 = h \cdot f\left(x_0 + \frac{1}{2}h,\ y_0 + \frac{1}{2}k_2\right) \\ \quad = h \cdot f(0.05, 1.055) = (0.1)(1.105) = 0.1105 \\ k_4 = h \cdot f(x_0 + h,\ y_0 + k_3) \\ \quad = h \cdot f(0.1, 1.1105) = (0.1)(1.2105) = 0.12105 \\ Y_1 = y_0 + \frac{1}{6}(k_1 + 2k_2 + 2k_3 + k_4) = 1 + \frac{1}{6}(0.66205) = 1.11034 \end{cases}$$

$$n = 1 \begin{cases} x_1 = 0.1,\ Y_1 = 1.11034 \\ k_1 = h \cdot f(x_1, Y_1) = (0.1)(1.21034) = 0.121034 \\ k_2 = h \cdot f\left(x_1 + \frac{1}{2}h,\ Y_1 + \frac{1}{2}k_1\right) \\ \quad = h \cdot f(0.15, 1.17086) = (0.1)(1.32086) = 0.132086 \\ k_3 = h \cdot f\left(x_1 + \frac{1}{2}h,\ Y_1 + \frac{1}{2}k_2\right) \\ \quad = h \cdot f(0.15, 1.17638) = (0.1)(1.32638) = 0.132638 \\ k_4 = h \cdot f(x_1 + h,\ Y_1 + k_3) \\ \quad = h \cdot f(0.2, 1.24298) = (0.1)(1.44298) = 0.144298 \\ Y_2 = Y_1 + \frac{1}{6}(k_1 + 2k_2 + 2k_3 + k_4) \\ \quad = 1.11034 + \frac{1}{6}(0.79478) = 1.24281 \end{cases}$$

Continuing in this manner through $n = 9$ would finally produce $y(1) = Y_{10} = 3.43656$.

The exact solution to the problem in Examples 1 and 2 can easily be obtained by analytical techniques. It is

$$y(x) = 2e^x - 1 - x$$

Its values are listed in Table 5.1 for comparison with the approximate results by various methods. With this simple differential equation and with the large step size $h = 0.1$, the results improve greatly when the Runge-Kutta methods are used. In fact, in this situation the fourth-order scheme produces the exact analytical result.

The programs to solve $y' = f(x, y)$, $y(x_0) = y_0$, by the higher-order Runge-Kutta techniques are listed under RUKU3 and RUKU4 in Appendix B. In each case the documentation on the program will advise you on how to use the program. A graphing routine for the numerical output is coupled to the program and may be used as desired.

The process of writing a program to work a particular problem or to display a particular routine is often greatly simplified if a flow chart is first constructed directly from the equations. Once the flow has been carefully outlined, the programming requires only a translation into a convenient language.

To solve the problem

$$y' = f(x, y); \quad y(x_0) = y_0$$

by Runge-Kutta IV, we use equation (2). To avoid more computer processing steps than are necessary, we have used algebra to slightly revise

Table 5.1

Solution to $y' = y + x$; $y(0) = 1$ on interval [0, 1], $h = 0.1$

x	Simple Euler	Improved Euler	Runge-Kutta III	Runge-Kutta IV	y (actual)
0.0	1.0	1.0	1.0	1.0	1.0
0.1	1.1	1.11	1.11033	1.11034	1.11034
0.2	1.22	1.24205	1.24279	1.24281	1.24281
0.3	1.362	1.39847	1.39969	1.39972	1.39972
0.4	1.5282	1.5818	1.5836	1.58365	1.58365
0.5	1.72102	1.79489	1.79738	1.79744	1.79744
0.6	1.94312	2.04086	2.04415	2.04424	2.04424
0.7	2.19743	2.32315	2.3274	2.3275	2.32751
0.8	2.48718	2.64558	2.65095	2.65108	2.65108
0.9	2.8159	3.01236	3.01904	3.0192	3.01921
1.0	3.18748	3.42816	3.43636	3.43656	3.43656

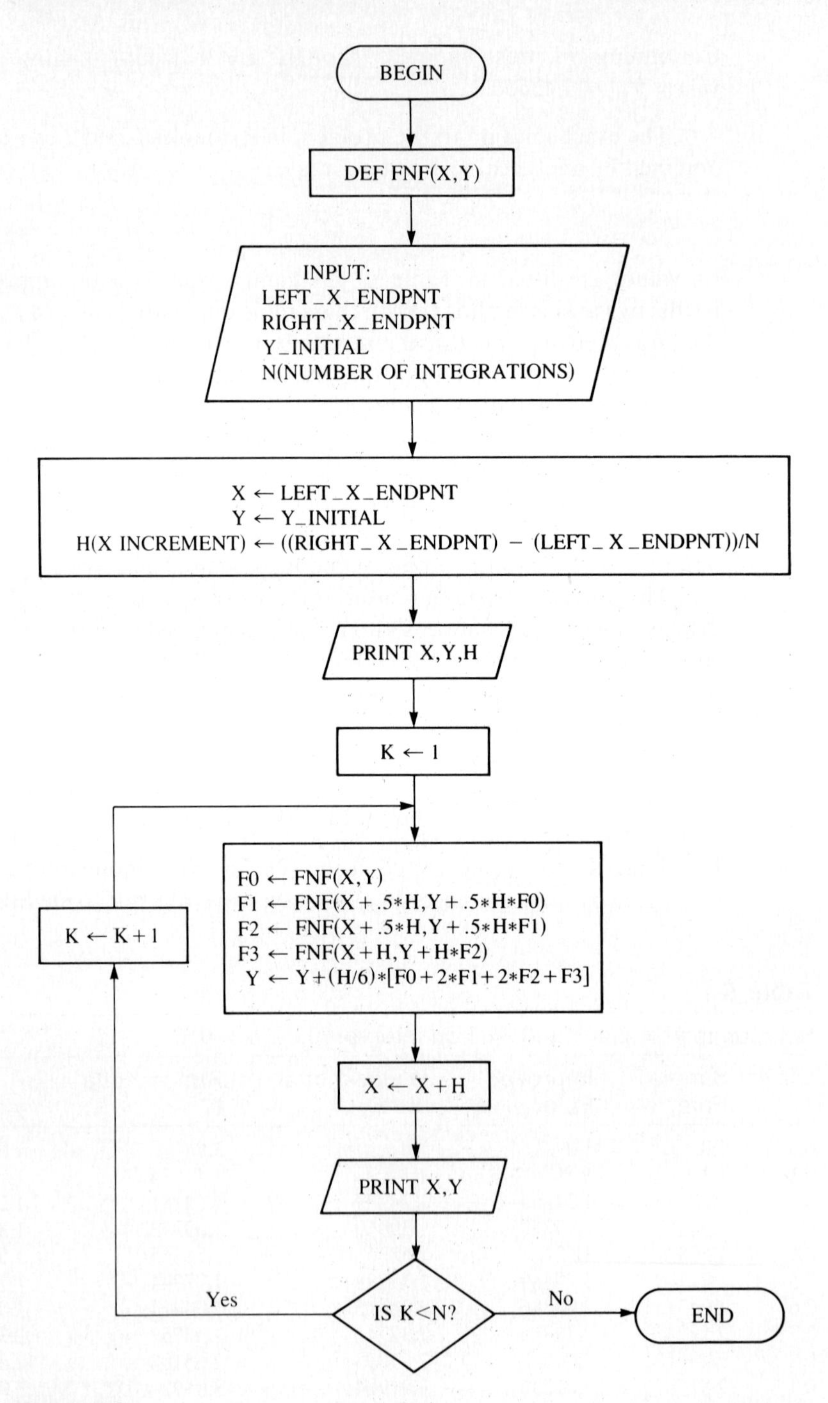
BEGIN
DEF FNF(X,Y)
INPUT:
LEFT_X_ENDPNT
RIGHT_X_ENDPNT
Y_INITIAL
N(NUMBER OF INTEGRATIONS)
X ← LEFT_X_ENDPNT
Y ← Y_INITIAL
H(X INCREMENT) ← ((RIGHT_X_ENDPNT) − (LEFT_X_ENDPNT))/N
PRINT X,Y,H
K ← 1
F0 ← FNF(X,Y)
F1 ← FNF(X+.5*H,Y+.5*H*F0)
F2 ← FNF(X+.5*H,Y+.5*H*F1)
F3 ← FNF(X+H,Y+H*F2)
Y ← Y+(H/6)*[F0+2*F1+2*F2+F3]
K ← K+1
X ← X+H
PRINT X,Y
IS K<N?
Yes
No
END

Figure 5.3

the formula as follows:

$$Y_{n+1} = Y_n + \frac{1}{6}h(K_1 + 2K_2 + 2K_3 + K_4)$$

where

$$K_1 = f(x_n, Y_n)$$

$$K_2 = f\left(x_n + \frac{1}{2}h, Y_n + \frac{1}{2}hK_1\right)$$

$$K_3 = f\left(x_n + \frac{1}{2}h, Y_n + \frac{1}{2}hK_2\right)$$

$$K_4 = f(x_n + h, Y_n + hK_3)$$

These equations and a carefully structured mental organization of what we are trying to do are used to construct a flow chart. The flow chart for integrating a first-order differential equation by using the fourth-order Runge-Kutta technique is shown in Figure 5.3.

A comparison of the program RUKU4 with Figure 5.3 will help you to understand the construction process. Further modifications to this (or any other) program could easily be made by a student who has programming experience.

5.3 Predictor-Corrector Methods

The Runge-Kutta methods for obtaining the approximate solution to

$$y' = f(x, y); \quad y(x_0) = y_0$$

were referred to as self-starting, because all the information needed to get the program started was contained in the initial value problem. The values (x_0, y_0) coupled with $f(x, y)$ produced (x_1, Y_1). This pair coupled with $f(x, y)$ produced (x_2, Y_2). The process continued to (x_3, Y_3), $(x_4, Y_4), \ldots,$ (x_n, Y_n). We will now describe a different approach, sometimes referred to as a **continuing** method, which was first developed by Milne (1890–1971) and published in 1926.

Suppose by referring to some other technique, possibly a Runge-Kutta process, we can find four starting values Y_n, Y_{n-1}, Y_{n-2}, and Y_{n-3} in the solution set. Of course, we know what we have for x_n, x_{n-1}, x_{n-2}, and x_{n-3}. These values (see Figure 5.4) give us four known points on the solution curve from which to launch our solution process for further points. Now using $y' = f(x, y)$, we can compute the functional values at these points and call these f_n, f_{n-1}, f_{n-2}, and f_{n-3} (see Figure 5.5).

To integrate the differential equation, we need an estimate for y'. We first construct the four points (x_{n-3}, f_{n-3}), (x_{n-2}, f_{n-2}), (x_{n-1}, f_{n-1}), and (x_n, f_n) that are in the x-y' pseudo space. Through these four points we

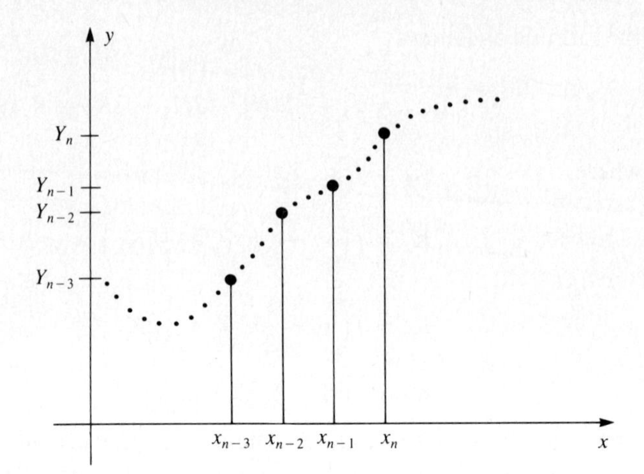

Figure 5.4

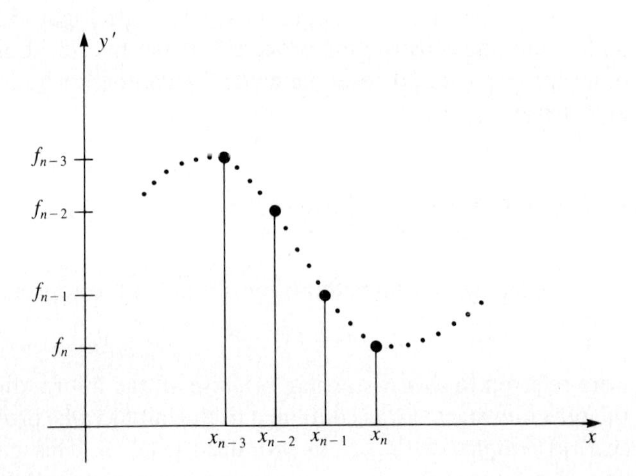

Figure 5.5

construct the unique cubic polynomial in (x, y') coordinates and use it to approximate $y'(x)$. Rewriting the differential equation as $dy = f(x, y)\,dx$ and integrating now from x_{n-3} to x_{n+1}, we get

$$\int_{x_{n-3}}^{x_{n+1}} dy = \int_{x_{n-3}}^{x_{n+1}} y'(x)\,dx$$

or
$$y(x_{n+1}) - y(x_{n-3}) = \int_{x_{n-3}}^{x_{n+1}} y'(x)\,dx$$

or
$$Y_{n+1} - Y_{n-3} = \int_{x_{n-3}}^{x_{n+1}} [\text{cubic polyn}]\,dx$$

or $$Y_{n+1} = Y_{n-3} + \int_{x_{n-3}}^{x_{n+1}} [\text{cubic polyn}]\, dx$$

By performing the integration of the cubic polynomial on the right we arrive at the approximate value for Y_{n+1}. (Note that everything on the right side is known.) We call this approximate value $\bar{y}_{n+1}$; and after some tedious integration work, we obtain the result

$$\bar{y}_{n+1} = Y_{n-3} + \frac{4h}{3}(2f_n - f_{n-1} + 2f_{n-2}) \tag{3}$$

This formula, the **predictor**, is now used to find a corrected value for Y_{n+1}. The derivation, which requires some complicated mathematics, produces the following **corrector** formula, which again is a weighted average of slopes:

$$Y_{n+1} = Y_{n-1} + \frac{h}{3}(f_{n-1} + 4f_n + f_{n+1}) \tag{4}$$

In this formula the f_{n+1} is calculated from $y' = f(x, y)$ by using the $\bar{y}_{n+1}$ obtained from the predictor formula for y and the prechosen x_{n+1}.

Formulas (3) and (4) are the equations for **Milne's method**, which is also a fourth-order method.

Example 3

Find $Y(1)$ for

$$y' = 2y - x; \quad y(0) = 1$$

by using Milne's method with $h = 0.1$. Here $f(x, y) = 2y - x$, $x_0 = 0$, and $y_0 = 1$. Using the values from a run of the same problem with Runge-Kutta IV, we obtain the other three required starting values $Y_1 = 1.21605$, $Y_2 = 1.46886$, and $Y_3 = 1.76658$. Thus

$$f_0 = 2y_0 - x_0 = 2$$

$$f_1 = 2Y_1 - x_1 = 2.3321$$

$$f_2 = 2Y_2 - x_2 = 2.73772$$

$$f_3 = 2Y_3 - x_3 = 3.23316$$

Now using (3) and (4) we will compute Y_4. In this case, $n = 3$.

$$\bar{y}_4 = y_0 + \frac{4h}{3}(2f_3 - f_2 + 2f_1) = 1 + \frac{0.4}{3}(8.3928) = 2.11904$$

Then $$f_4 = 2\bar{y}_4 - x_4 = 3.83808$$

So $$Y_4 = Y_2 + \frac{h}{3}(f_2 + 4f_3 + f_4) = 1.46886 + \frac{0.1}{3}(19.50844) = 2.11914$$

Computing Y_5, we will use $n = 4$. We also have $x_4 = 0.4$ and $f_4 = 2Y_4 - x_4 = 3.83828$.

$$\bar{y}_5 = Y_1 + \frac{4h}{3}(2f_4 - f_3 + 2f_2) = 1.21605 + \frac{0.4}{3}(9.91884) = 2.53856$$

Then
$$f_5 = 2\bar{y}_5 - x_5 = 4.57712$$

So
$$Y_5 = Y_3 + \frac{h}{3}(f_3 + 4f_4 + f_5) = 1.76658 + \frac{0.1}{3}(23.1634) = 2.53869$$

Continuing in this manner we obtain, after using $n = 5, 6, \ldots 9$, $Y(1) = Y_{10} = 6.29172$. The actual solution to the initial value problem is $y = \frac{3}{4}e^{2x} + \frac{1}{2}x + \frac{1}{4}$.

Using programs RUKU4 and MILNE, we can compare the results as demonstrated in Table 5.2. At the point $x = 1$, the MILNE predictor-corrector method reduced the error of 0.00012 with RUKU4 to an error of only 0.00007. In this case the MILNE method produced a significant improvement.

Another fourth-order method, which uses the Milne predictor formula but a modified corrector formula, is called **Hamming's method**.

$$\left.\begin{aligned} &\text{Predictor:} \quad \bar{y}_{n+1} = Y_{n-3} + \frac{4h}{3}(2f_n - f_{n-1} + 2f_{n-2}) \\ &\text{Corrector:} \quad Y_{n+1} = \frac{1}{8}(9Y_n - Y_{n-2}) + \frac{3h}{8}(f_{n+1} + 2f_n - f_{n-1}) \end{aligned}\right\} \qquad (5)$$

where again f_{n+1} is calculated by using the $\bar{y}_{n+1}$ from the predictor.

Table 5.2

Solution to $y' = 2y - x$; $y(0) = 1$, with step size $h = 0.1$

x	Y (RUKU4)	Y (MILNE)	y (actual)
0.0	1.0	1.0	1.0
0.1	1.21605	1.21605	1.21605
0.2	1.46886	1.46886	1.46887
0.3	1.76658	1.76658	1.76659
0.4	2.11914	2.11914	2.11916
0.5	2.53869	2.53869	2.53871
0.6	3.04005	3.04006	3.04009
0.7	3.64135	3.64137	3.6414
0.8	4.36471	4.36473	4.36477
0.9	5.23714	5.23718	5.23724
1.0	6.29167	6.29172	6.29179

Example 4

Find $Y(1)$ for

$$y' = 2y - x; \quad y(0) = 1$$

by using Hamming's method with $h = 0.1$. Again $f(x, y) = 2y - x, x_0 = 0$, and $y_0 = 1$. Using the same starting values as before ($Y_1 = 1.21605$, $Y_2 = 1.46886$, and $Y_3 = 1.76658$) and $f_0 = 2$, $f_1 = 2.3321$, $f_2 = 2.73772$, $f_3 = 3.23316$, and $n = 3$, we obtain from equations (5):

$$\bar{y}_4 = y_0 + \frac{4h}{3}(2f_3 - f_2 + 2f_1) = 2.11904$$

Then $$f_4 = 2\bar{y}_4 - x_4 = 3.83808$$

So $$Y_4 = \frac{1}{8}(9Y_3 - Y_1) + \frac{3h}{8}(f_4 + 2f_3 - f_2)$$

$$= \frac{14.68317}{8} + \frac{(0.3)(7.56668)}{8} = 2.11915$$

Now we use $n = 4$, $x_4 = 0.4$, and $f_4 = 3.83830$:

$$\bar{y}_5 = Y_1 + \frac{4h}{3}(2f_4 - f_3 + 2f_2) = 2.53856$$

Then $$f_5 = 2\bar{y}_5 - x_5 = 4.57712$$

So $$Y_5 = \frac{1}{8}(9Y_4 - Y_2) + \frac{3h}{8}(f_5 + 2f_4 - f_3)$$

$$= \frac{17.60349}{8} + \frac{(0.3)(9.02056)}{8} = 2.53871$$

Continuing in this manner, we obtain after using $n = 5, 6, \ldots, 9$, $Y(1) = Y_{10} = 6.29181$. Table 5.3 compares these results with the actual figures and results obtained by using programs RUKU4 and MILNE. Programs HAMMING, MILNE, RUKU4, and YACTUL are listed in Appendix B with documentation to describe the details of implementation for each program.

In comparing the two methods, we have discovered that when many values of x_n are required, Hamming's method is preferred to Milne's. If only a few values of x_n are needed, then Milne's technique is preferred, because it gives slightly better results.

A word concerning starting values is in order. The methods described above need more initial values than are available from the given first-order initial value problem. Both Milne's and Hamming's methods require y_1, y_2, and y_3 in addition to the given y_0. To generate the starting values for both of these, a fourth-order Runge-Kutta technique is generally used. Great

Table 5.3

Solution to $y' = 2y - x$; $y(0) = 1$, with step size $h = 0.1$.

x	Y (RUKU4)	Y (MILNE)	Y (HAMMING)	y (actual)
0.0	1.0	1.0	1.0	1.0
0.1	1.21605	1.21605	1.21605	1.21605
0.2	1.46886	1.46886	1.46886	1.46887
0.3	1.76658	1.76658	1.76658	1.76659
0.4	2.11914	2.11914	2.11915	2.11916
0.5	2.53869	2.53869	2.53871	2.53871
0.6	3.04005	3.04006	3.04008	3.04009
0.7	3.64135	3.64137	3.6414	3.6414
0.8	4.36471	4.36473	4.36477	4.36477
0.9	5.23714	5.23718	5.23724	5.23724
1.0	6.29167	6.29172	6.29181	6.29179

care must be exercised to make sure that the integration interval size in the RUKU4 problem is identical to the integration step size in the MILNE or HAMMING problem, because the first four number pairs in the two printouts must match. This is the only way that a valid comparison and/or solution can be generated.

5.4 Comparisons

After you have carefully digested these two higher-order approaches, make a comparison of their mechanisms and their respective merits and weaknesses. This section will help you with that task.

In a Runge-Kutta IV method, the function is evaluated four times for each step: once at the beginning of the step, twice in the middle of the step, and once again at the end. The contributions of these evaluations are computed by a weighted average, which produces the resultant increment in the dependent variable. Thus, at each step the new solution point is found from only new information with no benefit from prior work. In each case, the initial conditions for a given step are the results of the previous step. We noted earlier that reducing the step size required increased machine time and thus the attendant possibility of round-off error. These self-starting methods are easier to program, and in most cases the accuracy is easy to establish. Many higher-order Runge-Kutta techniques, which are beyond the scope of this text, are available.

In the predictor-corrector methods of Milne and Hamming and other more sophisticated extensions, the function is evaluated only once in the predictor and once in the corrector at each step. The functional values obtained earlier in the solution process are now used to estimate the solution point via a weighted average, and then the corrected solution point is found by again using a weighted average of functional values. Since only two function evaluations need to be done, machine time is saved in these processes, as the other inputs are already stored for usage.

The major disadvantage of the predictor-corrector methods is the possibility of an oscillating unstable solution that results from a round-off error, or some other inaccuracy, as it is multiplied down the line when successive steps are computed from prior material. Also programming is somewhat more complicated, and starting values for the process must be garnered from some other technique. In this alone, errors may crop up.

Detailed comparison studies have been conducted by numerical analysts on both Runge-Kutta self-starting methods and predictor-corrector methods. Maron* charts the initial value problem solvers and observes their local error, global error, function evaluations per step, and step-size control; and a detailed study of Adam's methods are contained in a text by Shampine and Gordon.** A book by Lambert† provides comparisons of explicit and implicit linear multistep methods, comparisons of Runge-Kutta methods with predictor-corrector methods, and comparisons of hybrid methods with linear multistep and Runge-Kutta methods. Also extensive investigations of many types of multistep, Runge-Kutta, and predictor-corrector methods are listed in a textbook by Lapidus and Seinfeld.††

Problems

1. For the initial value problem

$$y' - 2y = x + 3; \quad y(0) = -1$$

construct a table similar to Table 5.3 and list various solutions to the problem on the interval $[0, 1]$ with step size $h = 0.1$.

2. For the initial value problem

$$xy' + 2y = e^x; \quad y(1) = 2$$

construct a table showing solutions on $[1, 2]$ with $h = 0.1$, by using IMEUL, RUKU4, MILNE, HAMMING, and YACTUL.

3. For the initial value problem

$$y' = 3x^2 - 4; \quad y(0) = 0$$

---

* M. Maron, *Numerical Analysis: A Practical Approach* (New York: Macmillan, 1982), Chapter 8.

** L. F. Shampine and M. K. Gordon, *Computer Solution of Ordinary Differential Equations* (San Francisco: Freeman 1975).

† J. D. Lambert, *Computational Methods in Ordinary Differential Equations* (New York: John Wiley & Sons, 1973).

†† L. Lapidus and J.H. Seinfeld, *Numerical Solution of Ordinary Differential Equations* (Orlando, Fla.: Academic Press, 1971).

(a) Find the analytical solution on $[-3, 3]$ and sketch the graph.
(b) Solve on $[0, 3]$ and $[0, -3]$ by using RUKU4 with $h = 0.2$.
(c) Repeat part (b) with MILNE and compare all results.

4. For the initial value problem

$$y' = 4y^2 - 4x^2; \quad y(0) = \frac{1}{4}$$

(a) Solve on $[0, 1]$ with $h = 0.05$ and program RUKU4.
(b) Repeat part (a) with MILNE and HAMMING.

5. Use Milne's method to solve

$$y' = y^2 + 2; \quad y(0) = -6$$

on the interval $[0, 2]$. Obtain the starting values from RUKU4.

6. Solve the initial value problem

$$y' - 1 = y^{1.2}; \quad y(0) = 0$$

on the interval $[0, 1]$ by using $h = 0.1$ and $h = 0.01$ with IMEUL, RUKU4, and MILNE. Make a table of results.

7. Solvc

$$(x^2 + y^2)y' = 2y; \quad y(0) = 1$$

on the interval $[0, 1]$ with RUKU4 and HAMMING, obtaining three-figure accuracy.

8. Solve

$$y' = \sqrt{\cos y + \sin x}; \quad y(0) = 1$$

on the interval $[0, 2]$ by using RUKU4 and MILNE with $h = 0.2$.

9. Solve

$$y' = \frac{x - y}{x + y}; \quad y(3) = 2$$

on the interval $[2, 4]$ with $h = 0.1$. Choose any appropriate method.

10. Solve

$$y' = \frac{2}{x^2 + y^2 + 1}; \quad y(0) = 0$$

on the interval $[0, 2]$ by using $h = 0.05$ and any appropriate method.

11. Another predictor-corrector method enjoying wide usage is the Adams-Moulton method. It uses the predictor equation

$$\bar{y}_{n+1} = Y_n + \frac{h}{12}(23f_n - 16f_{n-1} + 5f_{n-2})$$

and the corrector equation

$$Y_{n+1} = Y_n + \frac{h}{12}(5f_{n+1} + 8f_n - f_{n-1})$$

where f_{n+1} is calculated from the differential equation by using $\bar{y}_{n+1}$ and x_{n+1}. Modify one of the existing programs (MILNE or HAMMING) to use these equations and run it on the problem

$$y' = x + y; \quad y(0) = 1$$

on the interval $[0, 2]$ with $h = 0.1$. Compare its accuracy against Milne, Hamming, and the actual solution.

12. To test the belief that all analytical methods can be discarded in favor of very accurate approximation schemes, solve the problem

$$y' = \frac{2xy}{x^2 - y^2}; \quad y(1) = 2$$

by using RUKU4 on the interval $[1, 1.5]$ with $h = 0.1$ and $h = 0.025$ and compare the results. Solve the problem analytically to discover why the discrepencies occur.

13. Solve

$$y' = 4x^3 y \cos(x^4); \quad y(0) = 1$$

by using RUKU4 on the interval $[0, 4]$ with $h = 0.1$. This is a separable differential equation. Obtain its solution analytically and explain the problem that appears in your numerical solution.

Applications Problems

Note: Any of the applications problems of Chapter 2 may be repeated now with a higher-order method, and a comparison of accuracy can be made.

A–1. **Biology: compartmental analysis.** A complicated physical or biological problem can often be separated into various distinct stages. The interaction between these individual stages then describes the collective process. Each stage is called a compartment (or cell or block) and the contents of each compartment are assumed to be well mixed. When material from a given cell is transferred to another, it is immediately incorporated into the receiver. Such a process is called a compartmental system (or a block diagram). If there are no inputs to or outputs from the system, it is called closed. Otherwise, it is referred to as an open system. The study of such systems is called compartmental analysis. In studying many diagnostic medical processes, researchers make compartmental analyses. For example, the blood is often treated as a container (or compartment).

Suppose a dose of p_0 milligrams of a drug is injected into the bloodstream and that the drug leaves the bloodstream and enters the urine at a rate proportional to the amount of drug present in the blood. Additionally, the drug is absorbed by a body organ at a rate proportional to the amount of drug present in the blood. Let the proportionality constant for the body organ absorption be 0.05, let the constant for the passage into the urine be 0.17, and let $p_0 = 10$.

(a) Set up the problem for this situation and solve for the amount of time at which the amount of drug in the bloodstream is 25% of its original value.

(b) Solve by using RUKU4 and run past the time found in part (a) until the amount in the bloodstream is 10% of its original value.

A–2. **Chemical reactions**. Consider a chemical mixture in which the concentration $v(t)$ of a single species confined to a closed volume has the same value at each point of the volume. In such a well-stirred system, the chemical law states that the rate at which the reaction is proceeding at any time is proportional to the concentration of chemical present, [i.e., $v'(t) = k(v(t))$, where the constant $k < 0$ is the reaction rate]. Typical reactions are the decomposition of dinitrogen pentoxide into oxygen and nitrogen dioxide and the decomposition of dimethyl ether into methane, carbon monoxide, and hydrogen. Suppose such a reaction is taking place with $k = -\frac{1}{4}$ per second. Let the initial concentration of the chemical be $v(0) = 20$.

(a) Set up the initial value problem for this situation.

(b) Solve it analytically; using YACTUL, run solutions from $t = 0$ to $t = 8$ in steps of $\frac{1}{2}$.

(c) Solve it by using RUKU4 on $[0, 8]$ with $h = 0.1$, but print only values at $0, 0.5, 1, 1.5, \ldots, 8$.

(d) Reduce h until three significant figures are obtained. Compare results with YACTUL.

A–3. **RL circuit**. Consider an electric circuit containing inductance and resistance, called an RL circuit. If no capacitance is present, then the differential equation for the current is $Ly' + Ry = 0$, which is coupled with an initial condition $y(t_0) = y_0$ to form a well-posed initial value problem. (Assume the applied voltage is $E = 0$.) Suppose that the inductance L in an RL circuit with constant resistance R is given by $L = L_0 + \beta^2 t$, where L_0 and β are given constants, and $t_0 = 0$.

(a) Show that the actual solution is given by

$$y = y_0\left(\frac{L_0}{L_0 + \beta^2 t}\right)^{R/\beta^2}$$

(b) Using $R = 2$, $\beta^2 = 1$, $L_0 = 3$, and $y_0 = 0.5$, graph this solution function on $[0, 5]$.

(c) Now use RUKU4 and MILNE on the differential equation over $[0, 5]$ with the constants from part (b). Obtain three significant figures; graph and compare results with parts (a) and (b).

(d) Suppose further that the resistance is not constant but that as current passes through the resistor it heats up and the resistance R increases. Assume $R(t) = 10 + 0.25t$. Find the current by using MILNE on $[0, 5]$ and obtain three significant figures.

A–4. **Physics.** If the magnetic characteristic of a coil wound on an iron core is assumed to be of cubic form, the sudden application of a periodic voltage $e = e_0 \sin \omega t$ across the coil gives rise to the initial value problem

$$e = iR + \frac{d\psi}{dt}$$

where $i = a\psi + b\psi^3$, with $\psi(0) = 0$ (ψ denotes magnetic flux). Define new variables

$$y = \frac{aR\psi}{e_0}$$

and

$$x = aRt$$

This becomes $y' = -y - 2y^3 + \sin 2x$, $y(0) = 0$, for the cases with $be_0^2 = 2a^3R^2$, $2aR = \omega$. (Check this out!) Run on $[-0.2, 0.6]$ by using RUKU4. (To get the initial conditions applied at $x = 0$ you need to be ingenious.)

A–5. **Population growth.** The Dutch mathematical biologist VerHulst did a considerable study of population growth. In 1837 his model for human populations consisted of the differential equation

$$\frac{dp}{dt} = aP - bP^2; \quad P(0) = P_0$$

where a and b are positive constants. To model the growth of the population in the United States he proposed the values $a = .03134$ and $b = (1.5887)10^{-10}$ and used time $t = 0$ as the year 1790 when the population P_0 was 3.9 million people. Using this model and the Runge-Kutta IV process, predict the population at 10-year intervals from 1790 to 1950.

A–6. **Blood analysis.** In studying the infusion of glucose into the bloodstream and its elimination by the body, if we let $c(t)$ denote the concentration of glucose in the blood, v denote the volume of distribution, p the rate of infusion, and k the velocity constant of elimination, then the glucose concentration satisfies

$$\frac{dc}{dt} = \frac{P}{v} - kc$$

Experimental studies show that $k = 0.0519$ and $p/kv = 53.8$. If $c(0) = 13.8$, find the values of $c(t)$ on $[0, 10]$ by using RUKU4.

A–7. **Societal mass behavior**. In studies of mass behaviors of humans, the effects of coercion of the majority by a small minority group (which poses the means for coercion) produce a sharp contrast to the imitative behavior exhibited when the majority of society voluntarily accepts a particular behavior (resulting in a band-wagon effect). We represent by φ the difference between the excitation ε_1 of a center responsible for behavior R_1 and excitation ε_2 of a center responsible for behavior R_2, with the two centers mutually inhibiting each other. Thus

$$\varphi = \varepsilon_1 - \varepsilon_2$$

When $\varphi = 0$, then $\varepsilon_1 = \varepsilon_2$ and there is no preference toward either behavior. A positive φ means preference for behavior R_1, and a negative φ means preference for behavior R_2. The preferences manifest themselves in the magnitude of φ. If ψ denotes the increments of the values of φ, we find that by using the equation of the mathematical theory of the central system that

$$\frac{d\psi}{dt} = C + Af(\psi) - a\psi$$

where C is a constant whose value is determined by the number of persons who always exhibit only R_1 and the number of persons who always exhibit only R_2; where the means available to them to coerce others, A and a, are fixed constants; and where

$$f(\psi) = N(1 - e^{-\alpha\psi})$$

where α is constant and N is the number of individuals who are affected by imitative behavior. Using the values $A = .080$, $a = 4.7$, $C = 3.4$, $\alpha = 5.1$, and $N = 40$, solve for $\psi(t)$ on $[0, 4]$ if $\psi(0) = 0.61$. Graph the result.

A–8. **Population growth**. An equation used by mathematical biologists to describe the growth of certain populations is

$$\frac{dy}{dt} = -ay\ln\left|\frac{y}{b}\right|$$

where a and b are positive constants. This is called the Gompertz growth equation. Assume that a specific species of fish has parameter values $a = 0.43$ and $b = 4.0$ and that $y(0) = 160$.

(a) Solve the initial value problem analytically. Put this solution into YACTUL and run on $[0, 6]$.

(b) Solve on $[0, 6]$ by using IMEUL and RUKU4 with $h = 0.01$ and compare results with part (a).

(c) Using a larger interval of t values, determine the limiting value y_∞ to which this population advances. Use your actual result to prove this.

Chapter 6

Solutions by Series Methods

6.1 Introduction

When we encounter a differential equation

$$y' = F(x, y) \qquad \text{or} \qquad y'' = G(x, y, y') \tag{1}$$

and are interested in an analytical solution, we sort through the special solution techniques studied in Chapters 1 and 3 for an applicable method. If this can be found, the solution is expressed as an n-parameter family of functions where n is the order of the equation. For example, we expressed the analytical solution to $y'' = G(x, y, y')$ in explicit form as $y(x) = c_1u_1(x) + c_2u_2(x)$ where u_1 and u_2 are linearly independent known functions of x. When initial conditions are given, this form allows easy evaluation of the solution functions at a specific point $x = x_1$, and we can obtain an exact result. Tables of values for the common functions often assist us in this evaluation.

Usually this kind of solution cannot be found, and frequently the solution functions defined by the differential equation are not any of the elementary functions that we encountered in calculus. We cannot assume that this limited collection of elementary functions will solve the general first- or second-order equation (1). The representation of a solution function must extend beyond the realm of polynomial, exponential, and trigonometric functions and their inverses. This versatility is obtained by representing the solution function in terms of a power series of the form

$$\sum_{n=0}^{\infty} a_n(x - x_0)^n$$

We are actually seeking the Taylor series expansion of the solution function by using this approach, and the one task that needs completion is the finding of that collection of undetermined coefficients $\{a_n\}, n = 0, 1, 2, \ldots$ so that this power-series function solves the given differential equation.

If we can find two such series that are linearly independent, we can couple them with given initial conditions and write the solution as

$$y(x) = y(x_0)\left(\sum_{n=0}^{\infty} a_n(x - x_0)^n\right) + y'(x_0)\left(\sum_{n=0}^{\infty} b_n(x - x_0)^n\right)$$

Now the two solutions, which formerly were known functions $u_1(x)$ and $u_2(x)$, become infinite-series defined functions whose properties and behavior may not be well known. Nevertheless, we at least have the functions. To now find the value of y at a given value of x, say x_1, we would need to substitute x_1 into the two infinite series and sum them. This is not an easy job. This is the spot at which the computer can become our strong ally, as it is capable of summing an infinite series to many decimals of accuracy in a short time. This technique will become our topic of study in Chapter 7. Thus we again can obtain a large collection of values of the solution function y at various x values in its domain of validity. Although we do not have the solution functions in a handy closed form, we at least have them.

A less convenient case results when we cannot find either an analytical solution or a power-series solution. In these cases we can get a numerical solution, which produces an enumeration of very accurate approximate values of the solution function $y(x)$. Unfortunately, in this case, we do not know anything about the function, but we can obtain its values and its graph as was witnessed by our work on numerical techniques in Chapters 2, 4, and 5. Thus the approach we are about to introduce is a sort of compromise. It is not as good as an analytical solution, but it is usually better than a purely numerical one.

From another point of view, the series solution method is important for applications problems. In many areas of mathematical physics and engineering, second-order linear differential equations with variable coefficients arise and often the series technique is the only analytical method available. In Section 6.4 we will consider some of the classical equations of this type, such as

$$x^2y'' + xy' + (x^2 - n^2)y = 0 \qquad \text{(Bessel's equation)}$$

$$(1 - x^2)y'' - 2xy' + n(n + 1)y = 0 \qquad \text{(Legendre's equation)}$$

In the power-series solution method, we will seek the solution to the differential equation in the form

$$y(x) = \sum_{n=0}^{\infty} a_n(x - x_0)^n$$

$$= a_0(x - x_0)^0 + a_1(x - x_0)^1 + a_2(x - x_0)^2 + \cdots$$

$$+ a_n(x - x_0)^n + \cdots$$

This is typically known as a Taylor series expansion of the function $y(x)$, centered at the given real value x_0, and the coefficients $a_0, a_1, a_2, \ldots, a_n, \ldots$ need to be determined to obtain the unique expansion for the function. This is an approximation to the function and is like an infinite-degree polynomial. Any truncation of it produces an actual polynomial approximation.

The point x_0 used in the trial solution will be a fixed point, generally determined by the given initial conditions, $y(x_0) = y_0$ and $y'(x_0) = v_0$.

This approach substitutes directly into the differential equation the predetermined trial solution and its derivatives and then determines the unknown coefficients a_n. When these have been found we have the power-series solution.

We are assuming here that the power-series solution converges in some interval containing x_0, and this interval can be verified by applying a convergence test to the series of the type generally studied in calculus. If this interval is found, we say that the solution function $y(x)$ is analytic at x_0. Another way to test a function for analyticity at a point is to calculate $\lim_{x \to x_0} f(x)$. If this limit exists the function is analytic at x_0.

Our work will be divided into two parts. First we will seek solutions to differential equations that are analytic at the initial point, and then we will undertake the more difficult task of determining the solution when it is not analytic at the given initial point. To set the tone for our work in these two cases, we will first consider an example where an easy first-order initial value problem is analyzed by the power-series technique.

Example 1

Consider the initial value problem

$$\left\{ \begin{array}{c} y' = xy \\ y(0) = 1 \end{array} \right\} \tag{2}$$

We assume that the solution function is analytic at the initial point $x = 0$ and use as our trial solution

$$y(x) = \sum_{n=0}^{\infty} a_n x^n \tag{3}$$

Thus, since differentiation of a series is allowed term by term, we get

$$y'(x) = \sum_{n=0}^{\infty} n a_n x^{n-1}$$

and our differential equation becomes

$$\sum_{n=0}^{\infty} na_n x^{n-1} = x\left(\sum_{n=0}^{\infty} a_n x^n\right)$$

or

$$\sum_{n=0}^{\infty} na_n x^{n-1} = \sum_{n=0}^{\infty} a_n x^{n+1}$$

Writing out the terms of these series both on the right and left sides, we obtain

$$\begin{aligned} 0 + 1a_1x^0 + 2a_2x^1 + 3a_3x^2 + 4a_4x^3 + 5a_5x^4 + \cdots \\ = a_0x^1 + a_1x^2 + a_2x^3 + a_3x^4 + a_4x^5 + \cdots \end{aligned}$$

By equating coefficients of like powers of x we get a set of relations concerning the unknown coefficients: $a_1 = 0$, $2a_2 = a_0$, $3a_3 = a_1$, $4a_4 = a_2$, $5a_5 = a_3$, and, in general, $na_n = a_{n-2}$ for $n \geq 2$.

Since $a_1 = 0$, it is easy to see that $a_3 = 0, a_5 = 0, \ldots, a_{2n+1} = 0, \ldots$ If we now solve the remaining collection of these recurrence equations for the higher subscripted constants in terms of the lower subscripted ones, we find

$$a_2 = \frac{1}{2} a_0, \quad a_4 = \frac{1}{4} a_2, \quad a_6 = \frac{1}{6} a_4, \ldots, a_{2n} = \frac{1}{2n} a_{2n-2}$$

These imply

$$a_4 = \frac{1}{4} \cdot \frac{1}{2} a_0, \quad a_6 = \frac{1}{6} \cdot \frac{1}{4} \cdot \frac{1}{2} a_0, \ldots,$$

$$a_{2n} = \frac{1}{2n} \cdot \frac{1}{2n-2} \cdot \frac{1}{2n-4} \cdot \cdots \cdot \frac{1}{2} a_0$$

Thus

$$a_{2_n} = \frac{1 \cdot a_0}{2 \cdot 4 \cdot 6 \cdot 8 \cdot \cdots \cdot 2n} = \frac{1}{2^n \cdot n!} a_0$$

We now go back to our original trial solution (3) and substitute these values back in. Thus

$$\begin{aligned} y(x) &= a_0 + a_1x + a_2x^2 + a_3x^3 + a_4x^4 + a_5x^5 \\ &\quad + a_6x^6 + \cdots + a_nx^n + \cdots \\ &= a_0 + 0 + \frac{1}{2} a_0x^2 + 0 + \frac{1}{2 \cdot 4} a_0x^4 + 0 \\ &\quad + \frac{1}{2 \cdot 4 \cdot 6} a_0x^6 + 0 + \cdots + \frac{a_0}{2^n n!} x^{2n} + \cdots \\ &= a_0\left(1 + \frac{1}{2}x^2 + \frac{1}{2 \cdot 4}x^4 + \frac{1}{2 \cdot 4 \cdot 6}x^6 + \cdots + \frac{1}{2^n n!}x^{2n} + \cdots\right) \\ &= a_0\left(\sum_{n=0}^{\infty} \frac{1}{2^n n!} x^{2n}\right) \end{aligned}$$

This is the power-series solution to the given differential equation. Now applying the initial condition that y is 1 when x is 0 we find

$$1 = a_0(1 + 0 + 0 + 0 + \cdots) = a_0$$

Thus the unique solution to the initial value problem (2) is

$$y(x) = \sum_{n=0}^{\infty} \frac{1}{2^n n!} x^{2n} \tag{4}$$

The ratio test will easily confirm that this series converges for all real x.

Since this problem was an easy setup we can solve it analytically for comparison. Separation of variables produces the solution

$$y(x) = e^{x^2/2}$$

Since from our calculus experience we know the expansion for $e^x = 1 + x + x^2/2! + x^3/3! + \cdots$, we can easily obtain the expansion for this solution function:

$$y(x) = e^{x^2/2} = 1 + \frac{x^2}{2} + \frac{x^4}{2^2 \cdot 2!} + \frac{x^6}{2^3 \cdot 3!} + \frac{x^8}{2^4 \cdot 4!} + \cdots$$

which is, of course, identical to the series solution (4).

It is important to emphasize that (4) would still be a solution of (2) if we were unable to recognize that the series in (4) converged to the function $e^{x^2/2}$. Generally one does not see that the series solutions converge to familiar elementary functions, because the power-series solutions are usually not elementary functions. Whenever we recognize the power-series solution as the expansion of a familiar function, we collect a bonus. By inserting the function, we have an analytical solution to a problem that originally did not appear to be solvable analytically.

Exercises 6.1

Find a power-series solution to the following problems.

1. $\dfrac{dy}{dx} = x + y$; $y(0) = 2$. Also solve analytically and compare answers.
2. $y' = x^2 + y$; $y(0) = 3$
3. $(1 + x^2)y' - y = 0$. Also solve analytically and compare answers.
4. $y'' = xy$
5. $x^{-1}y' = y + 1$; $y(0) = 2$
6. $y' = 4y + x$; $y(0) = 1$. Also solve analytically and compare answers.

6.2 Solution at an Ordinary Point

In this section we will study the linear homogeneous differential equations, which have the form

$$a_0(x)y'' + a_1(x)y' + a_2(x)y = 0$$

We will assume $a_0(x) \neq 0$ and divide by this function to obtain

$$y'' + \left[\frac{a_1(x)}{a_0(x)}\right] y' + \left[\frac{a_2(x)}{a_0(x)}\right] y = 0$$

We will then test the coefficient functions $a_1(x)/a_0(x)$ and $a_2(x)/a_0(x)$ for an important property, as outlined in the following definition.

Definition: The point $x = x_0$ is an **ordinary point** of the differential equation $y'' + p(x)\,y' + q(x)\,y = 0$ if the functions $p(x)$ and $q(x)$ are analytic at $x = x_0$.

In the original differential equation if the coefficient of y'' is not unity, we must divide by that coefficient and then inspect the resulting coefficient functions. All constant functions and all polynomial functions are analytic at all x values. In addition, the familiar functions $\sin x$, $\cos x$, $\tan x$, $\ln x$, e^x, $\arccos x$, and $\arctan x$, are analytic at all x values inside their domains.

It is easy to show that if functions f and g are analytic at x_0, then $f + g$ and fg are also analytic at x_0. Moreover, if $g(x_0) \neq 0$, then f/g is also analytic at x_0. Hence the quotient of two analytic polynomial functions is analytic at all points that are not zeros of the denominator polynomial.

Example 2

The differential equation $y'' + 2xy' - 4y = 0$ has coefficient functions that are analytic for all real x. Hence all x points are ordinary points.

Example 3

The equation $(x - 1)\,y'' + xy' - 3x^2y = 0$ must first be written as

$$y'' + \frac{x}{x-1}\,y' - 3\,\frac{x^2}{x-1}\,y = 0$$

Now it is evident that at $x = 1$ the coefficient functions are undefined and hence nonanalytic. Thus the point $x = 1$ is not an ordinary point, but all other x values are ordinary points.

If an initial point x_0 is defined for our problem by auxiliary conditions, we must determine first whether or not that point is an ordinary point. If it is, we may then proceed to an analytic solution function by using the trial function

$$y(x) = \sum_{n=0}^{\infty} a_n(x - x_0)^n$$

If not, we defer our work to Section 6.4.

Before proceeding with some sample problems that illustrate the technique to be used, we will briefly review some important properties of power series.

Property A

A power series in $(x - x_0)$ is an infinite series of the form

$$\sum_{n=0}^{\infty} a_n(x - x_0)^n$$

where the $a_0, a_1, \ldots$ are constants called the coefficients of the series and x_0 is a constant called the center of the series. In particular, if it is centered at $x = 0$, the series has the form

$$\sum_{n=0}^{\infty} a_n x^n$$

Property B

A series not centered at zero can be altered to the form

$$\sum_{n=0}^{\infty} a_n x^n$$

by the linear shift $X = x - x_0$. The behavior of the altered series near zero is exactly the same as the original series near $x = x_0$.

Property C

Every power series has an interval of convergence with the center point at its center and a nonnegative radius distance R such that the series converges inside this interval and diverges outside the interval.

Property D

A power series may be differentiated term by term, and the resultant series converges in the same interval of convergence.

Property E

The ratio test is an easy way to determine the interval of convergence of a power series:

$$\frac{1}{R} = \lim_{n \to \infty} \left| \frac{a_{n+1}}{a_n} \right|$$

Property F

A shift of the index for a power series does not alter the terms of the series, that is,

$$\sum_{n=k}^{\infty} a_n x^n = \sum_{n=k-L}^{\infty} a_{n+L} x^{n+L}$$

Property G

If a power series

$$\sum_{n=0}^{\infty} c_n(x - x_0)^n$$

has a sum that is identically zero for all $x_0 - R < x < x_0 + R$, then each coefficient is zero.

In those problems for which a power-series solution is appropriate, an initial point x_0 is not always provided. In that case we assume that $x = 0$ is the initial point, and we center our power-series solution at the point $x = 0$ and obtain an interval of convergence centered there.

Example 4

Consider the differential equation

$$y'' + xy' + y = 0 \tag{5}$$

Without initial conditions we will specify that $x = 0$ is our initial point and center the power-series solution at $x = 0$. Then our trial solution will be the same as equation (3):

$$y(x) = \sum_{n=0}^{\infty} a_n x^n$$

We will be finding values for the a_n coefficients as we complete this solution. By property D, we get

$$y' = \sum_{n=0}^{\infty} n a_n x^{n-1}$$

for which

$$xy' = \sum_{n=0}^{\infty} n a_n x^n$$

Also

$$y'' = \sum_{n=0}^{\infty} n(n-1) a_n x^{n-2}$$

Substituting these values into the differential equation gives

$$\sum_{n=0}^{\infty} n(n-1) a_n x^{n-2} + \sum_{n=0}^{\infty} n a_n x^n + \sum_{n=0}^{\infty} a_n x^n = 0$$

Applying property F to the first term by replacing n by $n + 2$ we obtain

$$\sum_{n=-2}^{\infty} (n+2)(n+1) a_{n+2} x^n + \sum_{n=0}^{\infty} n a_n x^n + \sum_{n=0}^{\infty} a_n x^n = 0$$

If we write out the terms for $n = -2$ and $n = -1$ in the first series we see that they are zero and thus may be discarded. Hence the differential equation becomes

$$\sum_{n=0}^{\infty} (n+2)(n+1) a_{n+2} x^n + \sum_{n=0}^{\infty} n a_n x^n + \sum_{n=0}^{\infty} a_n x^n = 0$$

We can now continue to the next step, because we have attained the following important properties: (1) The power of x is the same for each term, and (2) the lower summand is the same for each sum. We now combine the three sums into one "super sum":

$$\sum_{n=0}^{\infty} \{(n+2)(n+1) a_{n+2} + (n+1) a_n\} x^n = 0$$

Property G now allows us to write

$$(n+2)(n+1) a_{n+2} + (n+1) a_n = 0 \qquad \text{for all } n \geq 0 \tag{6}$$

This becomes what we will refer to as our **recurrence relation**. From this relation we will always solve for the highest subscripted a in terms of lower subscripted a's. Equation (6) can be reorganized to

$$a_{n+2} = -\frac{a_n}{n+2} \tag{7}$$

Thus we obtain a formula for finding any coefficient in our solution function in terms of a prior one for $n \geq 0$. Since the lowest subscripted a we can find from this relation is a_2, we must consider a_0 and a_1 to be arbitrary and find all other coefficients in terms of these two.

The theory blends together nicely at this point because these two arbitrary coefficients become the two parameters in the general solution to the given differential equation (5). If initial conditions had been specified at the beginning, the values of a_0 and a_1 would simply calculate to be the values of y and y' at $x = 0$.

Writing out some of the terms of (7) we find

$$a_2 = -\frac{a_0}{2}, \qquad a_4 = -\frac{a_2}{4} = \frac{a_0}{2 \cdot 4}, \qquad a_6 = -\frac{a_4}{6} = -\frac{a_0}{2 \cdot 4 \cdot 6}, \ldots$$

$$a_3 = -\frac{a_1}{3}, \qquad a_5 = -\frac{a_3}{5} = \frac{a_1}{3 \cdot 5}, \qquad a_7 = -\frac{a_5}{7} = -\frac{a_1}{3 \cdot 5 \cdot 7}, \ldots$$

We see the pattern that develops and this allows us to write the general terms:

$$a_{2n} = \frac{(-1)^n a_0}{2 \cdot 4 \cdot 6 \cdots (2n)} = \frac{(-1)^n a_0}{2^n n!}, \quad n \geq 0$$

and

$$a_{2n+1} = \frac{(-1)^n a_1}{3 \cdot 5 \cdot 7 \cdots (2n+1)}, \quad n \geq 0$$

Thus the solution (3) is now written

$$\begin{aligned} y(x) &= a_0 + a_1 x + a_2 x^2 + a_3 x^3 + a_4 x^4 + a_5 x^5 + a_6 x^6 + \cdots \\ &= a_0 + a_1 x - \frac{a_0}{2} x^2 - \frac{a_1}{3} x^3 + \frac{a_0}{2 \cdot 4} x^4 + \frac{a_1}{3 \cdot 5} x^5 + \cdots \\ &= a_0\left(1 - \frac{1}{2} x^2 + \frac{1}{2 \cdot 4} x^4 - \cdots + \frac{(-1)^n}{2^n n!} x^{2n} + \cdots\right) \\ &\quad + a_1\left(x - \frac{1}{3} x^3 + \frac{1}{3 \cdot 5} x^5 - \cdots + \frac{(-1)^n}{3 \cdot 5 \cdot 7 \cdots (2n+1)} x^{2n+1} + \cdots\right) \\ &= a_0\left(\sum_{n=0}^{\infty} \frac{(-1)^n}{2^n n!} x^{2n}\right) + a_1\left(\sum_{n=0}^{\infty} \frac{(-1)^n}{1 \cdot 3 \cdot 5 \cdots (2n+1)} x^{2n+1}\right) \end{aligned} \tag{8}$$

The two series converge for all x, as the ratio test will verify. Equation (8), which is equivalent to

$$y = a_0(u_1(x)) + a_1(u_2(x))$$

corresponds to the general solution in Chapter 3 for a second-order equation:

$$y(x) = c_1 u_1(x) + c_2 u_2(x)$$

Since the functions $u_1(x)$ and $u_2(x)$ in our present example are linearly independent (their powers of x are different), (8) is the general solution to (5).

In this problem we are able to collect a partial bonus. The solution function

$$u_1(x) = \sum_{n=0}^{\infty} \frac{(-1)^n}{2^n n!} x^{2n}$$

may be written as

$$u_1(x) = \sum_{n=0}^{\infty} \frac{1}{n!}\left(-\frac{x^2}{2}\right)^n$$

Since we know from our work in calculus with Maclaurin Expansions that

$$e^x = \sum_{n=0}^{\infty} \frac{1}{n!}(x)^n$$

We see that

$$u_1(x) = e^{-x^2/2}$$

The series for $u_2(x)$ is somewhat more complicated and does not provide the bonus. The graphs of $u_1(x)$ and $u_2(x)$ are illustrated in Figures 6.1 and 6.2.

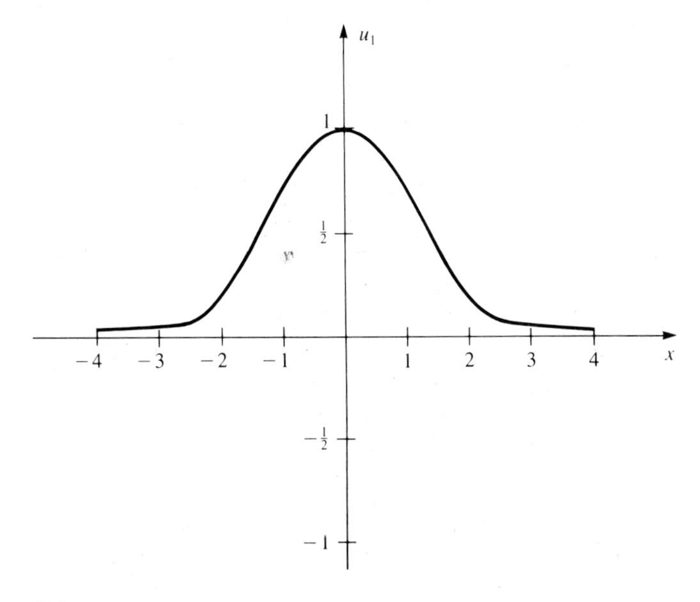

Figure 6.1

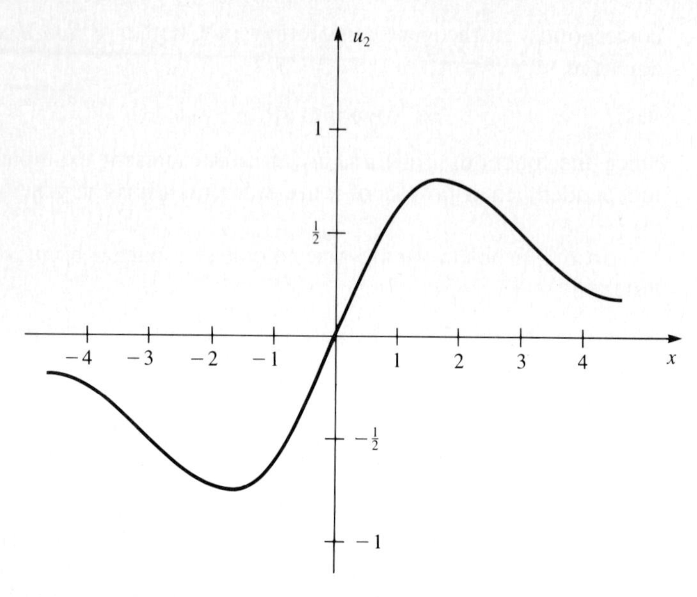

Figure 6.2

Example 5

As another example of this method consider

$$(x^2 + 3)y'' + 3xy' + y = 0 \tag{9}$$

Without specified initial conditions, we assume $x = 0$ is our point of interest. Since $x = 0$ is an ordinary point (the coefficient functions $3x/(x^2 + 3)$ and $1/(x^2 + 3)$ are not singular at $x = 0$) we use the trial solution

$$y(x) = \sum_{n=0}^{\infty} a_n x^n$$

As before,
$$y'(x) = \sum_{n=0}^{\infty} na_n x^{n-1}$$

so
$$3xy' = \sum_{n=0}^{\infty} 3na_n x^n$$

and
$$y''(x) = \sum_{n=0}^{\infty} n(n-1)a_n x^{n-2}$$

so
$$x^2 y'' = \sum_{n=0}^{\infty} n(n-1)a_n x^n$$

The differential equation now becomes

$$\sum_{n=0}^{\infty} n(n-1)a_n x^n + \sum_{n=0}^{\infty} 3n(n-1)a_n x^{n-2} + \sum_{n=0}^{\infty} 3na_n x^n + \sum_{n=0}^{\infty} a_n x^n = 0$$

Since the second sum is not matched in power of x to the others, we perform the index shift and replace n by $n + 2$.

$$\sum_{n=0}^{\infty} n(n-1)a_n x^n + \sum_{n=-2}^{\infty} 3(n+2)(n+1)a_{n+2}x^n + \sum_{n=0}^{\infty} 3na_n x^n + \sum_{n=0}^{\infty} a_n x^n = 0$$

In the second sum, the term for $n = -2$ is zero and the term for $n = -1$ is zero, so we may discard them. We can now combine the sums since the two earlier-listed criteria are met. The differential equation is now

$$\sum_{n=0}^{\infty} \{3(n+2)(n+1)a_{n+2} + [n(n-1) + 3n + 1]a_n\}x^n = 0$$

Property G allows us to set the bracketed coefficient to zero and solve for the higher subscripted a_{n+2} in terms of a_n to obtain

$$a_{n+2} = \frac{-(n^2 + 2n + 1)a_n}{3(n+2)(n+1)}$$

and in lowest terms this becomes

$$a_{n+2} = \frac{-(n+1)}{3(n+2)} a_n, \quad n \geq 0$$

Expanding these coefficients and combining them in terms of a_0 and a_1, the arbitrary coefficients, we get

$$a_2 = \frac{-1}{3\cdot 2} a_0, \quad a_4 = \frac{-3}{3\cdot 4} a_2 = \frac{3\cdot 1}{3^2\cdot 2\cdot 4} a_0,$$

$$a_6 = \frac{-5}{3\cdot 6} a_4 = \frac{-5\cdot 3\cdot 1}{3^3\cdot 2\cdot 4\cdot 6} a_0, \ldots$$

$$a_3 = \frac{-2}{3\cdot 3} a_1, \quad a_5 = \frac{-4}{3\cdot 5} a_3 = \frac{4\cdot 2}{3^2\cdot 3\cdot 5} a_1,$$

$$a_7 = \frac{-6}{3\cdot 7} a_5 = \frac{-6\cdot 4\cdot 2}{3^3\cdot 3\cdot 5\cdot 7} a_1, \ldots$$

Observing the patterns shown in developing these coefficients we are able to write the general forms

$$a_{2n} = \frac{(-1)^n\cdot 1\cdot 3\cdot 5\cdots(2n-1)}{3^n\cdot 2\cdot 4\cdot 6\cdots(2n)} a_0, \quad n \geq 1$$

$$a_{2n+1} = \frac{(-1)^n\cdot 2\cdot 4\cdot 6\cdots(2n)}{3^n\cdot 3\cdot 5\cdot 7\cdots(2n+1)} a_1, \quad n \geq 1$$

Now inserting these coefficients back into the trial solution gives

$$\begin{aligned}
y(x) &= a_0 + a_1 x + a_2 x^2 + a_3 x^3 + a_4 x^4 + a_5 x^5 + \cdots \\
&= a_0 + a_1 x + \frac{-1}{3 \cdot 2} a_0 x^2 + \frac{-2}{3 \cdot 3} a_1 x^3 + \frac{3 \cdot 1}{3^2 \cdot 2 \cdot 4} a_0 x^4 \\
&\quad + \frac{4 \cdot 2}{3^2 \cdot 3 \cdot 5} a_1 x^5 + \cdots \\
&= a_0\left(1 - \frac{1}{3 \cdot 2} x^2 + \frac{3 \cdot 1}{3^2 \cdot 2 \cdot 4} x^4 - \cdots\right) \\
&\quad + a_1\left(x - \frac{2}{3 \cdot 3} x^3 + \frac{4 \cdot 2}{3^2 \cdot 3 \cdot 5} x^5 - \cdots\right) \\
&= a_0\left(1 + \sum_{n=1}^{\infty} \frac{(-1)^n \cdot 1 \cdot 3 \cdot 5 \cdots (2n-1)}{3^n \cdot 2 \cdot 4 \cdot 6 \cdots (2n)} x^{2n}\right) \\
&\quad + a_1\left(x + \sum_{n=1}^{\infty} \frac{(-1)^n \cdot 2 \cdot 4 \cdot 6 \cdots (2n)}{3^n \cdot 3 \cdot 5 \cdot 7 \cdots (2n+1)} x^{2n+1}\right) \qquad (10) \\
&= a_0(u_1(x)) + a_1(u_2(x))
\end{aligned}$$

Again because they contain different powers of x, u_1 and u_2 are linearly independent. Hence (10) provides the two-parameter family of functions that is the general solution to (9). An ambitious student who can manipulate numbers would be able to show that (10) is equivalent to the slightly more compact form

$$y(x) = a_0\left(\sum_{n=0}^{\infty} \frac{(-1)^n(2n+1)!}{3^n \cdot 4^n(n!)^2} x^{2n}\right) + a_1\left(\sum_{n=0}^{\infty} \frac{(-1)^n 4^n (n!)^2}{3^n(2n+1)!} x^{2n+1}\right) \qquad (11)$$

The functions $u_1(x)$ and $u_2(x)$ in (11) are not recognizable as familiar functions. Their intervals of convergence could be checked by the ratio test, but a much easier method to find the common overlapping interval uses an elementary concept from complex-plane analysis. A complex number $z = a + bi$ is placed in the complex plane by moving a units along the horizontal axis and b units along the vertical axis. The distance D of such a point from the origin is measured by

$$D = \sqrt{a^2 + b^2}$$

When we are testing an infinite-series solution to a differential equation for convergence, the complex plane may be used to save some work. First locate and graph in the complex plane all of the singular points of the coefficient functions of the differential equation. By measuring the distance of each of these points from the origin, determine which point is closest to the origin. This distance is R, the radius of convergence on the real-number line of the power-series solution. The interval of convergence of the power-series solution is now $(-R, R)$.

Thus in Example 5, where the only singular points of the coefficient functions occur at $x = \pm\sqrt{3}i$ (the points where the denominator is zero), each is the same distance from the origin. On the x-axis the interval of convergence has radius $\sqrt{3}$. Our solution (11) is then valid in the interval $(-\sqrt{3}, \sqrt{3})$.

Example 6

The differential equation $(x^2 + 1)y'' - 5xy' + 3y = 0$ has a power-series solution valid in $(-1, 1)$ since the singularities occur at $x = \pm i$.

Example 7

The equation $(x - 2)(x^2 + 9)y'' - 2xy' - (4x + 1)y = 0$ has singularities at $x = 2$ and $x = \pm 3i$, so the minimal distance in the complex plane is 2 and the interval of convergence is $(-2, 2)$.

Example 8

The equation $y'' + 4xy' - 6y = 0$ has no singularities (every point x is an ordinary point) and hence the radius of convergence is infinite.

The next differential equation to be considered shows a little different behavior as its solution by power-series methods is obtained. We again point out that this problem cannot be solved by the analytical methods of Chapter 3.

Example 9

Consider

$$(4 + x^2)y'' + 2xy' - 2y = 0 \tag{12}$$

With $x = 0$ as the initial point and using the trial solution

$$y(x) = \sum_{n=0}^{\infty} a_n x^n$$

we obtain as before

$$y'(x) = \sum_{n=0}^{\infty} n a_n x^{n-1},$$

so

$$2xy' = \sum_{n=0}^{\infty} 2n a_n x^n$$

and

$$y''(x) = \sum_{n=0}^{\infty} n(n-1)a_n x^{n-2},$$

so

$$x^2 y'' = \sum_{n=0}^{\infty} n(n-1)a_n x^n$$

The differential equation becomes

$$\sum_{n=0}^{\infty} 4n(n-1)a_n x^{n-2} + \sum_{n=0}^{\infty} n(n-1)a_n x^n + \sum_{n=0}^{\infty} 2na_n x^n - \sum_{n=0}^{\infty} 2a_n x^n = 0$$

The index shift on the first sum and the combination of the sums give

$$\sum_{n=0}^{\infty} \{4(n+2)(n+1)a_{n+2} + [n(n-1) + 2n - 2]a_n\} x^n = 0$$

The recurrence formula that comes out of this is

$$a_{n+2} = \frac{-(n^2+n-2)}{4(n+2)(n+1)} a_n = \frac{-(n-1)}{4(n+1)} a_n, \quad n \geq 0$$

Inserting the various values of n gives us

$$a_2 = \frac{-(-1)}{4 \cdot 1} a_0, \quad a_4 = \frac{-1}{4 \cdot 3} a_2 = \frac{1(-1)}{4^2 \cdot 1 \cdot 3} a_0,$$

$$a_6 = \frac{-3}{4 \cdot 5} a_4 = \frac{1(-1)(-3)}{4^3 \cdot 1 \cdot 3 \cdot 5} a_0, \ldots$$

$$a_3 = \frac{-0}{4 \cdot 2} a_1 = 0, \quad a_5 = 0, \quad a_7 = 0, \ldots, a_{2n+1} = 0, \ldots$$

We see that a_1 is arbitrary but all higher odd subscripted a's are zero. The pattern on the evens leads us to

$$a_{2n} = \frac{1(-1)(-3)(-5)\cdots(-2n+3)}{4^n \cdot 1 \cdot 3 \cdot 5 \cdot 7 \cdots (2n-1)} a_0$$

This can be simplified if we cancel some terms:

$$a_{2n} = \frac{(-1)^{n+1}}{4^n(2n-1)} a_0$$

The trial solution $y = a_0 + a_1 x + a_2 x^2 + a_3 x^3 + a_4 x^4 + a_5 x^5 + \cdots$ now becomes

$$\begin{aligned} y &= a_0 + a_1 x + \frac{1}{4} a_0 x^2 + 0 + \frac{1(-1)}{4^2 \cdot 1 \cdot 3} a_0 x^4 + 0 + \cdots \\ &= a_1 x + a_0 \left(1 + \frac{1}{2} x^2 - \frac{1}{4^2 \cdot 1 \cdot 3} x^4 + \frac{1}{4^3 \cdot 5} x^6 \right. \\ &\qquad \left. + \cdots + \frac{(-1)^{n+1}}{4^n(2n-1)} x^{2n} + \cdots \right) \\ &= a_1 x + a_0 \left(1 + \sum_{n=1}^{\infty} \frac{(-1)^{n+1}}{4^n(2n-1)} x^{2n} \right) \end{aligned}$$

The new wrinkle that showed up produced a recognizable solution $u_1(x) = x$. Now if we go back to Section 3.11 we can use the technique there to produce the other solution; or, if you are a walking encyclopedia of power-series forms, you will recognize the series $u_2(x)$ as the expansion of

$$1 + \frac{1}{2} x \text{ Arctan } \frac{x}{2}$$

The general solution of equation (12) could then be expressed as

$$y(x) = a_1 x + a_0\left(1 + \frac{1}{2} \text{ Arctan } \frac{x}{2}\right)$$

The recognition in this example of the inverse tangent function provided an exceptional bonus. We arrived at a pair of elementary functions as our linearly independent solutions to a problem that we would have been unable to solve by *any* of our previously studied methods.

In each of the prior examples we used the point $x_0 = 0$ as our center or initial point. The trial solution under that arrangement is

$$y(x) = \sum_{n=0}^{\infty} a_n x^n$$

If, however, a point other than zero is the initial point, then our trial solution would have to take the form

$$y(x) = \sum_{n=0}^{\infty} a_n (x - x_0)^n$$

The mechanics of putting this trial solution through the differential equation is cumbersome, so we again turn to a change of variables as a way of simplifying the work. If we let $t = x - x_0$, which is nothing more than a linear shift, we may write our trial solution as

$$Y(t) = \sum_{n=0}^{\infty} a_n t^n$$

which again fits the form of our examples. But to use this change we must also switch the variables in the entire differential equation. But this is immediate since under the change of variables $dy/dx = dy/dt$ and $d^2y/dx^2 = d^2y/dt^2$.

We now solve the differential equation in the independent variable t with the solution centered at $t = 0$. This solution is switched back to the independent variable x by using $t = x - x_0$ again. We see then that one basic method covers many different cases.

Exercises 6.2

1. Find all the ordinary points of the following differential equations.
   (a) $xy'' + (x^2 - 1)y' - 2y = 0$
   (b) $y'' - (x + 3)y' + x^2y = 0$
   (c) $x(x + 1)y'' + 2y' - x^2y = 0$
   (d) $xy'' - (\sin x)y' + (e^x + 1)y = 0$

Find a power-series solution for each of the following initial value problems. In each case also find the interval of validity of the solution.

2. $y'' + y = 0;\ y(0) = 0,\ y'(0) = 1$
3. $(x^2 - 1)y'' + 2xy' - 6y = 0;\ y(0) = -1,\ y'(0) = 0$
4. $y'' + xy' + 2y = 0;\ y(0) = 2,\ y'(0) = 1$

Find the general solution near $x = 0$ by the power-series method for each of the following differential equations.

5. $y'' - xy = 0$ (This equation, known as Airy's equation, was originally studied in connection with light intensity calculations near surfaces of caustic materials.)
6. $(1 + x^2)y'' + xy' - y = 0$
7. $y'' + x^2y' + xy = 0$
8. $(1 - x^2)y'' + xy' - y = 0$
9. $(x^2 - 1)y'' - 2y = 0$
10. $y'' - xy' - (2 - 3x)y = 0$
11. $y''' - x^2y'' - 4xy' - 2y = 0$
12. $2(x^2 + 8)y'' + 2xy' + (x + 2)y = 0$
13. For Exercises 5–12 determine the interval of convergence for each equation by first establishing the circle of convergence in the complex plane.
14. In quantum mechanics, the theory of the oscillator uses the solutions of the differential equation

$$\frac{d^2u}{dx^2} - x^2u + (2n + 1)u = 0$$

for constant n, which remain finite as x grows unboundedly large. To find these solutions, we first show that when $n = 0$, the function $u = e^{-x^2/2}$ satisfies the equation. Then using the technique of Section 3.11 we set $u = ve^{-x^2/2}$ and substitute this into the differential equation. Show that we then get

$$\frac{d^2v}{dx^2} - 2x\frac{dv}{dx} + 2nv = 0$$

This should now be solved by infinite-series techniques to obtain solutions v_1 and v_2 and then obtain

$$u_1 = v_1e^{-x^2/2} \quad \text{and} \quad u_2 = v_2e^{-x^2/2}$$

We must now verify that these satisfy the finiteness requirement if and only if n is zero or a positive integer.

15. Consider the special differential equation of mathematical physics

$$(1 - x^2)y'' - 2xy' + n(n + 1)y = 0$$

called **Legendre's equation**. Show that if n is a positive integer, then one solution near $x = 0$ is a polynomial of degree n. Let the dummy index be k; choose the constant so that $y(1) = 1$ after you obtain the recurrence formula

$$a_{k+2} = \frac{-(n-k)(n+k+1)}{(k+2)(k+1)} a_k$$

Designating these polynomial solutions as $P_n(x)$ obtain

$$P_0(x) = 1, \qquad P_1(x) = x, \qquad P_2(x) = \frac{1}{2}(3x^2 - 1),$$

$$P_3(x) = \frac{1}{2}(5x^3 - 3x), \ldots$$

16. The differential equation

$$(1 - x^2)y'' - xy' + k^2 y = 0$$

is called **Tchebycheff's equation** and arises in many areas of mathematics and physics. Since $x = 0$ is an ordinary point, find two linearly independent solutions of the form

$$y = \sum_{n=0}^{\infty} a_n x^n$$

If k is chosen to be an integer n, then one of the above solutions is a polynomial of degree n. (See Exercise 10 in Section 1.1.)

17. The differential equation

$$y'' - 2xy' + ky = 0$$

is called **Hermite's equation** and has many applications in physics and engineering. Since all x are ordinary points, find two linearly independent solutions. If k is an even integer $2n$, then one of these solutions is a polynomial of degree n. (See Exercise 11 in Section 1.1.)

6.3 An Alternative Method: Taylor Series

For some problems, especially those which are nonhomogeneous and/or nonlinear, a method other than that described in Section 6.2 may prove to be simpler or necessary. In fact, there will be occasions when the method about to be described will be the only available method short of turning to the computer methods of Chapters 4 and 5. The slight advantage we gain here is that we get an approximation to the solution function, whereas other methods may not give us any idea of the function.

We will begin with a simple example, which can also be solved analytically, to demonstrate the procedure and then move to more complicated cases.

Example 10

Consider

$$y'' + y = 2x - 1; \quad y(1) = 1, \quad y'(1) = 3$$

This is a typical initial value problem that may be solved by the method of undetermined coefficients to obtain

$$y(x) = 2x - 1 + \sin(x - 1) \tag{13}$$

We suppose here that the solution to the differential equation is an analytic function that can be expressed in a power series in powers of $x - 1$. This form results from the given initial point $x = 1$. By Taylor's theorem for the expansion of a differentiable function, stated in most calculus books, we have

$$y(x) = y(1) + \frac{y'(1)}{1!}(x - 1) + \frac{y''(1)}{2!}(x - 1)^2 + \frac{y'''(1)}{3!}(x - 1)^3 + \cdots$$

The unknowns in this expansion are the values of y and its derivatives at $x = 1$. Our problem gives the values $y(1)$ and $y'(1)$. Values of the subsequent derivatives may be obtained from the differential equation and subsequent differentiations of that equation. For example, we can reorganize the differential equation to provide

$$y''(x) = -y + 2x - 1 \tag{14}$$

so that $\quad y''(1) = -y(1) + 2(1) - 1 = -1 + 2 - 1 = 0$

By now differentiating (14) we obtain

$$y'''(x) = -y'(x) + 2$$

so that $\quad y'''(1) = -y'(1) + 2 = -3 + 2 = -1$

Continuing to differentiate, we get

$$y''''(x) = -y''(x)$$

and $\quad y''''(1) = -y''(1) = 0$

We now use these values in Taylor's expansion to get

$$y(x) = 1 + \frac{3}{1}(x - 1) + \frac{0}{2!}(x - 1)^2 + \frac{-1}{3!}(x - 1)^3 + \frac{1}{5!}(x - 1)^5 + \cdots$$

$$= 1 + 3(x - 1) - \frac{1}{6}(x - 1)^3 + \frac{1}{120}(x - 1)^5 + \cdots \tag{15}$$

This can be extended to as many terms as desired for the degree of accuracy one wants. The convergence is most rapid for values near 1, but the series does converge to the solution function given earlier

$$y(x) = 2x - 1 + \sin(x - 1)$$

To see this, we reorganize (15) as follows:

$$y(x) = 1 + 2(x - 1) + \left\{(x - 1) - \frac{1}{3!}(x - 1)^3 + \frac{1}{5!}(x - 1)^5 + \cdots\right\}$$

$$= 2x - 1 + \{\sin(x - 1)\} \tag{16}$$

where we have used the known Taylor expansion of $\sin(x - x_0)$ derived in most calculus texts. Note that (16) agrees with (13).

While Example 10 illustrates the method, it is not of practical value for solving problems of the type proposed in the example. It is a plausible method, however, when the differential equation is *nonlinear*, for then the technique of Section 6.2 would involve the multiplication of power series, a very complicated process.

Example 11

Consider

$$yy'' + 2(y')^2 = 0; \quad y(0) = 1, \quad y'(0) = \frac{1}{2}$$

A nonproductive start would be to proceed as before and set

$$y = \sum_{n=0}^{\infty} a_n x^n$$

Since $(y')^2$ would then be a grand mess as a power series, we continue with the Taylor series method.

To write the Taylor series expansion in powers of x (the initial point designated by the initial conditions is $x = 0$) we will need the coefficient derivative evaluations. We are given $y(0) = 1$ and $y'(0) = \frac{1}{2}$. From the given differential equa ion we have $y(0)y''(0) + 2[y'(0)]^2 = 0$, so that $1y''(0) + 2(\frac{1}{2})^2 = 0$ or $y''(0) = -\frac{1}{2}$. By differentiating the given differential equation, we obtain

$$yy''' + y''y' + 4y'y'' = 0 \tag{17}$$

Evaluating this at $x = 0$, we get

$$1y'''(0) + \left(-\frac{1}{2}\right)\left(\frac{1}{2}\right) + 4\left(\frac{1}{2}\right)\left(-\frac{1}{2}\right) = 0$$

or

$$y'''(0) = \frac{5}{4}$$

Differentiation of equation (17) provides

$$(yy'''' + y'''y') + 5[(y'')^2 + y'y'''] = 0$$

Evaluating again at $x = 0$ gives

$$1y''''(0) + \left(\frac{5}{4}\right)\left(\frac{1}{2}\right) + 5\left(\frac{1}{4} + \frac{1}{2} \cdot \frac{5}{4}\right) = 0$$

or

$$y''''(0) = -5$$

This could continue on as far as one wished, but we will halt and substitute the coefficients into the series expansion. This produces

$$y(x) = y(0) + \frac{y'(0)}{1!}x + \frac{y''(0)}{2!}x^2 + \frac{y'''(0)}{3!}x^3 + \frac{y''''(0)}{4!}x^4 + \cdots$$

$$= 1 + \frac{1}{2}x - \frac{1}{4}x^2 + \frac{5}{24}x^3 - \frac{5}{24}x^4 + \cdots \tag{18}$$

To get a reasonably accurate set of solution values from this series would surely require quite a few more terms and the necessity of staying near the value $x = 0$. However, we now have some idea about the form of the solution function. A computer run on this problem would allow us to check the accuracy of equation (18).

Example 11 provides a differential equation that looks complicated but is easily solved by this series method. Nevertheless, you might have observed that the problem is solvable analytically by the method of Section 1.8, because it is a second-order equation that is reducible to first order. If one carries through that process, the analytical solution is

$$y = \left(\frac{3}{2}x + 1\right)^{1/3}$$

Comparing the value of $y(\frac{1}{2})$ in the two answers gives

$$y\left(\frac{1}{2}\right) = \left(\frac{3}{2} \cdot \frac{1}{2} + 1\right)^{1/3} = (1.75)^{1/3} = 1.20507$$

$$y\left(\frac{1}{2}\right) = 1 + \frac{1}{2} \cdot \frac{1}{2} - \frac{1}{4} \cdot \frac{1}{4} + \frac{5}{24} \cdot \frac{1}{8} - \frac{5}{24} \cdot \frac{1}{16} + \cdots = 1.2005$$

The first, of course, is the exact answer.

Exercises 6.3

Use the Taylor series techniques to find a power-series solution to each of the following problems. The series should contain at least five terms. Some answers can be checked by obtaining the analytical solution for comparison.

1. $y' = x + 2xy; y(0) = 1$
2. $y' = xy - x^2; y(0) = 3$

3. $y'' - 2y = e^{2x}$; $y(0) = 2$, $y'(0) = 2$
4. $y' = \sqrt{1 + xy}$; $y(0) = 1$
5. $y''' - y' - y^2 \ln|x + 2| = 0$; $y(0) = 2$, $y'(0) = 0$, $y''(0) = 1$
6. $(x^2 - 1)y'' - 6y = 0$; $y(0) = 2$, $y'(0) = 5$
7. $y''' - 4x^2y' + 12xy = 0$; $y(0) = y'(0) = 0$, $y''(0) = 2$
8. $y'' - e^{-x}y' = 1 + e^{-x}y^2$; $y(0) = y'(0) = 2$
9. $y'' + xy' - (1 - 2x)y = 0$; $y(-1) = 3$, $y'(-1) = -5$
10. $y'' + 2(y')^2 = e^x y^2$; $y(0) = -2$, $y'(0) = 2$

6.4 Solution at a Singular Point

We will now describe a method, attributed to Frobenius, which will cover the case described earlier when the initial point x_0 is not an ordinary point. Again the differential equation to be considered must first be put into the form

$$y'' + p(x)y' + q(x)y = f(x)$$

Our description of the method of Frobenius will be limited to homogeneous equations and thus $f(x) = 0$.

Definition: If the coefficient functions $p(x)$ and/or $q(x)$ are not analytic at the initial point x_0 (this condition is often equivalent to having either p or q become infinite at x_0), then the point x_0 is said to be a **singular point** of the differential equation. We further classify the singular point into two subcategories. If $p(x)$ and/or $q(x)$ is singular at the point x_0 but $(x - x_0)p(x)$ and $(x - x_0)^2 q(x)$ are analytic at x_0, then x_0 is called a **regular singular point**. Singular points which are not regular are called **irregular**.

Example 12

The differential equation $(x - 2)^3x^2y'' + 2(x - 2)xy' + 7y = 0$ is first written as

$$y'' + \left(\frac{2}{(x - 2)^2 x}\right)y' + \left(\frac{7}{(x - 2)^3 x^2}\right)y = 0$$

where it is clear that both $p(x)$ and $q(x)$ are singular at $x = 0$ and $x = 2$. Therefore, these two points are the singular points of this equation. All other points are ordinary points for this equation. Checking the singular point $x = 0$ for further classification, we compute according to the above definition,

$$xp(x) = \frac{2}{(x - 2)^2}$$

and

$$x^2q(x) = \frac{7}{(x - 2)^3}$$

Both of these expressions are now analytic at $x = 0$. Hence, $x = 0$ is a *regular* singular point.

Checking the point $x = 2$ we compute, again according to the definition,

$$(x - 2)p(x) = \frac{2}{x(x - 2)}$$

and

$$(x - 2)^2 q(x) = \frac{7}{x^2(x - 2)}$$

Both of these are nonanalytic at $x = 2$. Hence, $x = 2$ is an *irregular* singular point.

Example 13

The differential equation

$$y'' + \left[\frac{4x^2}{(x - 1)(x + 3)^2}\right]y' - \left[\frac{x - 3}{x(x - 1)^2}\right]y = 0$$

has singular points at $x = 1$, $x = -3$, and $x = 0$. The above method will easily show that $x = 1$ and $x = 0$ are regular singular points, that $x = -3$ is an irregular singular point, and that all other points are ordinary points.

Many of the classical differential equations used by engineers and physicists in applications areas possess singular points. Probably the two most famous are the **Bessel equation***

$$x^2y'' + xy' + (x^2 - n^2)y = 0$$

with a regular singular point at $x = 0$, and the **Legendre equation****

$$(1 - x^2)y'' - 2xy' + n(n + 1)y = 0$$

with regular singular points at $x = 1$ and $x = -1$.

The Method of Frobenius

We shall consider solutions of the homogeneous differential equation

$$y'' + p(x)y' + q(x)y = 0$$

near a regular singular point x_0. We shall assume this point to be the origin. If in a given problem $x_0 \neq 0$, we will shift linearly; and via the translation

---

* Named for Friedrich Wilhelm Bessel (1784–1846), a Prussian astronomer and mathematician.

** Named for Adrien-Marie Legendre (1752–1833), a French mathematician.

we can replace an equation in powers of $(x - x_0)$ by an equation in powers of x. (Also see property B in Section 6.2.)

The only modification we require in the method of Frobenius is to multiply our earlier assumed power-series solution $\sum_{n=0}^{\infty} a_n x^n$ by the term x^s, where s is a constant to be determined in addition to the a_i's. This is done by substituting the trial-solution series

$$y(x) = x^s \sum_0^\infty a_n x^n = \sum_0^\infty a_n x^{n+s} \tag{19}$$

into the equation to be solved and equating coefficients as in Section 6.2. For zero or positive integer values of s, the new series is the same or similar to the eariler used power series; but for negative or noninteger values of s, the series is of a new type. Since we will again assume $a_0 \neq 0$, the lowest power of x appearing in this series solution is x^s.

If the properties of general-power series listed in Section 6.2 are again employed, we will use

$$y'(x) = \sum_{n=0}^\infty (n + s)a_n x^{n+s-1} \tag{20}$$

and

$$y''(x) = \sum_{n=0}^\infty (n + s)(n + s - 1)a_n x^{n+s-2} \tag{21}$$

Substituting the expressions (19), (20), and (21) into the given differential equation, combining the coefficients into the new power-series expressions, and making a new arrangement of summing indices will again provide an opportunity to equate to zero the various coefficients of each power of x.

From these equations will emerge an expression called the **indicial equation** to easily find the applicable values of the index s as well as equations to find the values for the unknown coefficients $a_n, n = 0, 1, 2, \ldots$.

Suppose that two values of s are determined by solving the indicial equation. If for one value of s we can calculate a set of coefficients a_n involving only one that is arbitrary and if we substitute these values into (19), we will obtain an expression for $y(x)$, call it y_1, that is a particular solution to the original differential equation. If for the other value of s we can calculate a new set of coefficients a_n also involving only one that is arbitrary and if we substitute this collection into (19), we will obtain another particular solution, call it y_2. The sum $y = y_1 + y_2$, now involving two arbitrary constants, becomes the general solution to the original differential equation. Remember, however, that series solutions are valid only for those values of x for which both series converge.

The general case for the existence of series solutions at a regular singular point is covered by the following theorem whose proof may be found in more advanced level textbooks (see Ince* and Birkhoff and Rota**).

Theorem 6.1 (Existence Theorem):

If the roots s_1 and s_2 of the indicial equation corresponding to the differential equation

$$y'' + p(x)y' + q(x)y = 0$$

differ by a number that is not zero nor an integer, then there exist two linearly independent solutions of the form

$$y_1(x) = x^{s_1} \cdot \sum_0^\infty a_n x^n, \quad y_2(x) = x^{s_2} \cdot \sum_0^\infty b_n x^n$$

which are valid in some interval $0 < x < R$.

If the difference $s_1 - s_2$ is zero or an integer, then there exist linearly independent solutions of the form

$$y_1(x) = x^{s_1} \cdot \sum_0^\infty a_n x^n, \quad y_2(x) = x^{s_2} \cdot \sum_0^\infty b_n x^n + C y_1(x) \ln|x|$$

where the constant C is not zero when $s_1 = s_2$ and C may or may not be zero when the difference is an integer.

In any equation for which $x = 0$ is a regular singular point, there is at least one solution guaranteed which has the form y_1 of Theorem 6.1. We will now solidify Theorem 6.1 and the preceding comments through a collection of examples.

Example 14

Consider the differential equation

$$xy'' - \left(x - \frac{1}{2}\right)y' - \frac{1}{2}y = 0$$

One can easily verify that this equation has $x = 0$ as a regular singular point. We propose the trial solution

$$y(x) = x^s \sum_0^\infty a_n x^n = \sum_0^\infty a_n x^{n+s}$$

where the value of the real number s and the values of the constants $a_i, i = 1, 2, 3, \ldots,$ must be determined. Using the differentiation property, we

---

* E. L. Ince, *Ordinary Differential Equations* (New York: Dover, 1956), Chapter 15.
** G. Birkhoff and G. C. Rota, *Ordinary Differential Equations* (Boston: Ginn, 1962).

get

$$y'(x) = \sum_{0}^{\infty} (n + s)a_n x^{n+s-1}$$

so we have

$$xy' = \sum_{0}^{\infty} (n + s)a_n x^{n+s}$$

By a further application

$$y''(x) = \sum_{0}^{\infty} (n + s)(n + s - 1)a_n x^{n+s-2}$$

and thus

$$xy'' = \sum_{0}^{\infty} (n + s)(n + s - 1)a_n x^{n+s-1}$$

Substituting into the given differential equation produces

$$\sum_{0}^{\infty} (n + s)(n + s - 1)a_n x^{n+s-1} - \sum_{0}^{\infty} (n + s)a_n x^{n+s}$$
$$+ \sum_{0}^{\infty} \frac{1}{2}(n + s)a_n x^{n+s-1} - \sum_{0}^{\infty} \frac{1}{2} a_n x^{n+s} = 0$$

By index shifting we must produce like powers on the x terms of these sums. The second and fourth terms will undergo the shift $n \to n - 1$, producing

$$\sum_{0}^{\infty} (n + s)(n + s - 1)a_n x^{n+s-1} - \sum_{1}^{\infty} (n - 1 + s)a_{n-1} x^{n+s-1}$$
$$+ \sum_{0}^{\infty} \frac{1}{2}(n + s)a_n x^{n+s-1} - \sum_{1}^{\infty} \frac{1}{2} a_{n-1} x^{n+s-1} = 0 \quad (22)$$

Extra care must be taken while employing the method of Frobenius to keep the summation indices correct. No terms may be discarded here as was the case in earlier sections.

Before we are able to create our super sum, we must now write out the "zero'th" term of the first and third sums and thus get all four sums to start at $n = 1$. We actually let $n = 0$ in these terms and write down the results and group them up front. When this is done the sum begins at $n = 1$. Equation (22) now becomes

$$\left\{ s(s - 1)a_0 x^{s-1} + \sum_{1}^{\infty} (n + s)(n + s - 1)a_n x^{n+s-1} \right\}$$
$$- \sum_{1}^{\infty} (n + s - 1)a_{n-1} x^{n+s-1}$$
$$+ \left\{ \frac{1}{2} s a_0 x^{s-1} + \sum_{1}^{\infty} \frac{1}{2}(n + s)a_n x^{n+s-1} \right\} - \sum_{1}^{\infty} \frac{1}{2} a_{n-1} x^{n+s-1} = 0$$

or upon rearranging

$$\left(s(s-1)+\frac{1}{2}s\right)a_0x^{s-1}+\sum_1^{\infty}\left\{\left[(n+s)(n+s-1)+\frac{1}{2}(n+s)\right]a_n-\left(n+s-1+\frac{1}{2}\right)a_{n-1}\right\}x^{n+s-1}=0$$

Since a_0 is assumed to be nonzero (else we could begin summing our original trial solution at $n = 1$) and since property G states that the coefficient of each term on the left must vanish, we obtain

$$s(s-1)+\frac{1}{2}s=0$$

which is the *indicial equation*, and

$$\left[(n+s)(n+s-1)+\frac{1}{2}(n+s)\right]a_n-\left(n+s-1+\frac{1}{2}\right)a_{n-1}=0$$

which is the *general recurrence formula*. The roots of the indicial equation $s^2-\frac{1}{2}s=0$ are $s_1=0$ and $s_2=\frac{1}{2}$. Since they do not differ by zero or an integer, we are guaranteed by the existence theorem that two linearly independent series solutions will be obtained.

Working toward obtaining these two, we begin with the indicial root $s_1 = 0$ and the specialized recurrence formula obtained by using this s_1 value in the general expression:

$$\left[n(n-1)+\frac{1}{2}n\right]a_n-\left(n-1+\frac{1}{2}\right)a_{n-1}=0$$

This is written as

$$a_n=\frac{(n-1/2)a_{n-1}}{(n^2-1/2n)}=\frac{a_{n-1}}{n},\quad n\geq 1$$

We let a_0 be arbitrary and calculate from this formula the remaining a's:

$$a_1=a_0,\quad a_2=\frac{a_1}{2}=\frac{a_0}{2},\quad a_3=\frac{a_2}{3}=\frac{a_0}{3\cdot 2},$$

$$a_4=\frac{a_3}{4}=\frac{a_0}{4\cdot 3\cdot 2},\quad a_5=\frac{a_4}{5}=\frac{a_0}{5\cdot 4\cdot 3\cdot 2},\ldots,a_n=\frac{a_0}{n!},\ldots$$

Thus

$$y_1(x)\equiv x^{s_1}\sum_0^{\infty}a_nx^n=x^0(a_0+a_1x+a_2x^2+a_3x^3+\cdots)$$

$$=1\left(a_0+a_0x+\frac{a_0}{2}x^2+\frac{a_0}{3!}x^3+\cdots+\frac{a_0}{n!}x^n+\cdots\right)$$

$$=a_0\left(\sum_0^{\infty}\frac{x^n}{n!}\right)=a_0e^x$$

The indicial root $s_2 = \frac{1}{2}$, when substituted into the general recurrence formula, gives the new relation

$$\left\{\left(n+\frac{1}{2}\right)\left(n-\frac{1}{2}\right)+\frac{1}{2}\left(n+\frac{1}{2}\right)\right\}b_n - \left\{n+\frac{1}{2}-1+\frac{1}{2}\right\}b_{n-1} = 0$$

(where we have switched from letter a to letter b to avoid confusion). This is simplified to

$$n\left(n+\frac{1}{2}\right)b_n - nb_{n-1} = 0$$

or

$$b_n = \frac{b_{n-1}}{n+1/2} = \frac{2b_{n-1}}{2n+1}$$

Thus, by letting b_0 be arbitrary, we obtain the sequence

$$b_1 = \frac{2b_0}{3}, \qquad b_2 = \frac{2b_1}{5} = \frac{2^2 b_0}{3 \cdot 5}, \qquad b_3 = \frac{2b_2}{7} = \frac{2^3 b_0}{3 \cdot 5 \cdot 7},$$

$$b_4 = \frac{2b_3}{9} = \frac{2^4 b_0}{3 \cdot 5 \cdot 7 \cdot 9}, \ldots, b_n = \frac{2^n b_0}{3 \cdot 5 \cdot 7 \cdots (2n+1)}, \ldots$$

Using these values, we get

$$\begin{aligned} y_2(x) &\equiv x^{s_2} \sum_0^\infty b_n x^n = x^{1/2}(b_0 + b_1 x + b_2 x^2 + b_3 x^3 + \cdots) \\ &= x^{1/2}\left(b_0 + \frac{2}{3} b_0 x + \frac{2^2 b_0}{3 \cdot 5} x^2 + \cdots + \frac{2^n b_0}{3 \cdot 5 \cdots (2n+1)} x^n + \cdots\right) \\ &= b_0 x^{1/2}\left(\sum_0^\infty \frac{2^n}{3 \cdot 5 \cdot 7 \cdots (2n+1)} x^n\right) \end{aligned}$$

The ratio test will confirm that both $y_1(x)$ and $y_2(x)$ converge for all real x and thus the general solution is

$$\begin{aligned} y(x) &= a_0\left(\sum_0^\infty \frac{1}{n!} x^n\right) + b_0 x^{1/2}\left(\sum_0^\infty \frac{2^n}{3 \cdot 5 \cdot 7 \cdots (2n+1)} x^n\right) \\ &= a_0 e^x + b_0 x^{1/2} \sum_0^\infty \frac{(2x)^n}{3 \cdot 5 \cdot 7 \cdots (2n+1)} \\ &= a_0[e^x] + b_0\left[2x^{1/2} \sum_0^\infty \frac{(n+1)!}{(2n+2)!} (4x)^n\right] \end{aligned}$$

The bonus in Example 14 is that one of the power series was recognized as a familiar series and the function represented by that series could be put into the solution. This is not generally the case when the method of Frobenius is used.

Example 15

Solve $2x^2y'' + 5xy' - (2x^2 - 1)y = 0$.

The point $x = 0$ is a regular singular point. Using the trial solution

$$y(x) = x^s \sum_0^\infty a_n x^n$$

and its derivatives

$$y'(x) = \sum_0^\infty (n+s)a_n x^{n+s-1}$$

and

$$y''(x) = \sum_0^\infty (n+s)(n+s-1)a_n x^{n+s-2}$$

we substitute into the differential equation and obtain

$$\sum_0^\infty 2(n+s)(n+s-1)a_n x^{n+s} + \sum_0^\infty 5(n+s)a_n x^{n+s} - \sum_0^\infty 2a_n x^{n+s+2} + \sum_0^\infty a_n x^{n+s} = 0$$

Only the third sum needs an index shift, namely $n \to n-2$, making that term

$$\sum_{n=2}^\infty 2a_{n-2}\, x^{n+s}$$

Writing out the zero'th and first terms of the other three sums and the resulting super sum, we get

$$[2s(s-1)a_0 + 5sa_0 + a_0]x^s + [2(1+s)s + 5(1+s) + 1]a_1 x^{s+1} + \sum_{n=2}^\infty \{[2(n+s)(n+s-1) + 5(n+s) + 1]a_n - 2a_{n-2}\}x^{n+s} = 0$$

Equating the coefficient of each power of x to zero, we get

$$[2s(s-1) + 5s + 1]a_0 = 0 \tag{23}$$

$$[2(1+s)s + 5(1+s) + 1]a_1 = 0 \tag{24}$$

$$[2(n+s)(n+s-1) + 5(n+s) + 1]a_n - 2a_{n-2} = 0 \tag{25}$$

Since a_0 is assumed to be nonzero, we get from (23) the indicial equation

$$2s^2 + 3s + 1 = 0$$

whose roots are $s_1 = -\frac{1}{2}, s_2 = -1$. (Remember that only the equation resulting from the lowest power of x produces an indicial equation.) Using the first value $s_1 = -\frac{1}{2}$ in (24) we find

$$\left[2\left(\frac{1}{2}\right)\left(-\frac{1}{2}\right) + \frac{5}{2} + 1\right]a_1 = 0 \qquad \text{or} \qquad a_1 = 0$$

In equation (25), the value $s = -\frac{1}{2}$ produces the recurrence formula

$$\left[2\left(n-\frac{1}{2}\right)\left(n-\frac{3}{2}\right)+5\left(n-\frac{1}{2}\right)+1\right]a_n = 2a_{n-2}$$

$$(2n^2 + n)a_n = 2a_{n-2}$$

or

$$a_n = \frac{2a_{n-2}}{n(2n+1)}$$

Letting a_0 be arbitrary and noting from above that $a_1 = 0$, we obtain

$$a_2 = \frac{2a_0}{2 \cdot 5}, \qquad a_3 = 0, \qquad a_4 = \frac{2a_2}{4 \cdot 9} = \frac{2^2 a_0}{2 \cdot 4 \cdot 5 \cdot 9}, \qquad a_5 = 0,$$

$$a_6 = \frac{2a_4}{6 \cdot 13} = \frac{2^3 a_0}{2 \cdot 4 \cdot 6 \cdot 5 \cdot 9 \cdot 13}, \ldots,$$

$$a_{2n} = \frac{a_0}{n!\, 5 \cdot 9 \cdot 13 \cdots (4n+1)}, \qquad a_{2n+1} = 0, \ldots$$

Hence the first solution is

$$\begin{aligned} y_1(x) &\equiv x^{s_1} \sum_0^\infty a_n x^n \\ &= x^{-1/2}\left(a_0 + \frac{a_0}{1 \cdot 5}x^2 + \frac{a_0}{1 \cdot 2 \cdot 5 \cdot 9}x^4 + \frac{a_0}{1 \cdot 2 \cdot 3 \cdot 5 \cdot 9 \cdot 13}x^6 + \cdots\right) \\ &= a_0 x^{-1/2}\left(1 + \sum_1^\infty \frac{x^{2n}}{n!\, 5 \cdot 9 \cdot 13 \cdots (4n+1)}\right) \end{aligned}$$

Using the second root of the indicial equation $s_2 = -1$, we obtain in a similar manner the solution

$$y_2(x) = b_0 x^{-1}\left(1 + \sum_1^\infty \frac{x^{2n}}{n!\, 3 \cdot 7 \cdot 11 \cdots (4n-1)}\right)$$

Each of these series converges for all x, so we have a valid solution for all $x > 0$.

It is worth emphasizing again that if initial conditions had been specified with the original equation, we would be able to calculate the values of a_0 and b_0 and have a unique solution. The unfortunate case is that we still would know so little of the actual behavior of that unique function because of the complexity of having it expressed as a power series. In Chapter 7 we will investigate ways in which the computer can be an invaluable assistant in clearing up some of this built-in ambiguity.

The variety of cases involving various combinations of indicial roots and the resulting forms of the solutions dictated thereby according to our existence theorem are so numerous and so varied that an attempt to illustrate them by further examples would be pointless. The exercises for this section will include further work for uncovering some special cases; and Chapter 7 should give you a sufficient understanding of power-series solutions for an elementary course. For details of some of these advanced cases see Rabenstein* and Tenenbaum and Pollard.**

Because of the unquestioned usefulness in applications areas of the special functions of mathematical physics, we conclude this section with a study of the Bessel equation. This differential equation arises in a wide variety of problems, including those associated with aerodynamics, mechanics, electrical engineering, heat flow, and vibrations.

Example 16

Bessel's equation of order k is defined as

$$x^2y'' + xy' + (x^2 - k^2)y = 0$$

where k is any constant. Here we will restrict our study to real values of $k \geq 0$. By applying the test for classification of the point $x = 0$, we find that it is a regular singular point. Thus the trial solution

$$y(x) \equiv x^s \sum_0^{\infty} a_n x^n = \sum_0^{\infty} a_n x^{n+s}$$

will be applied with the method of Frobenius. Substitution into Bessel's equation produces

$$\sum_0^{\infty} (n+s)(n+s-1)a_n x^{n+s} + \sum_0^{\infty} (n+s)a_n x^{n+s} + \sum_0^{\infty} a_n x^{n+s+2} - \sum_0^{\infty} k^2 a_n x^{n+s} = 0$$

In the third sum we make the index shift $n \to n - 2$, producing the term

$$\sum_{n=2}^{\infty} a_{n-2} x^{n+s}$$

Writing out the first two terms of the other three series, we obtain the equation

$$[s(s-1) + s - k^2]a_0 x^s + [(1+s)s + (1+s) - k^2]a_1 x^{s+1} + \sum_{n=2}^{\infty} \{[(n+s)(n+s-1) + (n+s) - k^2]a_n + a_{n-2}\}x^{n+s} = 0$$

---

* A. L. Rabenstein, *Introduction to Ordinary Differential Equations* (Orlando, Fla.: Academic Press, 1972).

** M. Tenenbaum and H. Pollard, *Ordinary Differential Equations* (New York: Harper & Row, 1963).

Once again the coefficients must vanish for this to be an identity:

$$(s^2 - k^2)a_0 = 0$$

$$[(1 + s)^2 - k^2]a_1 = 0 \tag{26}$$

$$[(n + s)^2 - k^2]a_n + a_{n-2} = 0, \quad n = 2, 3, \ldots \tag{27}$$

Since a_0 is assumed to be nonzero, the indicial roots are $s^2 - k^2 = 0 \Rightarrow s_1 = k,\ s_2 = -k$. We will first find the solution corresponding to the nonnegative root $s_1 = k$. Putting $s = k$ in (26), we find that a_1 is zero. The recurrence formula (27) becomes

$$a_n = \frac{-a_{n-2}}{n^2 + 2kn}$$

From this we see that since $a_1 = 0$, we also have

$$a_3 = a_5 = a_7 = \cdots = a_{2n+1} = 0$$

With $s = k \geq 0$ we can write

$$a_n = \frac{-a_{n-2}}{n(n + 2k)}$$

Repeated applications of this formula will produce

$$a_2 = \frac{-a_0}{2(2 + 2k)} = \frac{-a_0}{4(k + 1)}$$

$$a_4 = \frac{-a_2}{4(4 + 2k)} = \frac{a_0}{2 \cdot 4(k + 2)4(k + 1)} = \frac{a_0}{2!\,4^2(k + 2)(k + 1)}$$

$$a_6 = \frac{-a_4}{6(6 + 2k)} = \frac{-a_0}{3!\,4^3(k + 3)(k + 2)(k + 1)}$$

$$\vdots$$

$$a_{2n} = \frac{(-1)^n a_0}{n!4^n(k + n)(k + n - 1)\cdots(k + 3)(k + 2)(k + 1)}$$

Thus one solution of Bessel's equation is

$$y_1(x) = a_0 \sum_0^\infty (-1)^n \frac{x^{k+2n}}{n!\,4^n(k + n)(k + n - 1)\cdots(k + 2)(k + 1)}$$

A particular choice of the arbitrary parameter a_0 results in a form of y_1 that is customarily used in applications problems. Let $a_0 = (2^k k!)^{-1}$ Then

$$y_1(x) \equiv J_k(x) = \sum_0^\infty \frac{(-1)^n x^{2n+k}}{2^{2n+k}\,n!\,(n + k)!} \tag{28}$$

is the solution and it is called the **Bessel function of the first kind of order *k*.**

The ratio test as shown in property E, as well as the recurrence formula, provides

$$\frac{1}{R} = \lim_{n\to\infty}\left|\frac{a_n}{a_{n-2}}\right| = \lim_{n\to\infty}\frac{1}{2kn+n^2} = 0$$

Therefore R is infinite. Thus the solution $J_k(x)$ converges for all x and $k \geq 0$. The tabulation of this special function showing the ease of its evaluation at any x value is considered in Chapter 7. In general the Bessel functions are nonelementary functions. The graph in Figure 6.3 shows the representation of the first three.

Now if k is nonzero and is not a positive integer, a second solution to Bessel's equation, linearly independent of $J_k(x)$, may be obtained by replacing k by $-k$. Then a general solution would be

$$y(x) = c_1 J_k(x) + c_2 J_{-k}(x)$$

valid for $x > 0$.

If k is zero or a positive integer, then the indicial equation has either equal roots or roots that differ by an integer. The existence theorem guarantees us the solution

$$J_k(x) = \sum_0^\infty \frac{(-1)^n x^{2n+k}}{2^{2n+k}\, n!\,(n+k)!}$$

and the second solution is

$$y_2(x) = x^{-k}\sum_0^\infty b_n x^n + KJ_k(x)\ln x$$

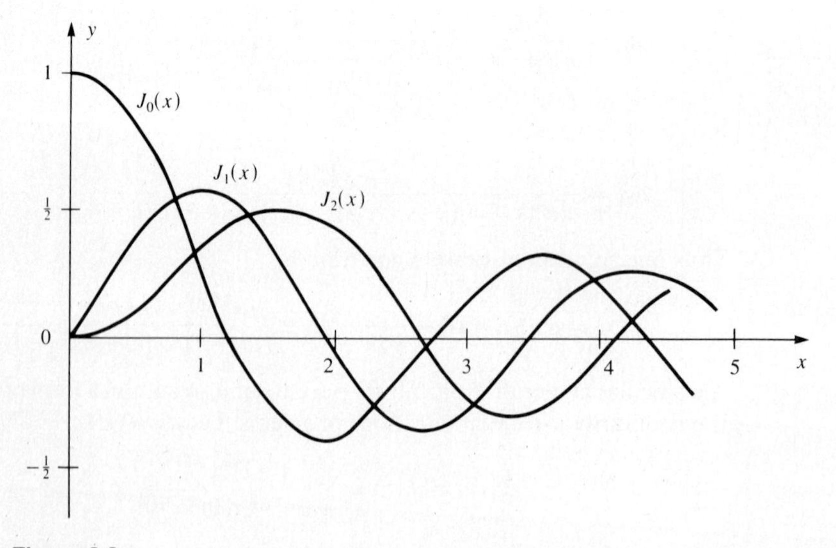

Figure 6.3

where the constants b_n and K may be obtained by substitution into Bessel's equation. The special cases in which $k = 0$ and $k = 1$ will be dealt with directly in the early part of Chapter 7.

A solution of Bessel's equation, called a **Bessel function of the second kind of order k**, that is linearly independent of $J_k(x)$, is denoted by $Y_k(x)$. To define this solution, we will use a function defined by

$$\varphi(p) = \begin{cases} 0 & \text{if } p = 0 \\ \sum_{j=0}^{p} j^{-1} & \text{if } p \text{ is a positive integer} \end{cases}$$

and a constant, called **Euler's constant**, defined by

$$\begin{aligned} \gamma &= \lim_{p \to \infty} \left(1 + \frac{1}{2} + \frac{1}{3} + \cdots + \frac{1}{p} - \ln p\right) \\ &= .5772157\cdots \end{aligned}$$

Then we can express the solution as

$$\begin{aligned} Y_k(x) = \frac{2}{\pi}\Bigg[\left(\ln\left|\frac{x}{2}\right| + \gamma\right)J_k(x) &- \frac{1}{2}\sum_{n=0}^{k-1} \frac{(k-n-1)!}{n!}\left(\frac{x}{2}\right)^{-k+2n} \\ &+ \frac{1}{2}\sum_{n=0}^{\infty} (-1)^{n+1}\left(\frac{\varphi(k+n) + \varphi(n)}{(n+k)!\,n!}\right)\left(\frac{x}{2}\right)^{k+2n}\Bigg] \end{aligned}$$

The graphs of Y_0 and Y_1 are illustrated in Figure 6.4.

The power-series method becomes an extremely valuable tool for second-order variable-coefficient equations of the type

$$(x-a)(x-b)y'' + (cx+d)y' + ey = 0$$

where a, b, c, d, and e are constants with $a \neq b$.

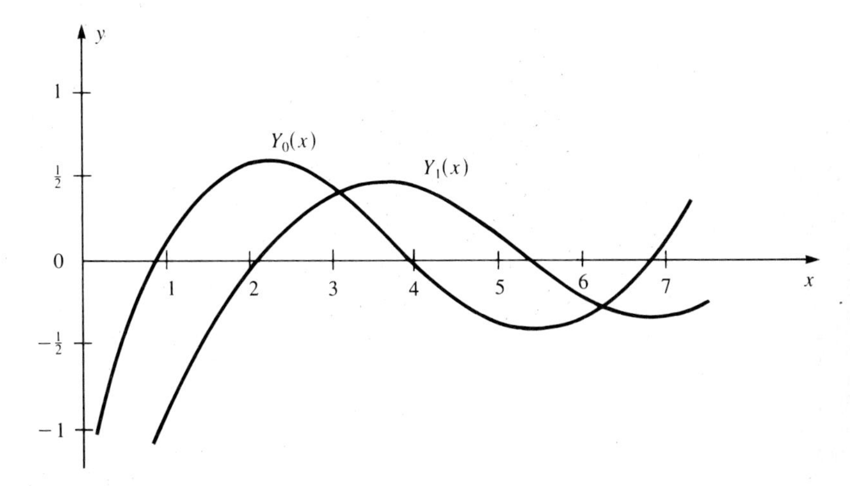

Figure 6.4

If we use the transformation $x = a + (b - a)t$ on an equation of this type, we change it into a differential equation of a type called **Gauss's hypergeometric equation** defined by

$$x(1 - x)y'' + [C - (A + B + 1)x]y' - ABy = 0$$

where A, B, and C are constants. This equation has many special functions related to it and also covers a wide selection of second-order equations.

We begin the solution around the regular singular point $x = 0$ with the trial solution

$$y = \sum_{0}^{\infty} a_n x^{n+s}$$

Using this, we see that the indicial equation is $s^2 + (C - 1)s = 0$, and its roots are 0 and $1 - C$. If C is not an integer, we will have two linearly independent solutions:

$$y_1(x) = \sum_{0}^{\infty} a_n x^n \qquad \text{and} \qquad y_2(x) = \sum_{0}^{\infty} b_n x^{n+1-C}$$

If we assume a_0 to be arbitrary, it can be verified that

$$a_n = \frac{A(A+1)\cdots(A+n-1)B(B+1)\cdots(B+n-1)}{n!\,C(C+1)(C+2)\cdots(C+n-1)}\,a_0, \qquad n = 1, 2, 3, \ldots$$

If we let $a_0 = 1$, we now obtain the first solution

$$y_1(x) = 1 + \frac{AB}{C}x + \frac{A(A+1)B(B+1)}{C(C+1)}\frac{x^2}{1\cdot 2} + \cdots$$

valid for $-1 < x < 1$. This is called the **hypergeometric series**, usually denoted by $F(A, B, C; x)$, and whose general form is

$$F(A, B, C; x) = 1 + \sum_{1}^{\infty} \frac{A(A+1)\cdots(A+n-1)B(B+1)\cdots(B+n-1)}{C(C+1)\cdots(C+n-1)} \cdot \frac{x^n}{n!}$$

With some special values of the constants we obtain some special functions:

$$F(A, B, C; 0) = 1$$

$$F(-A, B, B; -x) = (1 + x)^A$$

$$F(A, B, B; x) = (1 - x)^{-A}$$

$$F(1, 2, 2; -x) = \frac{1}{x}\ln(1 + x)$$

$$F(n + 1, -n, 1; x) = \text{polynomial of degree } n,\ n \geq 0$$

$$F\left(\frac{1}{2}, 1, \frac{3}{2}; x^2\right) = \frac{1}{2x}\ln\left(\frac{1 + x}{1 - x}\right)$$

$$F\left(\frac{k}{2}, \frac{-k}{2}, \frac{1}{2}; \sin^2 x\right) = \cos kx$$

$$xF\left(\frac{1}{2}, 1, \frac{3}{2}; -x^2\right) = \text{Arctan}\, x$$

$$xF\left(\frac{1}{2}, \frac{1}{2}, \frac{3}{2}; x^2\right) = \text{Arcsin}\, x$$

Exercises 6.4

1. Show that $x = 0$ is a singular point for each of the following equations. Then identify those that are regular singular points.
   (a) $4xy'' + 3y' - y = 0$
   (b) $x^2y'' + xy' + (x^2 - 9)y = 0$
   (c) $x^3y'' + 4x^2y' + (x + 2)y = 0$
   (d) $2x^2y'' - 3(\sin x)y' + y = 0$
   (e) $x^2y'' + x\left(x + \frac{1}{4}\right)y' - \left(x^2 + \frac{1}{2}\right)y = 0$
   (f) $3x^2y'' + x(x - 2)y' + (4 + x^2)y = 0$
   (g) $x(x + 1)^2y'' + x^2y' - (x + 1) = 0$

Use the method of Frobenius to solve the following differential equations. Use $x_0 = 0$.

2. $2x^2y'' + x(2x + 1)y' + 2xy = 0$
3. $2xy'' + y' - 2y = 0$
4. $2x^2y'' + xy' + (x - 1)(x + 1)y = 0$
5. $x(x - 2)y'' + y' - 2y = 0$
6. $x^2y'' + x(x^3 + 1)y' - y = 0$
7. $5xy'' - (x^2 - 2)y' - 4xy = 0$
8. $2x^2(x + 1)y'' + x(3x - 1)y' + y = 0$
9. $25x^2y'' + 2(x + 2)y = 0$
10. $3xy'' + (2 - x)y' - 2y = 0$
11. $x(x^2 - 1)y'' + 2(4x^2 - 1)y' + 12xy = 0$
12. $x^2y'' - 2x^3y' - 2y = 0$
13. $(1 + 3x^2)y'' + \left(\frac{2}{x}\right)y' - 6y = 0$
14. $x(x - 1)y'' + (x + 1)y' - y = 0$
15. $(2x^2 + x^3)y'' + (x + 3x^2)y' - (1 + 4x)y = 0$
16. Show that the power-series solution to $xy'' + 2y' - xy = 0$ can be expressed in finite form as

$$y(x) = a_0 \frac{\cosh x}{x} + b_0 \frac{\sinh x}{x}$$

17. Show that the solution to $4x^2y'' + 4xy' + (x^2 - 1)y = 0$ can be expressed in finite form as

$$y(x) = a_0 x^{1/2} \sin\frac{x}{2} + b_0 x^{-1/2} \cos\frac{x}{2}$$

18. Show that the solution to $x^2y'' - x(2x^2 - 1)y' - (1 + 2x^2)y = 0$ can be expressed in finite form as

$$y(x) = \frac{a_0}{x} + \frac{b_0 e^{x^2}}{x}$$

19. By writing out the expansions of $J_0(x)$ and $J_1(x)$ from equation (28), show that:

(a) $\frac{d}{dx} J_0(x) = -J_1(x)$

(b) $\frac{d}{dx}[xJ_1(x)] = xJ_0(x)$

20. A second way to find the indicial equation is to substitute $y = x^n$ into the differential equation and write the result as $f(n)x^{n-p} + g(n)x^{n-p+k}$,
where $p = 0$ or 1 and $k > 0$. Then $f(s) = 0$ is the indicial equation. Verify this for the equation of:
(a) Exercise 3
(b) Exercise 5
(c) Exercise 10
(d) Exercise 12
(e) Exercise 14

21. Yet another way to find the indicial equation for $y'' + p(x)y' + q(x)y = 0$ when $x = 0$ is a regular singular point is to create expressions $xp(x)$ and $x^2q(x)$ and expand them as power series:

$$xp(x) = p_0 + p_1x + p_2x^2 + \cdots$$

$$x^2q(x) = q_0 + q_1x + q_2x^2 + \cdots$$

The indicial equation is now $s(s - 1) + p_0s + q_0 = 0$. Verify this for the equations of
(a) Exercise 2
(b) Exercise 4
(c) Exercise 7
(d) Exercise 11
(e) Exercise 15

22. Show that $x^3y'' + x(2x + 1)y' - (1 + x)y = 0$ has an irregular singular point at $x = 0$. By attempting a solution of the form

$$y(x) = \sum_0^\infty a_n x^{n+s}$$

explain where the difficulties arise.

23. Show that

$$\sqrt{\frac{2}{\pi x}} \sin x$$

satisfies Bessel's equation with $k = \frac{1}{2}$.

24. Derive the formula

$$\frac{d}{dx}[x^2 J_2(x)] = x^2 J_1(x)$$

25. Find the solution of the equation $xy'' + y' + xy = 0$ for which $y(0) = 1$ and $y'(0) = 0$.

26. Show that a particular solution of the equation $xy'' + y' + a^2xy = 0$ can be written

$$y(x) = c_0\left(1 - \frac{a^2x^2}{2^2} + \frac{a^4x^4}{2^24^2} - \frac{a^6x^6}{2^24^26^2} + \cdots\right)$$

Then show that this is $y = c_0 J_0(ax)$.

In the case $a^2 = -i$, it can be written

$$J_0(\sqrt{-i}\,x) = J_0(i\sqrt{i}\,x) \equiv ber\, x + i \cdot bei\, x$$

where

$$ber\, x = 1 - \frac{x^4}{2^24^2} + \frac{x^8}{2^24^26^28^2} - \cdots$$

$$bei\, x = \frac{x^2}{2^2} - \frac{x^6}{2^24^26^2} + \frac{x^{10}}{2^24^26^28^210^2} - \cdots$$

These last two functions are the Bessel-real and Bessel-imaginary functions of importance in mathematical physics.

27. For each $n = 0, 1, 2, \ldots$ the **Laguerre differential equation** $xy'' + (1 - x)y' + ny = 0$, important in applications, has a polynomial solution which expressed in compact form is

$$L_n(x) = e^x \frac{d^n}{dx^n}(x^n e^{-x})$$

Find $L_n(x)$ in polynomial form for $n = 0, 1, 2, 3$ and verify that they solve the Laguerre differential equation.

28. For each nonnegative integer n, the **Legendre differential equation**, which plays an important role in many problems of engineering, is defined by $(1 - x^2)y'' - 2xy' + n(n + 1)y = 0$. A particular solution for each n is given by

$$y(x) = a_n\left[x^n - \frac{n(n-1)}{2(2n-1)}x^{n-2} + \frac{n(n-1)(n-2)(n-3)}{2 \cdot 4 \cdot (2n-1)(2n-3)}x^{n-4} - \cdots\right]$$

where
$$\begin{cases} a_0 = 1 \\ a_n = \dfrac{(2n)!}{2^n(n!)^2} \end{cases}$$

for $n = 1, 2, \ldots$. With these particular coefficients being used, the compact form

$$P_n(x) = \frac{1}{2^n n!} \frac{d^n}{dx^n}(x^2 - 1)^n$$

expresses the solutions, which are called the **Legendre polynomials**. Find the expressions for $P_0(x)$, $P_1(x)$, $P_2(x)$, and $P_3(x)$. Now show that

$$\int_{-1}^{1} P_i(x)P_j(x)\,dx = \begin{Bmatrix} 0 & \text{if} & i \neq j \\ 1 & \text{if} & i = j \end{Bmatrix} \qquad \text{for } i, j = 0, 1, 2, 3$$

This is an expression of the orthonormality of the functions, important in the Fourier series expansions of electronic-circuit theory.

29. In water-conservation projects, particularly in the building of river dams, concrete pipes of large diameters are needed. These pipes are strengthened with a device called a stiffener ring, which is inserted at an abrupt angle. This requires a careful study of the stresses in the circular pipe to accurately design it. The problem of determining the displacement in the radial direction of the stiffener-ring inserts due to water pressure inside may be reduced to the differential equation

$$r^2 \frac{d^4x}{dr^4} + 6r \frac{d^3x}{dr^3} + 6 \frac{d^2x}{dr^2} + x = 0$$

Find the recurrence formula and the indicial roots to potentially prepare the problem for a numerical solution.

Miscellaneous Exercises for Chapter 6

For each problem in Exercises 1–10 find all of the ordinary points, the regular singular points, and the irregular singular points.

1. $(2x - 1)y'' + y' - 5y = 0$
2. $x(x + 4)y'' + x^2y' - 2y = 0$
3. $2y'' + e^xy' + (\sin x)y = 0$
4. $(1 - x^2)y'' + 3xy' - 2y = 0$
5. $(2x^2 + 3x - 2)y'' + 7x^2y' + (\cos x)y = 0$
6. $3x^2(x - 1)y'' + (x^2 - 1)y' + (2x - 2)y = 0$
7. $(x^4 + x^2)y'' + (3x^3 - x)y' + (x^2 + 1)y = 0$
8. $(x^2 + 3x + 2)y'' + (x - 1)y' - x^2y = 0$

9. $x^2(4 + x)y'' + 9xy' - y = 0$
10. $x^2(x^2 - 5x + 6)y'' + x(3 - x)y' + (2x - 6)y = 0$

For Exercises 11–16 use the Taylor series method to find the first four nonzero terms in the solution to each initial value problem.

11. $y'' + y^2 = 0$; $y(0) = 1$, $y'(0) = 0$
12. $(x + 3)y'' + 2y = 0$; $y(0) = 0$, $y'(0) = 1$
13. $y'' - (x + 2)y' + 2y = 0$; $y(0) = 1$, $y'(0) = 3$
14. $y''' - y' = y^2 \ln x$; $y(1) = 1$, $y'(1) = 0$, $y''(1) = 1$
15. $y'' - y^2 = 4$; $y(0) = 0$, $y'(0) = 2$
16. $yy'' + 3(y')^2 = 0$; $y(0) = 1$, $y'(0) = \frac{1}{4}$

For Exercises 17–25 use series techniques to find the solution. Verify by analytical methods, if possible.

17. $y' = x + 3xy$; $y(0) = 1$
18. $y' = xy - x^2$; $y(0) = 2$
19. $x^2y'' + x^2y' - 2y = 0$
20. $(1 + x^2)y'' + xy' - y = 0$
21. $x^2y'' - 2xy' + 2y = 0$
22. $xy'' - (x + 2)y' + 2y = 0$
23. $xy'' + 2y' - xy = 0$
24. $x^2y'' + (x - 2x^3)y' - (1 + 2x^2)y = 0$
25. $(1 + 3x^2)y'' + \frac{2}{x}y' - 6y = 0$
26. Find the indicial roots used when solving $x^2y'' - 2x^2y' + (\frac{1}{4} + x^2)y = 0$ by series methods.
27. Find the indicial roots used when solving $x^2y'' + xy' + (x - 1)y = 0$ by series methods.
28. Find the indicial roots and the general recurrence formula when the method of Frobenius is applied to
$$3x^2y'' - (x^2 - 2x)y' - (x^2 + 2)y = 0$$
29. Find the indicial roots and the general recurrence formula when the method of Frobenius is applied to $x^2(x + 4)y'' + 7xy' - y = 0$.
30. Find the indicial roots and the general recurrence formula when the method of Frobenius is applied to
$$3x^2y'' - x(x - 5)y' + (2x^2 - 1)y = 0$$

Chapter 7

Numerical Methods for Series Solutions

As we have seen in our prior work, a considerable amount of algebraic manipulation is required to solve a second-order initial value problem by the methods of infinite series. The attendant opportunities for arithmetic errors and the unfamiliar appearance of the series result sometimes make this approach less than desirable. When the power series are not identified as particular functions, the calculation of the y-values that accompany a given interval of x-values is a formidable task. Now, as before, a numerical approach can produce what the problem proposer really desires. It enables us to obtain not only a table listing these x-y solution pairs, but also a graphical portrayal of the solution curve over any desired interval. This is often difficult to obtain from an analytical solution.

7.1 Bessel's Equations

To introduce the computational aspects, we will first examine two classical examples of second-order ordinary differential equations taken from mathematical physics. In these we will note how the evaluation of the series is done directly from the recurrence relation. This evaluation is especially important when the actual coefficients are impossible to find. We begin in Example 1 with a differential equation having a regular singular point at $x = 0$.

Example 1 (Bessel's Equation of Order Zero)

Consider $x^2y'' + xy' + x^2y = 0$. A few simple computations will produce the indicial equation $s^2 = 0$ and thus the double indicial root $s_1 = s_2 = 0$. The first solution obtained by using the method of Frobenius with $s_1 = 0$ is of the form

$$y_1(x) = \sum_{n=0}^{\infty} a_n x^n$$

It is made unique by choosing the arbitrary coefficient $a_0 = 1$, and its exact form by this method is

$$y_1(x) = 1 + \sum_{n=1}^{\infty} \frac{(-1)^n x^{2n}}{2^{2n}(n!)^2}$$

$$= 1 - \frac{1}{4}x^2 + \frac{1}{64}x^4 - \frac{1}{2304}x^6 + \cdots$$

The radius of convergence is easily found by the ratio test

$$\frac{1}{R} = \lim_{n\to\infty}\left|\frac{a_{n+1}}{a_n}\right| = \lim_{n\to\infty}\left\{\frac{2^{2n}(n!)^2}{2^{2n+2}[(n+1)!]^2}\right\} = \lim_{n\to\infty}\left[\frac{1}{4(n+1)^2}\right] = 0$$

implying that $R = \infty$. Hence this series converges for all x, and $y_1(x)$ solves the Bessel equation for all $x \neq 0$. (The point $x = 0$ is a singular point of the differential equation.) This solution, denoted by $J_0(x)$, is referred to as the **Bessel function of the first kind of order zero** and is important enough in applications to have extensive tables of values in print. We will now see how to produce such tables by evaluating the above series. The practice attained in this problem will assist us in subsequent work.

As we will see many times in this chapter, to evaluate directly from the series is much more difficult than to work from the recurrence relation itself. We are interested in finding the values of the solution $J_0(x)$ at a prechosen collection of x values in an interval $[A, B]$. We use the following approach:

The method of Frobenius used above has shown that all odd subscripted a's are zero. With arbitrary a_0 chosen to be 1, we calculate a_2, then a_4, etc., from the derived recurrence relation

$$a_{2n} = \frac{-a_{2n-2}}{(2n)^2}$$

noting at each step its absolute value and computing until the pre-established convergence tolerance is reached. By summing the series thus formed at a particular value of x, we get one solution value. This must then be done at each x value in $[A, B]$. This, of course, is a tedious and laborious project.

To produce these x-y solution pairs, we turn to the computer program BESSL0 (for Bessel-zero), which is written directly from the recurrence relation. This program, which is found in Appendix B, produces the values of $J_0(x)$ by the above-described process. It would be well worth your effort to study this program and then check the program by hand calculations for a few steps to get the feel of what it is doing. Then run the program and the associated graphing routine for practice. If you have programming expertise, try some modifications to fit the solution to an application problem (see the exercises).

A second linearly independent solution to the Bessel equation of order zero is

$$y_2(x) = y_1(x) \ln x + \sum_{n=1}^{\infty} \frac{(-1)^{n+1}}{2^{2n}(n!)^2} \left[1 + \frac{1}{2} + \frac{1}{3} + \cdots + \frac{1}{n} \right] x^{2n}$$

for $x > 0$. This series also converges for all $x > 0$. The program BESSL0 also computes simultaneously the values of y_2. The convergence tolerance allows the computation error to be less than the tolerance value. The values of the solutions y_1 and y_2 are printed simultaneously.

The graphing routine BESGRAF coupled to the program allows a pictorial representation of both solutions plotted on a common set of axes.

Example 2 (Bessel's Equation of Order One)

We now consider $x^2 y'' + xy' + (x^2 - 1)y = 0$. The method of Frobenius produces the indicial equation $s^2 - 1 = 0$. The roots $s_1 = 1$ and $s_2 = -1$ differ by an integer, and the solutions are somewhat more complicated. The first solution corresponding to $s_1 = 1$ is of the form

$$y_1(x) = \sum_{0}^{\infty} a_n x^{n+1}.$$

Putting this trial solution into the differential equation and performing the algebraic steps produce the recurrence relation

$$3a_1 = 0, \quad n(n-2)a_n + a_{n-2} = 0$$

Since $a_1 = 0$, we have $a_3 = a_5 \cdots = 0$. Additional algebraic work produces

$$a_{2n} = \frac{(-1)^n a_0}{2^{2n} n!(n+1)!}$$

and thus the first solution is

$$y_1(x) = a_0 x \left[1 + \sum_{n=1}^{\infty} \frac{(-1)^n x^{2n}}{2^{2n} n!(n+1)!} \right]$$

This series solution to Bessel's equation of order one converges for every x as can be shown by using the ratio test. With the arbitrary a_0 picked as $\frac{1}{2}$, this solution, denoted by $J_1(x)$, becomes the **Bessel function of the first kind of order one** and is also well tabulated numerically and widely used in applications.

A program to compute values of $J_1(x)$ and also the values of $y_2(x)$, the second linearly independent solution resulting from the use of $s_2 = -1$, appears in Appendix B under the name BESSL1. The automatic graphing option is once again included. Further mathematical computation will produce this second solution:

$$y_2(x) = -\frac{1}{2} y_1(x) \ln x + a_0 x^{-1}\left[1 - \sum_{1}^{\infty} \frac{(-1)^n}{2^{2n} n!(n-1)!}\left(2H - \frac{1}{n}\right)x^{2n}\right]$$

where $H = 1 + \frac{1}{2} + \cdots + \frac{1}{n}$, called the truncated harmonic series, and where we again pick the constant $a_0 = \frac{1}{2}$. Check this program also with a couple of hand calculations to appreciate its operation. Then run the program and the graphing routine to obtain the solution set with its graph.

In both of these examples, notice the behavior of the particular solutions near the origin. The first solution is bounded at $x = 0$ but the second is the unbounded singular solution. Of course, the logarithmic term in y_2 accounts for this behavior. In creating the graph via BESSL0 and/or BESSL1, do not include the point $x = 0$ in the solution set. The interval should begin near zero and run more than 10 units to the right (e.g., use [0.1, 15.1] with 30 printouts and a convergence tolerance of 0.0001).

7.2 Series Solution at an Ordinary Point

After having seen numerical evaluations for two specific differential equations, we now turn our attention to the process of obtaining the numerical values of the solution to an arbitrary second-order initial value problem via the methods of power-series manipulation. As mentioned earlier, the analytically obtained series solution gives little information on the properties of the solution. Insight into the behavior and nature of this solution function is gained by examining a graph of the function. To get a single point on the graph it is necessary to sum the infinite series in the solution at that value of x. To get the entire graph would be a sizable problem were it not for our speedy friend the computer.

For the computer to perform all these computations, the student must only obtain the recurrence formula by a few algebraic steps and input it into the program. However, the recurrence formula must be in a rather specific form: *It must be solved for the highest subscripted coefficient a_n in terms of lower subscripted ones, a_{n-1} and a_{n-2}, must be algebraically reduced to the* *simplest form, and should be broken into separate terms when possible.* The program SERSO is used when this recurrence formula contains series coefficients whose subscripts differ by at most two. If they differ by three, use SERSO3. Both programs are displayed in Appendix B. Carefully check the documentation for these programs to learn the correct method of inserting the recurrence formula into the program. For anything more complicated or sophisticated than this, one could follow the lead of those two programs and create an extended program capable of handling the particular problem. Coupled to these programs is a graphing routine that will present a pictorial representation of the unique solution to the initial value problem. In each case care must be exercised in determining the interval of convergence of the power series to know the values of x for which the solution being obtained is valid. One can then pick the desired interval on which the computer will calculate, print, and graph the solution function.

We will now illustrate the capabilities of these computer routines with a few examples.

Example 3

Consider the initial value problem

$$y'' + 4y = 0; \quad y(0) = 2, \quad y'(0) = 4$$

This problem, of course, can be solved analytically and the actual unique solution function is $y = 2\cos 2x + 2\sin 2x$. This will give us a good comparison for an accuracy check of our program SERSO. Using calculus to examine this function, we find it has zeros at $x = \frac{1}{2}\ \text{Arctan}(-1) = -0.392$ and 1.18 and 2.75, and it has relative maximum and minimum points at $[\frac{1}{2}\ \text{Arctan}\ 1,\ 2\cos(\text{Arctan}\ 1) + 2\sin(\text{Arctan}\ 1)]$ whose sampled values are (0.392, 2.83) and (3.53, 2.83) for maximum points and (1.96, −2.83) and (−1.18, −2.83) for minimum points. Since the initial point 0 is an ordinary point of the differential equation, we use the trial solution

$$y = \sum_{n=0}^{\infty} a_n x^n$$

and easily proceed to the recurrence formula

$$a_n = \frac{-4}{n(n-1)} a_{n-2}, \quad n \geq 2$$

We are given in the problem $a_0 = y(0) = 2$ and $a_1 = y'(0) = 4$. Using the program SERSO, we input the recurrence formula at line 800 in the form

```
800 DEF FNF(N, An_1, An_2) = -4 * An_2/(N * (N - 1))
```

Note that the coding in BASIC requires careful attention to the naming of variables. (The documentation in the program explains this.)

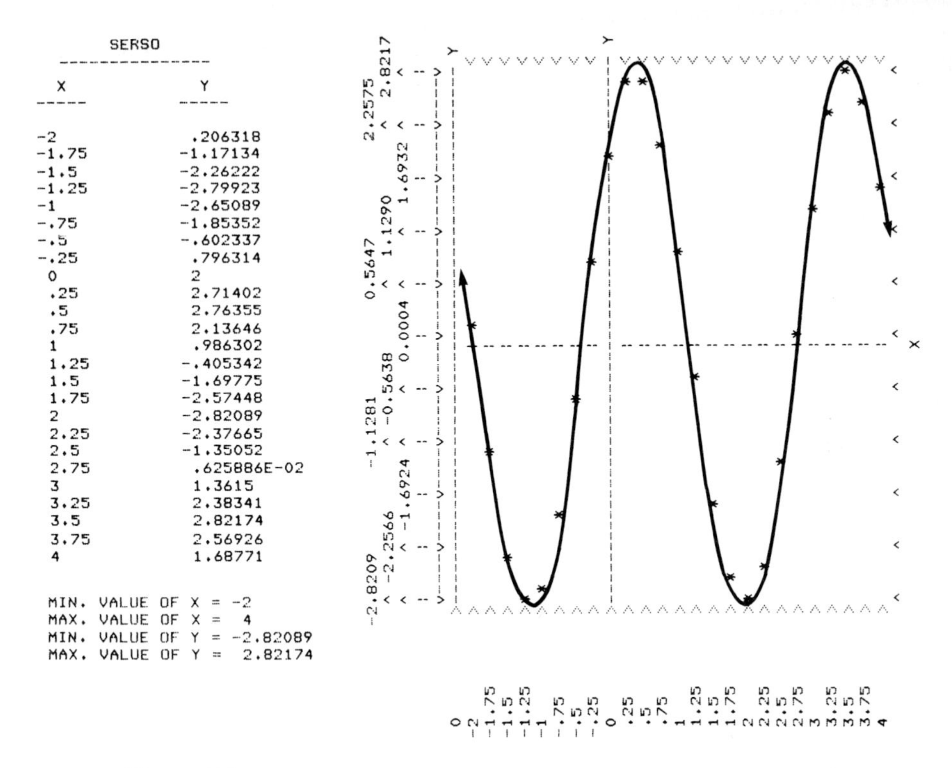

SERSO

X	Y
-2	.206318
-1.75	-1.17134
-1.5	-2.26222
-1.25	-2.79923
-1	-2.65089
-.75	-1.85352
-.5	-.602337
-.25	.796314
0	2
.25	2.71402
.5	2.76355
.75	2.13646
1	.986302
1.25	-.405342
1.5	-1.69775
1.75	-2.57448
2	-2.82089
2.25	-2.37665
2.5	-1.35052
2.75	.625886E-02
3	1.3615
3.25	2.38341
3.5	2.82174
3.75	2.56926
4	1.68771

MIN. VALUE OF X = -2
MAX. VALUE OF X = 4
MIN. VALUE OF Y = -2.82089
MAX. VALUE OF Y = 2.82174

Figure 7.1

Inspection of the differential equation shows no singular points, so we may choose our solution interval arbitrarily. Figure 7.1 shows the program output on $[-2, 4]$ in 24 evaluations and the associated graph. Observe the quality of the graph with respect to the zeros and maximum and minimum points of the actual solution calculated earlier.

Example 4

Consider

$$(x^2 + 4)y'' - (2x - 1)y' + 6y = 0; \quad y(0) = 2, \quad y'(0) = -2$$

Since 0 is an ordinary point of the differential equation, we proceed by the usual algebraic steps to the recurrence formula

$$a_n = \frac{(n^2 - 7n + 16)a_{n-2} + (n - 1)a_{n-1}}{-4n(n - 1)}$$

This relation is somewhat more complicated than are those we have studied thus far, but by no means is it harder for the computer to evaluate. Noting that potential singularities occur at $x = \pm 2i$, we proceed carefully again with our interval. We stretch it to its limits and understand

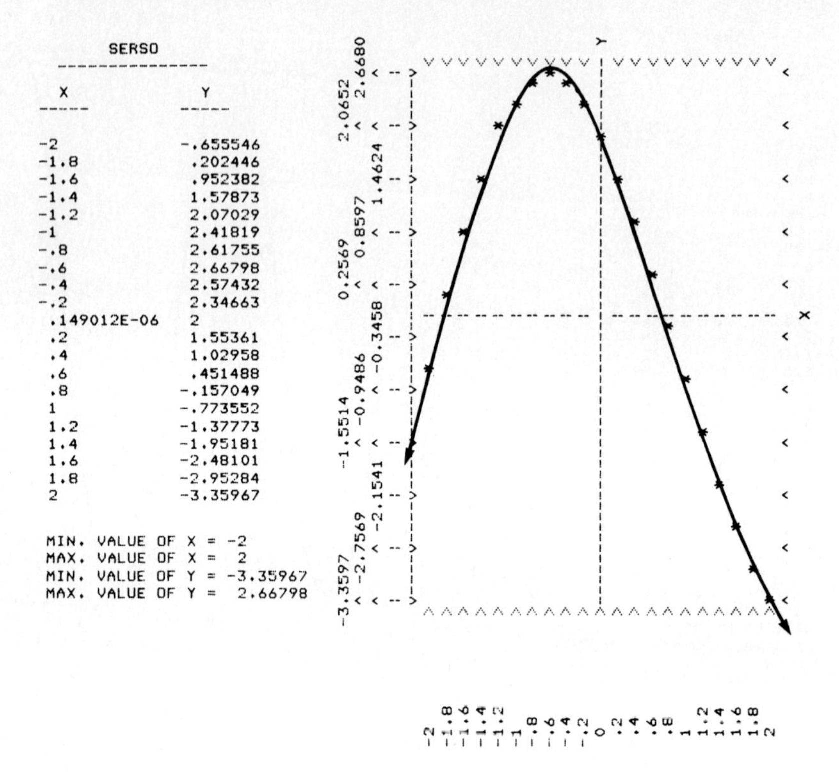
SERSO

X	Y
-2	-.655546
-1.8	.202446
-1.6	.952382
-1.4	1.57873
-1.2	2.07029
-1	2.41819
-.8	2.61755
-.6	2.66798
-.4	2.57432
-.2	2.34663
.149012E-06	2
.2	1.55361
.4	1.02958
.6	.451488
.8	-.157049
1	-.773552
1.2	-1.37773
1.4	-1.95181
1.6	-2.48101
1.8	-2.95284
2	-3.35967

MIN. VALUE OF X = -2
MAX. VALUE OF X = 2
MIN. VALUE OF Y = -3.35967
MAX. VALUE OF Y = 2.66798

Figure 7.2

then that the solution obtained is valid inside $[-2, 2]$. We use SERSO with statement 800 reading

```
800 DEF FNF(N, An_1, An_2)
    = ((N * N - 7 * N + 16) * An_2)/(-4 * N * (N - 1)) + An_1/(-4 * N)
```

and produce the results shown in Figure 7.2.

Example 5

Consider

$$(x^2 - 1)y'' + 6xy' + 4y = 0; \quad y(0) = 2, \quad y'(0) = 3$$

Before we solve the equation and apply the initial conditions, this example can be used to alert you about the potential pitfalls in computational work. To find the series solution to this differential equation by ordinary means, we would use the recurrence formula

$$a_n = \left(\frac{n+2}{n}\right) a_{n-2}$$

With a_0 and a_1 arbitrary, we would find

$$y = a_0(1 + 2x^2 + 3x^4 + 4x^6 + 5x^8 + \cdots)$$

$$+ a_1\left(x + \frac{5}{3}x^3 + \frac{7}{3}x^5 + \frac{9}{3}x^7 + \frac{11}{3}x^9 + \cdots\right) \tag{1}$$

The interesting patterns shown in the coefficients might lead us to a more complete investigation of these series. If we had at our disposal a large collection of infinite-series expansions, we might note the following:

$$\frac{1}{1-x} = 1 + x + x^2 + x^3 + x^4 + \cdots$$

$$\frac{1}{(1-x)^2} = 1 + 2x + 3x^2 + 4x^3 + 5x^4 + \cdots$$

$$\frac{1}{(1-x^2)^2} = 1 + 2x^2 + 3x^4 + 4x^6 + 5x^8 + \cdots$$

$$(3x - x^3)\cdot\frac{1}{(1-x^2)^2} = 3x + 5x^3 + 7x^5 + 9x^7 + 11x^9 + \cdots$$

Using these, the solution (1) could be written in a compact form:

$$y(x) = a_0 \cdot \frac{1}{(1-x^2)^2} + \frac{a_1}{3}\cdot\frac{3x - x^3}{(1-x^2)^2} \tag{2}$$

Now add the initial conditions $y(0) = 2$ and $y'(0) = 3$. Applying these to the original result (1) we obtain

$$y = 2(1 + 2x^2 + 3x^4 + 4x^6 + \cdots) + 3\left(x + \frac{5}{3}x^3 + \frac{7}{3}x^5 + \cdots\right)$$

$$= 2 + 3x + 4x^2 + 5x^3 + 6x^4 + \cdots + (n+2)x^n + \cdots$$

which because of the absence of fractions certainly appears to have a chance to converge only in $(-1, 1)$. (The ratio test confirms this.) This also would seem to agree with information from the differential equation showing singularities at $x = \pm 1$. However, if the initial conditions are put into the second result (2), we get the solution

$$y(x) = \frac{2 + 3x - x^3}{(1-x^2)^2} \tag{3}$$

This is a function displaying the property of vertical asymptotes of even powers at $x = \pm 1$. If the program SERSO were now run, it would give the results shown in Figure 7.3. Clearly the vertical asymptote at $x = 1$ shows, but at $x = -1$ it does not show. Checking the result (3) in

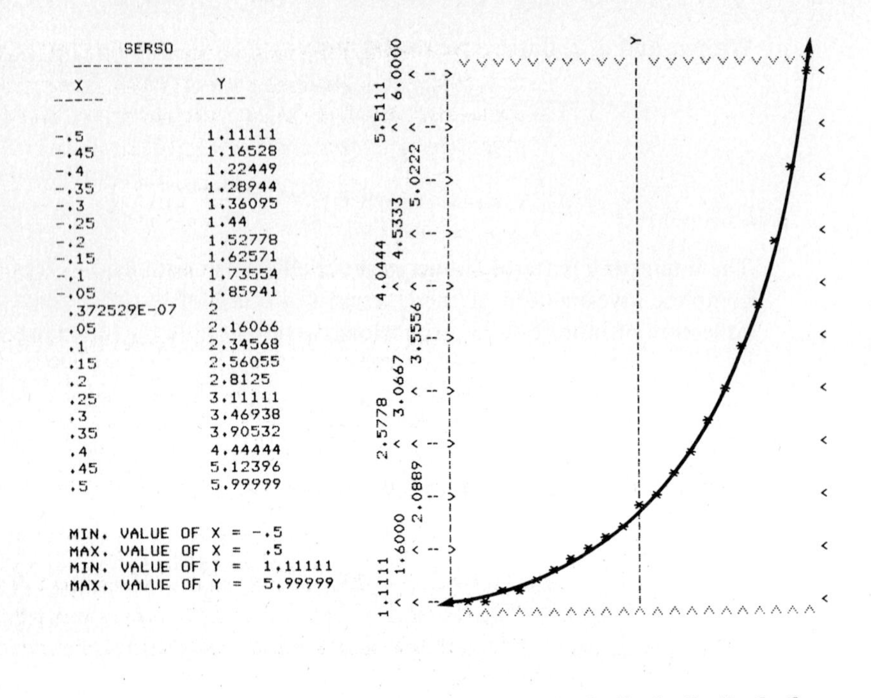

SERSO

X	Y
-.5	1.11111
-.45	1.16528
-.4	1.22449
-.35	1.28944
-.3	1.36095
-.25	1.44
-.2	1.52778
-.15	1.62571
-.1	1.73554
-.05	1.85941
.372529E-07	2
.05	2.16066
.1	2.34568
.15	2.56055
.2	2.8125
.25	3.11111
.3	3.46938
.35	3.90532
.4	4.44444
.45	5.12396
.5	5.99999

MIN. VALUE OF X = -.5
MAX. VALUE OF X = .5
MIN. VALUE OF Y = 1.11111
MAX. VALUE OF Y = 5.99999

Figure 7.3

YACTUL would confirm the correctness of the solution. Investigating further would show that (3) could be reduced to lower terms:

$$y(x) = \frac{2 + 3x - x^3}{(1 - x^2)^2} = \frac{(2 - x)(1 + x)^2}{(1 - x)^2(1 + x)^2} = \frac{2 - x}{(1 - x)^2}$$

Now everything checks: $y(-1) = \frac{3}{4} = 0.75$. Here we can clearly see why a series solution is not as desirable as an analytic solution. The interior graph shows the nature of the solution.

All the problems we have considered so far in this section have been initial value problems whose initial point is $x = 0$. This, as you probably realize, was done by design. When we encounter a problem posed at a different point, it is not difficult to modify our approach.

Suppose we want the solution to the posed problem near the initial point $x = a$. We make the change of variables $t = x - a$. The chain rule for derivatives ensures that $dy/dx = dy/dt$ and $d^2y/dx^2 = d^2y/dt^2$. Using these provides a new differential equation in the variables y and t with initial conditions at $t = 0$, which is solved for $y(t)$. When this process has been completed, the simple reverse substitution $t = x - a$ produces the unique solution $y = f(x)$ to the original problem. To illustrate this process, consider the following initial value problem in which the initial point is $x = 2$.

Example 6

Consider

$$y'' + (2 - x)y' + 2y = 0; \quad y(2) = 4, \quad y'(2) = 1$$

The substitution of $x = t + 2$, which is a linear shift of two units to the left, converts this to $y'' - ty' + 2y = 0$; $y(0) = 4$, $y'(0) = 1$. The recurrence formula for this initial value problem in the variables t and y is

$$a_n = \frac{n - 4}{n(n - 1)} a_{n-2}$$

Using the program SERSO on the t-interval $[-2, 2]$ we could easily obtain the printed results of y versus t; but since we need a linear shift of two units to the right to recover the originally intended solution, we modify the values of the independent variable to convert them to $x + 2$ values. This requires only three program-statement adjustments:

```
1400 LET X = X + 2
1405 PRINT X, Y
1415 LET X = X - 2
```

which would be inserted at the time statement 800 for the recurrence formula is input. The results of this problem (see Figure 7.4) now appear on $[0, 4]$, an interval whose center is the original initial value $x_0 = 2$. In the input list we still use the zero-centered values -2 and 2 for the independent variable, because the recurrence formula used is the formula for the shifted variable. The adjustments above automatically produce the final results over the correct domain.

Example 7

Consider

$$(x^2 - 1)y'' + 3xy' + xy = 0; \quad y(0) = -2, \quad y'(0) = 1$$

Noting that $x = 0$ is an ordinary point, we use the trial solution

$$y = \sum_{n=0}^{\infty} a_n x^n$$

and proceed to the recurrence formula

$$a_n = \frac{(n - 2)na_{n-2} + a_{n-3}}{n(n - 1)}$$

We notice that the subscripts differ by three and hence for this problem program SERSO3 must be used. However, program SERSO3 requires an initial value $y''(0)$, which is not given. It can be computed directly from the differential equation by solving for y'' and setting $x = 0$. For this

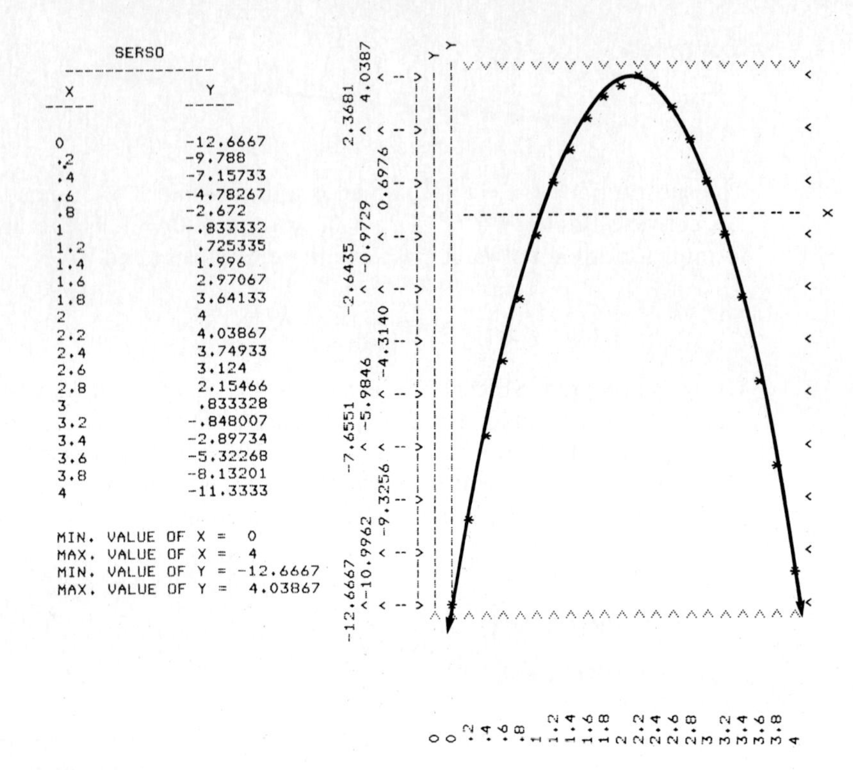
SERSO

X	Y
0	-12.6667
.2	-9.788
.4	-7.15733
.6	-4.78267
.8	-2.672
1	-.833332
1.2	.725335
1.4	1.996
1.6	2.97067
1.8	3.64133
2	4
2.2	4.03867
2.4	3.74933
2.6	3.124
2.8	2.15466
3	.833328
3.2	-.848007
3.4	-2.89734
3.6	-5.32268
3.8	-8.13201
4	-11.3333

MIN. VALUE OF X = 0
MAX. VALUE OF X = 4
MIN. VALUE OF Y = -12.6667
MAX. VALUE OF Y = 4.03867

Figure 7.4

problem we obtain $y''(0)=0$. Also observe that the differential equation is singular at $x = \pm 1$. We take care in choosing our interval and run this problem on $[-0.9, 0.9]$. The behavior at $x = \pm 1$ can be inferred from the graph (see Figure 7.5). Care must be taken in choosing the variable names; for SERSO3, a_n is denoted by An, a_{n-1} by An_1, a_{n-2} by An_2 and a_{n-3} by An_3. Check this in the listing of SERSO3 in Appendix B.

Example 8

Consider

$$y''' + x^2 y'' + 5xy' = 0; \quad y(0) = 1, \quad y'(0) = -2, \quad y''(0) = 1$$

This third-order equation has an ordinary point at $x = 0$, and we can proceed to a power-series solution by the usual steps. The easily derived recurrence formula is

$$a_n = \frac{-a_{n-3}}{(n-1)}$$

Using SERSO3, since the subscripts differ by three, we obtain the results in Figure 7.6.

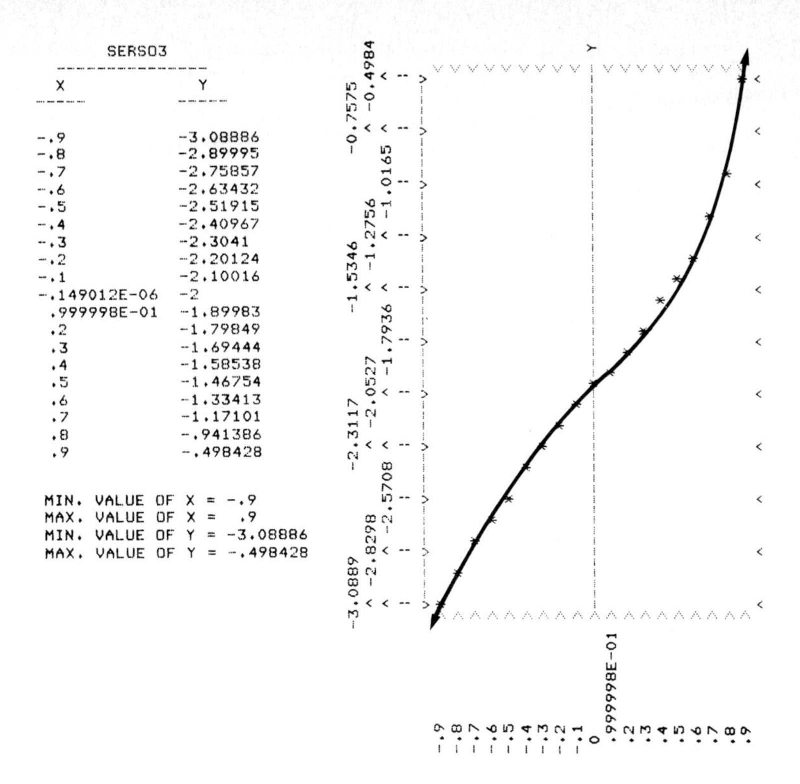

SERS03

X	Y
-.9	-3.08886
-.8	-2.89995
-.7	-2.75857
-.6	-2.63432
-.5	-2.51915
-.4	-2.40967
-.3	-2.3041
-.2	-2.20124
-.1	-2.10016
-.149012E-06	-2
.999998E-01	-1.89983
.2	-1.79849
.3	-1.69444
.4	-1.58538
.5	-1.46754
.6	-1.33413
.7	-1.17101
.8	-.941386
.9	-.498428

MIN. VALUE OF X = -.9
MAX. VALUE OF X = .9
MIN. VALUE OF Y = -3.08886
MAX. VALUE OF Y = -.498428

Figure 7.5

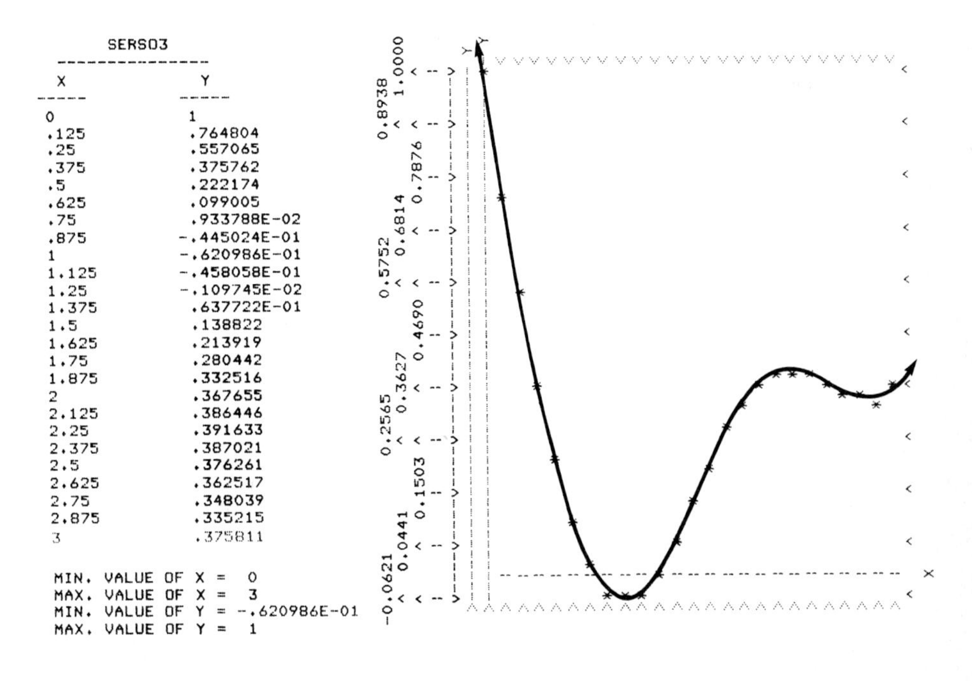

SERS03

X	Y
0	1
.125	.764804
.25	.557065
.375	.375762
.5	.222174
.625	.099005
.75	.933788E-02
.875	-.445024E-01
1	-.620986E-01
1.125	-.458058E-01
1.25	-.109745E-02
1.375	.637722E-01
1.5	.138822
1.625	.213919
1.75	.280442
1.875	.332516
2	.367655
2.125	.386446
2.25	.391633
2.375	.387021
2.5	.376261
2.625	.362517
2.75	.348039
2.875	.335215
3	.375811

MIN. VALUE OF X = 0
MAX. VALUE OF X = 3
MIN. VALUE OF Y = -.620986E-01
MAX. VALUE OF Y = 1

Figure 7.6

7.3 Series Solution at a Regular Singular Point

If a given differential equation has a regular singular point at an x-value near which a solution is desired, the method of Frobenius is usually used to produce the solution. Since that particular value of x is a singular point for the differential equation, we surely could not expect to have initial conditions at that point, which would define the value of the function and its derivative at the point. Thus we do not really have an initial value problem. The method of Frobenius does, however, allow us to make an arbitrary choice of the constant that appears in the particular solutions and this is often chosen to be 1. We are actually deciding arbitrarily to take one function from the family of solution functions. It is possible to gain some insight into the solution family by carefully inspecting one of its members. Hence, we are again interested in having the computer calculate the tedious series sums and print out results. The illustrations early in this chapter on Bessel's differential equations followed this format.

The user now must calculate the roots of the indicial equation as the differential equation is being processed en route to the recurrence formula. Once these have been determined, two separate runs of the computer program will be made. With a pick of the multiplying constants, these linearly independent solutions may be combined to the general solution. The three cases (in which the indicial roots are equal, differ by an integer, or neither of these) are not treated separately here since the computations of solution values are done directly from the recurrence formula. It should be clear that we could not input a repeated indicial root and expect distinct results. The case of the roots differing by an integer will in most cases produce linearly independent solutions anyway.

The program SERSING and its attached graphing routine will perform the above-mentioned operations. Reading the documentation of the program will help in its implementation. To get the feel of its mechanism, we turn to a couple more illustrations.

Example 9

Consider

$$x^2y'' + (x - x^2)y' - y = 0$$

By performing the necessary algebraic steps we obtain the indicial equation $s^2 - 1 = 0$, which gives the roots $s_1 = 1$ and $s_2 = -1$, and the concurrent recurrence formula

$$a_n = \frac{a_{n-1}}{n + s + 1}$$

Since we desire the solution in the neighborhood of the regular singular point $x = 0$, we must first look at the signs of the indicial roots to see if the origin could be included in our solution interval. In this problem the

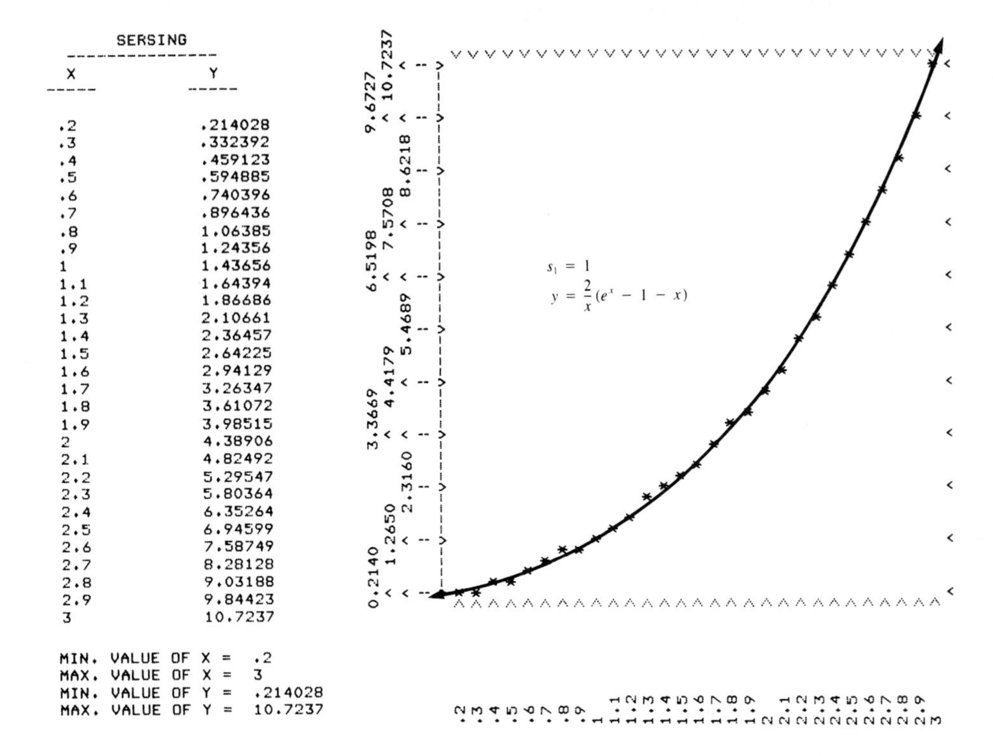

SERSING

X	Y
.2	.214028
.3	.332392
.4	.459123
.5	.594885
.6	.740396
.7	.896436
.8	1.06385
.9	1.24356
1	1.43656
1.1	1.64394
1.2	1.86686
1.3	2.10661
1.4	2.36457
1.5	2.64225
1.6	2.94129
1.7	3.26347
1.8	3.61072
1.9	3.98515
2	4.38906
2.1	4.82492
2.2	5.29547
2.3	5.80364
2.4	6.35264
2.5	6.94599
2.6	7.58749
2.7	8.28128
2.8	9.03188
2.9	9.84423
3	10.7237

MIN. VALUE OF X = .2
MAX. VALUE OF X = 3
MIN. VALUE OF Y = .214028
MAX. VALUE OF Y = 10.7237

Figure 7.7

solutions y_1 and y_2, which we obtain by setting $a_0 = 1$ and proceeding with the recurrence formula, would begin with x and x^{-1}. Thus y_1 would pass through the origin but y_2 would be singular there. We then should pick our solution interval of evaluation to exclude the origin. This problem is now run on $[0.2, 3]$ and its output and corresponding graphs are displayed in Figures 7.7 and 7.8. If we had carried out the analytic computations we would have obtained

$$y_1 = x\left(1 + \frac{1}{3}x + \frac{1}{12}x^2 + \frac{1}{60}x^3 + \frac{1}{360}x^4 + \cdots\right)$$

$$y_2 = x^{-1}\left(1 + x + \frac{1}{2!}x^2 + \frac{1}{3!}x^3 + \cdots\right)$$

The expression y_2 is easily seen to be the function e^x/x. The expression for y_1 may be rearranged:

$$\begin{aligned} y_1 &= x + \frac{1}{3}x^2 + \frac{1}{3\cdot 4}x^3 + \frac{1}{3\cdot 4\cdot 5}x^4 + \cdots \\ &= \frac{2}{x}\left(\frac{x^2}{2!} + \frac{x^3}{3!} + \frac{x^4}{4!} + \cdots\right) = \frac{2}{x}\left(e^x - 1 - x\right) \end{aligned}$$

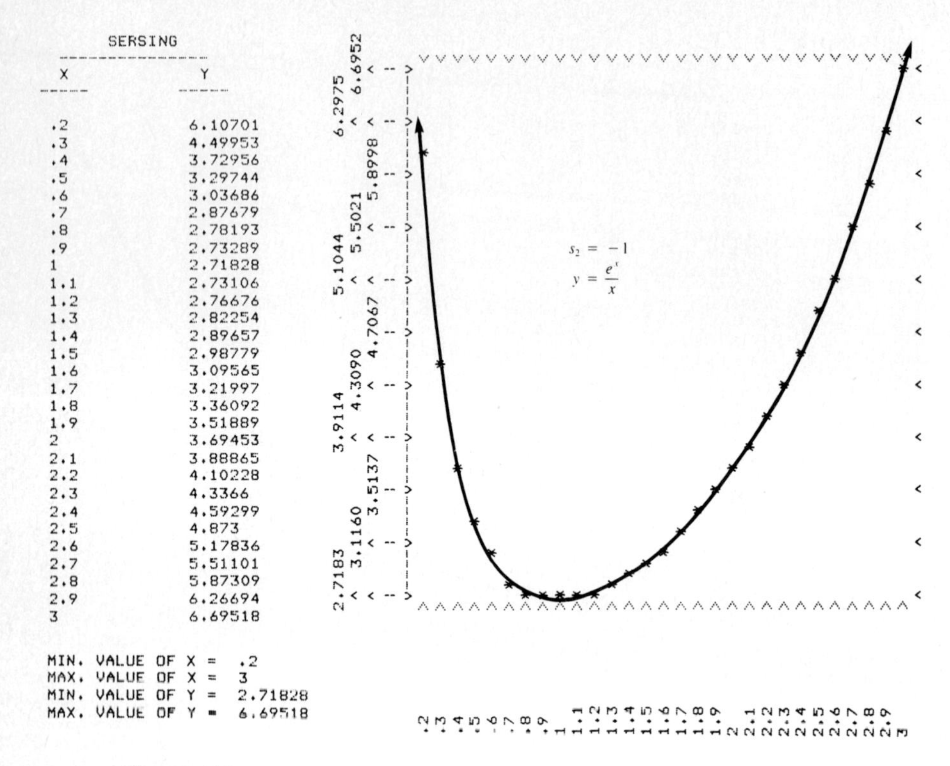

SERSING

X	Y
.2	6.10701
.3	4.49953
.4	3.72956
.5	3.29744
.6	3.03686
.7	2.87679
.8	2.78193
.9	2.73289
1	2.71828
1.1	2.73106
1.2	2.76676
1.3	2.82254
1.4	2.89657
1.5	2.98779
1.6	3.09565
1.7	3.21997
1.8	3.36092
1.9	3.51889
2	3.69453
2.1	3.88865
2.2	4.10228
2.3	4.3366
2.4	4.59299
2.5	4.873
2.6	5.17836
2.7	5.51101
2.8	5.87309
2.9	6.26694
3	6.69518

MIN. VALUE OF X = .2
MAX. VALUE OF X = 3
MIN. VALUE OF Y = 2.71828
MAX. VALUE OF Y = 6.69518

Figure 7.8

We gain the ultimate bonus in this problem, because we are now able to write the general solution as

$$y = c_1\left(\frac{e^x}{x}\right) + c_2\left[\frac{2}{x}(e^x - 1 - x)\right]$$

This provides a way of checking the accuracy of the power-series result.

Example 10

Consider

$$2x^2y'' + (x^2 - x)y' + y = 0$$

The usual algebraic operations on this equation produce the recurrence formula

$$a_n = \frac{-a_{n-1}}{2n + 2s - 1}$$

and the indicial equation $2s^2 - 3s + 1 = 0$, whose roots are $s_1 = 1$ and $s_2 = \frac{1}{2}$. Since both of these roots are positive, the solutions will not be singular at zero; the series for y_1 will lead with x, and the series for y_2 with

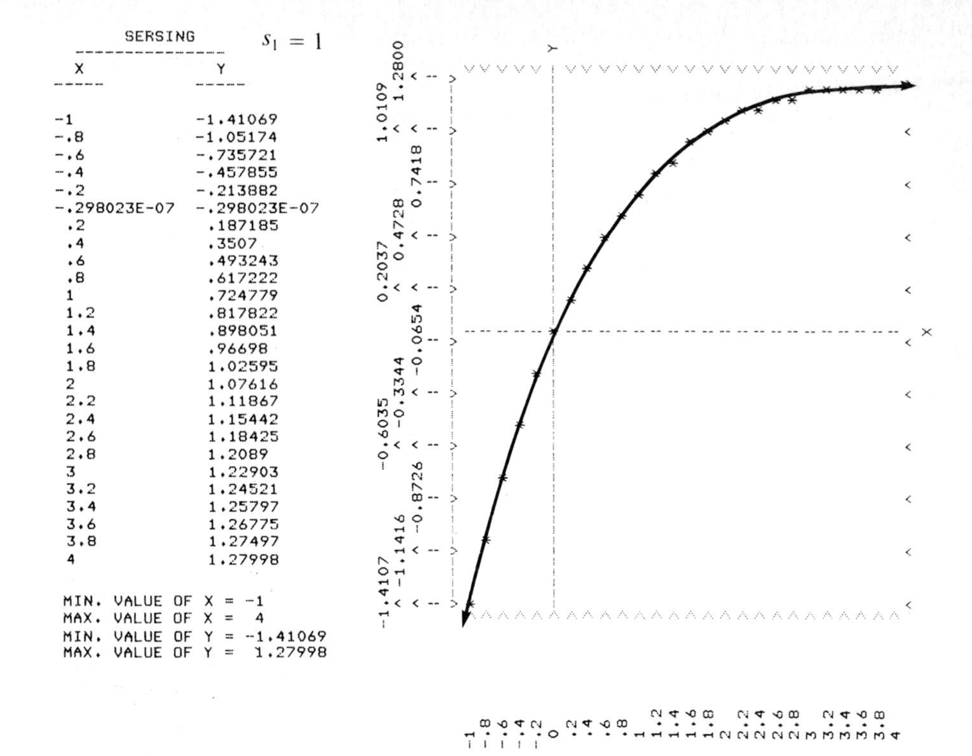

SERSING $s_1 = 1$

X	Y
-1	-1.41069
-.8	-1.05174
-.6	-.735721
-.4	-.457855
-.2	-.213882
-.298023E-07	-.298023E-07
.2	.187185
.4	.3507
.6	.493243
.8	.617222
1	.724779
1.2	.817822
1.4	.898051
1.6	.96698
1.8	1.02595
2	1.07616
2.2	1.11867
2.4	1.15442
2.6	1.18425
2.8	1.2089
3	1.22903
3.2	1.24521
3.4	1.25797
3.6	1.26775
3.8	1.27497
4	1.27998

MIN. VALUE OF X = -1
MAX. VALUE OF X = 4
MIN. VALUE OF Y = -1.41069
MAX. VALUE OF Y = 1.27998

Figure 7.9

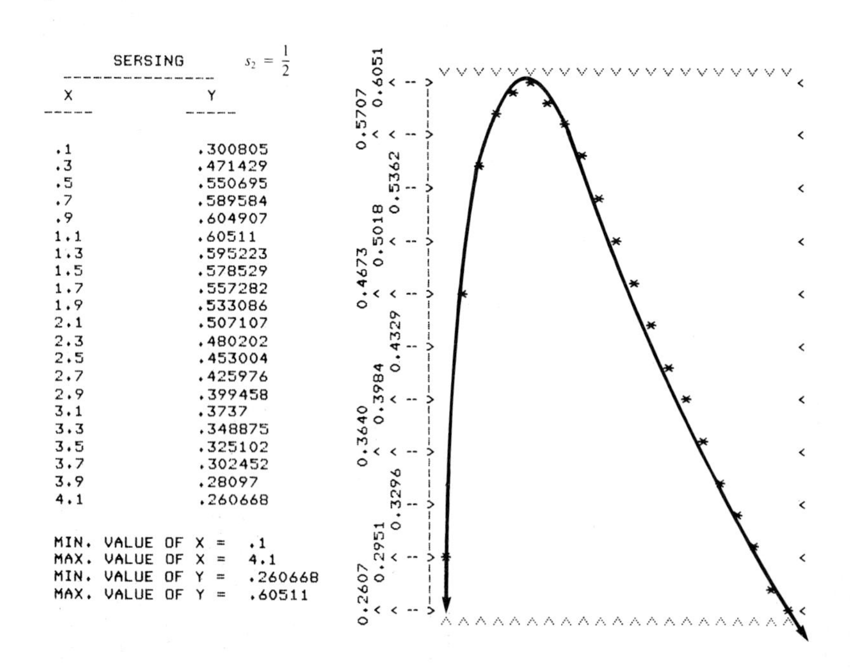

SERSING $s_2 = \frac{1}{2}$

X	Y
.1	.300805
.3	.471429
.5	.550695
.7	.589584
.9	.604907
1.1	.60511
1.3	.595223
1.5	.578529
1.7	.557282
1.9	.533086
2.1	.507107
2.3	.480202
2.5	.453004
2.7	.425976
2.9	.399458
3.1	.3737
3.3	.348875
3.5	.325102
3.7	.302452
3.9	.28097
4.1	.260668

MIN. VALUE OF X = .1
MAX. VALUE OF X = 4.1
MIN. VALUE OF Y = .260668
MAX. VALUE OF Y = .60511

Figure 7.10

$x^{1/2}$. Thus y_1 will pass through the origin and y_2 will stop at 0. We choose our interval of evaluation as $[-1, 4]$ for y_1 and pick the interval $[0, 4]$ for y_2. The actual solution power series are:

$$y_1 = x\left(1 - \frac{1}{3}x + \frac{1}{15}x^2 - \frac{1}{105}x^3 + \frac{1}{945}x^4 + \cdots\right)$$

$$y_2 = x^{1/2}\left(1 - \frac{1}{2}x + \frac{1}{8}x^2 - \frac{1}{48}x^3 + \cdots\right)$$

The output and graphs using SERSING are shown in Figures 7.9 and 7.10.

Problems

Using the programs SERSO, SERSO3, and SERSING, solve the following problems.

1. $(2x + 1)y'' + y' + 2y = 0$; $y(0) = 1$, $y'(0) = 1$
   (a) Noting that the first singularity occurs at $x = -\frac{1}{2}$, obtain the printed and graphed solution on $(-\frac{1}{2}, \frac{1}{2})$.
   (b) Using the program YACTUL, obtain for comparison the values of the actual y at $x = 0, \frac{1}{8}, \frac{1}{4}$, and $\frac{3}{8}$. The analytical solution is

$$y = \left(1 - x^2 + x^3 - \frac{13}{12}x^4 + \frac{17}{12}x^5 + \cdots\right)$$

$$+ \left(x - \frac{1}{2}x^2 + \frac{1}{6}x^3 - \frac{1}{8}x^4 + \frac{19}{120}x^5 - \cdots\right)$$

   To save work, combine these first. This series, when truncated, is a good approximation to the actual solution, especially for small x.

2. $2y'' - xy' - 2y = 0$; $y(0) = 1$, $y'(0) = 1$. Noting that there are no singularities, obtain the printed and graphed solution on $(-A, A)$ where you choose A.

3. $(2 + x^2)y'' + 5xy' + 4y = 0$; $y(0) = 1$, $y'(0) = 1$. Noting that the singularities occur at $\pm\sqrt{2}i$, obtain the printed and graphed solution on $(-1.4, 1.4)$.

4. Check the answer to Problem 3 at the points in $[0, 1]$ by obtaining the analytical solution.

5. $(x^2 - 1)y'' + xy' - y = 0$; $y(0) = 2$, $y'(0) = -3$. Solve on an appropriate interval.

6. $y'' - 2(x + 3)y' - 3y = 0$; $y(-3) = 1$, $y'(-3) = 0$. Run this on $[-5, -1]$. Also note what happens if $y'(-3)$ is changed slightly. (This problem requires a shift.)

7. $y'' - (2 - x)y = 0$; $y(0) = 3$, $y'(0) = 1$. Choose the solution interval $(-A, A)$.

8. $y'' - x^2y' - 2xy = 0$; $y(0) = 1$, $y'(0) = 1$. Find an interval on which $|y| < 5$.

9. $x^2y'' + (x^2 + 2x)y' - 2y = 0$. Run this on $[0.5, 2]$. In this problem

$$y_1 = \frac{3a_0}{x^2}(2 - 2x + x^2 - 2e^{-x})$$

$$y_2 = a_0 \frac{e^{-x}}{x^2}$$

Let $a_0 = 1$ as per the SERSING program. Check your answers by using YACTUL.

10. $2x^2y'' + 7x(x - 1)y' + 9y = 0$

11. $8x^2y'' + 10xy' + (x - 1)y = 0$

12. Pinpoint the first two positive zeros of $J_0(x)$ and $J_1(x)$ to four decimal places. (See definition of J_0 and J_1 in Section 7.1.)

13. $2xy'' - (2x^2 + 1)y' - xy = 0$. First find the indicial equation $2s^2 - 3s = 0$ and its associated recurrence formula. Using values $s_1 = 0$, obtain analytically the series solution and note that with $a_0 = 1$ we get $y_1 = e^{(1/2)x^2}$.
   (a) Compare the SERSING output and graph with the YACTUL output and graph.
   (b) Following the procedure studied in the text for finding a second solution when one solution has been found, make the change of dependent variables $y = ve^{x^2/2}$ and obtain the solution

$$y_2 = e^{x^2/2} \int_0^x t^{1/2} e^{-t^2/2}\, dt$$

   This nonelementary integral requires a numerical integration. A simple Euler approach will produce its values.
   (c) Now produce values for y_2 by computer routines and compare with the series solution:

$$y_2 = x^{3/2}\left(1 + \sum_{n=1}^{\infty} \frac{2^n x^{2n}}{7 \cdot 11 \cdot 15 \cdots (4n + 3)}\right)$$

Applications Problems (Programming Experience Required)

As mentioned earlier, numerous problems in mathematical physics produce various forms of Bessel's equations. We will enumerate a few and give the person with programming experience an opportunity to produce solutions.

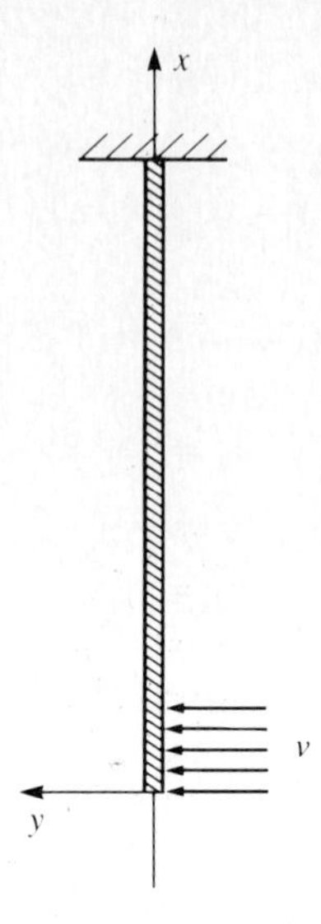

Figure 7.11

A–1. A uniform perfectly flexible cable of length $L = 10$ and weight-per-unit-length W hangs by one end from a frictionless hook. At $t = 0$, while the cable is at rest in a vertical position, a uniform and horizontal velocity v is applied to the lowest $\frac{1}{8}$ of the cable. Choose coordinates as shown in Figure 7.11. Vibrations will be set up in the cable. To study these, one must investigate solutions of the initial value problem that evolves:

$$(xX') + \lambda^2 X = 0; \quad X(L) = 0, \quad X(0) \text{ is finite}$$

where X is the displacement from the rest position. The Bessel function solution, when subjected to the initial conditions, produces

$$J_0(2\lambda\sqrt{L}) = 0$$

This is the frequency equation of the problem, and from it we can determine the natural frequencies of the cable. Find the values of $2\lambda\sqrt{L}$ by using BESSL0 and then calculate the natural frequency $\omega = \lambda\sqrt{g}$.

A–2. Certain equations closely resembling Bessel's equation occur so often that their solutions are also named and studied in their own right. The most important of these

$$x^2y'' + xy' - x^2y = 0$$

is known as the modified Bessel equation of order zero. Its real solution $I_0(x)$ is called the **modified Bessel function of the first kind of order zero**. Work through this problem to obtain the recurrence

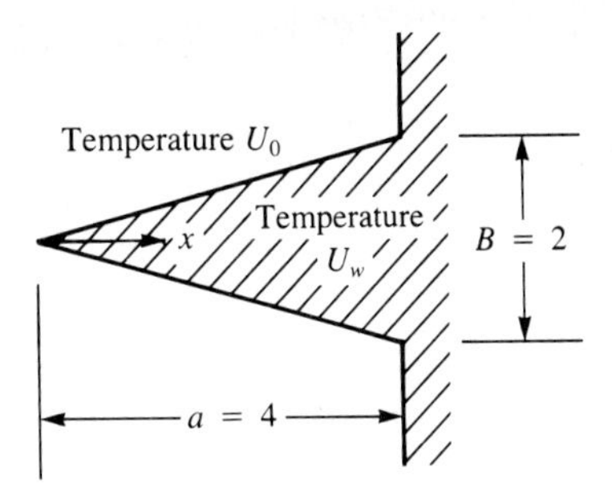

Figure 7.12

formula and write a program to tabulate its values. If you can also tabulate $I_0(\sqrt{x})$, then the following application problem can be solved.

A metal fin of a triangular cross section is attached to a plane surface to help carry off heat from the latter. Assuming dimensions and coordinates as shown in Figure 7.12, find the steady-state temperature distribution along the fin if the wall temperature is U_w and if the fin cools freely into air of constant temperature U_0. An analysis using a unit length of fin, thermal conductivity, and temperature gradients produces a solution

$$U = U_0 + (U_w - U_0) \cdot \frac{I_0(2\alpha\sqrt{x})}{I_0(2\alpha\sqrt{a})}$$

Choose units now so $\alpha = 0.5$. Having been given $a = 4$, obtain the solution on $[0, 4]$. Pick appropriate U_w and U_0.

Friedrich Bessel first used the now famous Bessel functions in his study of planetary motion in 1824.

The number of applications areas where Bessel's equations arise is astronomical. To impress you with their variety, we will list a few:

1. Stability of a vertical cylindrical rod.
2. Tidal waves in a long channel.
3. Vibrations of membranes.
4. Oscillations of a rotating liquid.
5. Propagation of electromagnetic waves along wires.
6. Potential theory.
7. Hydrostatic engineering.
8. Elasticity.

Chapter 8

The Laplace Transform

8.1 Introduction

This chapter introduces a special technique applicable especially to a wide range of initial value problems. The technique introduces the transformation of a function $f(x)$ of the real variable x into a function $F(s)$ of the real variable s, called the **Laplace transformation** (or **Laplace transform**) and denoted by $\mathscr{L}[f(x)] = F(s)$. This transformation acts as an operator, changing one function into another function. Its special usefulness with an initial value problem becomes evident by its ability to change such a problem into an algebraic problem in the variable s.

The techniques of using Laplace transforms are simple, and they give solutions to differential equations with initial conditions directly without the necessity of first finding the general solution and then using the initial conditions to find the parameter values. Because these are the types of problems that often arise in applications areas, transforms are important in mathematics, physics, and engineering.

Succeeding sections will show how this transformation operates on differential equations and their initial conditions, but first we must lay the groundwork.

8.2 Improper Integrals

To thoroughly understand Laplace transforms, we need a good comprehension of improper integrals. In the calculus, the convergence of an improper integral is defined as follows.

Definition: Let $f(x)$ be a continuous function on $0 \leq x \leq h$, $h > 0$. If the definite integral

$$\int_0^h f(x)\,dx$$

approaches a finite limit L as $h \to \infty$, the **improper integral**

$$\int_0^\infty f(x)\,dx$$

exists and **converges** to the value L. We write

$$\int_0^\infty f(x)\,dx = \lim_{h\to\infty} \int_0^h f(x)\,dx = L$$

If the limit does not exist, the improper integral **diverges**.

Example 1

Consider

$$\int_0^\infty \frac{1}{x+2}\,dx$$

We can write this as

$$\lim_{h\to\infty} \int_0^h \frac{1}{x+2}\,dx = \lim_{h\to\infty} \left[\ln|x+2| \Big|_0^h \right] = \lim_{h\to\infty} \Big[\ln|h+2| - \ln 2 \Big] \to \infty$$

Hence this improper integral diverges.

Example 2

Consider

$$\int_0^\infty e^{-2x}\,dx$$

Using the calculus approach, we write

$$\int_0^{\infty} e^{-2x}\,dx = \lim_{h\to\infty}\int_0^{h} e^{-2x}\,dx = \lim_{h\to\infty}\left[-\frac{1}{2}e^{-2x}\Big|_0^h\right]$$

$$= \lim_{h\to\infty}\left[-\frac{1}{2}e^{-2h} + \frac{1}{2}\right] = \frac{1}{2}$$

This improper integral converges.

Definition: If

$$\int_0^{\infty} |f(x)|\,dx$$

converges, then

$$\int_0^{\infty} f(x)\,dx$$

is said to **converge absolutely**.

The following calculus results, stated without proof, will be useful:

Proposition 1: If $\int_0^{\infty} |f(x)|\,dx$ converges, then $\int_0^{\infty} f(x)\,dx$ converges.

Proposition 2: If $0 \le f(x) \le g(x)$ on $0 \le x < \infty$ and if $\int_0^{\infty} g(x)\,dx$ converges, then $\int_0^{\infty} f(x)\,dx$ converges.

Proposition 3: If $0 \le f(x) \le g(x)$ on $0 \le x < \infty$, and if $\int_0^{\infty} f(x)\,dx$ diverges, then $\int_0^{\infty} g(x)\,dx$ diverges.

Proposition 4: If $\int_0^{\infty} e^{-sx} f(x)\,dx$ converges for a value $s = s_0$, then it converges for all $s > s_0$. (*Hint:* To see the validity of this result, apply Proposition 2.)

8.3 The Laplace Transform

Definition: The **Laplace transform** of a function $f(x)$, $0 \le x < \infty$, is defined by the improper integral.

$$\mathscr{L}[f(x)] = F(s) = \int_0^{\infty} e^{-sx} f(x)\,dx$$

where it is assumed that the integral converges for at least one value of s. The Laplace transform is an operator transforming the function $f(x)$ into its image $F(s)$. For many functions this is a simplification.

We begin our study of transforms by developing from the definition the Laplace transforms of some elementary functions. This will be followed by the development of a collection of theorems, which will allow further function work.

Our first result uses a function which is easily integrated.

Example 3

$$\mathscr{L}[e^{ax}] \equiv \int_0^\infty e^{-sx}e^{ax}\,dx = \lim_{h\to\infty}\int_0^h e^{-(s-a)x}\,dx$$

$$= \lim_{h\to\infty}\left[\frac{-1}{s-a}e^{-(s-a)x}\Big|_0^h\right]$$

$$= \lim_{h\to\infty}\left[\frac{-1}{s-a}(e^{-(s-a)h}-1)\right] = \frac{1}{s-a} \qquad \text{if} \qquad s > a \quad (1)$$

Since the integral diverges for $s \le a$, the Laplace transform does not exist for those values. Thus, the Laplace transform of the transcendental function e^{ax} is the algebraic function $1/(s-a)$.

We now apply the result of Example 3 to gain additional transforms.

Example 4

$$\mathscr{L}[e^{3x}] = \frac{1}{s-3}$$

$$\mathscr{L}[e^{-4x}] = \frac{1}{s-(-4)} = \frac{1}{s+4}$$

$$\mathscr{L}[1] = \mathscr{L}[e^{0x}] = \frac{1}{s-0} = \frac{1}{s}$$

Example 5

The transformation of the simple algebraic function $f(x) = x$ may also be obtained from the definition:

$$\mathscr{L}[x] \equiv \int_0^\infty e^{-sx}x\,dx = \lim_{h\to\infty}\int_0^h e^{-sx}x\,dx$$

The integration is done by parts, using

$$u = x \qquad dv = e^{-sx}\,dx$$

$$du = dx \qquad v = -\frac{1}{s}e^{-sx}$$

so that

$$\int_0^h e^{-sx}x\,dx = -\frac{1}{s}xe^{-sx}\Big|_0^h - \int_0^h \left(-\frac{1}{s}\right)e^{-sx}\,dx$$

$$= -\frac{h}{s}e^{-sh} + \frac{1}{s}\int_0^h e^{-sx}\,dx$$

$$= -\frac{h}{s}e^{-sh} + \frac{1}{s}\left(-\frac{1}{s}\right)\left[e^{-sx}\Big|_0^h\right]$$

$$= -\frac{h}{s}e^{-sh} - \frac{1}{s^2}e^{-sh} + \frac{1}{s^2}$$

$$\therefore \quad \mathscr{L}[x] = \lim_{h\to\infty}\left\{-\frac{h}{s}e^{-sh} - \frac{1}{s^2}e^{-sh} + \frac{1}{s^2}\right\} = \frac{1}{s^2} \qquad \text{if} \qquad s > 0$$

Before the development of other transforms using the definition, we obtain a very important and useful result.

Theorem 8.1:

*The operator $\mathscr{L}$ has the **linearity property**: For given real functions $f(x)$ and $g(x)$ and real numbers a and b,*

$$\mathscr{L}[a\cdot f(x) + b\cdot g(x)] = a\mathscr{L}[f(x)] + b\mathscr{L}[g(x)]$$
$$= aF(s) + bG(s)$$

for those values of s for which both transforms exist.

Proof:

By definition,

$$a\mathscr{L}[f(x)] = a\int_0^\infty e^{-sx}f(x)\,dx$$

which converges for $s > s_1$, and

$$b\mathscr{L}[g(x)] = b\int_0^\infty e^{-sx}g(x)\,dx$$

which converges for $s > s_2$. Thus both converge for $s > \max[s_1, s_2]$, and we write

$$a\mathscr{L}[f(x)] + b\mathscr{L}[g(x)] = \int_0^\infty ae^{-sx}f(x)\,dx + \int_0^\infty be^{-sx}g(x)\,dx$$

$$= \int_0^\infty e^{-sx}[af(x) + bg(x)]\,dx = \mathscr{L}[af(x) + bg(x)]$$

■

At this point it will be of value to consider the Laplace transform of a complex-valued function of the real variable x. Suppose

$f(x) = u(x) + i \cdot v(x)$ where $u(x)$ and $v(x)$ are continuous real functions. By using the limit process on complex functions, we can easily show that

$$\mathscr{L}[f(x)] = \mathscr{L}[u(x)] + i \cdot \mathscr{L}[v(x)]$$

By the identification property of complex numbers, we then have

$$\mathscr{L}[\text{real part of } f(x)] = \text{real part of } \mathscr{L}[f(x)] \qquad (2)$$

and $\qquad \mathscr{L}[\text{imaginary part of } f(x)] = \text{imaginary part of } \mathscr{L}[f(x)] \qquad (3)$

These formulas are immediately useful to generate additional transforms.

Example 6

Consider

$$\begin{aligned} f(x) &= u(x) + i \cdot v(x) \\ &= e^{ibx} \\ &= \cos bx + i \cdot \sin bx \end{aligned}$$

Then

$$\begin{aligned} \mathscr{L}[f(x)] &= \mathscr{L}[e^{ibx}] \\ &= \int_0^\infty e^{-sx} e^{ibx}\, dx = \lim_{h \to \infty} \int_0^h e^{(ib-s)x}\, dx \\ &= \lim_{h \to \infty} \left[\frac{e^{-(s-ib)x}}{ib - s} \bigg|_0^h \right] = \lim_{h \to \infty} \left[\frac{e^{-sh} e^{ibh} - 1}{-(s - ib)} \right] \\ &= \frac{-1}{-(s - ib)} = \frac{1}{s - ib} \qquad \text{if} \qquad s > 0 \end{aligned}$$

which corresponds to the complex case of formula (1). Using formulas (2) and (3), we have

$$\mathscr{L}[\text{real part of } e^{ibx}] = \text{real part of } \left(\frac{1}{s - ib} \right)$$

and

$$\mathscr{L}[\text{imaginary part of } e^{ibx}] = \text{imaginary part of } \left(\frac{1}{s - ib} \right)$$

Since

$$\frac{1}{s - ib} = \frac{s + ib}{(s - ib)(s + ib)} = \frac{s + ib}{s^2 + b^2} = \frac{s}{s^2 + b^2} + i \left(\frac{b}{s^2 + b^2} \right)$$

we obtain the useful relations

$$\begin{aligned} \mathscr{L}[\cos bx] &= \frac{s}{s^2 + b^2} \\ \mathscr{L}[\sin bx] &= \frac{b}{s^2 + b^2} \end{aligned} \qquad (4)$$

We will couple these results with the linearity property to further our knowledge of transforms.

Example 7

To find $\mathscr{L}[\sin^2 bx]$, we use a trigonometric identity to write

$$\begin{aligned}\mathscr{L}[\sin^2 bx] &= \mathscr{L}\left[\frac{1}{2} - \frac{1}{2}\cos 2bx\right]\\ &= \mathscr{L}\left[\frac{1}{2}\right] - \frac{1}{2}\mathscr{L}[\cos 2bx]\\ &= \frac{1}{2}\cdot\frac{1}{s} - \frac{1}{2}\cdot\frac{s}{s^2+4b^2} = \frac{2b^2}{s(s^2+4b^2)}\end{aligned}$$

Example 8

$$\begin{aligned}\mathscr{L}[2\sin 4x - 7\cos 3x] &= 2\mathscr{L}[\sin 4x] - 7\mathscr{L}[\cos 3x]\\ &= 2\cdot\frac{4}{s^2+16} - 7\frac{s}{s^2+9}\\ &= \frac{8}{s^2+16} - \frac{7s}{s^2+9}\end{aligned}$$

Another important property of Laplace Transforms is stated in the following theorem.

Theorem 8.2:

If $f(x)$ and $g(x)$ are continuous functions for $0 \leq x < \infty$ and if $\mathscr{L}[f(x)] = \mathscr{L}[g(x)]$, then $f(x) = g(x)$, and conversely.

The transform of a function is unique for continuous functions; that is, if $F(s)$ is the transform of $f(x)$, then no other continuous function has $F(s)$ for its transform. These facts imply the inverse property that if $\mathscr{L}[f(x)] = F(s)$, then $f(x) = \mathscr{L}^{-1}[F(s)]$, and we call $f(x)$ the **inverse Laplace transform of** $F(s)$.

Example 9

Using Example 4 with the continuous exponential functions, we find

$$\mathscr{L}^{-1}\left[\frac{1}{s-3}\right] = e^{3x} \quad \text{and} \quad \mathscr{L}^{-1}\left[\frac{1}{s}\right] = 1$$

and from Example 5 we have

$$\mathscr{L}^{-1}\left[\frac{1}{s^2}\right] = x$$

From Example 7 we find

$$\mathscr{L}^{-1}\left[\frac{2}{s^2+4}\right] = \sin 2x$$

We are now able to state a linearity property for the inverse transform, which parallels the property of Theorem 8.1.

Inverse linearity property: From the uniqueness of the inverse transform for continuous functions, we may write

$$\mathscr{L}^{-1}[a \cdot F(s) + b \cdot G(s)] = a\mathscr{L}^{-1}[F(s)] + b\mathscr{L}^{-1}[G(s)] = af(x) + bg(x)$$

Example 10

Since
$$\mathscr{L}[e^{ax}] = \frac{1}{s-a}$$

we have
$$e^{ax} = \mathscr{L}^{-1}\left[\frac{1}{s-a}\right]$$

and thus
$$\begin{aligned}\mathscr{L}^{-1}\left[\frac{2}{s} + \frac{5}{s-3}\right] &= 2\mathscr{L}^{-1}\left[\frac{1}{s}\right] + 5\mathscr{L}^{-1}\left[\frac{1}{s-3}\right] \\ &= 2e^{0x} + 5e^{3x} \\ &= 2 + 5e^{3x}\end{aligned}$$

There are continuous functions for which the Laplace transform does not exist. This situation will occur if the function grows too rapidly as x grows large. We will classify the functions for which we do have Laplace transforms as those of exponential order.

Definition: A function $f(x)$ that is continuous on $0 \leq x < \infty$ is said to be of **exponential order** α if there exist positive constants M and α such that $|f(x)| \leq Me^{\alpha x}$.

We see that a function $f(x)$ is of exponential order if it does not grow more rapidly than an exponential function grows as x gets unboundedly large.

Example 11

If a function is bounded, it is automatically a function of exponential order 0, since a bounded function $f(x)$ satisfies $|f(x)| \leq M = Me^{0x}$. Included in this set would be $\sin \beta x$ and $\cos \beta x$.

Example 12

The functions $e^{kx}\sin \beta x$ and $e^{kx}\cos \beta x$ are of exponential order k since

$$|e^{kx}\sin \beta x| = |e^{kx}| \cdot |\sin \beta x| \leq |e^{kx}| \leq 1e^{\alpha x}$$

for $\alpha = k$.

Example 13

The power function x^n is of exponential order α for any positive α, however small, since by the Maclaurin series expansion

$$e^{\alpha x} = \sum_{n=0}^{\infty} \frac{\alpha^n x^n}{n!} > \frac{\alpha^n x^n}{n!}$$

so that

$$x^n < \left(\frac{n!}{\alpha^n}\right) e^{\alpha x}$$

We are now able to prove the following important existence theorem.

Theorem 8.3:

If the function $f(x)$ is of exponential order α, then the Laplace transform of $f(x)$ exists and is absolutely convergent for all $s > \alpha$.

Proof:

First we can see that

$$|e^{-sx} f(x)| = e^{-sx}|f(x)| \le e^{-sx}(Me^{\alpha x}) = Me^{-(s-\alpha)x}$$

Then

$$\mathscr{L}[f(x)] \equiv \int_0^{\infty} e^{-sx} f(x)\, dx$$

converges absolutely if

$$\int_0^{\infty} |e^{-sx} f(x)|\, dx$$

converges, but

$$\int_0^{\infty} |e^{-sx} f(x)|\, dx \le M \int_0^{\infty} e^{-(s-\alpha)x}\, dx$$

which converges if $s > \alpha$. ■

Example 14

Find the inverse Laplace transform of $F(s) = (3s + 7)/(s^2 - 4)$.

$$\mathscr{L}^{-1}[F(s)] = \mathscr{L}^{-1}\left[\frac{3s+7}{s^2-4}\right] = \mathscr{L}^{-1}\left[\frac{3s+7}{(s-2)(s+2)}\right]$$

To apply the inverse linearity property, we need to perform a partial-fraction decomposition on the function $F(s)$:

$$\frac{3s+7}{(s-2)(s+2)} = \frac{A}{s-2} + \frac{B}{s+2}$$

If we multiply this identity by $s-2$ and then let $s=2$, we get

$$\left.\frac{3s+7}{s+2}\right|_{s=2} = A + B\left.\left(\frac{s-2}{s+2}\right)\right|_{s=2}$$

$$\frac{13}{4} = A + 0$$

Now we will go back to the original identity and multiply by $s+2$. Then we let $s=-2$ to obtain

$$\left.\frac{3s+7}{s-2}\right|_{s=-2} = A\left.\left(\frac{s+2}{s-2}\right)\right|_{s=-2} + B$$

$$\frac{1}{-4} = 0 + B$$

Hence

$$\begin{aligned}\mathscr{L}^{-1}\left[\frac{3s+7}{(s-2)(s+2)}\right] &= \mathscr{L}^{-1}\left[\frac{13}{4}\left(\frac{1}{s-2}\right) - \frac{1}{4}\left(\frac{1}{s+2}\right)\right]\\ &= \frac{13}{4}\mathscr{L}^{-1}\left[\frac{1}{s-2}\right] - \frac{1}{4}\mathscr{L}^{-1}\left[\frac{1}{s+2}\right]\\ &= \frac{13}{4}e^{2x} - \frac{1}{4}e^{-2x} = f(x)\end{aligned}$$

Example 15

Find
$$\mathscr{L}^{-1}\left[\frac{2s+3}{s^2+9}\right]$$

This rational function has an irreducible quadratic in the denominator, so factoring is impossible; but we have developed the transforms of the trigonometric functions, so we write

$$\begin{aligned}\mathscr{L}^{-1}\left[\frac{2s+3}{s^2+9}\right] &= \mathscr{L}^{-1}\left[\frac{2s}{s^2+9} + \frac{3}{s^2+9}\right]\\ &= 2\mathscr{L}^{-1}\left[\frac{s}{s^2+9}\right] + \mathscr{L}^{-1}\left[\frac{3}{s^2+9}\right]\end{aligned}$$

Since
$$\mathscr{L}[\cos 3x] = \frac{s}{s^2+9}$$

and
$$\mathscr{L}[\sin 3x] = \frac{3}{s^2+9}$$

we now get

$$\mathscr{L}^{-1}\left[\frac{2s+3}{s^2+9}\right] = 2\cos 3x + \sin 3x$$

Example 16

Find $$\mathscr{L}^{-1}\left[\frac{5}{3s^2+14}\right]$$

We first rearrange $F(s)$:

$$\frac{5}{3s^2+14} = \frac{5}{3(s^2+14/3)} = \frac{5}{3}\left(\frac{1}{s^2+14/3}\right)$$

$$= \frac{5}{3}\left(\frac{1}{s^2+(\sqrt{14/3})^2}\right) = \frac{5}{3}\sqrt{\frac{3}{14}}\left(\frac{\sqrt{14/3}}{s^2+(\sqrt{14/3})^2}\right)$$

Thus $$\mathscr{L}^{-1}\left[\frac{5}{3s^2+14}\right] = \frac{5}{\sqrt{42}}\mathscr{L}^{-1}\left[\frac{\sqrt{14/3}}{s^2+(\sqrt{14/3})^2}\right]$$

Using $$\mathscr{L}^{-1}\left[\frac{b}{s^2+b^2}\right] = \sin bx$$

we now have

$$\mathscr{L}^{-1}\left[\frac{5}{3s^2+14}\right] = \frac{5}{\sqrt{42}}\sin\sqrt{\frac{14}{3}}\,x$$

Example 17

Find $$\mathscr{L}^{-1}\left[\frac{4-5s}{s^3-3s^2-10s}\right]$$

First, we do the partial-fraction decomposition and write

$$\frac{4-5s}{s^3-3s^2-10s} = \frac{4-5s}{s(s+2)(s-5)} = \frac{A}{s} + \frac{B}{s+2} + \frac{C}{s-5}$$

Since the original denominator factored into distinct linear factors, we treat the decomposition as an identity and first multiply both sides by s and then let $s = 0$. This immediately gives a new identity $\frac{4}{-10} = \frac{2}{-5} = A$. Now we return to the original identity and multiply by $s + 2$ and then let $s = -2$. This produces the new identity $14/(-2)(-7) = 1 = B$. Now we multiply the original identity by $s - 5$ and let $s = 5$. This gives $\frac{-21}{35} = \frac{-3}{5} = C$. Thus our original decomposition has its constants determined and is:

$$\frac{4-5s}{s(s+2)(s-5)} = \frac{-2/5}{s} + \frac{1}{s+2} - \frac{3/5}{s-5}$$

Using the linearity property gives

$$\mathscr{L}^{-1}\left[\frac{4-5s}{s^3-3s^2-10s}\right] = -\frac{2}{5}\mathscr{L}^{-1}\left[\frac{1}{s}\right] + \mathscr{L}^{-1}\left[\frac{1}{s+2}\right]$$
$$-\frac{3}{5}\mathscr{L}^{-1}\left[\frac{1}{s-5}\right]$$
$$= -\frac{2}{5}\cdot 1 + e^{-2x} - \frac{3}{5}e^{5x}$$
$$= -\frac{2}{5} + e^{-2x} - \frac{3}{5}e^{5x}$$

Exercises 8–3

In Exercises 1–4 determine if the integral exists. If it converges, find the value.

1. $\displaystyle\int_0^\infty \frac{1}{1+x^2}\,dx$

2. $\displaystyle\int_0^\infty x^{-2}\,dx$

3. $\displaystyle\int_0^\infty (1+x)^{-4}\,dx$

4. $\displaystyle\int_0^\infty \frac{x\,dx}{(1+x^2)^2}$

In Exercises 5–9 show the validity of the Laplace transformations:

5. $\mathscr{L}\left[\dfrac{e^{ax}-e^{bx}}{a-b}\right] = \dfrac{1}{(s-a)(s-b)}$

6. $\mathscr{L}\left[\dfrac{1-\cos ax}{a^2}\right] = \dfrac{1}{s(s^2+a^2)}$

7. $\mathscr{L}[\cos^2 ax] = \dfrac{s^2+2a^2}{s(s^2+4a^2)}$

8. $\mathscr{L}[x^{-1/2}] = \sqrt{\dfrac{\pi}{s}}$ (*Hint:* First let $t = \sqrt{sx}$.)

9. $\mathscr{L}(\sin ax \sinh ax) = \dfrac{2a^2 s}{s^4+4a^4}$

10. Verify, in Exercises 5–9 that $\lim_{s\to\infty} F(s) = 0$.

11. Show that the function e^{t^2} is not of exponential order.
12. By using partial-fraction decomposition, show that

$$\mathscr{L}^{-1}\left[\frac{1}{s(s+1)^2}\right] = 1 - e^{-x} - xe^{-x}$$

13. Use partial-fraction decomposition to find

$$\mathscr{L}^{-1}\left[\frac{s}{s^2 - 4s + 3}\right]$$

14. Find

$$\mathscr{L}^{-1}\left[\frac{2s}{s^2 - 3s - 18}\right]$$

15. Find

$$\mathscr{L}^{-1}\left[\frac{2s+2}{s^3 - s}\right]$$

16. Show that $f(x) = 3\sin^2 4x$ is of exponential order. What order is it?
17. Show that $f(x) = 2e^{3x}\cos^2 2x$ is of exponential order 3.
18. Using the definition of the Laplace transform, find $\mathscr{L}[f(x)]$ if

$$f(x) = \begin{cases} -1, & 0 \le x \le 3 \\ 1, & x > 3 \end{cases}$$

19. Using the definition of the Laplace transform, find $\mathscr{L}[f(x)]$ if

$$f(x) = \begin{cases} 0, & 0 \le x < 4 \\ 2, & 4 \le x \le 7 \\ -1, & 7 < x \end{cases}$$

20. Prove that if $F(s)$ is the transform of $f(x)$ and $a > 0$, then $\frac{1}{a}F\left(\frac{s}{a}\right)$ is the transform of $f(ax)$.

8.4 Theoretical Considerations

It is now necessary to develop that portion of the theory that will permit us to solve linear differential equations with constant coefficients. The key concept is the transform of a derivative.

Theorem 8.4

If $f(x)$ is of exponential order α and $f'(x)$ is continuous, then $\mathscr{L}[f'(x)]$ exists for $s > \alpha$ and $\mathscr{L}[f'(x)] = s\mathscr{L}[f(x)] - f(0)$.

Proof:

By definition,

$$\mathscr{L}[f'(x)] = \int_0^\infty e^{-sx} f'(x)\,dx = \lim_{h\to\infty} \int_0^h e^{-sx} f'(x)\,dx$$

If we integrate by parts, setting

$$u = e^{-sx} \qquad dv = f'(x)\,dx$$

$$du = -se^{-sx}\,dx \qquad v = f(x)$$

we obtain $$\int_0^h e^{-sx} f'(x)\,dx = e^{-sx} f(x)\Big|_0^h + s\int_0^h e^{-sx} f(x)\,dx$$

Thus $$\lim_{h\to\infty}\int_0^h e^{-sx} f'(x)\,dx = \lim_{h\to\infty}[e^{-sh} f(h)] - f(0) + s\lim_{h\to\infty}\int_0^h e^{-sx} f(x)\,dx$$

$$= \lim_{h\to\infty}[e^{-sh} f(h)] - f(0) + s\mathscr{L}[f(x)]$$

$$\therefore \quad \mathscr{L}[f'(x)] = s\mathscr{L}[f(x)] - f(0) + \lim_{h\to\infty} e^{-sh}\cdot f(h)$$

However, since $f(x)$ is of exponential order α, the last limit goes to zero. Hence, $\mathscr{L}[f'(x)] = s\mathscr{L}[f(x)] - f(0)$. ■

Example 18

We have computed earlier that $\mathscr{L}[\sin\beta x] = \beta/(s^2 + \beta^2)$. Using Theorem 8.4 on the function $f(x) = \sin\beta x$, we have

$$s\mathscr{L}[\sin\beta x] - \sin(0) = \mathscr{L}[(\sin\beta x)']$$

$$s\cdot\frac{\beta}{s^2+\beta^2} - 0 = \mathscr{L}(\beta\cos\beta x) = \beta\mathscr{L}(\cos\beta x)$$

Hence $$\mathscr{L}[\cos\beta x] = \frac{s}{s^2+\beta^2}$$

This agrees with our earlier result (4). Now by repeated applications of Theorem 8.4, we may obtain the transforms of derivatives of any order.

Theorem 8.5

If $f, f', f'', f''', \ldots, f^{(n)}$ are of exponential order α, then for $s > \alpha$ we have:

$$\mathscr{L}[f'(x)] = s\mathscr{L}[f(x)] - f(0)$$

$$\mathscr{L}[f''(x)] = s^2\mathscr{L}[f(x)] - sf(0) - f'(0)$$

$$\mathscr{L}[f'''(x)] = s^3\mathscr{L}[f(x)] - s^2 f(0) - sf'(0) - f''(0)$$

$$\vdots$$

$$\mathscr{L}[f^{(n)}(x)] = s^n\mathscr{L}[f(x)] - s^{n-1}f(0) - s^{n-2}f'(0) - \cdots - f^{(n-1)}(0)$$

Proof:

We have from Theorem 8.4 that

$$\mathscr{L}[f'(x)] = s\mathscr{L}[f(x)] - f(0)$$

If we replace f by f', we obtain

$$\begin{aligned}\mathscr{L}[f''(x)] &= s\mathscr{L}[f'(x)] - f'(0) \\ &= s\{s\mathscr{L}[f(x)] - f(0)\} - f'(0) \\ &= s^2\mathscr{L}[f(x)] - sf(0) - f'(0)\end{aligned}$$

An induction argument using this process completes the proof. ■

Two additional theorems will allow us to construct a table of Laplace transforms, which is needed for the process of solving differential equations. The first (Theorem 8.6) allows us to take any function whose transform is known, multiply it by an exponential function and obtain the new transform. This greatly expands our ability to calculate transforms. The second (Theorem 8.7) is similar in that it allows us to multiply the known transformable function by any positive integer power of x and then to calculate the new transform.

Theorem 8.6 (Translation Property):

If $F(s)$, the Laplace transform of $f(x)$, exists for $s > \alpha$, then for any constant a,

$$\mathscr{L}[e^{-ax}f(x)] = F(s + a) \qquad \text{for} \qquad s > \alpha - a$$

Proof:

Since

$$F(s) = \mathscr{L}[f(x)] = \int_0^\infty e^{-sx}f(x)\,dx$$

we replace s by $s + a$, giving

$$\begin{aligned}F(s + a) &= \int_0^\infty e^{-(s+a)x}f(x)\,dx = \int_0^\infty e^{-sx}[e^{-ax}f(x)]\,dx \\ &= \mathscr{L}(e^{-ax}f(x)]\end{aligned}$$

and thus proving the theorem. ■

Example 19

Since $$\mathscr{L}[\sin \beta x] = \frac{\beta}{s^2 + \beta^2}, \quad s > 0$$

we have $$\mathscr{L}[e^{-ax}\sin \beta x] = \frac{\beta}{(s + a)^2 + \beta^2}, \quad s > -a$$

Example 20

Since
$$\mathscr{L}[x] = \frac{1}{s^2}, \quad s > 0$$

we have
$$\mathscr{L}[xe^{ax}] = \frac{1}{(s-a)^2}, \quad s > a$$

Theorem 8.7

If $f(x)$ is of exponential order α, then $F(s)$, the Laplace transform of $f(x)$, has derivatives of all orders, and

$$\begin{aligned}
F'(s) &= \mathscr{L}[-xf(x)] &\Leftrightarrow\quad& \mathscr{L}[xf(x)] = -F'(s)\\
F''(s) &= \mathscr{L}[x^2f(x)] &\Leftrightarrow\quad& \mathscr{L}[x^2f(x)] = F''(s)\\
F'''(s) &= \mathscr{L}[-x^3f(x)] &\Leftrightarrow\quad& \mathscr{L}[x^3f(x)] = -F'''(s)\\
&\vdots & & \vdots\\
F^{(n)}(s) &= \mathscr{L}[(-x)^nf(x)] &\Leftrightarrow\quad& \mathscr{L}[x^nf(x)] = (-1)^nF^{(n)}(s)
\end{aligned}$$

Proof:

Beginning with the definition
$$F(s) = \int_0^\infty e^{-sx}f(x)\,dx,$$

differentiate on both sides with respect to s, obtaining
$$F'(s) = \int_0^\infty (-xe^{-sx})f(x)\,dx = \int_0^\infty e^{-sx}[-xf(x)]\,dx = \mathscr{L}[-xf(x)]$$

Additional differentiation by the same process produces the remaining results. ■

Example 21

We found in equation (1) that
$$\mathscr{L}[e^{ax}] = \frac{1}{s-a}, \quad s > a$$

Now
$$\frac{d}{ds}\left(\frac{1}{s-a}\right) = -\frac{1}{(s-a)^2} = \mathscr{L}[-xe^{ax}] = -\mathscr{L}[xe^{ax}]$$

Hence
$$\mathscr{L}[xe^{ax}] = \frac{1}{(s-a)^2}$$

Note that this result agrees with Example 20. Also
$$\mathscr{L}[x^2e^{ax}] = \frac{d^2}{ds^2}\left(\frac{1}{s-a}\right) = \frac{2}{(s-a)^3}, \quad s > a$$

By repeating the process, we arrive at

$$\mathscr{L}[x^n e^{ax}] = \frac{n!}{(s-a)^{n+1}}, \quad s > a \tag{5}$$

and for the particular value $a = 0$,

$$\mathscr{L}[x^n] = \frac{n!}{s^{n+1}}, \quad s > 0$$

Table 8.1

Laplace Transforms*

	$\mathscr{L}[f(x)]$	$f(x)$
1.	$\frac{1}{s}$	1
2.	$\frac{1}{s^n}$, n = 1, 2, 3, ...	$\frac{x^{n-1}}{(n-1)!}$
3.	$\frac{1}{s^{1/2}}$	$\frac{1}{\sqrt{\pi x}}$
4.	$\frac{1}{s^{3/2}}$	$2\sqrt{\frac{x}{\pi}}$
5.	$\frac{1}{s+a}$	e^{-ax}
6.	$\frac{1}{s(s+a)}$	$\frac{1}{a}(1 - e^{-ax})$
7.	$\frac{1}{s^2(s+a)}$	$\frac{1}{a^2}(e^{-ax} + ax - 1)$
8.	$\frac{s}{s^2+a^2}$	$\cos ax$
9.	$\frac{1}{s^2+a^2}$	$\frac{1}{a}\sin ax$
10.	$\frac{s}{s^2-a^2}$	$\cosh ax$
11.	$\frac{1}{s^2-a^2}$	$\frac{1}{a}\sinh ax$
12.	$\frac{1}{s(s^2+a^2)}$	$\frac{1}{a^2}(1 - \cos ax)$
13.	$\frac{1}{s^2(s^2+a^2)}$	$\frac{1}{a^3}(ax - \sin ax)$

* Additional transforms are in tables preceding Appendix A.

We now tabulate the results we have obtained. To make this table most useful, we will need to algebraically reorganize some of the derived formulas. Equation (5) may be adjusted to

$$\frac{1}{n!}\mathscr{L}[x^n e^{ax}] = \frac{1}{(s-a)^{n+1}}$$

or

$$\frac{1}{(n-1)!}\mathscr{L}[x^{n-1} e^{ax}] = \frac{1}{(s-a)^{n}}$$

Table 8.1 *(Continued)*

	$\mathscr{L}[f(x)]$	$f(x)$
14.	$\dfrac{1}{(s+a)(s+b)},\ a \neq b$	$\left(\dfrac{1}{b-a}\right)(e^{-ax} - e^{-bx})$
15.	$\dfrac{s}{(s+a)(s+b)},\ a \neq b$	$\left(\dfrac{1}{a-b}\right)(ae^{-ax} - be^{-bx})$
16.	$\dfrac{1}{(s+a)^2}$	xe^{-ax}
17.	$\dfrac{1}{(s+a)^n},\ n = 1, 2, \cdots$	$\dfrac{x^{n-1}e^{-ax}}{(n-1)!}$
18.	$\dfrac{s}{(s+a)^2}$	$e^{-ax}(1-ax)$
19.	$\dfrac{1}{s(s+a)^2}$	$\dfrac{1}{a^2}[1-(1+ax)e^{-ax}]$
20.	$\dfrac{s+b}{(s+a)^2}$	$[(b-a)x+1]e^{-ax}$
21.	$\dfrac{1}{(s+a)^c},\ c > 0$	$\dfrac{x^{c-1}e^{-ax}}{\Gamma(c)}$
22.	$\dfrac{2as}{(s^2+a^2)^2}$	$x\sin(ax)$
23.	$\dfrac{s^2-a^2}{(s^2+a^2)^2}$	$x\cos(ax)$
24.	$\dfrac{b}{(s+a)^2+b^2}$	$e^{-ax}\sin(bx)$
25.	$\dfrac{s+a}{(s+a)^2+b^2}$	$e^{-ax}\cos(bx)$
26.	$\dfrac{1}{(s^2+a^2)^2}$	$\dfrac{\sin(ax) - ax\cos(ax)}{2a^3}$
27.	$\dfrac{s}{(s^2+a^2)^2}$	$\dfrac{x\sin(ax)}{2a}$

By taking inverse transforms, we obtain

$$\frac{x^{n-1}e^{ax}}{(n-1)!} = \mathscr{L}^{-1}\left[\frac{1}{(s-a)^n}\right]$$

Replacing a by $-a$, we get

$$\frac{x^{n-1}e^{-ax}}{(n-1)!} = \mathscr{L}^{-1}\left[\frac{1}{(s+a)^n}\right]$$

which is formula 17 in Table 8.1. This listing combines many of our derived formulas and examples into a reference table. The seeming reverse format will clarify itself as we move into the next section on solving differential equations.

Exercises 8.4

1. Find $\mathscr{L}[x^2 \sin ax]$.
2. Find $\mathscr{L}[x \sin^2 ax]$.
3. Find $\mathscr{L}[\sinh ax]$.
4. Find $\mathscr{L}[x \sinh ax]$.
5. Find $\mathscr{L}[x \cosh ax]$.
6. The Laplace transforms of some functions are difficult to calculate even with all the theorems. We may use the Taylor series to assist us if we assume that the transform of a power series may be computed term by term.

   (a) Use $e^x = \sum_0^\infty \frac{x^n}{n!}$ to find $\mathscr{L}[e^x]$.

   (b) Show that $\mathscr{L}\left[\frac{\sin x}{x}\right] = \text{Arctan}\left(\frac{1}{s}\right)$

   (c) In statistics a very important function is the error function defined by

   $$\text{Erf}\, x = \frac{2}{\sqrt{\pi}}\int_0^x e^{-t^2}\, dt$$

   Show that $\mathscr{L}[\text{Erf}\sqrt{x}] = \frac{1}{s\sqrt{s+1}}$.

   (d) We earlier used the Bessel function of order zero:

   $$J_0(x) = \sum_0^\infty \frac{(-1)^n}{(n!)^2}\left(\frac{x}{2}\right)^{2n}$$

   Show that $\mathscr{L}(J_0(x)] = 1/\sqrt{s^2+1}$.

7. Verify the formula for $\mathscr{L}[f'''(x)]$ in Theorem 8.5.
8. Verify the formula for $\mathscr{L}[x^3 e^{ax}]$ in Example 21.
9. Show the validity of

$$\mathscr{L}[e^{-ax}\sin bx] = \frac{b}{(s+a)^2 + b^2}$$

10. Verify that

$$\mathscr{L}^{-1}\left[\frac{s+3}{(s+3)^2+1}\right] = e^{-3x}\mathscr{L}^{-1}\left(\frac{s}{s^2+1}\right)$$

11. Find $\mathscr{L}^{-1}\left[\frac{4}{s^7} + \frac{2}{s-3}\right]$.

12. Find $\mathscr{L}^{-1}\left[\frac{s+k}{(s+k)^2+a^2}\right]$.

13. Using $\mathscr{L}[\sin x]$, find $\mathscr{L}[x^3 \sin x]$.
14. Calculate $\mathscr{L}[xe^{3x}]$ by three different ways.
15. Find $\mathscr{L}[xe^{-x}\cos 4x]$.
16. Find $\mathscr{L}[x^{9/2}]$.
17. Find $\mathscr{L}[\sqrt{x}\, e^{3x}]$.

8.5 Solving Linear Differential Equations with Constant Coefficients

To introduce the process of using Laplace transforms to solve an initial value problem, we consider the general second-order initial value problem

$$ay'' + by' + cy = f(x); \qquad y(0) = y_0, \qquad y'(0) = v_0$$

where a, b, c, y_0, and v_0 are real constants and where $a \neq 0$ and y is a function of exponential order.

Taking the Laplace transform of both sides of the differential equation and using the linearity property, we obtain

$$a\mathscr{L}[y''] + b\mathscr{L}[y'] + c\mathscr{L}[y] = \mathscr{L}[f(x)]$$

By Theorem 8.5, we then have

$$a\{s^2\mathscr{L}[y] - sy(0) - y'(0)\} + b\{s\mathscr{L}[y] - y(0)\} + c\mathscr{L}[y] = \mathscr{L}[f(x)]$$

Let $\mathscr{L}[y(x)]$ be denoted by $Y(s)$. Then

$$a[s^2 Y(s) - sy_0 - v_0] + b[sY(s) - y_0] + cY(s) = F(s)$$

$$(as^2 + bs + c)Y(s) - (as+b)y_0 - av_0 = F(s)$$

Thus
$$Y(s) = \frac{F(s) + (as + b)y_0 + av_0}{as^2 + bs + c} \tag{6}$$

The denominator in equation (6) is the characteristic polynomial of the differential equation. In those instances when $F(s)$ is a rational function, the entire expression for $Y(s)$ may then be expressed as a rational algebraic function. By factoring and rearranging the expression and employing the partial-fraction decomposition process, we will be able to use Table 8.1 to find the inverse transform of $Y(s)$. This inverse transform is the unique solution function $y(x)$ that solves the given initial value problem. Example 22 illustrates this process.

Example 22

Consider the following initial value problem with a polynomial right side. (Often in applications this function is called a **forcing function**.)

$$y' - y = 1 - x; \qquad y(0) = 3$$

Taking transforms on both sides, we obtain

$$\mathscr{L}[y' - y] = \mathscr{L}[1 - x]$$

$$\mathscr{L}[y'] - \mathscr{L}[y] = \mathscr{L}[1] - \mathscr{L}[x]$$

$$\{s\mathscr{L}[y] - y(0)\} - \mathscr{L}[y] = \frac{1}{s} - \frac{1}{s^2}$$

$$(s - 1)\mathscr{L}[y] = \frac{1}{s} - \frac{1}{s^2} + 3 = \frac{s - 1 + 3s^2}{s^2}$$

We now divide and use partial-fraction decomposition to obtain

$$\mathscr{L}[y] = \frac{3s^2 + s - 1}{s^2(s - 1)} = \frac{1}{s^2} + \frac{3}{s - 1}$$

Then

$$y = \mathscr{L}^{-1}\left[\frac{1}{s^2} + \frac{3}{s - 1}\right] = \mathscr{L}^{-1}\left[\frac{1}{s^2}\right] + 3\mathscr{L}^{-1}\left[\frac{1}{s - 1}\right]$$
$$= x + 3e^x$$

This is the unique solution to the given initial value problem.

The Laplace transform technique carried us directly to the solution without the necessity of finding the homogeneous solution or the particular solution, nor did we need to evaluate constants with the initial conditions. With the Laplace transform method we realize a definite saving of work.

Example 23

Consider now an initial value problem with an exponential forcing function:

$$y'' + y = 2e^x; \qquad y(0) = 2, \qquad y'(0) = 2$$

Following the above process we take transforms

$$\mathscr{L}[y'' + y] = \mathscr{L}[2e^x] \Rightarrow \mathscr{L}[y''] + \mathscr{L}[y] = 2\mathscr{L}[e^x]$$

$$\{s^2\mathscr{L}[y] - sy(0) - y'(0)\} + \mathscr{L}[y] = 2\left(\frac{1}{s-1}\right)$$

$$\{s^2\mathscr{L}[y] - 2s - 2\} + \mathscr{L}[y] = \frac{2}{s-1}$$

$$\therefore \quad \mathscr{L}[y] = \frac{[2/(s-1)] + 2s + 2}{s^2 + 1}$$

Separating the right side by partial-fraction decomposition, we get

$$\mathscr{L}[y] = \frac{1}{s-1} + \frac{1}{s^2+1} + \frac{s}{s^2+1}$$

$$\therefore \quad y = \mathscr{L}^{-1}\left[\frac{1}{s-1} + \frac{1}{s^2+1} + \frac{s}{s^2+1}\right]$$

$$= \mathscr{L}^{-1}\left[\frac{1}{s-1}\right] + \mathscr{L}^{-1}\left[\frac{1}{s^2+1}\right] + \mathscr{L}^{-1}\left[\frac{s}{s^2+1}\right]$$

$$= e^x + \sin x + \cos x$$

Example 24

The process remains the same if the equation is homogeneous or of higher order. Consider

$$\left\{\begin{array}{l} y''' + 3y'' - y' - 3y = 0 \\ y(0) = 1,\ y'(0) = 1,\ y''(0) = -1 \end{array}\right\}$$

We now take transforms:

$$\mathscr{L}[y'''] + 3\mathscr{L}[y''] - \mathscr{L}[y'] - 3\mathscr{L}[y] = \mathscr{L}[0] = 0$$

Using Theorem 8.5, we have

$$\{s^3\mathscr{L}[y] - s^2y(0) - sy'(0) - y''(0)\} + 3\{s^2\mathscr{L}[y] - sy(0) - y'(0)\}$$
$$- \{s\mathscr{L}[y] - y(0)\} - 3\mathscr{L}[y] = 0$$

$$\mathscr{L}[y](s^3 + 3s^2 - s - 3) = (s^2 \cdot 1 + s \cdot 1 - 1) + (3s \cdot 1 + 3 \cdot 1) - 1$$

or

$$\mathscr{L}[y] = \frac{s^2 + 4s + 1}{s^3 + 3s^2 - s - 3} = \frac{s^2 + 4s + 1}{(s-1)(s+1)(s+3)}$$

$$= \frac{3/4}{s-1} + \frac{1/2}{s+1} - \frac{1/4}{s+3}$$

$$\therefore \quad y = \frac{3}{4}\mathscr{L}^{-1}\left[\frac{1}{s-1}\right] + \frac{1}{2}\mathscr{L}^{-1}\left[\frac{1}{s+1}\right] - \frac{1}{4}\mathscr{L}^{-1}\left[\frac{1}{s+3}\right]$$

$$= \frac{3}{4}e^{x} + \frac{1}{2}e^{-x} - \frac{1}{4}e^{-3x}$$

Consider again the initial value problem at the beginning of this section and equation (6). We may express the transform of the solution $y(x)$ as

$$Y(s) = \frac{F(s)}{as^2 + bs + c} + y_0\frac{as + b}{as^2 + bs + c} + v_0\frac{a}{as^2 + bs + c}$$

We will refer to

$$T(s) \equiv \frac{1}{as^2 + bs + c}$$

as the **system function**, since its value is determined by the system under consideration. We can then calculate $y(x)$ in its general form by taking inverse transforms:

$$y(x) = \mathscr{L}^{-1}[Y(s)] = \mathscr{L}^{-1}[F(s)T(s)] + y_0\mathscr{L}^{-1}\left[\frac{as + b}{as^2 + bs + c}\right]$$

$$+ v_0\mathscr{L}^{-1}\left[\frac{a}{as^2 + bs + c}\right]$$

The values of the second and third terms are easy if we use partial-fraction decomposition, but the inverse transform of the product in term one still needs to be dealt with. This will be defined in the next section when the convolution theorem is stated. This convolution term becomes the particular solution to the differential equation, whereas the other two terms, which involve the initial conditions, are the homogeneous solution. If the differential equation had been homogeneous (i.e., if $f(x)$ equaled zero), then $F(s)$ would equal zero and the unique solution to the initial value problem would be

$$y(x) = y_0\mathscr{L}^{-1}\left[\frac{as + b}{as^2 + bs + c}\right] + v_0\mathscr{L}^{-1}\left[\frac{a}{as^2 + bs + c}\right]$$

The preceding process could not have been used if the initial value x_0 had not been zero. This requirement comes, of course, from the lower limit of the integral that defined the Laplace transform. If the initial conditions attach a value other than zero to x_0, we follow the lead of previous chapters and make a change of variables $t = x - x_0$. Now the initial value problem is declared in variables Y and t with initial conditions posed at $t = 0$. This is solved for $Y(t)$ by the current methods, and a reverse change of variables produces the result for the original problem.

Exercises 8.5

Use the Laplace transform methods to find solutions of the following initial value problems. Check some of your results by finding the solution by the method of an earlier chapter; check others by direct substitution.

1. $y' + 2y = e^{-t}$; $y(0) = 0$
2. $y' + ay = 1$; $y(0) = 0$
3. $x' - 5x = 1$; $x(0) = 3$
4. $y' + 2y = x$; $y(0) = -1$
5. $y' - y = \sin x$; $y(0) = -1$
6. $y'' - 5y' + 6y = \cos x$; $y(0) = 1$, $y'(0) = 0$
7. $y'' + 3y' + 2y = x + 1$; $y(0) = 1$, $y'(0) = -1$
8. $y'' + 4y' + 4y = e^{-2x}$; $y(0) = 0$, $y'(0) = 0$
9. $y'' - 2y' + 5y = 0$; $y(0) = 2$, $y'(0) = 3$
10. $y'' - 5y' - 6y = e^{3x} + 1$; $y(0) = 3$, $y'(0) = 1$
11. $y'' + 4y = 12$; $y(0) = 6$, $y'(0) = -3$
12. $y'' + 9y = 12\cos 3x$; $y(0) = 2$, $y'(0) = 5$
13. $(D^2 - 1)(D + 3)y = 0$; $y(0) = 1$, $y'(0) = 1$, $y''(0) = -1$
14. $(D + 1)^2 y = 6xe^{-x}$; $y(0) = 2$, $y'(0) = 2$
15. $y'' - 3y' + 2y = x$; $y(0) = 1$, $y'(0) = 2$
16. $y'' + y = 2\sin x$; $y(0) = 1$, $y'(0) = 3$
17. $y'' + y' = \cos x$; $y(0) = -\frac{1}{2}$, $y'(0) = -\frac{3}{2}$
18. $y''' + y' = x - 1$; $y(0) = 2$, $y'(0) = y''(0) = 0$
19. $y''' - 3y'' + 3y' - y = e^x$; $y(0) = 0$, $y'(0) = 1$, $y''(0) = -1$
20. $y'' - 2y' - 3y = e^x$; $y(0) = y'(0) = 1$
21. A particle passes through the point $(0, 1)$ with slope 1, and its acceleration is equal to the negative of twice the sum of the velocity and the y-coordinate. Find the path it follows.
22. Solve Exercise 27 in Section 3.10 by using Laplace transforms.
23. Solve Exercise 28 in Section 3.10 by using Laplace transforms.

8.6 Products of Transforms: Convolutions

We continually use the linearity properties of $\mathscr{L}$ as we solve problems. These properties assert the fact that $\mathscr{L}$ is a mapping that preserves the operations of addition of functions and multiplication of a function by a constant. There is also a type of multiplication of functions, denoted by $f(x) * g(x)$, for which $\mathscr{L}$ is a multiplicative mapping in the sense that

$$\begin{aligned}\mathscr{L}[f(x) * g(x)] &= \mathscr{L}[f(x)] \cdot \mathscr{L}[g(x)] \\ &= F(s) \cdot G(s)\end{aligned}$$

On the right we have the usual product of functions.

Definition: We call this new product the **convolution of the functions** $f(x)$ **and** $g(x)$, and it is defined by

$$f(x) * g(x) = \int_0^x f(x-t)g(t)\,dt$$

where $f(x)$ and $g(x)$ are continuous functions on $0 \leq x < \infty$ and are of exponential order α.

When these conditions are satisfied, the convolution of $f(x)$ and $g(x)$ exists for $s > \alpha$, is continuous, and is of exponential order α. It is easy to show that the convolution property follows the general properties of ordinary multiplication of functions:

$$f(x) * g(x) = g(x) * f(x)$$

$$[f(x) * g(x)] * h(x) = f(x) * [g(x) * h(x)]$$

$$f(x) * [g(x) + h(x)] = [f(x) * g(x)] + [f(x) * h(x)]$$

$$[cf(x)] * g(x) = f(x) * [cg(x)] = c[f(x) * g(x)]$$

Use the above definition to prove one or more of these properties.

The convolution property is most useful when solving differential equations with right sides that are somewhat complicated. Frequently we need to find the inverse transform of a product of two functions $F(s) \cdot G(s)$. If this cannot be decomposed by partial fractions, we often turn to the convolution property. The inverse linearity property again helps us. If $F(s) = G(s)H(s)$ and the inverse transforms $g(x)$ and $h(x)$ are known, then

$$\mathscr{L}^{-1}[F(s)] = \mathscr{L}^{-1}[G(s)H(s)] = g(x) * h(x)$$

We are now able to expand our table of inverse transforms by using this convolution property. Example 25 uses convolution of functions to find the inverse transform.

Example 25

To find

$$\mathscr{L}^{-1}\left[\frac{s}{(s^2+1)^2}\right]$$

we write it as

$$\mathscr{L}^{-1}\left[\frac{1}{s^2+1}\cdot\frac{s}{s^2+1}\right] = (\sin x) * (\cos x) = \int_0^x (\sin t)\cos(x-t)\,dt$$

$$= \frac{1}{2}x\sin x$$

The versatility of the Laplace transform method is illustrated in Example 26.

Example 26

Consider the following initial value problem with $f(x)$ arbitrary but assumed to have a Laplace transform:

$$y'' + 4y = f(x); \qquad y(0) = 0,\ y'(0) = 0$$

The function $f(x)$ could be an electromotive force in a circuit, a retarding force in a mechanical system, etc. Taking transforms gives

$$\mathscr{L}[y''] + 4\mathscr{L}[y] = \mathscr{L}[f(x)]$$

$$s^2\mathscr{L}[y] - sy(0) - y'(0) + 4\mathscr{L}[y] = \mathscr{L}[f(x)]$$

$$(s^2+4)\mathscr{L}(y) = \mathscr{L}[f(x)]$$

or

$$\mathscr{L}[y] = \left(\frac{1}{s^2+4}\right)\mathscr{L}[f(x)]$$

From the transform table we find that

$$\mathscr{L}^{-1}\left[\frac{1}{s^2+4}\right] = \frac{1}{2}\sin 2x$$

Hence, by the convolution property,

$$y = \mathscr{L}^{-1}\left[\frac{1}{s^2+4}\right] * \mathscr{L}^{-1}[\mathscr{L}[f(x)]]$$

$$= \left(\frac{1}{2}\sin 2x\right) * f(x)$$

$$= \frac{1}{2}\int_0^x (\sin 2t) f(x-t)\,dt$$

With the function $f(x)$ inserted in the form $f(x-t)$, we would integrate this to the unique solution.

As a more general statement that follows from the convolution definition, we obtain the following important property useful in two additional branches of applied mathematics: If $\mathscr{L}^{-1}[F(s)] = f(x)$ is known, then

$$\mathscr{L}^{-1}\left[\frac{1}{s}\cdot F(s)\right] = 1 * f(x) = \int_0^x f(t)\,dt$$

or

$$\mathscr{L}\left[\int_0^x f(t)\,dt\right] = \frac{1}{s}F(s)$$

This property will be useful in solving integral and integro-differential equations. (An integral equation is one that involves an unknown function and integrals of that function; an integro-differential equation is one that involves an unknown function, derivatives of that function, and integrals of that function.) Example 27 shows how this property is used in solving a problem regarding an electric circuit.

Example 27

In an RLC (R = resistance; L = inductance; C = capacitance) electric circuit, the oscillator equation is

$$L\frac{dI}{dt} + RI + \frac{1}{C}Q = f(t)$$

We could replace I by Q', and make this an equation involving Q (the electric charge) and its derivatives; but the item of interest is often the current $I(t)$, and we may proceed directly to the solution for $I(t)$ by first writing

$$Q(t) - Q(0) = \int_0^t I(\alpha)\,d\alpha$$

Substituting into the given differential equation, we obtain the integro-differential equation

$$LI' + RI + \frac{1}{C}\left(Q(0) + \int_0^t I(\alpha)\,d\alpha\right) = f(t)$$

Applying Laplace transforms to this and assuming the condition that the value of Q is zero at time $t = 0$ (i.e., $Q(0) = Q_0 = 0$), we get

$$\mathscr{L}[LI'] + \mathscr{L}[RI] + \mathscr{L}\left[\frac{Q_0}{C}\right] + \mathscr{L}\left[\frac{1}{C}\int_0^t I(\alpha)\,d\alpha\right] = \mathscr{L}[f(t)]$$

$$L\{s\mathscr{L}[I] - I(0)\} + R\mathscr{L}[I] + 0 + \frac{1}{C}\cdot\frac{1}{s}\mathscr{L}[I] = \mathscr{L}[f(t)]$$

$$\mathscr{L}[I]\left(Ls + R + \frac{1}{Cs}\right) = \mathscr{L}[f(t)] + LI(0)$$

or

$$\mathscr{L}[I(t)] = \left(\frac{1}{Ls + R + 1/Cs}\right)\{\mathscr{L}[f(t)] + LI(0)\}$$

Then

$$I(t) = \mathscr{L}^{-1}\left(\frac{1}{Ls + R + 1/Cs}\right) * \mathscr{L}^{-1}[\mathscr{L}[f(t)] + LI(0)]$$

When the values of L, R, and C and the forcing function $f(t)$ are specified, this function $I(t)$ may be calculated specifically. To carry this problem to its conclusion, assume $R = 4$ ohms, $C = \frac{1}{3}$ farad. $L = 1$ henry, and the forcing function $f(t) = e^t$ with $I(0) = 0$. Then

$$\begin{aligned}
I(t) &= \mathscr{L}^{-1}\left(\frac{1}{s + 4 + 3/s}\right) * \mathscr{L}^{-1}[\mathscr{L}[f(t)]] \\
&= \mathscr{L}^{-1}\left(\frac{s}{s^2 + 4s + 3}\right) * f(t) \\
&= \mathscr{L}^{-1}\left(\frac{s}{(s + 3)(s + 1)}\right) * f(t) \\
&= \frac{1}{2}(3e^{-3t} - e^{-t}) * e^t \\
&= \frac{3}{2}(e^{-3t} * e^t) - \frac{1}{2}(e^{-t} * e^t) \\
&= \frac{3}{2}\int_0^t e^{-3(t-x)} \cdot e^x\, dx - \frac{1}{2}\int_0^t e^{-(t-x)} \cdot e^x\, dx \\
&= \frac{3}{2}\int_0^t e^{-3t}e^{4x}\, dx - \frac{1}{2}\int_0^t e^{-t}e^{2x}\, dx \\
&= \frac{3}{8}e^{-3t}(e^{4t} - 1) - \frac{1}{4}e^{-t}(e^{2t} - 1) \\
&= -\frac{3}{8}e^{-3t} + \frac{1}{8}e^t + \frac{1}{4}e^{-t}
\end{aligned}$$

This function represents the behavior of the current $I(t)$ for any time t. This may be checked as the correct solution by substituting it into the original differential equation.

Example 28

We will apply the transform solution method to the integral equation

$$y(x) = \cos x + \int_0^x e^{-(x-t)} y(t)\, dt$$

where the unknown function y appears both in an integral and standing alone. First we take transforms of all terms:

$$\begin{aligned}
\mathscr{L}[y(x)] &= \mathscr{L}[\cos x] + \mathscr{L}\left[\int_0^x e^{-(x-t)} y(t)\, dt\right] \\
&= \mathscr{L}[\cos x] + \mathscr{L}[e^{-x} * y(x)]
\end{aligned}$$

or
$$Y(s) = \frac{s}{s^2+1} + \left(\frac{1}{s+1}\right) \cdot Y(s)$$

$$\therefore \quad Y(s) = \frac{s+1}{s^2+1} = \frac{s}{s^2+1} + \frac{1}{s^2+1}$$

Taking the inverse transform gives the solution

$$y(x) = \cos x + \sin x$$

This may be checked by direct substitution into the original integral equation.

Example 29 demonstrates the way in which the convolution property aids in evaluating the Euler Beta Function.

Example 29

The **Euler Beta Functions** are important in applications in mathematical physics. They are defined by

$$x^m * x^n = \int_0^x (x-t)^n t^m \, dt, \qquad m, n > -1$$

With the change of variable $t = xy$, we have

$$x^m * x^n = \int_0^1 x^m y^m \cdot x^n (1-y)^n x \, dy \equiv x^{m+n+1} \mathrm{B}(m+1, n+1)$$

The value of the beta function $\mathrm{B}(m+1, n+1)$ may now be evaluated by the convolution theorem.

$$\mathscr{L}[x^m * x^n] = \mathscr{L}[x^{m+n+1} \mathrm{B}(m+1, n+1)]$$

Since the left side is

$$\mathscr{L}[x^m] \cdot \mathscr{L}[x^n] = \frac{m!}{s^{m+1}} \cdot \frac{n!}{s^{n+1}} = \frac{m!\, n!}{s^{m+n+2}}$$

we have

$$\mathscr{L}[x^{m+n+1} \mathrm{B}(m+1, n+1)] = \frac{m!\, n!}{s^{m+n+2}}$$

or

$$x^{m+n+1} \mathrm{B}(m+1, n+1) = \mathscr{L}^{-1}\left[\frac{m!\, n!}{s^{m+n+2}}\right] = \frac{m!\, n!}{(m+n+1)!} x^{m+n+1}$$

by formula 2 of Table 8.1. Cancellation gives us

$$\mathrm{B}(m+1, n+1) = \frac{m!\, n!}{(m+n+1)!}$$

The gamma function of physics, defined by

$$\Gamma(k) = \int_0^{\infty} x^{k+1} e^{-x}\,dx, \qquad k > 0$$

has as its value for a positive integer $k = n$, $\Gamma(n + 1) = n!$ Therefore,

$$B(m + 1, n + 1) = \frac{\Gamma(m + 1)\Gamma(n + 1)}{\Gamma(m + n + 2)}$$

Exercises 8.6

1. Using the convolution property, find

   (a) $\mathscr{L}^{-1}\left[\dfrac{1}{s^2(s + a)}\right], \quad a \neq 0$

   (b) $\mathscr{L}^{-1}\left[\dfrac{s^2}{(s^2 + 9)^2}\right]$

   (c) $\mathscr{L}^{-1}\left[\dfrac{1}{s(s^2 + a^2)}\right], \quad a \neq 0$

2. Solve the initial value problem

   $y'' - 4y' = f(x); \qquad y(0) = 0, \qquad y'(0) = 0$

   Then insert $f(x) = e^{2x}$ into the solution and simplify.

3. Find the Laplace transform of the following convolution-type integrals:

   (a) $\displaystyle\int_0^x (x - t)\sin 3t\,dt$

   (b) $\displaystyle\int_0^x e^t(x - t)^2\,dt$

4. Assuming $y(x)$ has a transform, solve for $y(x)$ in the following integral equations:

   (a) $y(x) = x^2 + \displaystyle\int_0^x y(t)\sin(x - t)\,dt$

   (b) $y(x) = e^{-x} - 2\displaystyle\int_0^x y(t)\cos(x - t)\,dt$

   (c) $y(x) = 1 + \displaystyle\int_0^x 2y(x - t)e^{-2t}\,dt$

   (d) $4x^2 - y = \displaystyle\int_0^x y(t)e^t e^{-x}\,dt$

5. Suppose the electric circuit of Example 27 now has $L = 20$ henrys, $R = 30$ ohms, $C = 0.1$ farad, and $f(t) = \sin t$, with $I(0) = 0$. Solve for $I(t)$.
6. Prove that $f(x) * g(x) = g(x) * f(x)$.
7. Find the value of the Euler Beta Function $B(7, 4)$.

8.7 Discontinuous Functions and Periodic Functions

Thus far, the functions we have considered have been assumed to be continuous on the positive real axis. The transform method also works efficiently if the functions have finite or jump discontinuities or if the right-side function of a given differential equation is not continuous.

This type of function is prevalent in applications problems. For example, in many circuit problems you may encounter a pulsating source of the electromotive force (emf) or a source that varies with time in a discontinuous manner. Also, a particular part of some mechanical device may move during certain intervals and at other times be nonfunctional.

Of particular interest in applications problems is the **Heaviside unit-step function**

$$H_0(x) = \begin{cases} 1 & \text{if} \quad x > 0 \\ 0 & \text{if} \quad x < 0 \end{cases}$$

and extensions of it

$$H_A(x) = \begin{cases} 1 & \text{if} \quad x > A \\ 0 & \text{if} \quad x < A \end{cases}$$

and

$$H_{-A}(x) = \begin{cases} 0 & \text{if} \quad x > A \\ 1 & \text{if} \quad x < A \end{cases}$$

Figure 8.1 illustrates the Heaviside function. A more convenient and useful way of writing the $H_{-A}(x)$ function is to observe that (as illustrated in Figure 8.2)

$$H_0 - H_A = \begin{cases} 1 & \text{for} \quad 0 \le x \le A \\ 0 & \text{elsewhere} \end{cases}$$

On the positive axis, the region we are interested in for transform work, this combination is the same as $H_{-A}(x)$. In general, we can write

$$H_A - H_B = \begin{cases} 1 & \text{for} \quad A \le x \le B \\ 0 & \text{elsewhere} \end{cases}$$

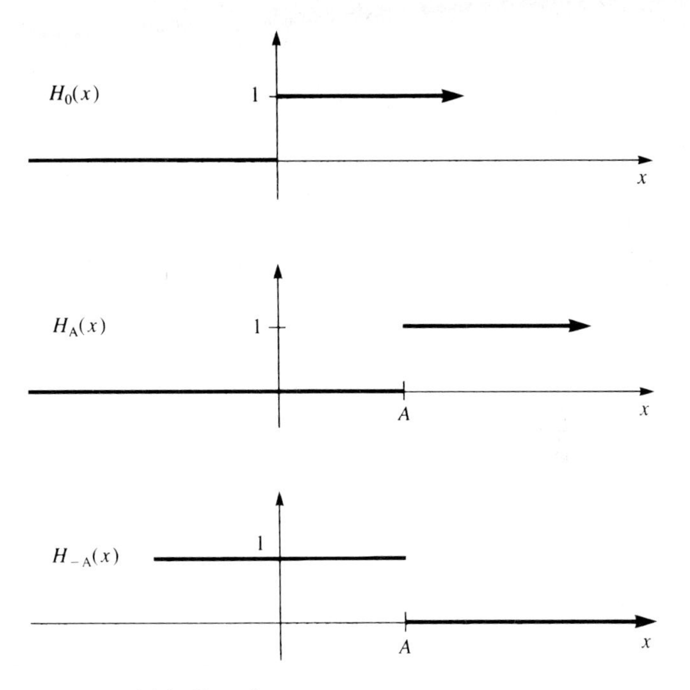

Figure 8.1 Heaviside Function

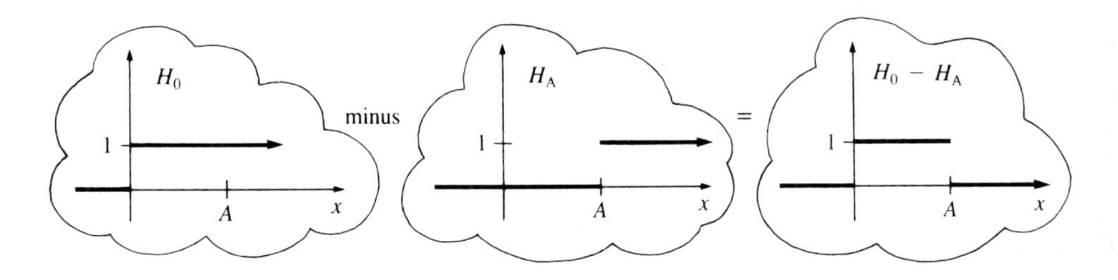

Figure 8.2

Trying an example of this type of combination step function, we can express $f(x)$ in terms of H:

$$f(x) = \begin{Bmatrix} 2 & \text{if} & 0 \le x \le 1 \\ -1 & \text{if} & 1 < x \le 3 \\ 4 & \text{if} & x > 3 \end{Bmatrix} = 2(H_0 - H_1) - 1(H_1 - H_3) + 4H_3$$

$$= 2H_0 - 3H_1 + 5H_3$$

Thus a function defined in a nonstandard way may be expressed as an additive combination of simple functions. The preceding function is illustrated in Figure 8.3.

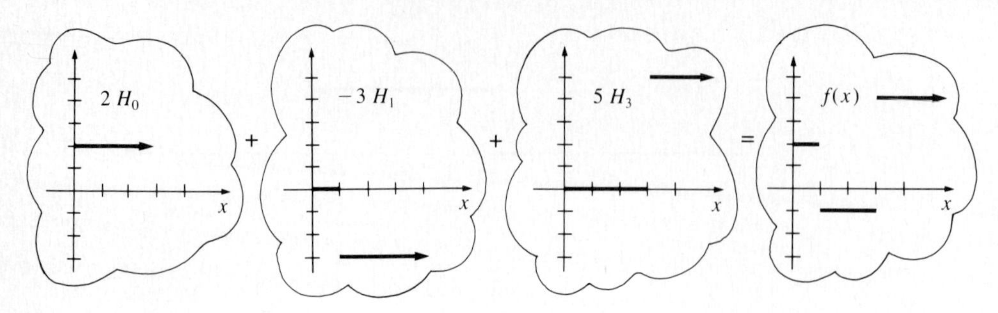

Figure 8.3

Example 30

If we have an initial value problem of the type

$$ay'' + by' + cy = f(t); \qquad y(0) = y_0, \qquad y'(0) = v_0$$

and $y(t)$ represents a charge in a circuit, then $f(t)$ represents a source of emf. If this source is not turned on until after A units of time and thereafter provides constant voltage E, it is represented as $f(t) = E \cdot H_A(x)$. If we find the Laplace transform of $H_A(x)$, we then need only to multiply it by E to obtain the transform of the step function $f(t)$.

By using the definition of the transform, we calculate the transform of the Heaviside function

$$\begin{aligned}\mathscr{L}[H_A(x)] &= \int_0^\infty e^{-sx} H_A(x)\,dx = \int_0^A e^{-sx} H_A(x)\,dx + \int_A^\infty e^{-sx} H_A(x)\,dx \\ &= 0 + \int_A^\infty e^{-sx} \cdot 1\,dx = -\frac{1}{s} e^{-sx}\Big|_A^\infty = -\frac{1}{s}\left[0 - e^{-sA}\right] = \frac{e^{-sA}}{s}\end{aligned} \tag{7}$$

which is valid for $A \geq 0$.

The Heaviside function, when used in combination with other functions, provides a realistic expression for a wide variety of discontinuous functions observed in applications. For example, consider the magnitude of the force exerted on a nail by a hammer as the nail is pounded into a plank, the charge on a capacitor if a spark jumps to it once each second, a sawtooth function applied as emf to a circuit, and the unit-impulse function of an impulse system.

If a function is given by different expressions in different domains, such as

$$f(x) = \begin{cases} u(x) & \text{if} \quad x < A \\ v(x) & \text{if} \quad x > A \end{cases}$$

then we conveniently write $f(x)$ by using the extended Heaviside function:

$$\begin{aligned} f(x) &= u(x)H_{-A}(x) + v(x)H_A(x) \\ &= u(x)[H_0 - H_A] + v(x)H_A \\ &= u(x)H_0 + [v(x) - u(x)]H_A \end{aligned}$$

Example 31

Let

$$f(x) = \begin{cases} x - 2 & \text{if} \quad 0 \leq x \leq 4 \\ 6 - x & \text{if} \quad x > 4 \end{cases}$$

Now $u(x) = x - 2$ and $v(x) = 6 - x$, so

$$\begin{aligned} f(x) &= (x - 2)H_0 + [(6 - x) - (x - 2)]H_4 \\ &= (x - 2)H_0 + (8 - 2x)H_4 \end{aligned}$$

This is illustrated in Figure 8.4.

We have seen how a function that is continuous but not differentiable everywhere and originally written in a nonstandard way is expressed as an algebraic sum. Its Laplace transform may now be computed via Theorem 8.7 and formula (7). We would use $\mathscr{L}[xf(x)] = -F'(s)$, so that

$$\mathscr{L}[xH_A(x)] = -\frac{d}{ds}\left(\frac{1}{s}e^{-sA}\right) = -\left[\frac{s(-A)e^{-sA} - e^{-sA}}{s^2}\right] = \frac{e^{-sA}(1 + sA)}{s^2} \tag{8}$$

This formula is easily extendable to higher powers of x.

If $f(x)$ is defined on $0 \leq x < \infty$ and we wish to shift $f(x)$ so that it acts only on $A \leq x < \infty$, we write

$$f_A(x) = f(x - A)H_A(x) = \begin{cases} f(x - A) & \text{if} \quad x > A \\ 0 & \text{if} \quad x < A \end{cases}$$

This expression leads us to an important property, which is stated as Theorem 8.8.

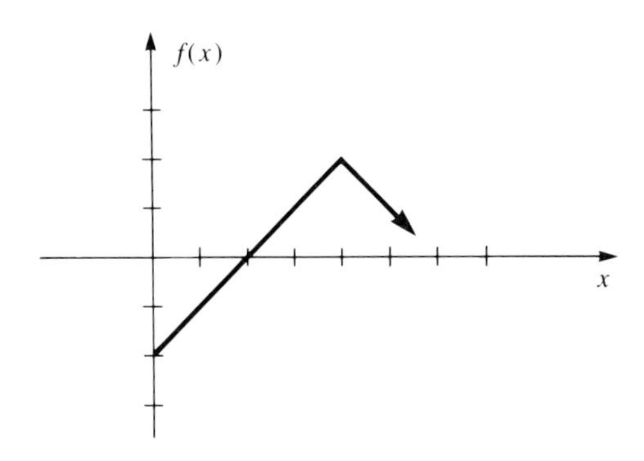

Figure 8.4

Theorem 8.8:

If $\mathscr{L}[f(x)]$ exists, then $\mathscr{L}[f_A(x)]$ exist and is expressed by

$$\mathscr{L}[f_A(x)] = \mathscr{L}[f(x-A)H_A(x)] = e^{-sA}\,\mathscr{L}[f(x)] = e^{-sA}F(s)$$

Proof:

$$\begin{aligned}\mathscr{L}[f_A(x)] &= \int_0^\infty e^{-sx}f_A(x)\,dx \\ &= \int_0^\infty e^{-sx}f(x-A)H_A(x)\,dx \\ &= \int_A^\infty e^{-sx}f(x-A)\,dx\end{aligned}$$

Now let $x - A = t$. Then $x = t + A$ and $dx = dt$. The integral becomes

$$\int_0^\infty e^{-s(t+A)}f(t)\,dt = e^{-sA}\int_0^\infty e^{-st}f(t)\,dt = e^{-sA}F(s) \qquad \blacksquare$$

Once this theorem is established, following naturally is the expression

$$\mathscr{L}^{-1}[e^{-sA}F(s)] = f(x-A)H_A(x) \tag{9}$$

which proves to be very valuable in applications.

Example 32

Consider

$$y'' + 4y' + 3y = f(x); \qquad y(0) = y'(0) = 0$$

where

$$f(x) = \begin{cases} 2 & \text{if} \quad 0 < x < 3 \\ -2 & \text{if} \quad x > 3 \end{cases}$$

We first write $f(x) = 2(H_0 - H_3) - 2H_3 = 2H_0 - 4H_3$ (as illustrated in Figure 8.5) and take Laplace transforms on both sides of the differential equation:

$$\mathscr{L}[y''] + 4\mathscr{L}[y'] + 3\mathscr{L}[y] = 2\mathscr{L}[H_0] - 4\mathscr{L}[H_3]$$

$$\{s^2\mathscr{L}[y] - sy(0) - y'(0)\} + 4\{s\mathscr{L}[y] - y(0)\} + 3\mathscr{L}[y] = 2\cdot\frac{1}{s} - 4\cdot\frac{e^{-3s}}{s}$$

$$(s^2 + 4s + 3)\mathscr{L}[y] = \frac{2 - 4e^{-3s}}{s}$$

$$\therefore \quad \mathscr{L}[y] = \frac{2}{s(s+3)(s+1)} - \frac{4e^{-3s}}{s(s+3)(s+1)}$$

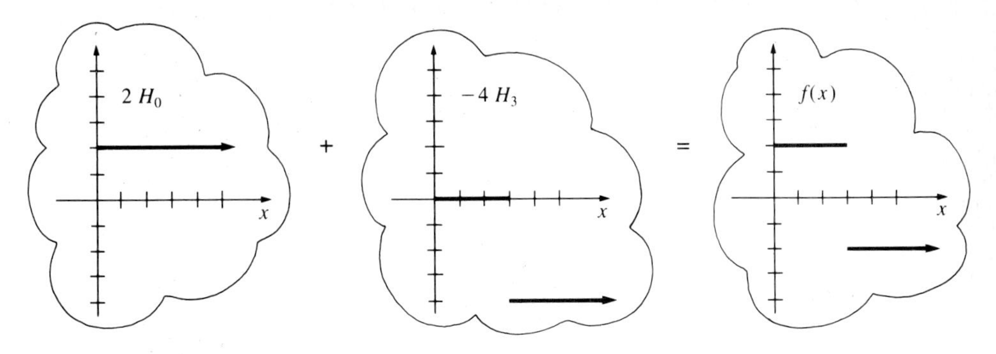

Figure 8.5

Note that an attempt must be made to get the function $F(s)$ isolated as well as multiplied by the exponential function. Then formula (9) can be used to compute the inverse transforms. By partial-fraction decomposition, we get the inverse transform of the first term:

$$\mathscr{L}^{-1}\left[\frac{2}{s(s+3)(s+1)}\right] = \mathscr{L}^{-1}\left[\frac{2/3}{s} + \frac{1/3}{s+3} - \frac{1}{s+1}\right]$$
$$= \frac{2}{3} + \frac{1}{3}e^{-3x} - e^{-x} \equiv T(x)$$

We can use (9) to find the inverse transform of the second term:

$$-2\mathscr{L}^{-1}\left[\frac{2e^{-3s}}{s(s+3)(s+1)}\right] = -2T(x-3)H_3(x)$$
$$= -2\left(\frac{2}{3} + \frac{1}{3}e^{-3(x-3)} - e^{-(x-3)}\right)H_3(x)$$

Hence the unique solution is

$$y(x) = \frac{2}{3} + \frac{1}{3}e^{-3x} - e^{-x} - \left(\frac{4}{3} + \frac{2}{3}e^{-3(x-3)} - 2e^{-(x-3)}\right)H_3(x)$$

Example 33

Consider the initial value problem

$$y'' - 3y' - 4y = f(x) = \begin{cases} 2x & \text{if} \quad 0 \le x \le 2 \\ \dfrac{x}{4} + \dfrac{7}{2} & \text{if} \quad x \ge 2 \end{cases}$$

$$y(0) = 0, \qquad y'(0) = 0$$

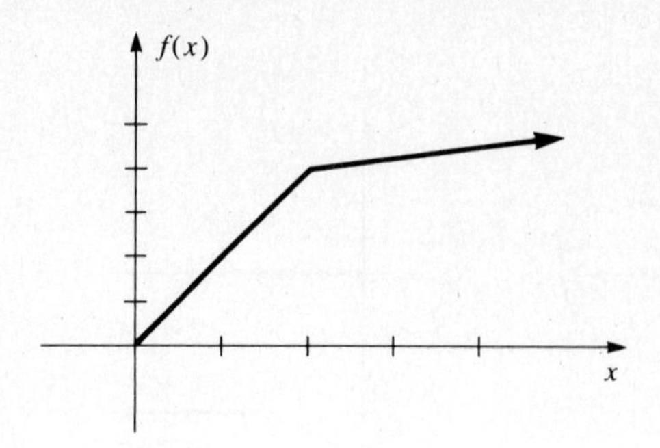

Figure 8.6

We first write the function $f(x)$ as

$$f(x) = 2x(H_0 - H_2) + \left(\frac{x}{4} + \frac{7}{2}\right)H_2(x)$$

$$= 2x\ H_0(x) + \left(-2x + \frac{x}{4} + \frac{7}{2}\right)H_2(x)$$

$$= 2x\ H_0(x) + \left(\frac{7}{2} - \frac{7}{4}x\right)H_2(x)$$

Taking transforms on both sides we get

$$\{s^2\mathscr{L}[y] - 0\} - 3\{s\mathscr{L}[y] - 0\} - 4\mathscr{L}[y]$$
$$= 2\mathscr{L}[xH_0] + \frac{7}{2}\mathscr{L}[H_2] - \frac{7}{4}\mathscr{L}[xH_2]$$

By formula (8) we get

$$\mathscr{L}[xH_0(x)] = \frac{1}{s^2}$$

$$\mathscr{L}[xH_2(x)] = \frac{e^{-2s}(1 + 2s)}{s^2}$$

Hence we have

$$(s^2 - 3s - 4)\mathscr{L}[y] = 2\left(\frac{1}{s^2}\right) + \frac{7}{2}\frac{e^{-2s}}{s} - \frac{7}{4}\left(\frac{e^{-2s}(1 + 2s)}{s^2}\right)$$

$$= \frac{2 + \frac{7}{2}se^{-2s} - \frac{7}{4}e^{-2s} - \frac{7}{2}se^{-2s}}{s^2}$$

$$= \frac{2}{s^2} - \frac{7}{4}\cdot\frac{e^{-2s}}{s^2}$$

$$\therefore \quad \mathscr{L}[y] = \frac{8 - 7e^{-2s}}{4s^2(s-4)(s+1)} = \frac{2}{s^2(s-4)(s+1)} - \frac{7}{8} \cdot \frac{2e^{-2s}}{s^2(s-4)(s+1)}$$

Then

$$y = \mathscr{L}^{-1}\left[\frac{2}{s^2(s-4)(s+1)}\right] - \frac{7}{8}\mathscr{L}^{-1}\left[\frac{2e^{-2s}}{s^2(s-4)(s+1)}\right]$$

$$= \mathscr{L}^{-1}\left[\frac{1/40}{s-4} - \frac{2/5}{s+1} + \frac{3/8}{s} - \frac{1/2}{s^2}\right]$$

$$- \frac{7}{8}\mathscr{L}^{-1}\left[e^{-2s}\left(\frac{1/40}{s-4} - \frac{2/5}{s+1} + \frac{3/8}{s} - \frac{1/2}{s^2}\right)\right]$$

$$= \left(\frac{1}{40}e^{4x} - \frac{2}{5}e^{-x} + \frac{3}{8} - \frac{1}{2}x\right)$$

$$- \frac{7}{8}\left[\frac{1}{40}e^{4(x-2)} - \frac{2}{5}e^{-(x-2)} + \frac{3}{8} - \frac{1}{2}(x-2)\right]H_2(x)$$

You are already familiar with trigonometric functions that are periodic. In many applications problems, particularly those of circuit analysis, the input function may be periodic.

Definition: A **periodic function** is a function for which $f(x + T) = f(x)$ for all x, and T is called the **period** of f.

Transforms of periodic functions may be found by applying the following theorem.

Theorem 8.9:

If a periodic function $f(x)$ has a transform and T is the period of f, then

$$\mathscr{L}[f(x)] = \frac{1}{1 - e^{-sT}}\int_0^T e^{-sx}f(x)\,dx, \qquad s > 0$$

Proof:

We have by definition,

$$F(s) = \int_0^\infty e^{-sx}f(x)\,dx = \int_0^T e^{-sx}f(x)\,dx + \int_T^\infty e^{-sx}f(x)\,dx$$

In the second integral we change variables by letting $x - T = y$. We then change

$$\int_T^\infty e^{-sx}f(x)\,dx$$

to

$$\int_0^\infty e^{-s(y+T)}f(y+T)\,dy$$

which equals

$$\int_0^\infty e^{-s(y+T)} f(y)\, dy$$

because of periodicity. Thus

$$\begin{aligned} F(s) &= \int_0^T e^{-sx} f(x)\, dx + e^{-sT} \int_0^\infty e^{-sy} f(y)\, dy \\ &= \int_0^T e^{-sx} f(x)\, dx + e^{-sT} \cdot F(s) \end{aligned}$$

Solving this for $F(s)$ we obtain

$$F(s) = \frac{1}{1 - e^{-sT}} \int_0^T e^{-sx} f(x)\, dx$$

Since $f(x)$ is periodic, we need to know the value of the function $f(x)$ only on $0 \leq x < T$ in order to compute its transform. It is sometimes useful to use a function $f_P(x)$ that equals $f(x)$ on $0 \leq x < T$ and is zero elsewhere. This may be written as

$$\begin{aligned} f_P(x) &= f(x) H_{-T}(x) + 0 \cdot H_T(x) \\ &= f(x)[H_0 - H_T] \end{aligned}$$

For a periodic $f(x)$, we can compute

$$\begin{aligned} \mathscr{L}[f(x)] &= \frac{1}{1 - e^{-sT}} \mathscr{L}[f_P(x)] = \frac{1}{1 - e^{-sT}} \int_0^\infty e^{-sx} f_P(x)\, dx \\ &= \frac{1}{1 - e^{-sT}} \int_0^T e^{-sx} f(x)\, dx \end{aligned}$$

Example 34

Consider the discontinuous sawtooth function illustrated in Figure 8.7. We can write this as

$$\begin{aligned} f_P(x) &= x(H_0 - H_1) \\ &= xH_0 - xH_1 \end{aligned}$$

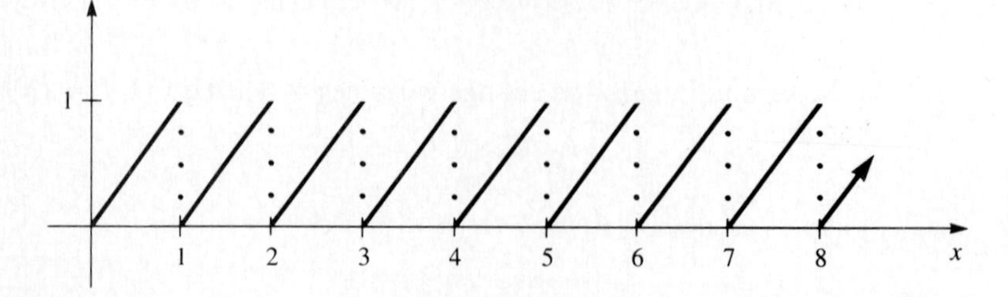

Figure 8.7

Then we can get the transform by differentiating:

$$\mathscr{L}[f(x)] = \frac{1}{1-e^{-s}}\mathscr{L}[f_P(x)] = \frac{1}{1-e^{-s}}\{\mathscr{L}[xH_0] - \mathscr{L}[xH_1]\}$$

$$= \frac{1}{1-e^{-s}}\left[-\frac{d}{ds}\left(\frac{1}{s}\right) - \frac{d}{ds}\left(\frac{1}{s}e^{-s}\right)\right]$$

$$= \frac{1}{1-e^{-s}}\left[\frac{1}{s^2} - \frac{e^{-s}}{s} - \frac{e^{-s}}{s^2}\right] = \frac{1}{s^2} - \frac{e^{-s}}{s(1-e^{-s})}$$

The same result may be obtained by integrating:

$$\mathscr{L}[f(x)] = \frac{1}{1-e^{-s}}\int_0^1 e^{-sx}x\,dx \overset{\text{By}}{\underset{\text{parts}}{\Longrightarrow}} \frac{1}{1-e^{-s}}\left\{-\frac{x}{s}e^{-sx}\Big|_0^1 + \frac{1}{s}\int_0^1 e^{-sx}\,dx\right\}$$

$$= \frac{1}{1-e^{-s}}\left\{-\frac{1}{s}e^{-s} - \frac{1}{s^2}(e^{-s}-1)\right\}$$

$$= \frac{1}{1-e^{-s}}\left\{\frac{1}{s^2} - \frac{e^{-s}}{s} - \frac{e^{-s}}{s^2}\right\}$$

$$= \frac{1}{s^2} - \frac{e^{-s}}{s(1-e^{-s})}$$

Example 35

To calculate the Laplace transform of the square wave illustrated in Figure 8.8, we note that the function is periodic with period 2 and defined by

$$f(x) = \begin{cases} 1 & \text{on} \quad 0 < x \le 1 \\ -1 & \text{on} \quad 1 < x \le 2 \end{cases}$$

Using Theorem 8.9, we obtain

$$F(s) = \frac{\int_0^2 e^{-sx}f(x)\,dx}{1-e^{-2s}}$$

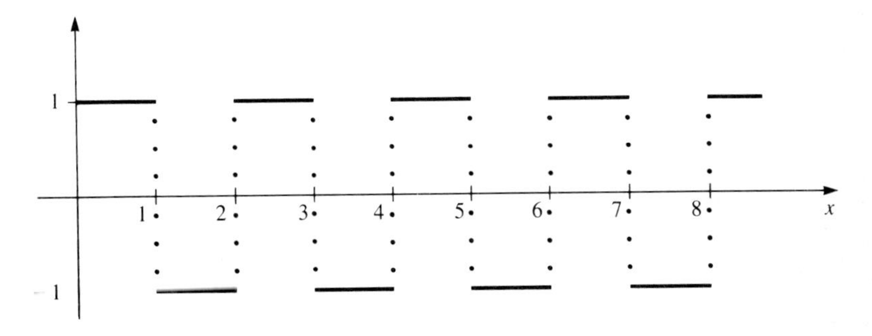

Figure 8.8

However,

$$\int_0^2 e^{-sx} f(x)\, dx = \int_0^1 e^{-sx}(1)\, dx + \int_1^2 e^{-sx}(-1)\, dx$$

$$= -\frac{1}{s}\left\{ e^{-sx}\Big|_0^1 \right\} + \frac{1}{s}\left\{ e^{-sx}\Big|_1^2 \right\}$$

$$= -\frac{1}{s}(e^{-s} - 1) + \frac{1}{s}(e^{-2s} - e^{-s})$$

$$= \frac{1}{s}(e^{-2s} - 2e^{-s} + 1)$$

$$\therefore \quad F(s) = \frac{e^{-2s} - 2e^{-s} + 1}{s(1 - e^{-2s})} = \frac{(e^{-s} - 1)(e^{-s} - 1)}{s(1 - e^{-s})(1 + e^{-s})} = \frac{1 - e^{-s}}{s(1 + e^{-s})}$$

This may also be written as

$$F(s) = \frac{1 - e^{-s}}{s(1 + e^{-s})} = \frac{1}{s} \cdot \frac{e^{-s/2}(e^{s/2} - e^{-s/2})}{e^{-s/2}(e^{s/2} + e^{-s/2})}$$

$$= \frac{1}{s} \tanh\left(\frac{s}{2}\right)$$

Exercises 8.7

1. Solve the initial value problem

$$y'' + y = f_1(t); \qquad y(0) = 0, \qquad y'(0) = 0$$

where
$$f_1(t) = \begin{cases} 1, & 0 \le t < 4 \\ 0, & t > 4 \end{cases}$$

2. Solve the initial value problem

$$y'' + y = f_2(t); \qquad y(0) = 0, \qquad y'(0) = 1$$

where
$$f_2(t) = \begin{cases} 1, & 0 \le t < 1 \\ 0, & 1 \le t < 2 \\ 1, & t \ge 2 \end{cases}$$

3. Solve the initial value problem

$$y'' + y = f_3(t); \qquad y(0) = 0, \qquad y'(0) = 0$$

where
$$f_3(t) = \begin{cases} 8, & 0 \le t \le 2 \\ 6 + t, & t > 2 \end{cases}$$

4. Solve the initial value problem

$$y'' + 4y = f_4(t); \qquad y(0) = 1, \qquad y'(0) = 1$$

where
$$f_4(t) = \begin{cases} t, & 0 \le t \le 1 \\ 1, & t > 1 \end{cases}$$

5. Find the Laplace transform of the sawtooth wave illustrated in Figure 8.9.

6. Find the Laplace transform of the clipped sawtooth wave illustrated in Figure 8.10.

7. Find the Laplace transform of the rectified sine wave $f(x) = \sin x$, $0 \le x < \pi$, with $f(x + \pi) = f(x)$, which is illustrated in Figure 8.11.

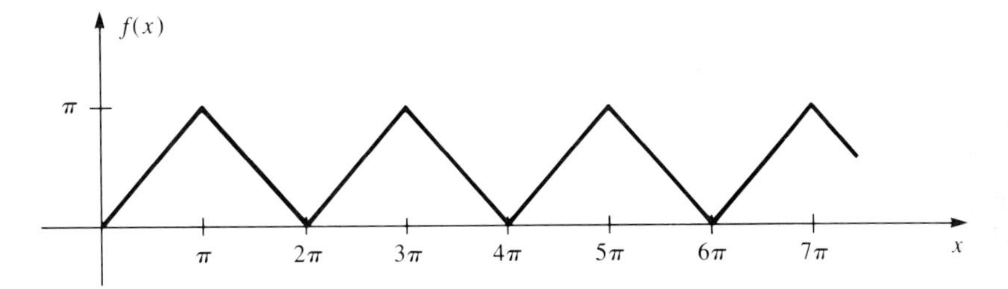

Figure 8.9

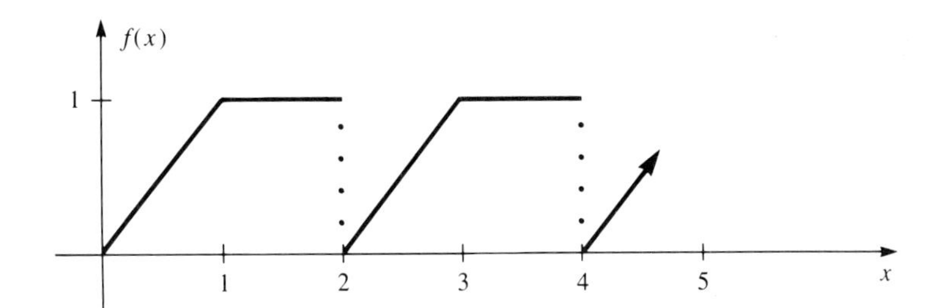

Figure 8.10

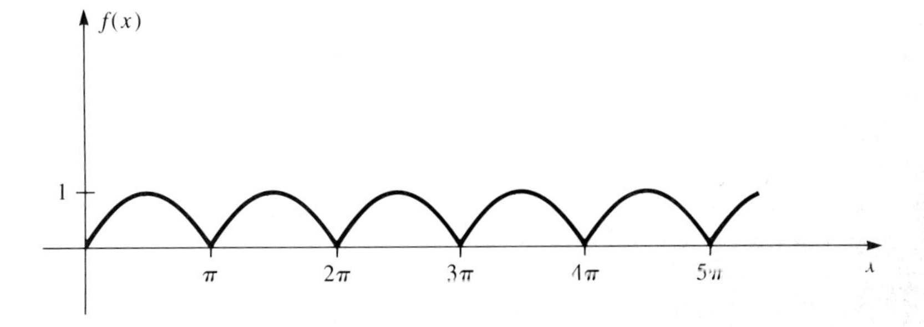

Figure 8.11

8.8 Equations with Variable Coefficients

To solve linear differential equations with variable coefficients by Laplace transforms, we need to use the formula from Theorem 8.7:

$$\mathscr{L}[(-x)^n f(x)] = \frac{d^n}{ds^n} F(s)$$

For example, when $y(0) = a$ and $y'(0) = b$ are the given initial conditions, then combining this formula with

$$\mathscr{L}\left[\frac{d^n}{dx^n} f(x)\right] = s^n \mathscr{L}[f(x)] - s^{n-1} f(0) - s^{n-2} f'(0)$$
$$- \cdots - s f^{(n-2)}(0) - f^{(n-1)}(0)$$

from Theorem 8.5, we obtain

$$\mathscr{L}[xy''] = -\frac{d}{ds}\{s^2 \mathscr{L}[y] - as - b\} = -s^2 \frac{d\mathscr{L}[y]}{ds} - 2s\mathscr{L}[y] + a$$

This method is illustrated in Example 36.

Example 36

Consider

$$y'' + xy' - 2y = 4; \qquad y(0) = y'(0) = 0$$

Taking transforms on both sides, we obtain

$$s^2 \mathscr{L}[y] - \frac{d}{ds}\{s\mathscr{L}[y] - y(0)\} - 2\mathscr{L}[y] = \frac{4}{s}$$

or

$$s^2 \mathscr{L}[y] - s \cdot \frac{d\mathscr{L}[y]}{ds} - \mathscr{L}[y] - 2\mathscr{L}[y] = \frac{4}{s}$$

$$-s \cdot \frac{d\mathscr{L}[y]}{ds} - (3 - s^2)\mathscr{L}[y] = \frac{4}{s}$$

$$\frac{d}{ds}\mathscr{L}[y] + \frac{3 - s^2}{s}\mathscr{L}[y] = -\frac{4}{s^2}$$

or

$$\mathscr{L}'[y] + \frac{3 - s^2}{s}\mathscr{L}[y] = -\frac{4}{s^2}$$

(*Note*: $\mathscr{L}'[y]$ is the notation for $\frac{d}{ds}\mathscr{L}[y]$.) Since $\mathscr{L}(y) \equiv F(s)$, this is a linear equation that may be solved for $\mathscr{L}(y)$ by first finding the integrating factor.

$$I(s) = e^{\int \frac{3 - s^2}{s} ds} = e^{3\ln|s| - s^2/2} = s^3 e^{-s^2/2}$$

$$\therefore \quad \frac{d}{ds}\left[(s^3e^{-s^2/2})\mathscr{L}[y]\right] = -\frac{4}{s^2}(s^3e^{-s^2/2}) = -4se^{-s^2/2}$$

Then $$s^3e^{-s^2/2}\cdot\mathscr{L}[y] = 4e^{-s^2/2} + C$$

or $$\mathscr{L}[y] = \frac{4}{s^3} + \frac{Ce^{s^2/2}}{s^3}$$

However, $$\lim_{s\to\infty}\left(\frac{e^{s^2/2}}{s^3}\right) \neq 0$$

so we must have $C = 0$. Hence $\mathscr{L}[y] = 4/s^3$ provides

$$y(x) = \mathscr{L}^{-1}[4s^{-3}] = 4\left(\frac{x^2}{2!}\right) = 2x^2$$

This result is easily checked to be the correct function.

Example 37

Consider the linear nonhomogeneous differential equation with variable coefficients

$$xy'' - (2 + x)y' + 3y = x - 1; \qquad y(0) = 0$$

Taking transforms we get

$$\{-s^2\mathscr{L}'[y] - 2s\mathscr{L}[y]\} - 2s\mathscr{L}[y]$$

$$+ \{s\mathscr{L}'[y] + \mathscr{L}[y]\} + 3\mathscr{L}[y] = \frac{1}{s^2} - \frac{1}{s}$$

$$(-s^2 + s)\mathscr{L}'[y] - (4s - 4)\mathscr{L}[y] = \frac{1-s}{s^2}$$

$$s\mathscr{L}'[y] + 4\mathscr{L}[y] = \frac{1}{s^2}$$

$$\mathscr{L}'[y] + \frac{4}{s}\mathscr{L}[y] = \frac{1}{s^3}$$

The integrating factor is s^4, so we obtain

$$\{s^4\mathscr{L}[y]\}' = s$$

Hence $$\mathscr{L}[y] = \frac{1}{2s^2} + \frac{C}{s^4}$$

and thus

$$y = \frac{1}{2}x + Cx^3$$

If $y'(0)$ had also been specified, we could evaluate C.

Miscellaneous Exercises for Chapter 8

1. Find $\mathscr{L}^{-1}\left[\dfrac{5s}{s^2+8}\right]$.
2. Find $\mathscr{L}[x \sin \beta x]$.
3. Find $\mathscr{L}^{-1}\left[\dfrac{6s+2}{s^2+4}\right]$.
4. Find $\mathscr{L}^{-1}\left[\dfrac{e^{-4s}}{s}\right]$.
5. Find $\mathscr{L}^{-1}\left[\dfrac{s}{(s-2)^3}\right]$.
6. Find $\mathscr{L}^{-1}\left[\dfrac{1}{(s^2-1)^2}\right]$.
7. Find $\mathscr{L}^{-1}\left[\dfrac{s+3}{s^2+6s+13}\right]$.
8. Find $\mathscr{L}^{-1}\left[e^{-2s}\left(\dfrac{s}{s^2+4s+3}\right)\right]$.
9. Find $\mathscr{L}\left[\int_0^x \sin 2t\, dt\right]$.
10. Find $\mathscr{L}^{-1}\left[\dfrac{16s^2}{(s-3)(s+1)^2}\right]$.
11. Find $\mathscr{L}^{-1}\left[\dfrac{1}{(s+1)^3}\right]$.
12. Find $\mathscr{L}^{-1}\left[\dfrac{1}{s^2+6s+13}\right]$.

In Exercises 13–23 use the methods of Laplace transforms to solve the initial value problem.

13. $y' + y = e^{-t}$; $y(0) = 0$
14. $y' + 2y = e^{-t}$; $y(0) = 0$
15. $y' - 3y = 0$; $y(0) = -4$
16. $y'' + 2y' + 5y = 0$; $y(0) = 1$, $y'(0) = 5$
17. $y'' + 4y = \sin t$; $y(0) = 0$, $y'(0) = 0$
18. $y'' + 0.01y' + y = 2$; $y(0) = 0$, $y'(0) = 0$
19. $D^2(D^2 + 1)y = 3e^{-t}$; $y(0) = 0$, $y'(0) = 0$
20. $y'' + 2y' + y = 2e^{-x}$; $y(0) = 0$, $y'(0) = 1$

21. $y'' + 2y' + 10y = 0;\ y(0) = 0,\ y'(0) = 1$
22. $4y'' + 7y = 0;\ y(0) = 5,\ y'(0) = 1$
23. $y'' + \alpha^2 y = 0;\ y(0) = 1,\ y'(0) = 0$
24. In the circuit equation

$$L\frac{d^2i}{dt^2} + \frac{1}{C}i = \begin{cases} -16 & \text{if} \quad 0 < t \le 3 \\ 0 & \text{if} \quad 3 < t \end{cases}$$

if $L = 1$ henry, $C = \frac{1}{16}$ farad and $i(0) = i'(0) = 0$, find $i(t)$.

25. Find $\mathscr{L}[f(x)]$ if $f(x) = \begin{cases} 0 & \text{if} \quad x \le 1 \\ 4 & \text{if} \quad 1 < x \le 3 \\ 2 & \text{if} \quad 3 < x \end{cases}$

26. Find $\mathscr{L}[f(x)]$ if $f(x) = \begin{cases} 0 & \text{if} \quad x \le 3 \\ x & \text{if} \quad x > 3 \end{cases}$

27. Find $\mathscr{L}[f(x)]$ if $f(x) = \begin{cases} x & \text{if} \quad 0 < x \le 2 \\ 4 - x & \text{if} \quad 2 < x \le 4 \\ 0 & \text{if} \quad x > 4 \end{cases}$

28. Use the convolution theorem in the solution of $y'' + 4y = \cos 2x$; $y(0) = 1,\ y'(0) = -2$.

29. Let

$$f(x) = \begin{cases} 1 & \text{if} \quad 0 \le x < 1 \\ 2 & \text{if} \quad 1 \le x < 2 \\ \vdots & \\ n & \text{if} \quad n - 1 \le x < n \\ \vdots & \end{cases}$$

Find $\mathscr{L}[f(x)]$.

30. Define $K(t) = H_1(t) + H_2(t) + H_3(t) + \cdots$.
    (a) Graph $K(t)$.
    (b) Graph $t - K(t - 1)$.
    (c) Graph $K(t) - t$.
    (d) Find the Laplace transform of each of the functions in (a), (b), and (c).
31. Use the definition of the Laplace transform to find $\mathscr{L}[\sin ax]$. (Use integration by parts twice.)
32. Use the formula for the Laplace transform of a periodic function to find $\mathscr{L}[\cos x]$.
33. If $\mathscr{L}[f(x)] = F(s)$ and if

$$\lim_{x \to 0^+} \frac{f(x)}{x}$$

exists, then

$$\mathscr{L}\left[\frac{f(x)}{x}\right] = \int_s^\infty F(t)\,dt$$

Use this to find

$$\mathscr{L}\left[\frac{\sin x}{x}\right]$$

and

$$\mathscr{L}\left[\frac{\sin 2x}{x}\right]$$

34. Prove that $[f(x) + g(x)] * h(x) = [f(x) * h(x)] + [g(x) * h(x)]$.

35. Find the Laplace transform of each of the following functions:

(a) $\int_0^x \sin(x - t)\cos t\,dt$

(b) $\int_0^x t\cos(x - t)\,dt$

(c) $\int_0^x t^2 e^{-t}\sin(x - t)\,dt$

(d) $\int_0^x (x - t)^2 \cos 4t\,dt$

36. Solve the following integral equations for $y(x)$:

(a) $y(x) = x - 3\int_0^x y(t)\sin(x - t)\,dt$

(b) $y(x) = e^{-2x} + 4\int_0^x y(t)\cos 2(x - t)\,dt$

(c) $y(x) = \sin 2x - \int_0^x (x - t)y(t)\,dt$

(d) $y(x) = \sin x - \int_0^x (x - t)y(t)\,dt$

37. Given $\mathscr{L}[e^{ax}] = 1/(s - a)$ and $\mathscr{L}[e^{-ax}] = 1/(s + a)$, find

$$\mathscr{L}[\cosh ax] = \mathscr{L}\left[\frac{e^{ax} + e^{-ax}}{2}\right]$$

and

$$\mathscr{L}[\sinh ax] = \mathscr{L}\left[\frac{e^{ax} - e^{-ax}}{2}\right]$$

38. Use the results of Exercise 37 and the identities $\cosh(i \cdot ax) = \cos ax$ and $\sinh(i \cdot ax) = i \sin ax$ to find $\mathscr{L}[\cos ax]$ and $\mathscr{L}[\sin ax]$.
39. Find $\mathscr{L}[xe^{ax}]$ and $\mathscr{L}[xe^{-ax}]$. Combine these to find $\mathscr{L}[x \cosh ax]$ and $\mathscr{L}[x \sinh ax]$.
40. Use the results of Exercise 39 and replace a by ia to obtain the formulas for $\mathscr{L}[x \cos ax]$ and $\mathscr{L}[x \sin ax]$.

Chapter 9

Systems of Differential Equations

9.1 Introduction: Model Formulation

Many of the applications problems in science and engineering lead to differential equations involving more than one unknown function. If the motion of a particle in the plane is being studied, we must determine both x and y as functions of time t. In most cases we would have distance, velocity, and acceleration as functions of time in both directions. This would provide more than one equation in more than one dependent variable.

In previous chapters we studied the methods of solution for one differential equation in one unknown function. In this chapter we will extend our capabilities to a system (a collection) of differential equations consisting of n equations in n unknown functions. We will limit our attention to linear systems and reserve for Chapter 10 the numerical solutions to the more complicated nonlinear cases. In our solution techniques we will return to differential operators and operator methods that were studied earlier.

We will first illustrate the model formulation of the systems of differential equations for some application situations.

Example 1

A particle moving in the plane is constrained by forces such that its horizontal velocity at a point (x, y) is equal to the sum of its coordinate values, and its vertical velocity at that point is twice the first coordinate value minus the second coordinate value. There is no acceleration.

Model Formulation: Using t as the single independent variable, we may write

$$\frac{dx}{dt} = x + y$$

$$\frac{dy}{dt} = 2x - y$$

This is a system of two first-order equations in two unknown functions.

Example 2

In a community water purification system, two holding tanks are interconnected by inlet and outlet pipes as shown in Figure 9.1: Each tank holds 10^6 gallons and the flow capacity of each pipe is indicated. The impure water from the city sanitary system flows in through pipe A and after circulation through the purification process flows into Lake Pleasant through pipe B. Suppose at time $t = 0$, a spill of 40 pounds of toxic chemical C enters tank A. Suppose additionally that the chemical cannot be filtered out but may be slowly released through outlet B by a diverter valve to a holding pond. How long must the outlet stream be diverted before the concentration of chemical C falls below the level of .001 pounds/gallon and once again the water flows into Lake Pleasent?

Model Formulation: Let $x(t)$ denote the amount of chemical C in tank A and $y(t)$ denote the amount of C in tank B at any time t. The

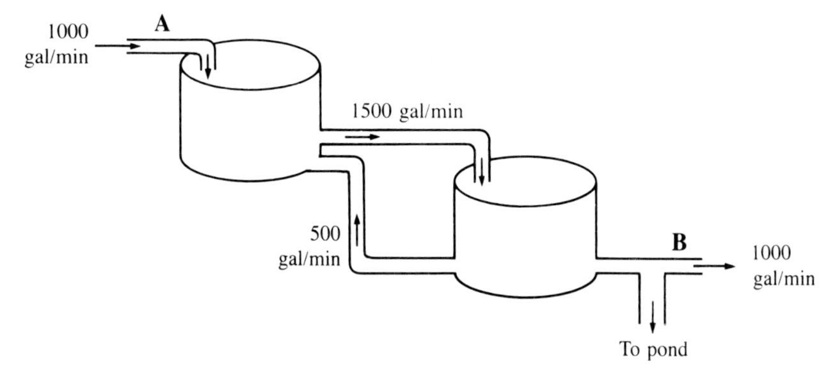

Figure 9.1

concentration of chemical C in each tank is $x/10^6$ and $y/10^6$, respectively, so the equations that are used to solve for x and y are derived from these facts. To tank A is added 500 gallons of concentration $(y/10^6)$ each minute and from tank A is deleted 1500 gallons at concentration $(x/10^6)$. The other 1000 gallons of chemical-free water flows in through pipe A. To tank B is added 1500 gallons at concentration $(x/10^6)$ and deleted is 500 gallons into tank A and 1000 gallons into pipe B both at concentration $(y/10^6)$. Therefore

$$\frac{dx}{dt} = 500\left(\frac{y}{10^6}\right) - 1500\left(\frac{x}{10^6}\right)$$

$$\frac{dy}{dt} = 1500\left(\frac{x}{10^6}\right) - 1500\left(\frac{y}{10^6}\right)$$

Initial conditions are that $x = 40$ and $y = 0$ at time $t = 0$. When this system has been solved for $x(t)$ and $y(t)$ uniquely, the value of y would be set at .001 and that equation solved for time t.

This system with its initial conditions is a simple mathematical model whose solution helps to control water quality. The model will accept varying pollutant amounts and can be changed to reflect varying flow rates.

Example 3

A pair of second-order linear equations is obtained by considering the motion of a mechanical system of two masses M_1 and M_2 and three springs with spring constants k_1, k_2, and k_3, as illustrated in Figure 9.2.

Model Formulation: If we assume the motion is on a smooth, frictionless plane, the only forces on the masses are the spring forces. If x_1 and x_2 denote displacement from the equilibrium position, then Newton's second law provides

$$M_1 \frac{d^2x_1}{dt^2} = -k_1x_1 + k_2(x_2 - x_1)$$

$$M_2 \frac{d^2x_2}{dt^2} = -k_2(x_2 - x_1) - k_3x_2$$

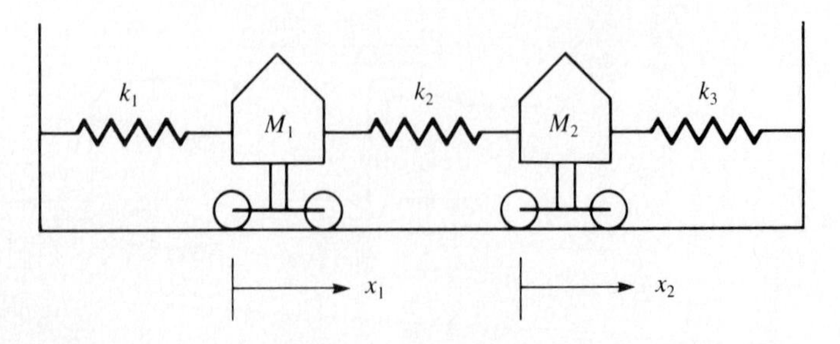

Figure 9.2

This system may be solved if the displacement and velocity for each mass is given at time $t = 0$.

Example 4

In an effort to better understand the functions of the various parts of the human heart and their interrelations, an electrical model of the heart and arteries was constructed by Sidney Roston. The electric circuit is advantageous as a model because its behavior is easily measured and understood.

First we will look at a model of the heart and its arteries, which is diagramed in Figure 9.3. The battery B, resistance R_1, and capacitance C_1 represent the heart. At the end of the rhythmic expansion of the heart cavities, during which they fill with blood, the voltage across C_1 is taken as V_0. No current i flows through B during the expansion of the heart. In an actual model a simple relay switch with a repetitive timing device will easily simulate this feature. As the heart begins its contraction to pump the blood, the battery B is activated, i_5 becomes positive, and the voltage V_1 across C_1 increases. When this voltage reaches adequate magnitude, the special thyratron tube becomes conductive and i_2 becomes positive. The inductance L and accompanying resistance R_3 prevent any initial surge in the value of i_2, and their function is in agreement with the corresponding gradual increase in blood flow from the heart when the aortic valve opens. Because i_2 increases slowly, V_1 continues to increase. The nature of this increase depends on B and L. A healthy heart represented by high B and low L has a peak value of V_1 early in the contraction stage and then a rapid decline. A failing heart cannot significantly increase V_1 and hence the peak occurs later in the contraction. This may be detected by an instrument and the strength of the heart determined. Also in the healthy heart, the outward push would occur early in the contraction, providing a stronger pulse;

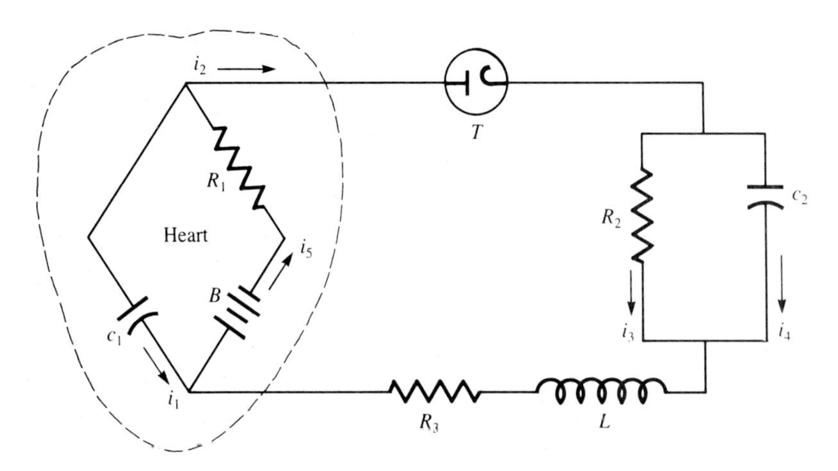

Figure 9.3

whereas in the failing heart the flow occurs more uniformly over the contraction period.

The swelling of the large arteries is represented by C_2. Because of peripheral resistance R_2, the blood that leaves the heart partially expands the artery and partially flows through it. The accumulation of charge in C_2 and the increase in i_3 represent these aspects of the blood flow.

When the voltage across C_1 eventually decreases to V_0, the current flow through the thyratron abruptly ceases; and during the pause between flows, C_2 discharges exponentially through R_2 just as the extended artery gives rise to blood flow during the expansion phase via peripheral resistance.

Model Formulation: The mathematical representation of the circuit after the flow through the thyratron begins is straightforward via the circuit laws:

$$\frac{q_1}{C_1} + R_1 i_5 = B \tag{1}$$

where q_1 is the charge in C_1.

$$L\frac{di_2}{dt} + R_3 i_2 + R_2 i_3 - \frac{q_1}{C_1} = 0 \tag{2}$$

$$R_2 i_3 - \frac{q_2}{C_2} = 0 \tag{3}$$

where q_2 is the charge in C_2.

$$i_1 + i_2 - i_5 = 0 \tag{4}$$

$$i_3 + i_4 - i_2 = 0 \tag{5}$$

The differentiation of equations (1), (2), and (3) to remove the charge values yields

$$\frac{i_i}{C_1} + R_1\frac{di_5}{dt} = 0 \tag{6}$$

$$L\frac{d^2 i_2}{dt^2} + R_3\frac{di_2}{dt} + R_2\frac{di_3}{dt} - \frac{i_1}{C_1} = 0 \tag{7}$$

$$R_2\frac{di_3}{dt} - \frac{i_2}{C_2} = 0 \tag{8}$$

Equations (4) through (8) now become a system of differential equations to be solved for the currents i_1, i_2, i_3, i_4, and i_5 and thus a prediction of the blood flow.

Example 5

Parasitic infections are among the most important world health problems. Malaria leads the list, followed by schistosomiasis and other worm infections. These diseases are natural occurrences in most of the tropical and subtropical areas of the world. George Macdonald* derived a mathematical model that described the complicated interaction between parasitic worms, human hosts, and secondary hosts (snails). These schistosomes have a complicated life history. Male and female worms must mate within a human host. Thereafter, the female deposits eggs, some of which leave the host in the feces. Other eggs may be trapped in various organs of the body and these create the symptoms of the disease. If an egg comes into contact with fresh water, a larva hatches. To continue the cycle, the larva must penetrate the body of a snail. Once a snail is infected, a new variety of larvae is quickly produced. As these swim freely, they seek to penetrate the skin of a human in order to complete the cycle.

Model Formulation: The quantities of interest are the mean number of worms infecting a human, called m, and the number of infected snails, denoted by I. If we assume the death rate of snails, denoted by δ, is constant and the rate at which snails are infected is proportional to the number of healthy snails, then

$$\frac{dI}{dt} = -\delta I + B \cdot P(m)(S - I)$$

where S is the total number of snails (which is considered constant), B is a proportionality factor independent of m, and $P(m)$ is the number of paired worms per human host. Then by similar analysis, we get

$$\frac{dm}{dt} = -rm + \frac{A}{S} I$$

where r is the death rate for worms inside the human host and A is a factor taking into account the number of secondary larvae produced per infected snail. Scientific investigation notes that m changes very slowly since the life span of a worm inside a human host is about three years.

This pair of first-order equations may now be carefully studied by mathematical methods. We would use an expression for $P(m)$ based on the Poisson probability distribution for the worms and proceed to analyze the equation for equilibrium solutions (i.e., those for which $dm/dt = 0$ and $dI/dt = 0$).

---

* G. Macdonald, "The Dynamics of Helminth Infections, with Special Reference to Schistosomes," *Transactions of the Royal Society of Tropical Medicine and Hygiene*, 59 (1965), pp. 489–506.

These five problems serve as typical examples of how systems of differential equations are used in research in many phases of science and engineering. As was made evident in the examples, the model formulation lies mainly in the particular scientific field, but the solution process and the analysis of the solution lie in the realm of expertise of the differential-equations student. We hope to sharpen this skill in this chapter.

9.2 Linear Systems

To begin our study of the process of solving a system of differential equations, we first need to define some terms and introduce some special techniques.

Definition: A **general linear system** of two first-order differential equations in unknown functions $x(t)$ and $y(t)$ is of the form

$$\begin{cases} A_1(t)\dfrac{dx}{dt} + A_2(t)\dfrac{dy}{dt} + A_3(t)x + A_4(t)y = f(t) \\ \\ B_1(t)\dfrac{dx}{dt} + B_2(t)\dfrac{dy}{dt} + B_3(t)x + B_4(t)y = g(t) \end{cases} \tag{9}$$

This system has **constant coefficients** if $A_i(t)$ and $B_i(t)$, $i = 1, 2, 3, 4$, are constants. This system is called **homogeneous** if $f(t) = g(t) = 0$.

Example 6

An example of a linear nonhomogeneous system of first-order equations with constant coefficients would be

$$\begin{cases} 2\dfrac{dx}{dt} + 4\dfrac{dy}{dt} - 5x + 3y = t + 1 \\ \\ 4\dfrac{dx}{dt} - 3\dfrac{dy}{dt} + x - 2y = e^t \end{cases}$$

The above definitions can easily be extended to systems of n equations in n unknown functions.

Definition: A system of first-order equations will be called a **standardized first-order system** if it can be put into the special form

$$\begin{cases} \dfrac{dx_1}{dt} = f(x_1, x_2, t) \\ \\ \dfrac{dx_2}{dt} = g(x_1, x_2, t) \end{cases}$$

If f and g are linear combinations of x_1 and x_2, the system would be first order and linear; and if t does not specifically appear, it would be homogeneous and would have constant coefficients.

Example 7

The system

$$\frac{dx_1}{dt} = 3x_1 - 4x_2$$

$$\frac{dx_2}{dt} = 4x_1 + x_2$$

is a standardized first-order system that is linear and homogeneous and has constant coefficients.

An important fundamental property of a first-order system is its relationship to a higher-order differential equation in one unknown function. An nth-order linear differential equation may be broken down into a linear first-order system with n equations. The only requirement is that the highest derivative can be isolated on one side of the equation, as in

$$\frac{d^{(n)}y}{dx^n} = A_1(x)\frac{d^{(n-1)}y}{dx^{n-1}} + \cdots + A_{n-1}(x)\frac{dy}{dx} + A_n(x)y + F(x)$$

The technique requires us to make the following variable changes:

$$u_1 = y, \quad u_2 = \frac{dy}{dx}, \quad u_3 = \frac{d^2y}{dx^2},$$

$$u_4 = \frac{d^3y}{dx^3}, \ldots, u_{n-1} = \frac{d^{(n-2)}y}{dx^{n-2}}, \quad u_n = \frac{d^{(n-1)}y}{dx^{n-1}}$$

By combining these we obtain

$$\frac{du_1}{dx} = \frac{dy}{dx} = u_2, \quad \frac{du_2}{dx} = \frac{d}{dx}\left(\frac{dy}{dx}\right) = \frac{d^2y}{dx^2} = u_3,$$

$$\frac{du_3}{dx} = u_4, \ldots, \frac{du_{n-1}}{dx} = u_n$$

and the differential equation becomes

$$\frac{d^{(n)}y}{dx^n} = \frac{d}{dx}\left(\frac{d^{(n-1)}y}{dx^{n-1}}\right) = \frac{d}{dx}(u_n) = A_1u_n + A_2u_{n-1} + \cdots + A_nu_1 + F(x)$$

We now write these in an organized way and obtain

$$\left\{\begin{aligned}
\frac{du_1}{dx} &= u_2 \\
\frac{du_2}{dx} &= u_3 \\
\frac{du_3}{dx} &= u_4 \\
&\vdots \\
\frac{du_{n-1}}{dx} &= u_n \\
\frac{du_n}{dx} &= A_n u_1 + A_{n-1} u_2 + \cdots + A_2 u_{n-1} + A_1 u_n + F(x)
\end{aligned}\right\}$$

which is a standardized first-order system. If $F(x) = 0$ in this equation, the system is linear and homogeneous. Note that we have converted an nth-order differential equation into a system of n first-order differential equations.

Example 8

Consider

$$3y''' - 2xy'' + (x + 2)y' + 2y = e^x$$

This third-order equation is reduced to a system of three first-order equations by using the variable change $u_1 = y$, $u_2 = y'$, $u_3 = y''$. With this set we get $u_3' = y'''$ and thus from the given differential equation we obtain

$$3u_3' = 2xu_3 - (x + 2)u_2 - 2u_1 + e^x$$

Combining this with the earlier variable-change equations provides

$$\begin{aligned}
u_1' &= u_2 \\
u_2' &= u_3 \\
u_3' &= -\frac{2}{3}u_1 - \left(\frac{x+2}{3}\right)u_2 + \frac{2}{3}xu_3 + \frac{1}{3}e^x
\end{aligned}$$

which is a standardized first-order system. The solution techniques of systems of equations, which will be studied shortly, will give us $u_1(x)$, $u_2(x)$, and $u_3(x)$, which actually are the solutions $y(x)$, $y'(x)$, and $y''(x)$ for the original differential equation. If the given equation had resulted from motion of a particle, we would then have the position, velocity, and acceleration at any time t.

Definition: A solution of a system of equations (9) is an ordered set of functions $\{x(t), y(t)\}$, each defined on a common interval I, such that all equations in the system become identities when the functions are appropriately inserted into the system.

In order to put the studies in this chapter on a sound theoretical basis, we now state without proof a basic theorem.

Theorem 9.1 (Existence and Uniqueness):

Consider the system

$$\frac{dy_1}{dt} = f_1(t, y_1, y_2)$$

$$\frac{dy_2}{dt} = f_2(t, y_1, y_2)$$

with the initial conditions (see Figure 9.4)

$$y_1(t_0) = A_1, \quad y_2(t_0) = A_2$$

Let the functions f_1 and f_2 be continuous in a rectangular region of space R defined by

$$\{-k_0 < t - t_0 < k_0, -k_1 < y_1 - A_1 < k_1, -k_2 < y_2 - A_2 < k_2\}$$

In R, let each function satisfy a Lipschitz condition:

$$|f_1(t, y_1, y_2) - f_1(t, \hat{y}_1, \hat{y}_2)| \leq K(|y_1 - \hat{y}_1| + |y_2 - \hat{y}_2|)$$

and

$$|f_2(t, y_1, y_2) - f_2(t, \hat{y}_1, \hat{y}_2)| \leq K(|y_1 - \hat{y}_1| + |y_2 - \hat{y}_2|)$$

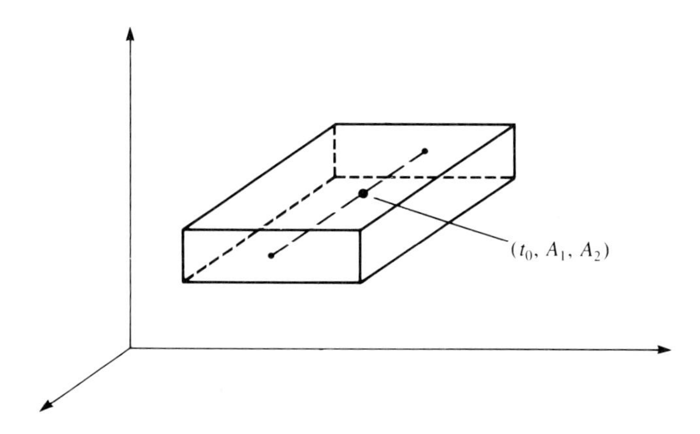

Figure 9.4

where K is a positive constant and (t, y_1, y_2) and $(t, \hat{y}_1, \hat{y}_2)$ are any two points in R. Then there exists an interval I: $-\alpha < t - t_0 < \alpha$ in which there is one and only one set of continuous functions $\{y_1(t), y_2(t)\}$ with continuous derivatives in I that identically satisfies the given system with initial conditions.

We now have the available knowledge to adequately develop the solution technique for a simple first-order system.

The operator notation introduced in Section 3.2 proves to be a handy laborsaving device for systems. We replace each derivative by its operator notation and collect the multipliers of each unknown function. The process closely parallels the high-school algebra technique of a simultaneous solution of algebraic equations.

Example 9

The standardized first-order system

$$\frac{dx_1}{dt} = 2x_1 + 3x_2$$

$$\frac{dx_2}{dt} = \frac{1}{3}x_1 + 2x_2$$

is rewritten as

$$Dx_1 = 2x_1 + 3x_2$$

$$Dx_2 = \frac{1}{3}x_1 + 2x_2$$

and then, collecting terms and transposing in order to line up the variables, we write it as

$$(D - 2)x_1 - 3x_2 = 0$$

$$-\frac{1}{3}x_1 + (D - 2)x_2 = 0$$

We now solve this pair by the familiar elimination process of elementary algebra for homogeneous equations. Using our operators as algebraic quantities, which is valid following earlier studies, we multiply the first equation by $\frac{1}{3}$ and the second by $(D - 2)$ to allow elimination of the variable $x_1(t)$:

$$\frac{1}{3}(D - 2)x_1 - x_2 = 0$$

$$-\frac{1}{3}(D - 2)x_1 + (D - 2)(D - 2)x_2 = 0$$

Adding these gives

$$[(D^2 - 4D + 4) + (-1)]x_2 = 0$$

$$(D^2 - 4D + 3)x_2 = 0$$

$$(D - 3)(D - 1)x_2 = 0$$

This is a familiar second-order linear homogeneous differential equation with constant coefficients, which is easily solved to

$$x_2(t) = C_1 e^{3t} + C_2 e^t$$

From the second equation of the original pair we have

$$\frac{1}{3} x_1 = \frac{dx_2}{dt} - 2x_2$$

or

$$x_1 = 3\frac{dx_2}{dt} - 6x_2$$

$$= 3(3C_1 e^{3t} + C_2 e^t) - 6(C_1 e^{3t} + C_2 e^t)$$

$$= 3C_1 e^{3t} - 3C_2 e^t$$

This pair of functions $\{3C_1 e^{3t} - 3C_2 e^t, C_1 e^{3t} + C_2 e^t\}$ is easily checked as the valid solution to the original system.

In Example 9 we had a homogeneous system of equations. While these are usually easier to solve than the nonhomogeneous case, we proceed to the more difficult cases without altering our approach.

Example 10

Consider the system

$$\frac{dy}{dx} = x^2 + \frac{1}{3} z$$

$$\frac{dz}{dx} = 3x^2 + 6y + z$$

Writing this in operator notation gives

$$Dy - \frac{1}{3} z = x^2$$

$$-6y + (D - 1)z = 3x^2$$

We may now eliminate either y or z by algebra; we will arbitrarily choose z. Multiply the first equation by $(D - 1)$ and the second by $\frac{1}{3}$.

$$(D - 1)Dy - \frac{1}{3}(D - 1)z = (D - 1)x^2 = Dx^2 - x^2 = 2x - x^2$$

$$-2y + \frac{1}{3}(D - 1)z = x^2$$

In the first operation, because of the nonhomogeneous nature of the system, we needed to exercise *extra care* when multiplying on the right by the operator. Don't fail to do this important step! We now add these to obtain

$$[D(D-1)-2]y = 2x$$

$$(D^2 - D - 2)y = 2x$$

$$(D-2)(D+1)y = 2x$$

This is a second-order linear nonhomogeneous equation that is solvable by the method of undetermined coefficients to

$$y(x) = C_1 e^{2x} + C_2 e^{-x} - x + \frac{1}{2}$$

Now using the first of the original pair, we may most easily obtain the solution function for the other dependent variable:

$$\frac{1}{3} z = \frac{dy}{dx} - x^2$$

$$z = 3\frac{dy}{dx} - 3x^2$$

$$= 3(2C_1 e^{2x} - C_2 e^{-x} - 1) - 3x^2$$

or

$$z(x) = 6C_1 e^{2x} - 3C_2 e^{-x} - 3 - 3x^2$$

The elimination method usually works and provides us with a solution. It will, in fact, work even if one or more system equation is not first order.

Example 11

In the late nineteenth century the physicist J. J. Thomson experimented to determine the ratio m/e of the mass m of the electron to the charge e of an electron. (Do not confuse this e with the base of natural logarithms.) The electrons were subjected to an electric field of intensity E and a magnetic field of intensity H. The motion equations used were

$$m\frac{d^2x}{dt^2} + He\frac{dy}{dt} = Ee$$

$$m\frac{d^2y}{dt^2} - He\frac{dx}{dt} = 0$$

The initial conditions at $t = 0$ are $x = y = dx/dt = dy/dt = 0$. To determine the path of the electron, we solve this system. In the operator form we have

$$mD^2x + HeDy = Ee$$

$$-HeDx + mD^2y = 0$$

Multiplying the second equation by mD and the first by He, we get

$$mHeD^2x + H^2e^2Dy = HEe^2$$
$$-mHeD^2x + m^2D^3y = 0$$

Adding gives

$$m^2D^3y + H^2e^2Dy = HEe^2$$

or

$$D^3y + \left(\frac{He}{m}\right)^2 Dy = HE\left(\frac{e}{m}\right)^2$$
$$\left[D^3 + \left(\frac{He}{m}\right)^2 D\right]y = HE\left(\frac{e}{m}\right)^2$$
$$D\left[D^2 + \left(\frac{He}{m}\right)^2\right]y = HE\left(\frac{e}{m}\right)^2$$

Solving this third-order nonhomogeneous differential equation by methods of Chapter 3 gives

$$y = C_1\cos\left(\frac{He}{m}\right)t + C_2\sin\left(\frac{He}{m}\right)t + C_3 + \frac{E}{H}t$$

The initial conditions $y = dy/dt = 0$ when $t = 0$ give

$$C_2 = -\frac{Em}{eH^2}, \quad C_1 = -C_3$$

The second equation of the original system is now revised to

$$Dx = \frac{m}{He}D^2y$$
$$= \frac{m}{He}\left[-\left(\frac{He}{m}\right)^2 C_1\cos\left(\frac{He}{m}\right)t - \left(\frac{He}{m}\right)^2 C_2\sin\left(\frac{He}{m}\right)t\right]$$

Since $dx/dt = 0$ when $t = 0$ gives $C_1 = 0$, we have

$$x(t) \equiv \int_0^t Dx = -\left(\frac{He}{m}\right)C_2\int_0^t \sin\left(\frac{He}{m}\right)t\,dt$$
$$= \left(\frac{He}{m}\right)C_2\left[\left(\frac{m}{He}\right)\cos\left(\frac{He}{m}\right)t\Big|_0^t\right] = C_2\left[\cos\left(\frac{He}{m}\right)t - 1\right]$$

Hence the unique results for the parametric equations of the path are

$$x(t) = \frac{Em}{eH^2}\left[1 - \cos\left(\frac{He}{m}\right)t\right]$$

and

$$y(t) = \frac{E}{H}t - \frac{Em}{eH^2}\sin\left(\frac{He}{m}\right)t = \frac{Em}{eH^2}\left[\left(\frac{He}{m}\right)t - \sin\left(\frac{He}{m}\right)t\right]$$

This is the parametric pair of a cycloid, the path traced out by a point on the rim of a rolling wheel.

The elimination process also works on systems of more than two equations and sometimes does not involve much elimination.

Example 12

In the successive radioactive changes of thorium 228 to radium 224 to radon 220, let x_1, x_2, and x_3 denote the number of atoms of the three successive substances at time t. Let the numbers at time $t = 0$ be $x_1(0) = A_1$, $x_2(0) = 0$, $x_3(0) = 0$. The law of radioactive decay gives the system

$$\begin{cases} \dfrac{dx_1}{dt} = -k_1x_1 \\ \dfrac{dx_2}{dt} = k_1x_1 - k_2x_2 \\ \dfrac{dx_3}{dt} = k_2x_2 \end{cases}$$

where the values of the k's are determined from the half-life values:

Th-228 $\xrightarrow{\alpha}$ Ra-224 in 1.913 years

Ra-224 $\xrightarrow{\alpha}$ Rn-220 in 3.64 days

Rn-220 $\rightarrow$ Po-216 in 55.6 seconds

Adding the equations in this system gives

$$\frac{dx_1}{dt} + \frac{dx_2}{dt} + \frac{dx_3}{dt} = D(x_1 + x_2 + x_3) = 0$$

This implies $x_1 + x_2 + x_3 = C$, a constant, but

$$x_1(0) + x_2(0) + x_3(0) = A_1 + 0 + 0 = C$$

$$\therefore \quad x_1(t) + x_2(t) + x_3(t) = A_1$$

which expresses conservation of mass.

Arranging these equations in operator notation, we obtain the equivalent system

$$\begin{aligned} (D + k_1)x_1 \qquad\qquad\qquad &= 0 \\ -k_1x_1 + (D + k_2)x_2 \qquad &= 0 \\ -k_2x_2 + Dx_3 &= 0 \end{aligned}$$

The first differential equation involves only one unknown function, and its solution by methods of Chapter 1 is

$$x_1(t) = C_1e^{-k_1t}$$

Since $x_1(0) = A_1$, $C_1 = A_1$; thus $x_1(t) = A_1 e^{-k_1 t}$. Putting this value into the second equation gives

$$(D + k_2)x_2 = k_1 A_1 e^{-k_1 t}$$

This is a first-order linear differential equation in the variable $x_2(t)$. We again use a method of Chapter 1 to obtain

$$x_2(t) = \left(\frac{k_1 A_1}{k_2 - k_1}\right) e^{-k_1 t} + C_2 e^{-k_2 t}$$

The initial condition gives

$$0 = \left(\frac{k_1 A_1}{k_2 - k_1}\right) + C_2$$

Hence
$$x_2(t) = \left(\frac{k_1 A_1}{k_2 - k_1}\right)\left(e^{-k_1 t} - e^{-k_2 t}\right)$$

Now equation three of the system may be written as

$$\begin{aligned} Dx_3 &= k_2 x_2 \\ &= \left(\frac{k_1 k_2 A_1}{k_2 - k_1}\right)\left(e^{-k_1 t} - e^{-k_2 t}\right) \end{aligned}$$

Hence

$$\begin{aligned} x_3 \equiv \int_0^t Dx_3 &= \frac{k_1 k_2 A_1}{k_2 - k_1} \int_0^t (e^{-k_1 t} - e^{-k_2 t})\, dt \\ &= \frac{k_1 k_2 A_1}{k_2 - k_1}\left(\frac{1 - e^{-k_1 t}}{k_1} - \frac{1 - e^{-k_2 t}}{k_2}\right) \\ &= A_1\left(1 + \frac{k_1 e^{-k_2 t} - k_2 e^{-k_1 t}}{k_2 - k_1}\right) \end{aligned}$$

To again verify conservation of mass, we add the newly found results to obtain

$$\begin{aligned} x_1(t) + x_2(t) + x_3(t) &= \left[A_1 e^{-k_1 t}\right] + \left[\frac{k_1 A_1}{k_2 - k_1}(e^{-k_1 t} - e^{-k_2 t})\right] \\ &\quad + \left[A_1 + \frac{A_1 k_1 e^{-k_2 t}}{k_2 - k_1} - \frac{A_1 k_2 e^{-k_1 t}}{k_2 - k_1}\right] = A_1 \end{aligned}$$

The graph of this system of equations is shown in Figure 9.5.

A valuable aid in determining the correct form of the solution set to a system of equations is expressed in the following test.

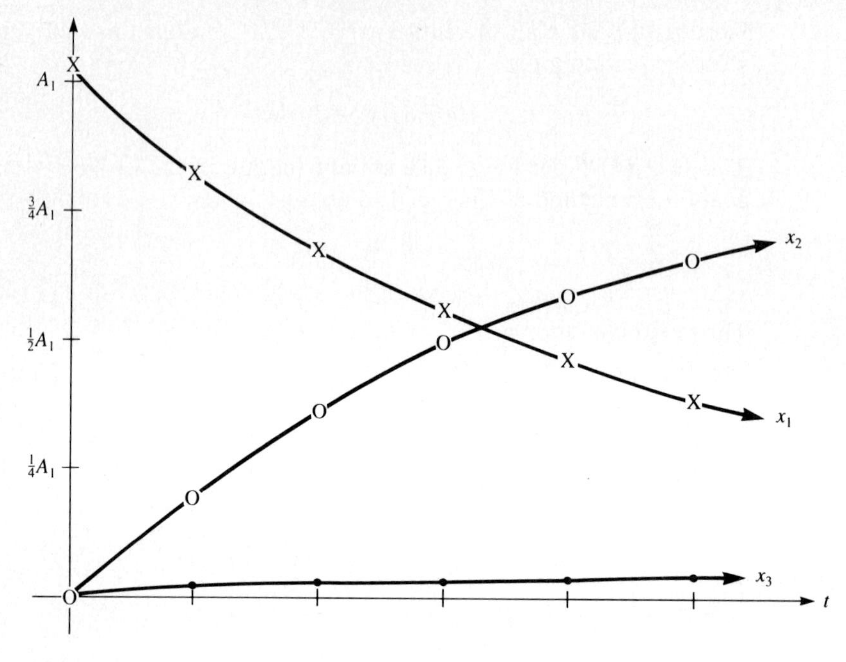

Figure 9.5

Operator Coefficient Determinant Test

It can be shown that for a system of differential equations written in the operator notation and aligned according to variables, such as

$$(D - a)x_1 + (D - b)x_2 = f_1$$
$$(D - c)x_1 + (D - k)x_2 = f_2$$

that the number of arbitrary constants appearing in the general solution to the system is equal to the exponent of the highest power of D in the expansion of the determinant of the operator coefficients of the dependent variables x_1 and x_2. Should this exponent be zero, the system is called degenerate and has either no solution or an infinite number of solutions. Examples 13 and 14 demonstrate how this works.

Example 13

For the system

$$\frac{dy}{dx} = y + z + 2\cos 2x$$

$$\frac{dz}{dx} = 3y + z$$

the equations are rearranged to

$$(D-1)y - \qquad z = 2\cos 2x$$

$$-3y + (D-1)z = 0$$

and the determinant of operator coefficients is

$$\begin{vmatrix} D-1 & -1 \\ -3 & D-1 \end{vmatrix} = (D-1)^2 - 3 = D^2 - 2D - 2$$

so *two* arbitrary constants will appear in the solution. If additional constants appear in the solution, they must be expressed in terms of only two.

Example 14

Consider the system

$$\frac{dx}{dt} + \frac{dy}{dt} - 2x - 2y = t + 1$$

$$\frac{dx}{dt} + \frac{dy}{dt} + 3x + 3y = t + 1$$

In operator notation this system is

$$(D-2)x + (D-2)y = t + 1$$

$$(D+3)x + (D+3)y = t + 1$$

and the operator coefficient determinant is

$$\begin{vmatrix} D-2 & D-2 \\ D+3 & D+3 \end{vmatrix} = 0$$

Therefore the system is degenerate and has no solutions. This implies that there is no pair $\{x(t), y(t)\}$ that will satisfy the two given differential equations simultaneously.

If you find that your solution contains more arbitrary constants than this test dictates, then select a proper number of constants and substitute the solution found into the differential equations and solve for the extra constants in terms of the chosen collection.

Exercises 9.2

1. For Examples 1 through 5 of Section 9.1, use the operator coefficient determinant test to find the number of arbitrary constants that should appear in the general solution.

In Exercises 2–12, solve the system of equations. Use the test to help get the correct form for the general solution.

2. $\dfrac{dx_1}{dt} = -3x_1 + 4x_2, \quad \dfrac{dx_2}{dt} = -2x_1 + 3x_2$

3. $\dfrac{dx}{dt} = 3x - y, \quad \dfrac{dy}{dt} = x + y$

4. $\dfrac{dy}{dx} = 3y - z, \quad \dfrac{dz}{dx} = 3y$

5. $\dfrac{dr}{dt} = \dfrac{1}{2}r + \dfrac{1}{2}s, \quad \dfrac{ds}{dt} = -\dfrac{3}{2}r + \dfrac{5}{2}s$

6. $\dfrac{dx}{dt} = 3x + 2y, \quad \dfrac{dy}{dt} = -x + y$

7. $x_1' = 0.5x_1 + x_2 + t - 1, \quad x_2' = -1.5x_1 + 2.5x_2 - 2t$

8. $\dfrac{dx}{dt} = 2x + y - 2z, \quad \dfrac{dy}{dt} = 3y - 2z, \quad \dfrac{dz}{dt} = 3x + y - 3z$

9. $\dfrac{dx}{dt} = 3x - 3y + z, \quad \dfrac{dy}{dt} = 2x - y, \quad \dfrac{dz}{dt} = x - y + z$

10. $\dfrac{d^2x_1}{dt^2} - \dfrac{dx_2}{dt} + t = 1, \quad \dfrac{dx_1}{dt} + 2\dfrac{dx_2}{dt} = 4e^t + x_1$

11. $\dfrac{d^2y}{dx^2} - y + z - \dfrac{d^2z}{dx^2} + \cos 2x = 0, \quad 2\dfrac{dy}{dx} - \dfrac{dz}{dx} - z = 0$

12. $\begin{cases} Dx_1 - x_2 + x_3 = 0 \\ -x_1 + (D-1)x_2 = 0 \\ -x_1 + (D-1)x_3 = 0 \end{cases}$

13. If x denotes the number of hermit crabs on a one-mile stretch of Florida panhandle beach and y is the number of sea anemones in the same area, and if

$$\frac{dx}{dt} = k_1x(a - x), \quad \frac{dy}{dt} = k_2xy$$

find the relation between the crabs and anemones. If $k_1 = 10^{-5}$, $k_2 = 10^{-6}$, and $a = 10^5$, find $x(t)$ and $y(t)$.

14. The inductance of the primary coil of a transformer is 3 henrys and its resistance is 7 ohms; corresponding figures for the secondary are 6 henrys and 10 ohms; the neutral inductance is 4 henrys. The free oscillations in the two circuits are given by

$$3\frac{di_1}{dt} + 4\frac{di_2}{dt} + 7i_1 = 0; \quad 4\frac{di_1}{dt} + 6\frac{di_2}{dt} + 10i_2 = 0$$

$$i_1(0) = \frac{15}{7}, \quad i_2 = 7$$

where i_1 and i_2 (amp) are the respective currents in the primary and secondary at time t.

(a) Show i_2 continuously decreases but i_1 increases to a maximum and then decreases toward zero as t becomes large.

(b) Find the maximum value of i_1 and the time it takes to reach the maximum.

(c) Find the values of i_1 and i_2 at $t = 0.01$ second.

15. (a) A study, begun during World War I by Frederick W. Lanchester, on the dynamics of combat led to a system of differential equations for the numbers of opposing combatants at time t. An oversimplified version is

$$\frac{dx}{dt} = -\frac{1}{1+E}, \quad \frac{dy}{dt} = -\frac{E}{1+E}$$

for constant E. If $x(0) = x_0$ and $y(0) = y_0$, find $x(t)$ and $y(t)$.

(b) A more sophisticated study by B. O. Koopman yields

$$\frac{dx}{d\tau} = -yE^{-1/2}, \quad \frac{dy}{d\tau} = -xE^{1/2}$$

where τ is a function of t and E. Solve for $x(\tau)$ and $y(\tau)$.

16. A significant portion of medical research involves systems of linear differential equations. Examples include hormone concentration adjustments, gland interreactions, and models for the detection of diabetes. In these uses the system is

$$\frac{dx_1}{dt} = A_1 - A_2x_1 - A_3x_2$$

$$\frac{dx_2}{dt} = B_1 + B_2x_1 - B_3x_2$$

where $A_i, B_i \geq 0$. Solve for $x_1(t)$ and $x_2(t)$ if $A_1 = 2$, $A_2 = A_3 = 1$, $B_1 = 3$, $B_2 = 1$, $B_3 = 2$.

17. In compartmental analysis of the concentration of a drug in adjacent cells of volume V_1 and V_2, respectively, through which diffusion takes place, the concentrations in the cells at any time t may be represented by the solution of the system

$$\frac{dx_1}{dt} = \frac{x_2 - x_1}{v_1}$$

$$\frac{dx_2}{dt} = \frac{x_1 - x_2}{v_2}$$

Solve for $x_1(t)$ and $x_2(t)$.

18. In the study of a cat's spinal-cord motoneuron, the following first-order system appears

$$\frac{dx_1}{dt} = x_2$$

$$\frac{dx_2}{dt} = -x_2\left[\left(\frac{T_2 + T_3}{T_2 T_3}\right) - \left(\frac{K T_1}{T_2 T_3}\right) f'(x)\right] - \frac{1}{T_2 T_3} x_1 + \frac{K}{T_2 T_3} f(x)$$

where K is a scaling factor, T_1, T_2, and T_3 are time constants, and $f(x)$ is the transmembrane current best described by

$$f(x) = (x_1 + 2x_2)(x_1^2 + 1)$$

Let $K = 2.5$, $T_1 = T_2 = 1$, $T_3 = 0.5$. Write this system as a standardized first-order system, then ascribe the correct terminology to the system.

19. Attempts have been made to find a mathematical model to describe periodicities in the concentration of bodily substances that enhance or inhibit each other's rate of production (or dissipation). The nature of the model demands that the solutions of the differential equations be nonnegative at all times (i.e., the steady states are all positive).

Lamport* proposed a model for the push-pull theory of hormone interaction, which maintains that the follicle-stimulating hormone of the pituitary gland enhances the production of estrogen while estrogen inhibits the production of the follicle-stimulating hormone. Denoting the hormones by F and E, respectively, we obtain

$$\frac{dF}{dt} = a_{10} - a_{11}F - a_{12}E$$

$$\frac{dE}{dt} = a_{21}F - a_{22}E$$

In deriving these (check the terms), we assume each hormone dissipates at a rate proportional to its own concentration, and we include the enhancing and inhibitory effects attributed to the hormones (i.e., that F stimulates the production of E and E inhibits the production of F and also that the pituitary is able to produce its hormone independently at the rate a_{10} but that the ovary is unable to produce estrogen independently).

---

* H. Lamport, "Periodic Changes in Blood Estrogen." *Endocrinology*, 27(1940), pp. 673–80.

Solve for the periodic solutions $F(t)$ and $E(t)$ if they each have a unit value at time $t = 0$ and if $a_{11} = -1, a_{12} = a_{22} = 1, a_{21} = 2$, and $a_{10} = 1$.

20. Consider the third-order push-pull system used to describe the thyroid-pituitary system where x_1 is the concentration of the pituitary hormone thyrotrophin, x_2 is the concentration of an activated enzyme within the thyroid gland, and x_3 is the concentration of thyroid hormone. The periodic variation in metabolic rate observed in the mental disorder *periodic catatonia* was explained by the following push-pull system:

$$\frac{dx_1}{dt} + x_1 = 1 - 9x_3$$

$$\frac{dx_2}{dt} + 0.5x_2 = x_1$$

$$\frac{dx_3}{dt} + 1.5x_3 = 7x_2$$

where we see that:
(a) only x_1 is produced independently,
(b) x_1 stimulates production of x_2,
(c) x_2 stimulates production of x_3,
(d) x_3 inhibits production of x_1,
(e) the components dissipate in proportion to their respective concentrations.

Show that $x_3(t) = 0.1098 + C_1e^{-5t} + C_2e^{t}\cos(3.43)t + C_3e^{t}\sin(3.43)t$ with the value of $x_3 < \frac{1}{9}$ producing a periodic cycle.

21. In the study of specialized types of coupled reactions, the diffusion equations are created as follows:

Suppose the first substance is produced at a rate a_1c_1 gm/cm^3 per second, proportional to its own concentration c_1. The second substance catalyzes the decomposition of the first so the decomposition happens at rate b_1c_2, proportional to the concentration c_2 of the second. Production of the second is catalyzed by the first at rate a_2c_1 and also decomposes at rate b_2c_2. If h_1 and h_2 are the permeabilities of the substances and k_1 and k_2 are the corresponding external concentrations, we have

$$\frac{dc_1}{dt} = \left(a_1 - \frac{3h_1}{r_0}\right)c_1 - b_1c_2 + \frac{3h_1}{r_0}k_1$$

$$\frac{dc_2}{dt} = a_2c_1 - \left(b_2 + \frac{3h_2}{r_0}\right)c_2 + \frac{3h_2}{r_0}k_2$$

Values from an actual experiment are as follow:

$$h_1 = 2 \times 10^{-4} \text{ cm/sec}$$

$$h_2 = 10^{-4} \text{ cm/sec}$$

$$a_1 = 3 \times 10^{-1} = a_2$$

$$k_1 = k_2 = 10^{-4} \text{ gm/cm}^3$$

$$b_1 = 10^{-1}$$

$$b_2 = 0$$

$$r_0 = 3 \times 10^{-3} \text{ cm}$$

These should give a period of cell life on the order of one minute. Solve for $c_1(t)$ and $c_2(t)$.

9.3 Solving a System of Equations by Using Laplace Transforms

As we saw in Section 9.1, many applications problems are sufficiently complicated to produce differential equations in more than one dependent variable. The initial conditions that usually accompany such situations should be used to solve the problem completely. In Chapter 8, the methods of Laplace transforms used these initial conditions in the solution process as an integral part of the operation; the unique solution was produced immediately, and later application of the initial conditions was unnecessary. This same process may now be used on a system of differential equations with initial conditions. As the Laplace transform process converts each differential equation to an algebraic equation, we will now be faced with the simpler process of solving sets of simultaneous algebraic equations in the unknowns $\mathscr{L}[x]$, $\mathscr{L}[y]$, etc., to ultimately produce the unique solution to the system. This process is often most easily done by using the technique called **Cramer's rule**, in which determinants are used.

We will first define Cramer's rule for a purely algebraic pair of equations

$$a_1 x + a_2 y = c_1$$

$$b_1 x + b_2 y = c_2$$

Let D be the determinant of coefficients of the unknowns on the left side:

$$D \equiv \begin{vmatrix} a_1 & a_2 \\ b_1 & b_2 \end{vmatrix} = a_1 b_2 - a_2 b_1$$

Let D_x be the determinant obtained when the column of coefficients of x is replaced by the column of constants from the right side of the equations:

$$D_x \equiv \begin{vmatrix} c_1 & a_2 \\ c_2 & b_2 \end{vmatrix} = c_1 b_2 - c_2 a_2$$

The determinant D_y is obtained from D by replacing the column of coefficients of y by the column of constants:

$$D_y \equiv \begin{vmatrix} a_1 & c_1 \\ b_1 & c_2 \end{vmatrix} = a_1 c_2 - b_1 c_1$$

The pair is uniquely solved if $D \neq 0$, and the solution is

$$x = D_x/D, \quad y = D_y/D$$

An extension to three equations in three unknowns is similarly made and uses determinants of order three. Example 15 demonstrates Cramer's rule.

Example 15

Use Cramer's rule to solve

$$\begin{aligned} 2x - y &= 1 \\ -4x + 3y &= 5 \end{aligned}$$

We will first find D, D_x, and D_y:

$$D = \begin{vmatrix} 2 & -1 \\ -4 & 3 \end{vmatrix} = 2(3) - (-4)(-1) = 2$$

$$D_x = \begin{vmatrix} 1 & -1 \\ 5 & 3 \end{vmatrix} = 1 \cdot 3 - 5(-1) = 8$$

$$D_y = \begin{vmatrix} 2 & 1 \\ -4 & 5 \end{vmatrix} = 2 \cdot 5 - (-4)(1) = 14$$

The unique solution is $x = D_x/D = \frac{8}{2} = 4$, $y = D_y/D = \frac{14}{2} = 7$.

We will now solve a system of first-order differential equations by using Laplace transforms and Cramer's rule.

Example 16

Consider the system

$$\begin{aligned} y' &= y + 2z \\ z' &= 2y + z \end{aligned}$$

with $y(0) = 2$, $z(0) = 0$.

Taking Laplace transforms of both sides of each equation, we obtain

$$\begin{aligned} s\mathscr{L}[y] - y(0) &= \mathscr{L}[y] + 2\mathscr{L}[z] \\ s\mathscr{L}[z] - z(0) &= 2\mathscr{L}[y] + \mathscr{L}[z] \end{aligned}$$

Using the initial conditions and reorganization produces

$$\begin{aligned} (s-1)\mathscr{L}[y] - \quad 2\mathscr{L}[z] &= 2 \\ -2\mathscr{L}[y] + (s-1)\mathscr{L}[z] &= 0 \end{aligned}$$

We will now solve the problem as an algebraic system of two equations in the two unknowns $\mathscr{L}[y]$ and $\mathscr{L}[z]$. Using Cramer's rule on this pair we first obtain the determinant of coefficients of the left side.

$$\begin{aligned} D \equiv \begin{vmatrix} s-1 & -2 \\ -2 & s-1 \end{vmatrix} &= (s-1)^2 - 4 = s^2 - 2s + 1 - 4 \\ &= s^2 - 2s - 3 \\ &= (s-3)(s+1) \end{aligned}$$

Then the determinant D_y is obtained by replacing the y-column in D by the column of constants from the right side of the equations and leaving the z-column unchanged.

$$D_y \equiv \begin{vmatrix} 2 & -2 \\ 0 & s-1 \end{vmatrix} = 2(s-1)$$

We similarly obtain D_z by replacing the z-column of D by the column of constants.

$$D_z \equiv \begin{vmatrix} s-1 & 2 \\ -2 & 0 \end{vmatrix} = 4$$

Then $$\mathscr{L}[y] = \frac{D_y}{D} = \frac{2(s-1)}{(s-3)(s+1)}$$

and $$\mathscr{L}[z] = \frac{D_z}{D} = \frac{4}{(s-3)(s+1)}$$

Partial-fraction decomposition of these rational functions gives

$$\mathscr{L}[y] = \frac{1}{s-3} + \frac{1}{s+1}$$

and $$\mathscr{L}[z] = \frac{1}{s-3} - \frac{1}{s+1}$$

The table of inverse transforms now produces the unique solution pair,

$$y = e^{3x} + e^{-x}$$

$$z = e^{3x} - e^{-x}$$

The technique works with equal ease on systems of three or more linear equations.

Example 17

Consider

$$\begin{aligned}
\frac{dx}{dt} &= -2x - y + 1 \\
\frac{dy}{dt} + \frac{dz}{dt} &= y - 2z \\
\frac{dx}{dt} + \frac{dz}{dt} &= x - 1
\end{aligned}$$

with $x(0) = 1$, $y(0) = 0$, $z(0) = 0$.

We first reorganize the equations to the form

$$\begin{aligned}
\frac{dx}{dt} + 2x + y &= 1 \\
\frac{dy}{dt} - y + \frac{dz}{dt} + 2z &= 0 \\
\frac{dx}{dt} - x + \frac{dz}{dt} &= -1
\end{aligned}$$

and apply the Laplace transform to both sides of each equation.

$$\begin{aligned}
s\mathscr{L}[x] - x(0) + 2\mathscr{L}[x] + \mathscr{L}[y] &= \frac{1}{s} \\
s\mathscr{L}[y] - y(0) - \mathscr{L}[y] + s\mathscr{L}[z] - z(0) + 2\mathscr{L}[z] &= 0 \\
s\mathscr{L}[x] - x(0) - \mathscr{L}[x] + s\mathscr{L}[z] - z(0) &= -\frac{1}{s}
\end{aligned}$$

We insert the initial conditions in these equations and reorganize them:

$$\begin{aligned}
(s+2)\mathscr{L}[x] + \mathscr{L}[y] &= \frac{s+1}{s} \\
(s-1)\mathscr{L}[y] + (s+2)\mathscr{L}[z] &= 0 \\
(s-1)\mathscr{L}[x] + s\mathscr{L}[z] &= \frac{s-1}{s}
\end{aligned}$$

For this set of three algebraic equations in the three unknowns $\mathscr{L}[x]$, $\mathscr{L}[y]$, and $\mathscr{L}[z]$ we have

$$D = \begin{vmatrix} s+2 & 1 & 0 \\ 0 & s-1 & s+2 \\ s-1 & 0 & s \end{vmatrix} = (s+1)(s-1)(s+2)$$

$$D_x = \begin{vmatrix} \dfrac{s+1}{s} & 1 & 0 \\ 0 & s-1 & s+2 \\ \dfrac{s-1}{s} & 0 & s \end{vmatrix} = \left(\frac{s-1}{s}\right)(s^2+2s+2)$$

$$D_y = \begin{vmatrix} s+2 & \dfrac{s+1}{s} & 0 \\ 0 & 0 & s+2 \\ s-1 & \dfrac{s-1}{s} & s \end{vmatrix} = (s+2)\left(\frac{-1(s-1)}{s}\right)$$

$$D_z = \begin{vmatrix} s+2 & 1 & \dfrac{s+1}{s} \\ 0 & s-1 & 0 \\ s-1 & 0 & \dfrac{s-1}{s} \end{vmatrix} = (s-1)\left(\frac{s-1}{s}\right)$$

$$\therefore \quad \mathscr{L}[x] = \frac{D_x}{D} = \frac{s^2+2s+2}{s(s+1)(s+2)} = \frac{1}{s} - \frac{1}{s+1} + \frac{1}{s+2}$$

$$\mathscr{L}[y] = \frac{D_y}{D} = \frac{-1}{s(s+1)} = -\frac{1}{s} + \frac{1}{s+1}$$

$$\mathscr{L}[z] = \frac{D_z}{D} = \frac{s-1}{s(s+1)(s+2)} = \frac{-1/2}{s} + \frac{2}{s+1} - \frac{3/2}{s+2}$$

Inverse tranforms of these equations give the unique solution

$$x(t) = 1 - e^{-t} + e^{-2t}$$

$$y(t) = -1 + e^{-t}$$

$$z(t) = -\frac{1}{2} + 2e^{-t} - \frac{3}{2}e^{-2t}$$

We will now employ the systems method on a multiple-spring oscillator problem of the type discussed in Example 3 of Section 9.1.

Example 18

Suppose the spring constants are $k_1 = 3$, $k_2 = 2$, $k_3 = 0$ (this one is temporarily disconnected) and $M_1 = M_2 = 1$ in suitable units. Let the second mass be initially stretched $\frac{1}{2}$ unit and released. The equations become

$$D^2 x_1 = -3x_1 + 2(x_2 - x_1)$$
$$D^2 x_2 = -2(x_2 - x_1) - 0x_2$$

Since at $t = 0$ the mass M_1 is in equilibrium, so that $5x_1(0) = 2x_2(0)$, and $x_2 = \frac{1}{2}$, we have $x_1(0) = \frac{1}{5}$, $x_2(0) = \frac{1}{2}$, $x_1'(0) = 0$, $x_2'(0) = 0$. We will write the differential equations as

$$D^2 x_1 + 5x_1 - 2x_2 = 0$$
$$D^2 x_2 + 2x_2 - 2x_1 = 0$$

and take Laplace transforms of both equations:

$$s^2 \mathscr{L}[x_1] - sx_1(0) - x_1'(0) + 5\mathscr{L}[x_1] - 2\mathscr{L}[x_2] = 0$$
$$s^2 \mathscr{L}[x_2] - sx_2(0) - x_2'(0) + 2\mathscr{L}[x_2] - 2\mathscr{L}[x_1] = 0$$

or
$$(s^2 + 5)\mathscr{L}[x_1] - 2\mathscr{L}[x_2] = \frac{1}{5} s$$
$$-2\mathscr{L}[x_1] + (s^2 + 2)\mathscr{L}[x_2] = \frac{1}{2} s$$

For this system the determinants to be used in Cramer's rule are

$$D = s^4 + 7s^2 + 6$$
$$D_{x_1} = \frac{s^3 + 7s}{5}$$
$$D_{x_2} = \frac{5s^3 + 29s}{10}$$

which give
$$\mathscr{L}[x_1] = \frac{D_{x_1}}{D} = \frac{s^3 + 7s}{5(s^2 + 1)(s^2 + 6)}$$

and
$$\mathscr{L}[x_2] = \frac{D_{x_2}}{D} = \frac{5s^3 + 29s}{10(s^2 + 1)(s^2 + 6)}$$

Partial-fraction decomposition, along with inverse transforms, gives

$$x_1 = \frac{6}{25} \cos t - \frac{1}{25} \cos \sqrt{6}\, t$$
$$x_2 = \frac{12}{25} \cos t + \frac{1}{50} \cos \sqrt{6}\, t$$

This pair gives the displacements for each weight at all times $t > 0$. Note the two competing periods.

One more example should give you sufficient experience to solve initial value problems of systems of equations.

Example 19

Consider

$$\frac{dx}{dt} - y = e^t$$

$$\frac{dy}{dt} + x = \sin t$$

with $x(0) = 1$, $y(0) = 0$. Taking transforms on both sides produces

$$s\mathscr{L}[x] - \mathscr{L}[y] = \frac{1}{s-1} + 1$$

$$\mathscr{L}[x] + s\mathscr{L}[y] = \frac{1}{s^2+1}$$

Solving these algebraically gives

$$\mathscr{L}[x] = \frac{1}{2}\left[\frac{1}{s-1} + \frac{1}{s^2+1} + \frac{s}{s^2+1} + \frac{2}{(s^2+1)^2}\right]$$

and
$$\mathscr{L}[y] = \frac{1}{2}\left[\frac{-1}{s-1} - \frac{1}{s^2+1} + \frac{s}{s^2+1} + \frac{2s}{(s^2+1)^2}\right]$$

Inverse transforms give the following solution pair:

$$x(t) = \frac{1}{2}(e^t + 2\sin t + \cos t - t\cos t)$$

and
$$y(t) = \frac{1}{2}(-e^t - \sin t + \cos t + t\sin t)$$

Exercises 9.3

Solve the following systems with given initial conditions by using Laplace transform methods. Check answers to Exercises 1 and 2 by direct substitution. Check your answers to Exercises 3–8 by reworking the problems with operator methods.

1. $(D-1)x + 2y = 0$
   $3x + (D-2)y = 0$
   $x(0) = 1$, $y(0) = 0$
2. $y' = 3y - 2z$
   $z' = 2y - z$
   $y(0) = 0$, $z(0) = -1$

3. $$\frac{dx}{dt} + \frac{dy}{dt} - y = 3e^t$$

$$2\frac{dx}{dt} + \frac{dy}{dt} + 2y = 0$$

$x(0) = 1$, $y(0) = 0$

4. $$\begin{aligned} (D-1)y + (D+1)z &= e^{-x} \\ D^2y + Dz &= 2e^{-x} \end{aligned}$$

$y(0) = 1$, $z(0) = 2$, $y'(0) = 0$

5. $$\frac{dy}{dx} - y + z = e^x$$

$$\frac{dz}{dx} - 2y + z = 1$$

$y(0) = 0$, $z(0) = 0$

6. $$\begin{aligned} y'' + 2z' + 3z &= 3x \\ y' + z' &= x - 1 \end{aligned}$$

$y(0) = 1$, $z(0) = -1$, $y'(0) = -2$

7. $$\begin{aligned} y' + z'' - y - z &= 3e^{2x} \\ y'' + z' + z &= 0 \end{aligned}$$

$$y(0) = -1,\ z(0) = \frac{4}{3}$$

$$y'(0) = -2,\ z'(0) = \frac{8}{3}$$

8. $$\begin{aligned} x'' - y + 2z &= 3e^{-t} \\ -2x' + 2y' + z &= 0 \\ 2x' + 2y'' - 2y' + z' &= 0 \end{aligned}$$

$x(0) = x'(0) = 1$, $y(0) = 2$, $z(0) = 2$, $z'(0) = -2$

9. A set of stiff differential equations that came out of the Saturn space program is

$$y' = -6y + 5z + 2\sin t$$

$$z' = 94y + 95z$$

At $t = 0$, $y = z = 0$. Solve by any available method.

9.4 Applications Problems

Numerous applications problems give rise to systems of differential equations. This section lists only a few samples, which produce both easy and difficult systems (i.e., systems that may be solved by analytical techniques and those that need a numerical solution by the computer).

Example 20

Consider the electrical network of Figure 9.6 containing resistors R_1, R_1, and R_2, two capacitors C, two inductors L, and a source of electromotive force $E(t)$. Assume that at time $t = 0$ the currents are zero and the charges on the capacitors are zero. The components, not including $E(t)$, are assumed to be invarient with time. We are interested in finding the currents through the inductance coils for all time $t > 0$.

Since the current i is the rate of change of charge q in a resistor, $i = dq/dt$, the voltage drop across the resistor is $R(dq/dt)$. If q is the charge on a capacitor, then the voltage drop across the capacitor is $\frac{1}{c} \cdot q$. The voltage necessary to induce the rate of change di/dt in the inductor is $L(di/dt)$. Thus Kirchhoff's network laws provide our circuit equations.

Let q_1 be the charge on the capacitor in the left circuit, q_2 in the right circuit. Then the currents through the inductance coils are dq_1/dt and dq_2/dt, and the current through resistor R_2 is $\frac{d}{dt}(q_1 - q_2)$. Since around any closed circuit in a network the sum of the instantaneous voltage drops in the specific direction is zero, we get

$$L\frac{d^2q_1}{dt^2} + R_1\frac{dq_1}{dt} + \frac{1}{c}q_1 + R_2\frac{d}{dt}(q_1 - q_2) = E(t)$$

$$L\frac{d^2q_2}{dt^2} + R_1\frac{dq_2}{dt} + \frac{1}{c}q_2 - R_2\frac{d}{dt}(q_1 - q_2) = 0$$

with $q_1 = q_2 = dq_1/dt = dq_2/dt = 0$ when $t = 0$. Writing these in operator form preparing for transformation, we get

$$\left[LD^2 + (R_1 + R_2)D + \frac{1}{c}\right]q_1 - \qquad R_2Dq_2 = E(t)$$

$$-R_2q_1 + \left[LD^2 + (R_1 + R_2)D + \frac{1}{c}\right]q_2 = 0$$

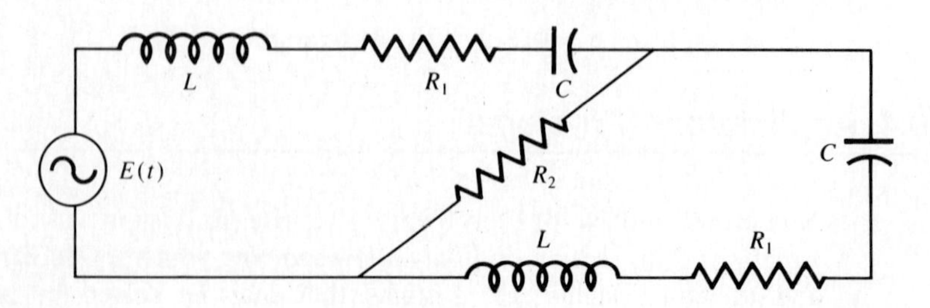

Figure 9.6

Now taking transforms and solving for $\mathscr{L}[q_1]$ and $\mathscr{L}[q_2]$, we get

$$\mathscr{L}[q_1] = \frac{Ls^2 + (R_1 + R_2)s + 1/c}{\mathscr{D}(s)} \cdot \mathscr{L}[E(t)]$$

and
$$\mathscr{L}[q_2] = \frac{R_2 s}{\mathscr{D}(s)} \cdot \mathscr{L}[E(t)]$$

where
$$\mathscr{D}(s) = \left(Ls^2 + R_1 s + \frac{1}{c}\right)\left[Ls^2 + (R_1 + 2R_2)s + \frac{1}{c}\right]$$

Using inverse transforms on these would give $q_1(t)$ and $q_2(t)$. Since the current $i = dq/dt$ we could first get the Laplace transforms of the currents as

$$\mathscr{L}[i_1] = \frac{Ls^3 + (R_1 + R_2)s^2 + (1/c)s}{\mathscr{D}(s)} \cdot \mathscr{L}[E(t)]$$

$$\mathscr{L}[i_2] = \frac{R_2 s^2}{\mathscr{D}(s)} \cdot \mathscr{L}[E(t)]$$

and then find the currents in the two circuits of the network by using inverse transforms.

Example 21

Often differential equations in biological problems create a system to be solved. For example, the egg of a certain parasite is deposited on a host, and when hatched it kills the host. We are interested in the variation of host and parasite populations with time.

Let $h(t)$ and $p(t)$ denote the number of hosts and parasites at any time t. The number of eggs deposited per unit time depends on the probability of hosts and parasites coming together and is proportional to the population product (i.e., Khp). If b is the birth rate of the hosts and d is their death rate when no parasites are present, and if $\hat{d}$ is the death rate of parasites, then

$$\frac{d}{dt} h(t) = b \cdot h(t) - d \cdot h(t) - K \cdot h(t)p(t)$$

describes the host-population changes, and

$$\frac{d}{dt} p(t) = K \cdot h(t)p(t) - \hat{d} \cdot p(t)$$

describes the parasite-population changes. Letting $b - d = \alpha$, we get

$$\frac{dh}{dt} = \alpha h - Khp$$

$$\frac{dp}{dt} = Khp - \hat{d}p$$

We assume $b > d$ so that $\alpha > 0$. There are populations $h(t)$ and $p(t)$ whose size remains stationary. No growth implies $dh/dt = dp/dt = 0$, so that

$$\alpha h - Khp = h(\alpha - Kp) = 0$$

$$Khp - \hat{d}p = p(Kh - \hat{d}) = 0$$

Since h and p are not zero we get an equilibrium point

$$h = \frac{\hat{d}}{K}$$

$$p = \frac{\alpha}{K} = \frac{b-d}{K}$$

This equilibrium host-population size is dependent on the death rate of parasites, and the equilibrium parasite population is dependent on birth and death rates of the host.

The solution of the system in which both equations are nonlinear is an exercise to be discussed in Chapter 10, where the computer is used as a tool for solving the more complicated problems.

Example 22

A problem of current interest in medical research concerns how a drug accumulates in body cells as the drug diffuses across cell walls. A model for this process, on which compartmental analysis is now performed, consists of n compartments (modeling the cells) placed in linear juxtaposition. We assume that within each cell the compound distribution is uniform at all times t and that the rate of diffusion from a cell A to its neighboring cell B is a constant r_{AB} multiplied by the compound concentration in cell A. The reverse process would have a rate (r_{BA}) (concentration B). We additionally assume that the units of dimension are chosen so that cell volume equals the area of the cell wall through which diffusion takes place.

If we denote the concentration in cell j at time t by $C_j(t)$, the diffusion in and out of cell j causes C_j to increase at rate

$$r_{j-1,j}C_{j-1} + r_{j+1,j}C_{j+1} \tag{10}$$

and decrease at rate

$$r_{j,j-1}C_j + r_{j,j+1}C_j \tag{11}$$

The rate of change for the jth cell may now be expressed as $(d/dt)C_j(t)$ and is the sum of expressions (10) and (11). This, of course, may be expanded to a multicell analysis, providing a large collection of simultaneous equations. Suppose we simplify our model to contain two cells and consider, for instance, a blood cell A and a liver cell B; and suppose at time $t = 0$ a unit amount of drug is contained in cell A and none in cell B. If the rates are r_A and r_B, we get

$$\frac{d}{dt} C_A = -r_A C_A + r_B C_B$$

$$\frac{d}{dt} C_B = r_A C_A - r_B C_B$$

with $C_A(0) = 1$, $C_B(0) = 0$. Using the methods of the previous sections, we can readily solve these to yield

$$C_A(t) = \frac{r_B}{r_A + r_B} + \frac{r_A}{r_A + r_B} e^{-(r_A + r_B)t}$$

$$C_B(t) = \frac{r_A}{r_A + r_B} - \frac{r_A}{r_A + r_B} e^{-(r_A + r_B)t}$$

As t grows large, the second terms will begin to dissolve, leaving a steady-state value for

$$\bar{C}_A = \frac{r_B}{r_A + r_B}$$

and

$$\bar{C}_B = \frac{r_A}{r_A + r_B}$$

Biomathematics is a rapidly expanding field and the study of ecosystems is an important area. The ecology of a system of four species with interaction between them would be modeled by the system in Example 23.

Example 23

Consider two prey populations H_1 and H_2 and two predator populations P_1 and P_2. If $a_1, a_2, b_1, b_2, \alpha_{ij}$, and β_{ij} are positive parameters representing the interactions, then

$$\frac{dH_1}{dt} = H_1[a_1 - \alpha_{11} P_1 - \alpha_{12} P_2]$$

$$\frac{dH_2}{dt} = H_2[a_2 - \alpha_{21} P_1 - \alpha_{22} P_2]$$

$$\frac{dP_1}{dt} = P_1[-b_1 + \beta_{11} H_1 + \beta_{12} H_2]$$

$$\frac{dP_2}{dt} = P_2[-b_2 + \beta_{21} H_1 + \beta_{22} H_2]$$

Some realistic parameter values would include $a_1 = a_2 = 3$, $\alpha_{11} = \alpha_{22} = 2$, $\alpha_{12} = \alpha_{21} = 1$, $b_1 = 40$, $b_2 = 20$, $\beta_{11} = 3$, and $\beta_{12} = \beta_{21} = \beta_{22} = 1$, with

$H_1(0) = \frac{1}{2}$, $H_2(0) = 1$, and $P_1(0) = P_2(0) = 1$. Try to draw a graph of the subsequent time path for $H_1(t)$.

If we changed the parameters to $\alpha_{11} = \alpha_{22} = 1$ and $\alpha_{12} = \alpha_{21} = 2$ and left the rest unchanged and if we changed the initial value $H_1(0)$ to $\frac{9}{10}$ and left the others the same, we would obtain an unstable two-predator/two-prey system. If we then added explicit competition between the two prey populations via ε_{ij} parameters, then

$$\frac{dH_1}{dt} = H_1[a_1 - \varepsilon_{11}H_1 - \varepsilon_{12}H_2 - \alpha_{11}P_1 - \alpha_{12}P_2]$$

$$\frac{dH_2}{dt} = H_2[a_2 - \varepsilon_{21}H_1 - \varepsilon_{22}H_2 - \alpha_{21}P_1 - \alpha_{22}P_2]$$

$$\frac{dP_1}{dt} = P_1[-b_1 + \beta_{11}H_1 + \beta_{12}H_2]$$

$$\frac{dP_2}{dt} = P_2[-b_2 + \beta_{21}H_1 + \beta_{22}H_2]$$

where $\varepsilon_{11}\varepsilon_{22} - \varepsilon_{12}\varepsilon_{21} > 0$.

If this system were analyzed, one would find that for relatively large values of ε_{11} or ε_{12} the ecosystem would be stable whereas large values of ε_{12} or ε_{21} would produce instability.

In nearly all real-world applications of the type in Example 23, the systems are solved on high-speed computers. This will be our topic of study in Chapter 10.

Example 24 extends some of the theory of Section 1.10 on the spread of an infection through a population.

Example 24

If a third class in the population is created, called the removal class R, and if no vital dynamics are considered (i.e., if no births or deaths are counted), then the model is symbolized as $S \to I \to R$ (where S denotes the susceptibles and I denotes the infectives). We may consider the members of R as having recovered from the disease and having acquired immunity and thus are not eligible for class S again, as was the case in Section 1.10.

Since the population size is the constant value N, we may write

$$\frac{d(Ns)}{dt} = -\lambda Nsi$$

This represents the removals from S that are due to contact with an infective. Also

$$\frac{d(Ni)}{dt} = \lambda Nsi - \gamma Ni$$

The terms on the right represent, respectively, the influx to I that is due to those in S contracting the disease and the exit from I that is due to gaining immunity. In addition

$$\frac{d(Nr)}{dt} = \gamma Ni$$

represents the influx to class R that is due to those in I who recover and gain immunity.

Since $s(t)$, $i(t)$, and $r(t)$ represent fractions of the population in the respective classes S, I, and R, we have $s(t) + i(t) + r(t) = 1$. We also acquire a handy way to check our differential equations since by differentiating this identity with respect to t we get $ds/dt + di/dt + dr/dt = 0$. The three differential equations, when the constant N has been canceled, become

$$\frac{ds}{dt} = -\lambda si$$

$$\frac{di}{dt} = \lambda si - \gamma i$$

$$\frac{dr}{dt} = \gamma i$$

Appropriate initial conditions would be $s(0) = s_0 > 0$, $i(0) = i_0 > 0$, and $r(0) = r_0 = 0$. When appropriate values for the parameters and the initial conditions are specified, these equations are usually solved by approximation techniques.

A theoretical analysis to determine the behavior of the various classes can be conducted by a phase-plane analysis (see Coddington and Levinson*), which produces the following result:

If $\sigma s_0 \leq 1$, then $i(t)$ approaches zero monotonically by continually decreasing and no epidemic results. If $\sigma s_0 > 1$, then $i(t)$ increases to a maximum value,

$$i_{\text{max}} = 1 - \frac{1}{\sigma} - \frac{1}{\sigma}\log(\sigma s_0)$$

and then decreases to zero and $s(t)$ decreases to a fixed value that is the unique root of

$$x = 1 + \frac{1}{\sigma}\log\left(\frac{x}{s_0}\right)$$

in the interval $(0, 1/\sigma)$.

---

* E. Coddington and N. Levinson, *Theory of Ordinary Differential Equations* (New York: McGraw-Hill, 1955).

Another case models a situation in which the disease is periodic. The symbolic representation is $S \to I \to R \to S$, with vital dynamics, and for which $s(t) + i(t) + r(t) = 1$. We will assume that there is a temporary immunity with the rate of loss of immunity per immune individual per unit time denoted by α. The differential equation for $s(t)$ is

$$\frac{ds}{dt} = -\lambda is + \delta - \delta s + \alpha r$$

(Note that the constant N has already been canceled.) The terms on the right are due, respectively, to:
(a) the decrease in S due to individuals' contracting the disease;
(b) the births into class S;
(c) the deaths from class S; and
(d) the influx to S of persons who lost their immunity.
The differential equation for $i(t)$ is

$$\frac{di}{dt} = \lambda is - \delta i - \gamma i$$

The right-side terms are due, respectively, to:
(a) those coming from S by having contracted the disease;
(b) those dying while in class I; and
(c) the outflow that is due to recovery from the disease.
By using $s(t) + i(t) + \mathrm{r(t)} = 1$, we can remove r from this pair and thus obtain the equation pair

$$\frac{ds}{dt} = -\lambda is + (\delta + \alpha) - (\delta + \alpha)s - \alpha i$$

$$\frac{di}{dt} = \lambda i\left(s - \frac{\gamma + \delta}{\lambda}\right)$$

This is a nonlinear pair of differential equations that must be solved for $s(t)$ and $i(t)$ by a numerical scheme once appropriate values for the parameters and the initial conditions are specified.

Exercise 9.4

1. Consider the network in Figure 9.7. At time $t = 0$ there is no current flowing and no charge on C. Determine the charge $q(t)$ if $C = 10^{-5}$ farad, $L = 0.1$ henry, $R_1 = 50$ ohms, $R_2 = 200$ ohms, and $E = 100$ volts.

Miscellaneous Exercises for Chapter 9

In the following exercises solve the system of equations by either the elimination calculus technique or the Laplace transform technique.

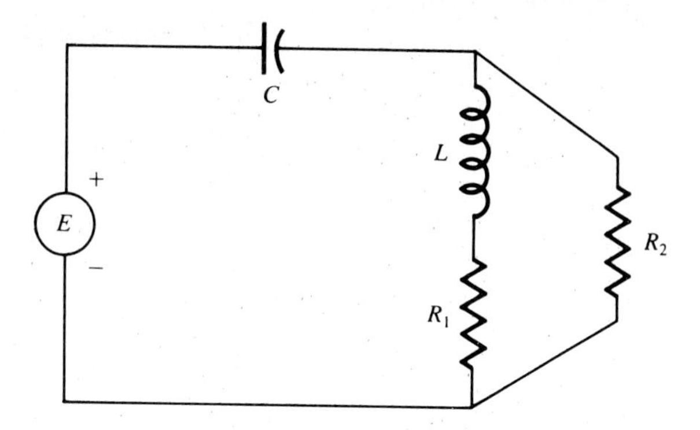

Figure 9.7

1. $\dfrac{dy}{dt} = 2t, \quad \dfrac{dz}{dt} = (y + z)/t; \quad y(1) = 0, z(1) = 4$

2. $(D + 4)y - 6z = 2$
   $3y + (D - 5)z = e^x$

3. $\dfrac{dx}{dt} = y + 10, \quad \dfrac{dy}{dt} = -5x - 4y;$
   $x(0) = 7, y(0) = -5$

4. $\dfrac{dx}{dt} + 2x + y = t; \quad x(0) = \frac{3}{2}$

   $\dfrac{dx}{dt} + \dfrac{dy}{dt} + y = 1; \quad y(0) = -\frac{1}{2}$

5. $\dfrac{dx_1}{dt} = \frac{1}{2}(x_1 + x_2); \quad x_1(0) = 1$

   $\dfrac{dx_2}{dt} = -\frac{1}{2}(3x_1 - 5x_2); \quad x_2(0) = 5$

6. $\dfrac{dx}{dt} = 3x + 2y, \quad \dfrac{dy}{dt} = -x + y$

7. $\dfrac{dx_1}{dt} = 2x_1 + x_2, \quad \dfrac{dx_2}{dt} = 3x_1 + 4x_2;$
   $x_1(0) = -2, x_2(0) = 6$

8. $\dfrac{dx}{dt} + \dfrac{dy}{dt} = 4x + 1; \quad x(0) = \frac{5}{4}$

   $\dfrac{dx}{dt} + y - 3x = t^2; \quad y(0) = \frac{9}{4}$

9. $\dfrac{dx}{dt} - 2x - 4y = -8; \quad x(0) = 1$

$\dfrac{dy}{dt} - 3x - 6y = 0; \quad y(0) = 8$

10. $4\dfrac{dx}{dt} - \dfrac{dy}{dt} = x - y$

$-8\dfrac{dx}{dt} + 2\dfrac{dy}{dt} = x - y$

11. $t^2\dfrac{d^2y}{dt^2} + t\dfrac{dy}{dt} + z = 1 + \ln t$

$t\dfrac{dy}{dt} + t\dfrac{dz}{dt} - y - z = \ln t$

12. Find a curve through the point (1, 1, 1) satisfying the equations:

$\dfrac{dy}{dx} = -2xz, \qquad \dfrac{dz}{dx} = 2xy$

13. $D^2x = 2x - 3y$

$D^2y = x - 2y$

14. $\dfrac{dy}{dx} - 9y = 8z; \quad y(0) = 1$

$\dfrac{dz}{dx} + 19z = 24y; \quad z(0) = 0$

15. $\dfrac{dx}{dt} = x + y, \quad \dfrac{dy}{dt} = -x + z, \quad \dfrac{dz}{dt} = y + z$

$x(0) = -1, \quad y(0) = 4, z(0) = 1$

16. $t\dfrac{dx}{dt} = y + z, \quad t\dfrac{dy}{dt} = z + x, \quad t\dfrac{dz}{dt} = x + y$

Chapter 10

Numerical Methods for Systems

10.1 Creating a System

The numerical techniques that we applied to first-order initial value problems in the earlier chapters will now be extended in two ways. We will derive techniques to solve: (1) a system of first-order initial value problems and (2) a high-order initial value problem that is reduced to a first-order system.

The elimination calculus techniques, studied in Chapter 9, are usually limited to systems in which the coefficients are constant. Chapter 10 allows the differential equations to be nonlinear and to have variable coefficients. These provisions greatly expand our capabilities.

Earlier we learned how to reduce a high-order differential equation to a system of first-order equations, premised on the restriction that the term $d^n x/dt^n$ may be isolated in the differential equation. This technique provides additional flexibility for applications.

Chapter 10 examines four numerical methods for solving systems of differential equations. In each of these we will restrict ourselves to a system of two first-order differential equations:

$$\left\{\begin{array}{l} y' = f(x, y, z) \\ z' = g(x, y, z) \\ y(x_0) = y_0, \quad z(x_0) = z_0 \end{array}\right\} \tag{1}$$

Extensions to systems of more than two equations will be discussed in the problems at the end of this chapter.

An important property of system study is that the second-order initial value problem

$$y'' = g(x, y, y'); \quad y(x_0) = y_0, \quad y'(x_0) = z_0$$

is transformed to the form of (1) if we set $y' = z$ and $y'' = z' = g$. Thus we arrive at an alternate way to solve the second-order initial value problem. By a substitution we obtain a system of two differential equations. The experience we gain here with the second-order equation will help in solving the higher-order problems.

Example 1

Reduce the initial value problem

$$y'' = 3y + x; \quad y(0) = 0, y'(0) = z_0 = 2$$

to a system of two first-order differential equations with appropriate initial conditions.

Define $z = y'$. Then $z(0) = y'(0) = 2$ and $z' = y''$. Therefore the differential equation becomes $z' = 3y + x$, so the system is

$$\left\{\begin{array}{l} y' = z \\ z' = 3y + x \\ y(0) = 0, \quad z(0) = 2 \end{array}\right.$$

Example 2

Reduce the initial value problem

$$2y'' - 3xy' - 2y = 0; \quad y(0) = 3, \quad y'(0) = -2$$

to a system.

Again we define $z = y'$ and solve for y''. We get $y'' = \frac{3}{2}xy' + y = \frac{3}{2}xz + y$. Since $z' = y''$, we get the system

$$\left\{\begin{array}{l} y' = z \\ z' = \frac{3}{2}xz + y \\ y(0) = 3, \quad z(0) = -2 \end{array}\right.$$

10.2 Euler's Method

It is easy to extend the Euler technique of Chapter 2 to a system of two first-order equations as in (1). Formula (5) of Section 2.2 is generalized to become

$$\left\{\begin{aligned} Y_{n+1} &= Y_n + h\cdot f(x_n, Y_n, Z_n) \\ Z_{n+1} &= Z_n + h\cdot g(x_n, Y_n, Z_n) \end{aligned}\right\} \tag{2}$$

This pair is now coupled with appropriate initial conditions $y(x_0) = y_0$, $z(x_0) = z_0$ to formulate **Euler's system method.**

Example 3

Find the approximate numerical solution on $[0, 1]$ for

$$y'' = 3y + x; \quad y(0) = 0, \quad y'(0) = 2$$

by using Euler's system method with $h = 0.1$.

From Example 1 we obtain the equivalent set:

$$\begin{cases} y' = z \\ z' = 3y + x \\ y(0) = 0, \quad z(0) = 2 \end{cases}$$

Now using (2) and (1) we see that

$$\begin{cases} f(x, y, z) = z, \\ g(x, y, z) = 3y + x \\ x_0 = 0, \quad y_0 = 0, \quad z_0 = 2 \end{cases}$$

Computing gives us

$$n = 0 \begin{cases} y'_0 = f(x_0, y_0, z_0) = z_0 = 2 \\ z'_0 = g(x_0, y_0, z_0) = 3y_0 + x_0 = 0 \\ Y_1 = y_0 + h\cdot f(x_0, y_0, z_0) = 0 + (0.1)(2) = 0.2 \\ Z_1 = z_0 + h\cdot g(x_0, y_0, z_0) = 2 + (0.1)(0) = 2 \end{cases}$$

We calculate the values of the functions. (first two lines)

We use (2) to get new points in the solution set. (last two lines)

$$n = 1 \begin{cases} Y'_1 = f(x_1, Y_1, Z_1) = Z_1 = 2 \\ Z'_1 = g(x_1, Y_1, Z_1) = 3Y_1 + x_1 = 0.6 + 0.1 = 0.7 \\ Y_2 = Y_1 + hY'_1 = 0.2 + (0.1)(2) = 0.4 \\ Z_2 = Z_1 + hZ'_1 = 2 + (0.1)(0.7) = 2.07 \end{cases}$$

New points

$$n = 2\begin{cases} Y'_2 = f(x_2, Y_2, Z_2) = Z_2 = 2.07 \\ Z'_2 = g(x_2, Y_2, Z_2) = 3Y_2 + x_2 = 1.2 + 0.2 = 1.4 \\ Y_3 = Y_2 + hY'_2 = 0.4 + (0.1)(2.07) = 0.607 \\ Z_3 = Z_2 + hZ'_2 = 2.07 + (0.1)(1.4) = 2.21 \end{cases} \quad \text{New points}$$

Continuing in this manner for $n = 3, 4, \ldots, 9$, we obtain the result

$$y(1) = Y_{10} = 2.89368$$

The program EULSYS, used to work problems of this type, is listed with its documentation in Appendix B. The functions $f(x, y, z)$ and $g(x, y, z)$, as defined in (1), are put into the program with statements 800 and 810, respectively. The initial values are included in the data input. As you will find, this program simply performs the exact steps of Euler's method as just outlined. Table 10.1 shows the computer results of this program and compares them with the actual solution

$$y = \frac{7}{18}\sqrt{3}(e^{\sqrt{3}x} - e^{-\sqrt{3}x}) - \frac{1}{3}x$$

Table 10.1

Solution to $y'' = 3y + x$; $y(0) = 0$, $y'(0) = 2$ on [0, 1] with $h = 0.1$

x	Y (EULSYS)	y (actual)
0.0	0.0	0.0
0.1	0.200000	0.201168
0.2	0.400000	0.409389
0.3	0.607000	0.631928
0.4	0.828000	0.876479
0.5	1.070210	1.151400
0.6	1.341260	1.465960
0.7	1.649420	1.830630
0.8	2.003810	2.257370
0.9	2.414690	2.760020
1.0	2.893680	3.354700

From this example we again see that the Euler method is simple to understand and easy to program, but it is rather inaccurate. The accuracy could be improved with a smaller step size, but that is usually not a fruitful path to follow. We turn instead to the more powerful methods.

10.3 A Fourth-Order Runge-Kutta Method

The natural extension of the fourth-order Runge-Kutta scheme for a single first-order equation listed in equation (2) of Section 5.1 leads us to the equations:

$$Y_{n+1} = Y_n + \frac{1}{6}(k_1 + 2k_2 + 2k_3 + k_4)$$

$$Z_{n+1} = Z_n + \frac{1}{6}(l_1 + 2l_2 + 2l_3 + l_4)$$

where:

$$\begin{aligned}
k_1 &= h \cdot f(x_n, Y_n, Z_n) \\
l_1 &= h \cdot g(x_n, Y_n, Z_n) \\
k_2 &= h \cdot f\left(x_n + \frac{1}{2}h, Y_n + \frac{1}{2}k_1, Z_n + \frac{1}{2}l_1\right) \\
l_2 &= h \cdot g\left(x_n + \frac{1}{2}h, Y_n + \frac{1}{2}k_1, Z_n + \frac{1}{2}l_1\right) \\
k_3 &= h \cdot f\left(x_n + \frac{1}{2}h, Y_n + \frac{1}{2}k_2, Z_n + \frac{1}{2}l_2\right) \\
l_3 &= h \cdot g\left(x_n + \frac{1}{2}h, Y_n + \frac{1}{2}k_2, Z_n + \frac{1}{2}l_2\right) \\
k_4 &= h \cdot f(x_n + h, Y_n + k_3, Z_n + l_3) \\
l_4 &= h \cdot g(x_n + h, Y_n + k_3, Z_n + l_3)
\end{aligned} \tag{3}$$

In applying this method to a pair of first-order differential equations, one first calculates all the k's and l's and then computes the new Y and Z values for each step. This is a self-starting method requiring only the functions $f(x, y, z)$ and $g(x, y, z)$ and the initial values x_0, y_0, and z_0.

As we saw in Chapter 5, this method may also be used to provide starting values of Milne's and Hamming's predictor-corrector methods, which will be applied to systems in subsequent sections. Meanwhile we will apply the fourth-order Runge-Kutta method in Examples 4 and 5.

Example 4

Find the numerical approximate solution on [0, 1] to

$$y'' = 3y + x; \qquad y(0) = 0, \qquad y'(0) = 2$$

by using the fourth-order Runge-Kutta method with $h = 0.1$. After reducing the given problem to the first-order system, we have

$$f(x, y, z) = z; \qquad g(x, y, z) = 3y + x;$$

$$x_0 = 0, y_0 = 0, z_0 = 2$$

Using equations (3) we compute:

$$\text{For } n = 0 \begin{cases} k_1 = h \cdot f(x_0, y_0, z_0) = h \cdot z_0 = (0.1)(2) = 0.2 \\ l_1 = h \cdot g(x_0, y_0, z_0) = h(3y_0 + x_0) = (0.1)(0) = 0 \\ k_2 = h \cdot f(0.05, 0.1, 2) = (0.1)(2) = 0.2 \\ l_2 = h \cdot g(0.05, 0.1, 2) = (0.1)(0.3 + 0.05) = 0.035 \\ k_3 = h \cdot f(0.05, 0.1, 2.0175) = (0.1)(2.0175) = 0.20175 \\ l_3 = h \cdot g(0.05, 0.1, 2.0175) = (0.1)(0.3 + 0.05) = 0.035 \\ k_4 = h \cdot f(0.1, 0.20175, 2.035) = (0.1)(2.035) = 0.2035 \\ l_4 = h \cdot g(0.1, 0.20175, 2.035) = (0.1)(0.70525) = 0.070525 \end{cases}$$

Then $$Y_1 = y_0 + \frac{1}{6}(k_1 + 2k_2 + 2k_3 + k_4)$$

$$= 0 + \frac{1}{6}(0.2 + 0.4 + 0.4035 + 0.2035) = 0.201167$$

$$Z_1 = z_0 + \frac{1}{6}(l_1 + 2l_2 + 2l_3 + l_4)$$

$$= 2 + \frac{1}{6}(0 + 0.07 + 0.07 + 0.070525) = 2.035088$$

$$\text{For } n = 1 \begin{cases} k_1 = hz_1 = (0.1)(2.035088) = 0.203509 \\ l_1 = h(3y_1 + x_1) = (0.1)(0.7035) = 0.07035 \\ k_2 = h \cdot f(0.15, 0.30292, 2.07026) = (0.1)(2.07026) = 0.207026 \\ l_2 = h \cdot g(0.15, 0.30292, 2.07026) = (0.1)(1.05876) = 0.105876 \\ k_3 = h \cdot f(0.15, 0.30468, 2.08803) = (0.1)(2.08803) = 0.208803 \\ l_3 = h \cdot g(0.15, 0.30468, 2.08803) = (0.1)(1.06404) = 0.106404 \\ k_4 = h \cdot f(0.2, 0.40997, 2.14149) = (0.1)(2.14149) = 0.214149 \\ l_4 = h \cdot g(0.2, 0.40997, 2.14149) = (0.1)(1.42991) = 0.142991 \end{cases}$$

Then $$Y_2 = Y_1 + \frac{1}{6}(k_1 + 2k_2 + 2k_3 + k_4)$$

$$= 0.201167 + \frac{1}{6}(1.249316) = 0.409386$$

$$Z_2 = Z_1 + \frac{1}{6}(l_1 + 2l_2 + 2l_3 + l_4)$$

$$= 2.035088 + \frac{1}{6}(0.637901) = 2.14141$$

Continuing in this manner for $n = 2, 3, \ldots, 9$, we would obtain the result $y(1) = Y_{10} = 3.35466$.

The program to run problems of this type is stored under RKU4S—the Runge-Kutta fourth-order method for a system—and requires the functions $f(x, y, z)$ and $g(x, y, z)$ as defined in (1) to be put in as statements 800 and 810, respectively. The results of using this program for Example 4 and the actual solution are listed in Table 10.2. A dramatic improvement in accuracy must be noted! A relative error of almost 14% has been reduced to 0.001%.

Table 10.2

Solution to $y'' = 3y + x$; $y(0) = 0$, $y'(0) = 2$, on [0, 1] with $h = 0.1$

x	Y (EULSYS)	Y (RKU4S)	y (actual)
0.0	0.0	0.0	0.0
0.1	0.200000	0.201167	0.201168
0.2	0.400000	0.409386	0.409389
0.3	0.607000	0.631922	0.631928
0.4	0.828000	0.876471	0.876479
0.5	1.070210	1.151390	1.151400
0.6	1.341260	1.465950	1.465960
0.7	1.649420	1.830610	1.830630
0.8	2.003810	2.257340	2.257370
0.9	2.414690	2.759990	2.760020
1.0	2.893680	3.354660	3.354700

Although our development and examples here have assumed a second-order equation, one main usage of the systems technique covers the solutions of higher-ordered equations that are first broken down into a first-order system. The Runge-Kutta equations above can easily be extended to systems of more than two equations. In addition, complicated coupled systems, which often arise in applications, can be treated in a similar manner. Both types are featured in the problems for this chapter.

Example 5

The movement of substances through the gastral tract is usually influenced by the tension in the circumferential muscles. If the canal is a surface of revolution defined by $r = r(z)$ and hydrostatic forces are considered, then the equilibrium equations are

$$\frac{dr}{dz} = \tan\varphi$$

$$T = p\cos\varphi\left[r(\sec^2\varphi) + \frac{r^2}{2}\cdot\frac{d}{dz}(\tan\varphi)\right]$$

The change of variables $x = pz/T$ and $y = pr/T$ and some algebraic manipulations produce

$$\frac{d\varphi}{dx} = \frac{2}{y}\left(\frac{\cos\varphi}{y} - 1\right)$$

$$\frac{dy}{dx} = \tan\varphi$$

Initial conditions are $y(0) = 2$, $\varphi(0) = \varphi_0 = 0$. Solve for $\varphi(x)$ and $y(x)$ on $[0, \frac{5}{2}]$ using an appropriate step size.

Preparing this problem for computer solution, we obtain for insertion into the RKU4S program the statements

```
800 DEF FNF(X,Y,Z) = TAN(Z)
810 DEF FNG(X,Y,Z) = (2/Y)*((COS(Z)/Y) - 1)
```

with input 0, 2.5, 2, 0, 250, 10. The results of this run are shown in Table 10.3. From these results one would then calculate $r(z)$ and T.

Table 10.3

x	y	φ
0.0	2.0	0.0
0.3	1.90788	−0.150568
0.6	1.90788	−0.304693
0.9	1.78606	−0.466765
1.2	1.60000	−0.643501
1.5	1.32288	−0.848062
1.8	0.87178	−1.11977
2.1	−0.937964	−2.01115
2.2	−0.681848	−1.87508
2.3	0.207130	−1.50111
2.4	−0.537666	0.950925
2.5	0.093534	2.77669

10.4 Milne's Method

To provide an algorithm for solving systems of two equations by the predictor-corrector technique, we perform the natural extension of Milne's previous results from equations (3) and (4) of Section 5.3 to obtain the equations for numerically solving a system of two first-order equations.

Predictors:

$$\left.\begin{aligned} \bar{Y}_{n+1} &= Y_{n-3} + \frac{4h}{3}(2f_n - f_{n-1} + 2f_{n-2}) \\ \bar{Z}_{n+1} &= Z_{n-3} + \frac{4h}{3}(2g_n - g_{n-1} + 2g_{n-2}) \end{aligned}\right\} \quad (4a)$$

Correctors:

$$\left.\begin{aligned} Y_{n+1} &= Y_{n-1} + \frac{h}{3}(f_{n+1} + 4f_n + f_{n-1}) \\ Z_{n+1} &= Z_{n-1} + \frac{h}{3}(g_{n+1} + 4g_n + g_{n-1}) \end{aligned}\right\} \qquad (4b)$$

where f_{n+1} and g_{n+1} are obtained, respectively, from the predictor formulas $\bar{Y}_{n+1}$ and $\bar{Z}_{n+1}$; $f_{n+1} = f(x_{n+1}, \bar{Y}_{n+1}, \bar{Z}_{n+1})$; and $g_{n+1} = g(x_{n+1}, \bar{Y}_{n+1}, \bar{Z}_{n+1})$. When using this method we need starting values y_1, y_2, y_3, and z_1, z_2, z_3, which are obtained from some other method previously run. The functions f and g from the problem allow the computation of other required values.

A program to solve initial value problems by this method is listed in Appendix B under the name MLNS2—Milne's method on a system of two equations. The functions $f(x, y, z)$ and $g(x, y, z)$, as defined by (1), are put in as statements 800 and 810, respectively.

10.5 Hamming's Method

As an alternative to Milne's method, we present the extensions to one more predictor-corrector scheme by expanding equations (5) of Section 5.3 to formulate the Hamming method for solving a system of two equations.

Predictors:

$$\bar{Y}_{n+1} = Y_{n-3} + \frac{4h}{3}(2f_n - f_{n-1} + 2f_{n-2})$$

$$\bar{Z}_{n+1} = Z_{n-3} + \frac{4h}{3}(2g_n - g_{n-1} + 2g_{n-2})$$

Correctors:

$$Y_{n+1} = \frac{1}{8}(9Y_n - Y_{n-2}) + \frac{3h}{8}(f_{n+1} + 2f_n - f_{n-1})$$

$$Z_{n+1} = \frac{1}{8}(9Z_n - Z_{n-2}) + \frac{3h}{8}(g_{n+1} + 2g_n - g_{n-1}) \qquad (5)$$

where f_{n+1} and g_{n+1} are obtained, respectively, from the predictor formulas $\bar{Y}_{n+1}$ and $\bar{Z}_{n+1}$. A program to use these equations is easy to construct by modifying the MLNS2 statements.

As was mentioned earlier, a person with programming experience can construct the formulas to extend to three or more first-order equations in a system and then modify the existing programs to handle these new cases. One should consider the equations given here as springboards to solutions of a large variety of differential equations and initial value problems that would normally not be encountered in a traditional course in undergraduate differential equations.

Although much care must still be exercised in applying numerical methods, they clearly expand our capabilities manyfold.

Problems

1. Find $y(1)$ for

$$y'' - 7y' + 6y = 0; \qquad y(0) = 1, \qquad y'(0) = 0$$

by using Euler's method with $h = 0.1$.
(a) Obtain the steps for $n = 0, 1, 2$ by hand.
(b) Use the computer program EULSYS to complete the solution on $[0, 1]$ and check the computations of part (a).
(c) Find the analytical solution and compare with the solution from above.

2. (a) Solve Problem 1 by the fourth-order Runge-Kutta method. Compare your answer with actual values.
(b) Use MLNS2 to solve Problem 1 by Milne's method. Find the starting values first. Compare your answer with actual values.

3. Find $y(1)$ for

$$y'' + 4y = 0; \qquad y(0) = 1, \qquad y'(0) = 0$$

by using Euler's method with $h = 0.1$ and $h = 0.05$. Find the analytical solution and compare these values with the numerical Euler values.

4. (a) Solve Problem 3 by a fourth-order Runge-Kutta method.
(b) Solve Problem 3 by the Milne method.
(c) Compare the results of (a) and (b) with the actual result and calculate the percentage of error in each.

5. Extend the Euler equations (2) to handle a system of three equations. Use these in a modified EULSYS program to solve

$$y''' - xy' + y^2 = 0; \qquad y(0) = 1, \qquad y'(0) = 0, \qquad y''(0) = 1$$

on $[0, 1]$ with $h = 0.1$.

6. (a) Write down the additional equations needed to extend the Runge-Kutta set (3) to solve a system of three equations.
(b) Modify the given program RKU4S to accommodate your new set and work a problem of your choice.
(c) Modify the Milne set (4) to solve a system of three equations.

(d) Construct a program to use Hamming's equations (5). Rework any of the above problems to make a comparison of accuracies.

7. Using Milne's method, tabulate the solution of the system

$$2\frac{dy}{dx} = y^2 + xz$$

$$2\frac{dz}{dx} = x^2 + yz$$

$$y(0) = 0, \qquad z(0) = 1$$

on $[0, 2]$ with $h = 0.1$.

Applications Problems

A–1. **An RLC circuit.** If a circuit contains resistance, inductance, and capacitance but no voltage is applied, then the differential equation governing the determination of the charge $Q(t)$ for all $t > 0$ is

$$Q'' + \frac{R}{L}Q' + \frac{1}{LC}Q = 0; \qquad Q(0) = Q_0, \qquad Q'(0) = 0$$

(a) Given the circuit illustrated in Figure 10.1 with an initial charge in the capacitor of 1, determine the actual solution. Use YACTUL and determine the maximum value of the charge and the time at which the charge first reaches zero.
(b) Repeat (a) but use RKU4S and obtain four significant figures of accuracy. Compare these results with those of part (a).
(c) See if MLNS2 gives more accurate results.

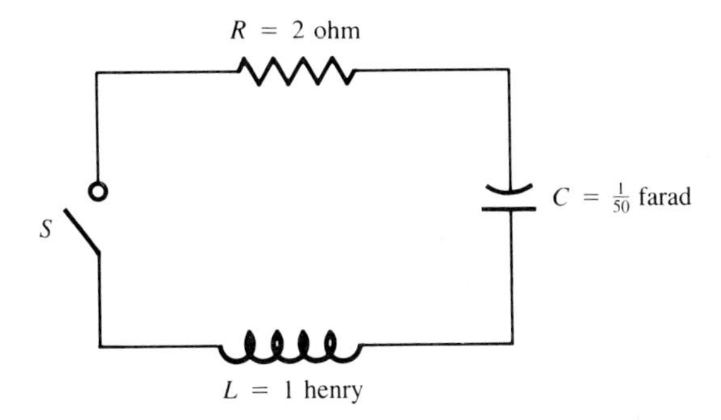

Figure 10.1

A–2. **Oscillatory motion.** As a pendulum oscillates, its position at any time t is determined by frictional forces such as resistance of the medium through which the pendulum swings and the mechanical friction at the pivot point. These damp the pendulum motion by dissipation of energy. The differential equation that describes such motion is

$$\frac{d^2\theta}{dt^2} + \frac{c}{M} \cdot \frac{d\theta}{dt} + \frac{g}{L}\theta = 0$$

where c is the friction coefficient; M, the mass of the pendulum; L, its length; and g, the gravitational constant. Typical initial conditions are $\theta(0) = \theta_0$, $\theta'(0) = 0$.

(a) A pendulum for which $L = 4$ feet and $M = 2$ slugs is oscillating in a viscous fluid for which $c = 8$. Suppose $\theta(0) = 0.1$ radian and $\theta'(0) = 0$. Using RKU4S, determine the oscillation pattern and graph on $[0, 2\pi]$.

(b) Repeat part (a) if $\theta'(0) = -0.1$ radian/second. Estimate the maximum amplitude.

A–3. **Economics-market model**. In the simplified formulation of a dynamic market model, the demand and supply functions, denoted by Q_d and Q_s, respectively, are generally taken to be functions of the current price P alone. More realistically, however, buyers and sellers may base their market behavior not only on the current price, but also on the price trend prevailing at the time. The price trend is likely to lead them to certain expectations regarding price level in the future; and these expectations will, in turn, influence the demand and supply situations. With these considerations we express the demand and supply functions as

$$Q_d = j_1 + k_1 P + m_1 P' + n_1 P'' \qquad (j_1 > 0, \qquad k_1 < 0)$$

$$Q_s = j_2 + k_2 P + m_2 P' + n_2 P'' \qquad (j_2 < 0, \qquad k > 0)$$

The unspecified parameters play a significant role in this model. If $m_1 > 0$, then a rising price will cause Q_d to rise, and hence buyers and sellers will prefer to increase their transactions while the price is relatively low. If $m_1 < 0$, then the expectation is a reversal of price trend; hence buyers will cut purchases and wait for the price to go down again. The other parameters m_2, n_1, and n_2 can be similarly interpreted.

Suppose the adjustment process is such that the price is always set at a level that will clear the market (i.e., $Q_d = Q_s$). What kind of price-time path will emerge? Equating the demand and supply functions and combining terms, we get

$$P'' + \left(\frac{m_1 - m_2}{n_1 - n_2}\right)P' + \left(\frac{k_1 - k_2}{n_1 - n_2}\right)P = -\left(\frac{j_1 - j_2}{n_1 - n_2}\right), \qquad n_1 \neq n_2$$

Suppose for a particular situation we are given

$$Q_d = 10 - P - 4P' + P''$$

$$Q_s = -2 + 2P + 5P' + 10P''$$

with $P(0) = 5$, $P'(0) = 0.5$.

(a) Create the second-order initial value problem and solve analytically. Put the result into YACTUL on [0, 100].

(b) Solve by using IMEUL2, RKU4S, and MLNS2 on the interval [0, 100] with $h = 1$ and look for damped periodic fluctuations of period $(2\pi) \div (\sqrt{3}/6)$. Note also that the price is converging to a stationary equilibrium price of 4. Explain why this should happen in light of the form of the actual solution.

A–4. **Predator-prey**. The original theory describing predator-prey interactions was derived and studied independently by V. Volterra, a mathematician, and A. Lotka, a biologist. Their theory followed from these considerations: If each of two species existed alone, they would obey simple growth and/or decay via the earlier model $y' = c_1 y$ and $w' = c_2 w$, where the parameters c_1 and c_2 are the respective differences in birth and death rates for the species. If, however, the predators eat the prey, then c_1 will decrease while c_2 increases. The speed of decrease of c_1 depends on the quantity of predators in existence, while the rapidity of increase for c_2 depends on the size of the prey population. These facts are incorporated into the simple mathematical model for the parameters $c_1 = (a_1 - a_2 w)$ and $c_2 = (-b_1 + b_2 y)$, where a_1, a_2, b_1, and b_2 are all positive. The following Volterra-Lotka equations are then obtained by inserting these values of c into the previous simple equations:

$$y' = (a_1 - a_2 w)y$$

and

$$w' = (-b_1 + b_2 y)w$$

These are a coupled pair of nonlinear differential equations. We cannot solve the pair analytically, but to learn something of the nature of the solutions we analyze the equations as follows.

When $a_1 - a_2 w = 0 = -b_1 + b_2 y$, we have $y' = w' = 0$; therefore, the values $y_c = b_1/b_2$ and $w_c = a_1/a_2$ are called **critical values.** If a study is made of the signs of y' and w' when these elements are near the critical values, it can be shown that the solutions $y(t)$ and $w(t)$ will oscillate about these critical values y_c and w_c. This analysis provides a first insight into the nature of the solutions. To provide further analysis we linearize the system about $y_c = b_1/b_2$ and $w_c = a_1/a_2$ by first changing variables with the linear shift

$$s = y - \frac{b_1}{b_2} = y - y_c$$

and

$$q = w - \frac{a_1}{a_2} = w - w_c$$

and obtaining the system

$$s' = -\frac{a_2 b_1}{b_2} q - a_2 sq$$

$$q' = \frac{a_1 b_2}{a_2} s + b_2 sq$$

The nonlinear product terms in this system may be dropped when s and q are small, and we obtain the linearized system

$$s' = -\frac{a_2 b_1}{b_2} q$$

$$q' = \frac{a_1 b_2}{a_2} s$$

By combining these two equations into a single equation, show that a solution to this linear system in s and q also satisfies $z'' + a_1 b_1 z = 0$ (whose solution you should study) and that as a consequence s and q will probably oscillate. Explain why this suggests that $y(t)$ and $w(t)$ will probably oscillate around b_1/b_2 and a_1/a_2, respectively. Sketch a graph. Now confirm these findings numerically by using the following problem:

$$y' = 6y - yw = (6 - w)y$$

$$w' = -20w + 2yw = (-20 + 2y)w$$

(a) Set $y' = 0$ and $w' = 0$ to calculate the critical values.
(b) Using EULSYS with a very small h value (to avoid machine overflow), solve this nonlinear problem on $0 \le t \le 5$, with initial values of 12 and 2 for y and w, respectively. By graphing, confirm the expected oscillations around y_c and w_c.
(c) Hand plot the y and w values against each other. What type of life cycle results?

A–5. **Extended predator-prey.** Consider an isolated intertidal marine community where two species live in the same environment and compete for the same food resources, but neither is a predator nor a prey for the other. Volterra proposed a mathematical model for this situation. If $y(t)$ and $z(t)$ represent the number present in each population, then

$$y' = [a_1 - a_2(y + z)]y$$

$$z' = [b_1 - b_2(y + z)]z$$

where $a_1, a_2, b_1, b_2 > 0$. The values of these parameters portray the strength of the species, their eating habits, appetites, birth and death rates, etc. Let $y(0) = 100$ and $z(0) = 50$. Investigate the colony sizes of the species under the following circumstances.

(a) If the food supply is limited compared to the species' appetites, we might find $a_1 = 2, a_2 = 0.2, b_1 = 1, b_2 = 0.1$. Solve on $[0, 5]$ by using RKU4S and $h = 0.01$. Interpret your results.

(b) If both species are tough enough to survive on what is available, we might find $a_1 = 4$, $a_2 = 0.1$, $b_1 = 2$, $b_2 = 0.05$. Solve on $[0, 5]$. Interpret again.

(c) Let $a_1 = 12, a_2 = 0.1, b_1 = 2, b_2 = 0.05$. Interpret the results.

(d) Let $a_1 = 12, a_2 = 0.1, b_1 = 2, b_2 = 0.01$. Interpret the results.

(e) Try to find interesting cases by varying not only these parameters but the initial values as well.

A–6. **Chemical reactors.** A nonlinear system of equations can be obtained from studying a first-order irreversible chemical reaction carried out under nonisothermal conditions in a continuously stirred tank reactor. Control of this reactor is to be achieved by manipulation of the flow of cooling fluid through a cooling coil inserted in the reactor. The dynamic-mass and heat-balance equations are

$$y' = -(1 - e^A)y - \frac{1}{2}(e^A - 1)$$

$$z' = e^A y - 8.9z^2 - 4.225 + \frac{1}{2}(e^A - 1)$$

where

$$A = \frac{25z}{z + 2}$$

and

$$y(0) = -.1111889$$

$$z(0) = .0323358$$

This represents a self-sustained oscillation of an energy-mass system. It is termed a "limit cycle"; and on a yz-plane, a closed elliptic curve will be achieved. Starting from the given initial conditions, the values of y and z will change and eventually return to the initial point. Solve the system by using RKU4S; run it until at least one cycle has been achieved. Do a hand plot of the output points for four cycles and see if you get the predicted ellipse.

A–7. **Reversible chemical reaction**. A chemical reaction in which one molecule of product A is transformed eventually into one molecule of product C may actually consist of two successive reactions in which first one molecule of A is transformed into one molecule of B and then one of B goes into one of C. Moreover, the intermediate product B and the final product C may spontaneously revert to A and B, respectively.

Assume such a reaction is started at $t = 0$ with $n = 2000$ molecules of A. Determine the amount of A present at subsequent time t. Let $k_1 = 0.6$ be the speed coefficient for $A \to B$, $k_2 = 0.4$ for $B \to A$, $k_3 = 0.3$ for $B \to C$, and $k_4 = 0.15$ for $C \to B$. Let x, y, and z denote the number of molecules of A, B, and C, respectively, present at time t. Using $x + y + z = n$, the total-rate equations become

$$\frac{dx}{dt} + k_1 x - k_2 y = 0$$

$$\frac{dy}{dt} - (k_1 - k_4)x + (k_2 + k_3 + k_4)y = k_4 n$$

with $x(0) = n$, $y(0) = 0$. Solve by using RKU4S on $[0, 10]$ and graph both x versus t and y versus t. (This may require program modifications.)

An example of this type of chemical reaction (but not producing the above differential equation) is a process used in the laboratory for producing carbon dioxide:

$$CaCO_3 + 2HCl \rightleftarrows CaCl_2 + H_2CO_3$$

$$H_2CO_3 \rightleftarrows H_2O + CO_2$$

Another example of a reversible chemical reaction is found in the manufacture of mercurous chloride, Hg_2Cl_2, used in medicine, from mercuric chloride, $HgCl_2$, and mercury, Hg:

$$\underbrace{HgCl_2 + Hg}_{A} \rightleftarrows \underbrace{2HgCl}_{B} \rightleftarrows \underbrace{Hg_2Cl_2}_{C}$$

Let's look at terms A and B of this process. If the forward and backward speed coefficients are k_1 and k_2, respectively, and we have n_1 and n_2 molecules of mercuric chloride and mercury, respectively, then the differential equation for the number of molecules of HgCl is

$$2\frac{dx}{dt} = k_1(n_1 - x)(n_2 - x) - 2k_2 x, \qquad x(0) = 0$$

A–8. **Electric transformer**. An iron-core transformer used in communications circuits has a primary coil with resistance R_1 and self-inductance L_1 and a secondary coil with resistance R_2 and self-inductance L_2. Let the windings be such that the mutual inductance M is positive. If we apply the emf $E_0 \sin \omega t$ to the primary and we short the secondary at $t = 0$, then Kirchhoff's voltage law gives

$$L_1 \frac{di_1}{dt} + R_1 i_1 + M \frac{di_2}{dt} = E_0 \sin \omega t$$

$$M \frac{di_1}{dt} + L_2 \frac{di_2}{dt} + R_2 i_2 = 0$$

with $i_1(0) = 0$, $i_2(0) = 0$. Suppose $R_1 = 3$ ohms, $R_2 = 1$ ohm, $L_1 = \frac{5}{2}$ henry, $L_2 = \frac{1}{2}$ henry, $M = \frac{1}{2}$ henry, $E_0 = 10$ volts, $\omega = 2$ radians/second. Set up the above equations, simplify, and write as a first-order system:

$$i_1' = f(t, i_1, i_2), \qquad i_1(0) = 0$$

$$i_2' = g(t, i_1, i_2), \qquad i_2(0) = 0$$

Solve this system by using RKU4S on $[0, 4]$ and compare your answer with the actual analytical solution:

$$i_1 = e^{-t} + \frac{5}{13} e^{-3t} - \frac{18}{13} \cos 2t + \frac{14}{13} \sin 2t$$

$$i_2 = e^{-t} - \frac{15}{13} e^{-3t} + \frac{2}{13} \cos 2t - \frac{16}{13} \sin 2t$$

Note that the terms containing the exponentials represent the transient part while the trigonometric terms represent the steady-state part of the solution.

A–9 **Electric network**. The simple RLC circuit studied earlier is an inadequate model for applications in many electrical phenomena. The study of the simple transformer requires more complicated circuits, similar to those of the television receiver and the computer networks. The generalization is to interconnect RLC circuits into an electric network. A typical two-loop network is shown in Figure 10.2, in which $E(t)$ is an electromotive force applied to the left loop. Using Kirchhoff's circuit laws on the loops, we are able to determine the currents i_1, i_2, and i_3 in the portions of the network. The differential equations for a no-capacitance network are the simplest. First we see at the junction point J_1 that $i_1 - i_2 - i_3 = 0$. Also the circuit laws produce

$$\begin{cases} L_1 i_2' + (R_1 + R_3) i_2 + (R_1 + R_3) i_3 = E \\ L_2 i_3' + R_2 i_3 - L_1 i_2' = 0 \end{cases}$$

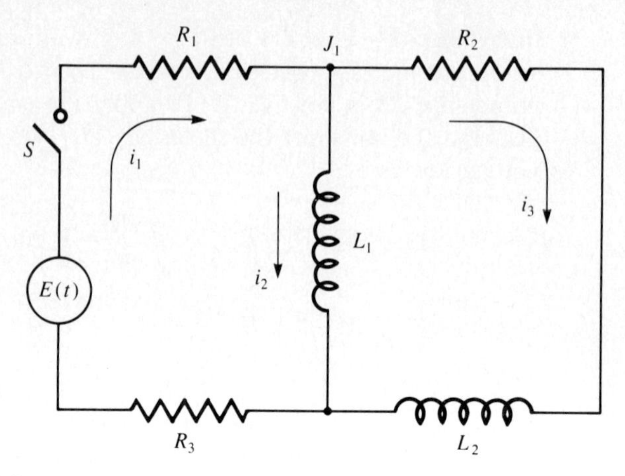

Figure 10.2

Let $L_1 = 1$ henry, $L_2 = \frac{1}{4}$ henry, $R_1 = 1$ ohm, $R_2 = 2$ ohms, $R_3 = \frac{1}{2}$ ohm, and $E(t) = \sin t$. Determine the currents i_1, i_2, and i_3 by using RKU4S on an eight-second time interval after closing the switch S. Graph and interpret the results.

A–10. **Radiation heat transfer**. As technology expands and man reaches for the stars, the temperature extremes encountered require a more thorough understanding of radiative heat transfer. Since radiation is a fourth-power phenomenon, the basic equations are nonlinear and thus numerical integrations are mandatory. A typical case of this is the transient heat flow in an electron tube filled with an inert gas (see Figure 10.3). Before the current is applied the whole system is at temperature T_1. At $t = 0$ the filament temperature is suddenly raised to $T_2 > T_1$ by the electric current. Heat is convected to the

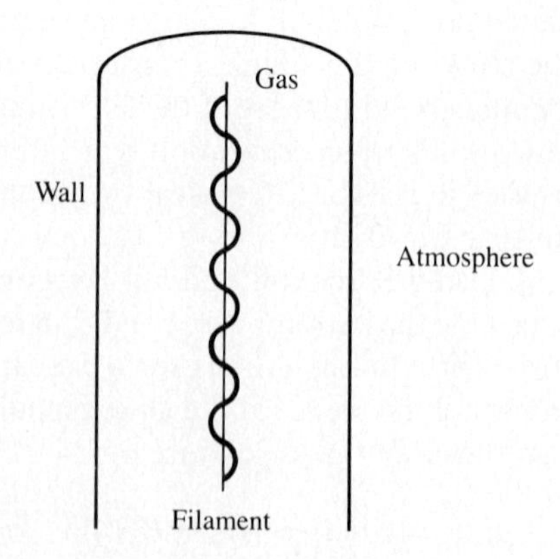

Figure 10.3

surrounding gas and radiated to the tube walls. The wall receives heat by convection from the gas and by radiation from the filament. Finally we note that the wall transfers heat by convection to the surrounding atmosphere, which is at temperature T_1. If we let $T_2 = 2T_1$ and if the heat capacities C_1 and C_2 of the gas and wall, respectively, are related by $C_2 = 2C_1$ and if the radiation coefficients are known, then we can formulate the system of differential equations for the propogation problem in dimensionless variables:

$$\frac{dy}{dt} = -2y + z + 2, \qquad y(0) = 1$$

$$\frac{dz}{dt} = \frac{1}{2}y - z + \frac{16 - z^4}{10}, \qquad z(0) = 1$$

Solve on [0, 3] with $h = 0.1$ in MLNS2.

A–11. **Dynamic dampers** (programming experience required). If a heavy piece of machinery of mass M is subjected to a pulsating vertical force, $F = F_0 \sin \omega t$, due to the unbalanced action of an internal combustion engine, the vibrations thus created could be transmitted to the supporting structure. To avoid this feature, the machine is supported on a collection of vertical springs having spring constant K. When the installation operates at certain speeds, the vibrational pulsating effects are multiplied (resonance occurs). If the system so operates for any length of time, the support system could be severely damaged, since the amplitude of the vibration increases linearly with time under resonance. To avoid such damage, a dynamic damper consisting of a mass m is attached to the machine M by a spring of constant k (see Figure 10.4). Let $x(t)$ and $y(t)$ denote the positive vertically downward displacements of M and m respectively, with the origin at the position of static equilibrium. Neglecting friction and applying Newton's laws to the system, we arrive at the differential equations

$$m\frac{d^2y}{dt^2} = k(y - x)$$

$$M\frac{d^2x}{dt^2} = k(y - x) - Kx + F_0 \sin \omega t$$

with $x(0) = x'(0) = y(0) = y'(0) = 0$. Let $W = Mg = 10{,}000$ pounds, $w = mg = 1500$ pounds, $K = 2 \cdot 10^5$ pounds/foot, $k = 2 \cdot 10^4$ pounds/foot, $F_0 = 300$ pounds, $\omega = 25$ radians/second. Find the steady-state vibration amplitudes of M and m.

Remember that decaying exponential terms (from the homogeneous solution) are transient terms and die out while the

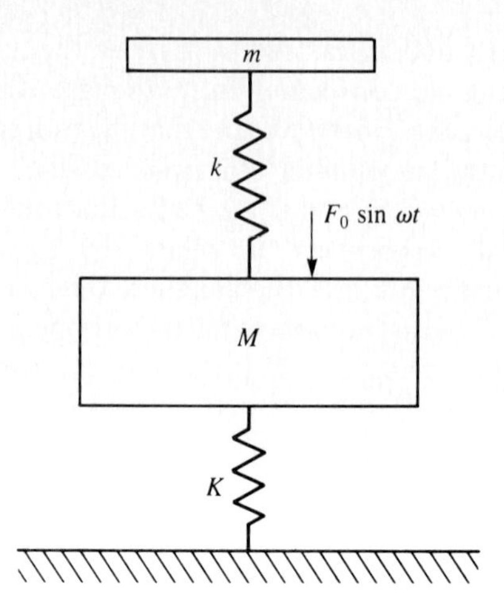

Figure 10.4

sinusoidal terms (from the particular solution) are steady-state terms and remain. A modification of RKU4S must be made to solve this coupled pair of second-order equations. If in addition to choosing the correct weight w, the value of the spring constant k is chosen so that

$$\sqrt{\frac{k}{m}} = \sqrt{\frac{K}{M}} = \omega$$

(this is called a tuned damper), the entire vibration of M is killed. This, of course, assumes the engine runs at $\omega = \sqrt{K/M}$ radians/second. Thus the damper eliminates resonance at speed $\omega = \sqrt{K/M}$, but two new resonance speeds are introduced. If these are sufficiently far apart and are outside of the operating speed of the engine, resonance is avoided except for the momentary passage through the lower speed, which produces only momentary vibration of small amplitude. Tortional oscillation dampers based on the same principle are used on most airplane engines and many stationary reciprocating engines.

A–12. **Hydraulic surge tank**. Consider a conduit supplying a hydraulic turbine (see Figure 10.5). When the valve is open, the flow velocity is V_0. Because of the fluid friction in the conduit, the pressure p_2 is less than p_1, and the fluid level in the surge tank does not rise to full height H. Thus the head loss is

$$H_L = F \cdot \frac{L}{p} \frac{V_0^2}{2g}$$

where F is the friction factor and g is gravity. We will study the propagation problem that evolves when, at time $t = 0$, the valve is

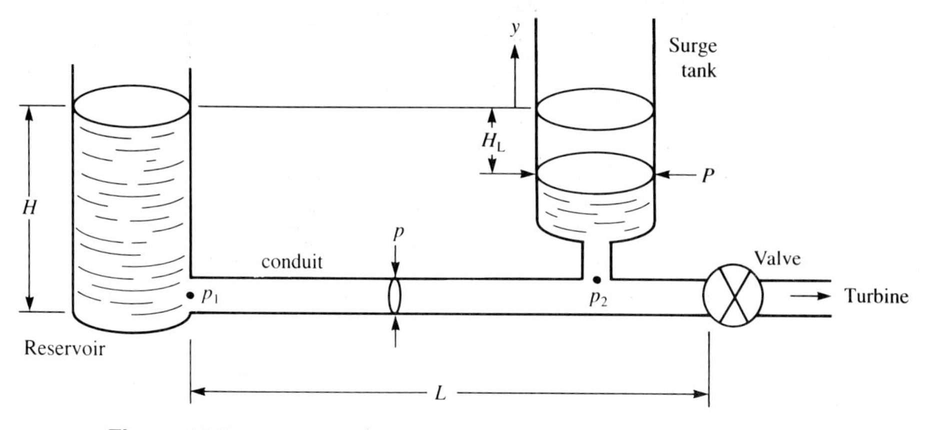

Figure 10.5

suddenly closed. The level in the surge tank will suddenly rise and fall until the excess kinetic energy is dissipated by fluid friction. Of practical interest is the maximum value of y reached in the first surge. The laws governing the motion are Newton's second law and the law of continuity of flow. We obtain

$$p^2 \frac{dy}{dt} = P^2 V$$

where V is instantaneous velocity in the conduit and where p and P are the diameters of the conduit and surge tank, respectively, and we obtain

$$-\rho L \frac{\pi}{4} P^2 \frac{dV}{dt} = (p_2 - p_1) \frac{\pi P^2}{4} \left(1 + H_L g \frac{V^2}{V_0^2}\right)$$

where ρ is mass density of the liquid. The pressure difference may be approximated by $\rho g y$. Then the second equation becomes

$$L \frac{dV}{dt} = -gy - F \cdot \frac{L}{P} \cdot \frac{V^2}{2}$$

The initial conditions are

$$V(0) = V_0$$

$$y(0) = -F \frac{L}{P} \frac{V_0^2}{2g} = -H_L$$

This initial value problem would not be difficult to solve numerically if we were given the values of p, P, L, F, and V_0. A convenient way of solving the problem but waiting to input the constants (so that we may later see what happens when one or more constants are altered) is to go to a nondimensional formulation by introducing new variables:

$$x_1 = \frac{y}{H_L} = \frac{y}{F \cdot \frac{L}{P} \cdot \frac{V_0^2}{2g}}, \qquad x_2 = \frac{V}{V_0}, \qquad t_1 = \frac{t}{\frac{2P}{F \cdot V_0}}$$

We then get for our problem

$$\frac{dx_1}{dt_1} = \beta x_2$$

$$\frac{dx_2}{dt_1} = -x_1 - x_2^2$$

where
$$\beta = \frac{g}{L}\left(\frac{2P^2}{F \cdot p \cdot V_0}\right)^2$$

and $x_1(0) = -1$ (note that $y = -H_L$) and $x_2(0) = 1$ (note that $V(0) = V_0$). Solve this system for x_1 and x_2 on $[0, 10]$ by using RKU4S with $h = 0.05$ and a value 0.8 for β. Note that this allows adjustment of L, P, F, p, or V_0. The values of y and V, of course, are then easily obtained from the equations of variable change.

Now run the problem again after doubling the length of L and doubling the diameter P of the surge tank (i.e., use a new β value). From an engineering point of view it is desirable to have $y < H_L$, so one would design the surge-tank system to keep $x_1 < 1$.

A–13. **Extended predator-prey**. The simple Volterra-Lotka equations for interaction between a predator (or parasite) population $P(t)$ and a prey (or host) population $H(t)$ are

$$\frac{dH(t)}{dt} = H(t)[a - \alpha P(t)]$$

$$\frac{dP(t)}{dt} = P(t)[-b + \beta H(t)]$$

where a (the per capita birth rate of prey), b (the death rate of the predator), and α and β (interaction parameters between the species) are positive constants. In nature, the responses and interactions between predators and prey rarely happen immediately; they usually involve time delays, which tend to produce destabilization. To involve these, we replace a (the birth rate) by a function of their population density, using the **Verhulst-Pearl logistic form**

$$a \to r\left[1 - \frac{H}{K}\right]$$

where r is a proportionality constant and K is a carrying capacity set by environmental resources. The rate at which predators remove prey is described by the predators functional response, the term $-\alpha P(t)H(t)$ above, and it corresponds to unlimited attack capacity per predator, which increases linearly with prey density. A more realistic functional response would have the predators' consumption increasing less than linearly as prey population size increases. The Russian marine biologist Ivlev used the replacement

$$\alpha P(t)H(t) \to kP(t)[1 - e^{-cH(t)}]$$

where k and c are constants. The differential equations then become

$$\frac{dH}{dt} = rH\left(1 - \frac{H}{K}\right) - kP[1 - e^{-cH}]$$

$$\frac{dP}{dt} = P[-b + \beta(1 - e^{-cH})]$$

Find how the size of the population $H(t)$ varies as t ranges over $[0, 70]$. Use parameter values $c = 10^{-3}$, $\beta = 1.5$, $k = b = r = 1$, $K = 4000$, and the initial value $H(0) = 1.8$. Use $P(0) = 1$.

A–14. **Extended predator-prey**. Other variations of the replacements used in problem A-13 produce slightly different models. Often traits of the particular species involved dictate the variation to be used. We replace a by $r[1 - (H/K)]$ as before, but now we also replace $\alpha P(t)H(t)$ by $kHP/(H + D)$, where D is a given value of the prey population beyond which the predators attack capability begins to be saturated. This particular replacement is often used in invertebrate ecology. The rate equation for the predators in this situation replaces the conventional K, which measures the predator resources, by γH, a value proportional to prey abundance. We are thus led to the new differential equation pair

$$\frac{dH}{dt} = rH\left[1 - \frac{H}{K}\right] - \frac{kPH}{H + D}$$

$$\frac{dP}{dt} = sP\left[1 - \frac{P}{\gamma H}\right]$$

Convert these to a dimensionless pair to avoid size considerations by using the variable change equations $X = H/D$ and $Y = P/\gamma D$ and use rt to represent the time variable. Then solve the equations for $X(t)$ on $[0, 8]$ and $Y(t)$ on $[0, 4]$ by using $0 \le t \le 80$. The parameter values that arise from actual data are $r/s = 6$, $K/D = 10$, and $k\gamma/r = 1$, and initial values are $X(0) = 5.5$ and $Y(0) = 3.9$.

A–15. **Crime control**. The contemporary social problem of crime control may be investigated through a system of differential equations. If we regard m_1 as the segment of the total population M committing criminal acts and m_2 as the segment engaged in crime prevention and law enforcement, then $M - m_1 - m_2$ is the remaining portion, which may also have an influence. We will let γ_1 be the influence by this largest group on the rate of change of m_1 and γ_2 be the influence on the rate of change of m_2. For example, these changes may be produced by varying the degree of enforcement support, altering the parole policy, and enacting or rescinding legislation. We then obtain

$$\frac{dm_1}{dt} = \gamma_1 + \alpha_1 m_1 - \beta_1 m_1 m_2$$

$$\frac{dm_2}{dt} = \gamma_2 - \alpha_2 m_2 + \beta_2 m_1 m_2$$

(This pair embraces both the theories of social behavior and group conflict in crime.) Note that the coefficient of autoincrease α_1 is positive and $-\alpha_2$ is negative (i.e., in the absence of a control force, the number of violators would increase, whereas, if there were no crime, prevention and enforcement personnel would decay in number). The signs of the product terms also have direct interpretation: incarceration of violators and intrinsic growth of an organization. The effects are proportional to the number of encounters $m_1 m_2$. Solve this pair numerically by using $\alpha_1 = 0.1$, $\alpha_2 = 0.4$, $\beta_1 = 0.08$, $\beta_2 = 0.05$, $\gamma_1 = 1.6$, and $\gamma_2 = 0.7$ on the interval $[0, 2]$; also use $m_1(0) = 40$ and $m_2(0) = 6$.

A–16. **Foucault pendulum**. The lobby of the United Nations headquarters in New York displays a Foucault pendulum, which automatically keeps the correct time. This pendulum, which is free to swing in both the x and y directions, exhibits the rotation of the earth as the direction of oscillation varies continuously. When the damping of friction has been accounted for, the equations of motion are

$$\frac{d^2x}{dt^2} - 2\omega(\sin\lambda)\frac{dy}{dt} + \frac{g}{L}x = 0$$

$$\frac{d^2y}{dt^2} + 2\omega(\sin\lambda)\frac{dx}{dt} + \frac{g}{L}y = 0$$

where λ is the latitude on earth of the pendulum; ω is the angular velocity of the earth's rotation, which equals $0.729 \times 10^{-4}\ \text{sec}^{-1}$; g is the gravitational constant 9.8 meters/second^2; and L is the length of the pendulum.

(a) Minneapolis is at 45° north latitude. If the pendulum is 20 meters long, how long will it take to rotate the plane of swing by 90° at this location.
(b) Find the latitude where you live. Determine how long the pendulum must be to rotate the plane of swing by 360° in 24 hours.

A–17. **Transformer flux**. In the study of the variation of flux F in a large transformer a second-order nonlinear differential equation is used. If a sinusoidal source voltage $E = 150 \sin \omega t$, where $\omega = 120\pi$ is used to drive the transformer, then the flux will vary according to

$$\frac{d^2F}{dt^2} + k_1 F + k_2 F^3 = \frac{\omega}{N} E$$

where k_1 and k_2 are design parameters and N is the number of turns in the primary winding. A particular application has parameter values $N = 450$, $k_1 = 74$, and $k_2 = 0.22$. Using RKU4S, solve for flux F as a function of time on the interval $[0, 2.5]$ if initial values are $F(0) = 0.5$ and $dF/dt(0) = 0.1$. Graph the results.

A–18. **Heartbeat**. When we develop a mathematical model for a biological phenomenon, our object should be to describe qualitatively the observed biological process. The process of interest in this case is the beating action of the heart. The heart is an elastic muscle that contracts and relaxes in a regular rhythm regulated by the pacemaker. When the chemical control present in the tissue of the pacemaker reaches a specific value, called the threshold, an electrochemical wave is triggered, thus causing the heart to contract. As the chemical control decreases, the heart muscle relaxes. The model of the heartbeat developed by E. C. Zeeman is based on specific features of the heart. First, the model should exhibit an equilibrium state corresponding to the relaxed state of the heartbeat cycle. The model should also reflect a rapid return to this equilibrium state. The rate of change of the muscle-fiber length depends at any particular instant on the tension of the fiber and the value of the chemical control. The chemical control changes at a rate proportional to muscle-fiber extension. The model by Zeeman is of the form:

$$\frac{dx}{dt} = f(x, b)$$

$$\frac{db}{dt} = g(x, b)$$

where $x(t)$ represents the change in muscle-fiber length and is dependent on time t, and $b(t)$ represents a chemical control variable that governs the electrochemical wave. Based on the generalizations above concerning the heartbeat cycle, the following equations were developed:

$$\varepsilon \frac{dx}{dt} = -(x^3 - Tx + b)$$

$$\frac{db}{dt} = (x - b_0)$$

T represents the tension in the system and must be greater than zero and is determined by $3x_0^2 - a$. The initial value of b at time $t = 0$ must be negative and is determined by $2x_0^3 - ax_0$. The constant a is chosen such that T and $b(0)$ are in the correct range. The parameter x_0 is the initial muscle-fiber length and ε is a small positive constant. The resulting solution curves oscillate about $x = b = 0$ such that the minimum x-value corresponds to diastolic relaxation and while b is negative, x increases to a maximum representing systolic contraction.

(a) For $a = 0.45$, $x_0 = 0.45$, and $\varepsilon = 0.025$, solve for T and b_0 and code the equations in BASIC for use in RKU4S.

(b) Solve on the interval $[0, 5]$, printing values for $0, 0.1, 0.2, \ldots$. Graph the results and trace the oscillations for the heartbeat and the pacemaker. Label the diastolic and systolic equilibrium points.

(c) Graph over the same interval for $a = 0.45$ and $\varepsilon = 0.0125$ and then again with $\varepsilon = 0.005$. How do these changes affect the model of the heartbeat?

A–19. **Nerve impulse transmission**. The firing of the axon of a nerve cell is the result of the change in concentration of potassium ions within the cell and sodium ions outside the cell. Although the firing process is complex, basically it begins when this difference in concentration is large enough to cause sodium ions to flow through the cell walls into the interior of the axon. As sodium ions are introduced into the axon, the cell walls become more permeable, resulting in even more sodium ions flowing into the axon. Once a critical level of membrane permeability is reached, a rapid rise in the potential difference occurs and the axon fires. Following the firing, the sodium ions gradually flow outward and the potassium ions rapidly flow into the axon, and the system returns to its initial state of rest.

Many mathematical models have been developed for this phenomenon, including one proposed in 1952 by physiologists A. L. Hodgkin and A. F. Huxley, who were awarded the Nobel prize for this work. A simplified model of much interest at present was proposed by R. Fitzhugh in 1961 and developed by J. Nagumo in 1962 and takes the form

$$\frac{\partial u}{\partial t} = \frac{\partial^2 u}{\partial x^2} + u(1 - u)(u - a) - w$$

$$\frac{\partial w}{\partial t} = bu - \gamma w$$

where a, b, and γ are positive constants with $0 < a < 1$. In this model, u represents membrane potential, with x measuring the distance along the axon and t referring to time. The cubic term $u(1 - u)(u - a)$ represents the flow inward of sodium ions, and w is a recovery variable analogous to the flow of potassium ions following the firing. The solutions depend on $x + ct$ and are of the form

$$u(x, t) = \phi(x + ct) = \phi(\varepsilon)$$

$$w(x, t) = \psi(x + ct) = \psi(\varepsilon)$$

By substituting the solution forms into the system, a coupled system of ordinary differential equations is obtained:

$$\phi'' = c\phi' - \phi(1 - \phi)(\phi - a) + \psi$$

$$c\psi' = b\phi - \gamma\psi$$

where primes denote differentiation with respect to ε. By setting $\theta = \phi'$, the equations can be rewritten as

$$\theta' = c\theta - \phi(1 - \phi)(\phi - a) + \psi$$

$$\phi' = \theta$$

$$\psi' = \frac{b}{c}\phi - \frac{\gamma}{c}\psi$$

Mathematical investigation and analysis of the differential equations reveal that $\psi \equiv 0$. Thus the system simplifies to

$$\theta' = c\theta - \phi(1 - \phi)(\phi - a)$$

$$\phi' = \theta$$

Experimental evidence shows that when $c = \sqrt{2}(\frac{1}{2} - a)$ and $0 < a < \frac{1}{2}$, then a solution curve exists in the (ϕ, θ) plane that leaves $(0, 0)$ and approaches $(1, 0)$.

(a) Analytical solutions can be developed for the system such that

$$\phi(\varepsilon) = 1/(1 + e^{-\varepsilon/\sqrt{2}})$$

$$\theta(\varepsilon) = e^{-\varepsilon/\sqrt{2}}/\sqrt{2}(1 + e^{-\varepsilon/\sqrt{2}})^2$$

Using YACTUL, obtain values for ϕ and θ over the interval $[-15, 15]$ and graph the results.

(b) Code the system into BASIC for use in RKU4S and use a value of $a = 0.35$. Solve over $[-10,10]$ with values determined from YACTUL for $Y(-10)$ and $Z(-10)$. Graph the results.

A–20. **Population cycles.** In the study of the population cycles in rodents, the following model was created:

$$\frac{dn_1}{dt} = n_1[a_1(t) - (b_1 - c_1)n_2 - c_1(n_1 + n_2)] \qquad (1)$$

$$\frac{dn_2}{dt} = n_2[-a_2(t) + b_2 n_1] \qquad (2)$$

In this model n_1 is the density (in animals per acre) of so-called "emigrants," and n_2 is the density of the second type, which will be denoted as "tolerants." If there were no tolerants, the number of emigrants, starting with low density, would first increase exponentially as $e^{a_1 t}$; but as density increases, the emigrants show a high tendency to leave the habitat. This phenomenon, generated by total density, is the term $-c_1 n_1(n_1 + n_2)$ in equation (1). Emigrants have high reproductive potential. Young lactating females constitute the major portion of this class. The tolerants are older and less active. However, their interaction, both sexual and social, with emigrants is expressed by the term $b_2 n_2 n_1$ in equation (2). Thus we see a possible flow of individuals between the two populations.

Experimental values for the parameters consist of the following: $b_1 = 1.75 \times 10^{-3}$, $b_2 = 1.50 \times 10^{-3}$, and $c_1 = 1.0 \times 10^{-3}$. Initial values are $n_1(0) = n_2(0) = \frac{1}{2}$. Consider the variable t as months and compute the population values over the interval $0 \le t \le 50$ with step size $h = 1$ month. Now the functions $a_1(t)$ and $a_2(t)$ need to be determined to complete the model. Assuming that the net reproduction rate is minimal in late winter and maximal in late summer, we choose

$$a_1(t) = 1 + 0.35 \sin\frac{\pi}{6}t \qquad \text{and} \qquad a_2(t) = 0.14 - 0.075 \sin\frac{\pi}{6}t$$

where $t = 0$ represents mid-June in the northern hemisphere.

Table A

Additional Laplace Transforms

	$\mathscr{L}[f(x)]$	$f(x)$
28.	$\dfrac{1}{1+as}$	$\dfrac{1}{a}e^{-x/a}$
29.	$\dfrac{1}{s(1+as)}$	$1-e^{-x/a}$
30.	$\dfrac{1}{(1+as)^2}$	$\dfrac{1}{a^2}xe^{-x/a}$
31.	$\dfrac{1}{(1+as)(1+bs)}$	$\dfrac{e^{-x/a}-e^{-x/b}}{a-b}$
32.	$\dfrac{s}{(1+as)(1+bs)}$	$\dfrac{ae^{-x/b}-be^{-x/a}}{ab(a-b)}$
33.	$\dfrac{1}{\sqrt{s+a}}$	$\dfrac{e^{-ax}}{\sqrt{\pi x}}$
34.	$\dfrac{s\sin b+a\cos b}{s^2+a^2}$	$\sin(ax+b)$
35.	$\dfrac{s\cos b-a\sin b}{s^2+a^2}$	$\cos(ax+b)$
36.	$\dfrac{a^2}{s^3+a^3}$	$\dfrac{e^{-ax}}{3}-\dfrac{e^{ax/2}}{3}\left(\cos\dfrac{\sqrt{3}}{2}ax-\sqrt{3}\sin\dfrac{\sqrt{3}}{2}ax\right)$
37.	$\dfrac{a^2s}{s^4+a^4}$	$\sin\dfrac{a}{\sqrt{2}}x\sinh\dfrac{a}{\sqrt{2}}x$
38.	$\dfrac{as^2}{s^2+a^4}$	$\dfrac{1}{\sqrt{2}}\left(\cos\dfrac{a}{\sqrt{2}}x\sinh\dfrac{a}{\sqrt{2}}x+\sin\dfrac{a}{\sqrt{2}}x\cosh\dfrac{a}{\sqrt{2}}x\right)$
39.	$\dfrac{s^3}{s^4+a^4}$	$\cos\dfrac{a}{\sqrt{2}}x\cosh\dfrac{a}{\sqrt{2}}x$

	$\mathscr{L}[f(x)]$	$f(x)$
40.	$\dfrac{a^3}{s^4 - a^4}$	$\dfrac{1}{2}(\sinh ax - \sin ax)$
41.	$\dfrac{a^2 s}{s^4 - a^4}$	$\dfrac{1}{2}(\cosh ax - \cos ax)$
42.	$\dfrac{as^2}{s^4 - a^4}$	$\dfrac{1}{2}(\sinh ax + \sin ax)$
43.	$\dfrac{s^3}{s^4 - a^4}$	$\dfrac{1}{2}(\cosh ax + \cos ax)$
44.	$\dfrac{as^2}{(s^2 + a^2)^2}$	$\dfrac{1}{2}(\sin ax + ax \cos ax)$
45.	$\dfrac{s^3}{(s^2 + a^2)^2}$	$\cos ax - \dfrac{ax}{2} \sin ax$
46.	$\dfrac{a^3}{(s^2 - a^2)^2}$	$\dfrac{1}{2}(ax \cosh ax - \sinh ax)$
47.	$\dfrac{as}{(s^2 - a^2)^2}$	$\dfrac{x}{2} \sinh ax$
48.	$\dfrac{as^2}{(s^2 - a^2)^2}$	$\dfrac{1}{2}(\sinh ax + ax \cosh ax)$
49.	$\dfrac{s^3}{(s^2 - a^2)^2}$	$\dfrac{1}{2}(2 \cosh ax + ax \sinh ax)$
50.	$\dfrac{ab}{(s^2 + a^2)(s^2 + b^2)}$	$\dfrac{a \sin bx - b \sin ax}{a^2 - b^2}$
51.	$\dfrac{s}{(s^2 + a^2)(s^2 + b^2)}$	$\dfrac{\cos bx - \cos ax}{a^2 - b^2}$
52.	$\dfrac{s^2}{(s^2 + a^2)(s^2 + b^2)}$	$\dfrac{a \sin ax - b \sin bx}{a^2 - b^2}$
53.	$\dfrac{s^3}{(s^2 + a^2)(s^2 + b^2)}$	$\dfrac{a^2 \cos ax - b^2 \cos bx}{a^2 - b^2}$
54.	$\dfrac{ab}{(s^2 - a^2)(s^2 - b^2)}$	$\dfrac{b \sinh ax - a \sinh bx}{a^2 - b^2}$
55.	$\dfrac{s}{(s^2 - a^2)(s^2 - b^2)}$	$\dfrac{\cosh ax - \cosh bx}{a^2 - b^2}$
56.	$\dfrac{s^2}{(s^2 - a^2)(s^2 - b^2)}$	$\dfrac{a \sinh ax - b \sinh bx}{a^2 - b^2}$
57.	$\dfrac{s^3}{(s^2 - a^2)(s^2 - b^2)}$	$\dfrac{a^2 \cosh ax - b^2 \cosh bx}{a^2 - b^2}$

	$\mathscr{L}[f(x)]$	$f(x)$
58.	$\dfrac{a^2}{s^2(s^2+a^2)}$	$x - \dfrac{1}{a}\sin ax$
59.	$\dfrac{a^2}{s^2(s^2-a^2)}$	$\dfrac{1}{a}\sinh ax - x$
60.	$\dfrac{a^4}{s(s^2+a^2)^2}$	$1 - \cos ax - \dfrac{a}{2}x\sin ax$
61.	$\dfrac{a^4}{s(s^2-a^2)^2}$	$1 - \cosh ax + \dfrac{a}{2}x\sinh ax$

Table B

Integrals and Identities

Integrals

$$\int xe^{ax}\,dx = \frac{axe^{ax} - e^{ax}}{a^2} + C$$

$$\int \log x\,dx = x\log x - x + C$$

$$\int x\log x\,dx = \frac{1}{2}x^2\log x - \frac{1}{4}x^2 + C$$

$$\int x^2\log x\,dx = \frac{x^3}{3}\log|x| - \frac{x^3}{9} + C$$

$$\int x\sin x\,dx = \sin x - x\cos x + C$$

$$\int x\cos x\,dx = \cos x + x\sin x + C$$

$$\int x^2\sin^2 x\,dx = \frac{x^3}{6} - \left(\frac{x^2}{4} - \frac{1}{8}\right)\sin 2x - \frac{x\cos 2x}{4} + C$$

$$\int x\cos^2 x\,dx = \frac{x^2}{4} + \frac{x\sin 2x}{4} + \frac{\cos 2x}{8} + C$$

$$\int x^2\cos^2 x\,dx = \frac{x^3}{6} + \left(\frac{x^2}{4} - \frac{1}{8}\right)\sin 2x + \frac{x\cos 2x}{4} + C$$

$$\int \frac{dx}{(x+a)(x+b)} = \frac{1}{a-b}\log\left|\frac{x+b}{x+a}\right| + C$$

$$\int \frac{x\,dx}{(x+a)(x+b)} = \frac{1}{a-b}(a\log|x+a| - b\log|x+b|) + C$$

$$\int \frac{dx}{(x+a)^2(x+b)} = \frac{1}{a-b}\left(\frac{1}{x+a} + \frac{1}{a-b}\log\left|\frac{x+b}{x+a}\right|\right) + C$$

$$\int \frac{x\,dx}{(x+a)^2(x+b)} = \frac{-a}{(a-b)(x+a)} - \frac{b}{(a-b)^2}\log\left|\frac{x+b}{x+a}\right| + C$$

$$\int \sin^2 x\,dx = \frac{1}{2}(x - \sin x\cos x) + C$$

$$\int \cos^2 x\,dx = \frac{1}{2}(x + \sin x\cos x) + C$$

$$\int \sin^2 x\cos^2 x\,dx = \frac{1}{8}x - \frac{1}{32}\sin 4x + C$$

$$\int \frac{1}{x^2+a^2}\,dx = \frac{1}{a}\operatorname{Arctan}\frac{x}{a} + C$$

$$\int \frac{1}{x^2-a^2}\,dx = \frac{1}{2a}\log\left|\frac{x-a}{x+a}\right| + C$$

Identities

$$\sin(x \pm y) = \sin x \cos y \pm \cos x \sin y$$

$$\cos(x \pm y) = \cos x \cos y \mp \sin x \sin y$$

$$\sin x \sin y = \frac{1}{2}\cos(x - y) - \frac{1}{2}\cos(x + y)$$

$$\cos x \cos y = \frac{1}{2}\cos(x - y) + \frac{1}{2}\cos(x + y)$$

$$\sin x \cos y = \frac{1}{2}\sin(x - y) + \frac{1}{2}\sin(x + y)$$

$$\sinh x = \frac{e^x - e^{-x}}{2}$$

$$\cosh x = \frac{e^x + e^{-x}}{2}$$

Appendix A

Computer Documentation

This appendix contains the documentation (as listed below) for the Digital Equipment Corporation VAX 11/750 computer system. However, most of the information is portable to other computer installations. The software manufacturer (CONDUIT) has agreed to supply details on program availability. The firm can be contacted at the address furnished in the Preface.

- 1.0 System Overview
- 1.1 System Resources
- 1.2 Program Resources
- 1.3 File Structure
- 1.4 Screen Handling
- 1.5 Special Program Interface
- 1.6 Assembling the Programs
- 1.7 Program Instructions
- 1.8 Sample Program Output
- 1.9 Program Errors

Hope College Documentation Form
Prepared by Mike Ely and Joellyn Shull

```
1.0      SYSTEM OVERVIEW - DIFFEQLIB

     Located in account 'VANIWAARDEN' is a subdirectory, 'DIFFEQLIB',
which contains programs that are used to numerically solve differential
equations.  These programs are designed specifically for the use of
students in the Differential Equations class taught by Dr. VanIwaarden.

     In the DIFFEQLIB is a subdirectory named 'PRINTGRAF', which
contains the programs necessary to graphically display data points
and/or create files for printouts.  The student need not access these
programs directly unless he wishes to make changes to the graph or
printout.  The execution modules within PRINTGRAF are 'chained' to
by the programs in DIFFEQLIB.


     The following is a list of all the programs necessary to complete
the Differential Equations problems:

     DIRECTORY [VANIWAARD]

  DEFILES.COM
```

```
   DIRECTORY [VANIWAARD.DIFFEQLIB]

 BESSLO.BAS      BESSL1.BAS      EULER.BAS       EULSYS.BAS
 HAMMING.BAS     IMEUL.BAS       IMEUL2.BAS      IMEUL3.BAS
 MLNS2.BAS       MILNE.BAS       RKU4S.BAS       RUKU3.BAS
 RUKU4.BAS       SERSING.BAS     SERSO.BAS       SERSO3.BAS
 YACTUL.BAS

   DIRECTORY [VANIWAARD.DIFFEQLIB.PRINTGRAF]

 BESGRAF.BAS     BESGRAF.EXE     BESPRINT.BAS    BESPRINT.EXE
 DEGRAF.BAS      DEGRAF.EXE      DEPRINT.BAS     DEPRINT.EXE
 PRINTOUT.BAS    PRINTOUT.EXE    PROGPRINT.DAT
```

1.1 SYSTEM RESOURCES

The programs in the 'DIFFEQLIB' and 'PRINTGRAF' directories are all written in BASIC and are designed to be run in the interactive mode of BASIC. The 'PRINTOUT', 'BESPRINT', and 'DEPRINT' progams will return the user to the DCL mode when completed. All other programs leave the user in the interactive BASIC mode.

After requesting a printout, a file named 'PRINTFILE.DAT' is created in the user's account. It contains the output from running a 'DIFFEQLIB' program and/or a graphic representation of the data. The user may have the file printed out on the printer at Durfee Computer Center or on any print queue located elsewhere on the campus.

The DCL command module 'DEFILES.COM' assigns logical names to allow users easier access to programs in the DIFFEQLIB. It also allows use of the 'MATHHELP' help library.

1.2 PROGRAM RESOURCES

Files in directory [VANIWAARD]:

```
        ****  DEFILES.COM ****

$ DEFINE BESSLO [VANIWAARD.DIFFEQLIB]BESSLO.BAS
$ DEFINE BESSL1 [VANIWAARD.DIFFEQLIB]BESSL1.BAS
$ DEFINE EULER [VANIWAARD.DIFFEQLIB]EULER.BAS
$ DEFINE EULSYS [VANIWAARD.DIFFEQLIB]EULSYS.BAS
$ DEFINE HAMMING [VANIWAARD.DIFFEQLIB]HAMMING.BAS
$ DEFINE IMEUL [VANIWAARD.DIFFEQLIB]IMEUL.BAS
$ DEFINE IMEUL2 [VANIWAARD.DIFFEQLIB]IMEUL2.BAS
$ DEFINE IMEUL3 [VANIWAARD.DIFFEQLIB]IMEUL3.BAS
$ DEFINE MILNE [VANIWAARD.DIFFEQLIB]MILNE.BAS
$ DEFINE MLNS2 [VANIWAARD.DIFFEQLIB]MLNS2.BAS
$ DEFINE RKU4S [VANIWAARD.DIFFEQLIB]RKU4S.BAS
$ DEFINE RUKU3 [VANIWAARD.DIFFEQLIB]RUKU3.BAS
$ DEFINE RUKU4 [VANIWAARD.DIFFEQLIB]RUKU4.BAS
$ DEFINE SERSING [VANIWAARD.DIFFEQLIB]SERSING.BAS
$ DEFINE SERSO [VANIWAARD.DIFFEQLIB]SERSO.BAS
$ DEFINE SERSO3 [VANIWAARD.DIFFEQLIB]SERSO3.BAS
$ DEFINE YACTUL [VANIWAARD.DIFFEQLIB]YACTUL.BAS
$ ASSIGN DISK$ACD1OF1:[VANIWAARD.DIFFEQLIB]MATHHELP.HLB HLP$LIBRARY
```

1.3 FILE STRUCTURE

This is a diagram of the possible routes a user might take while working with the Differential Equations programs. 'BASIC' and 'DCL' refer to the mode that the user will return to upon completion of the sequence.

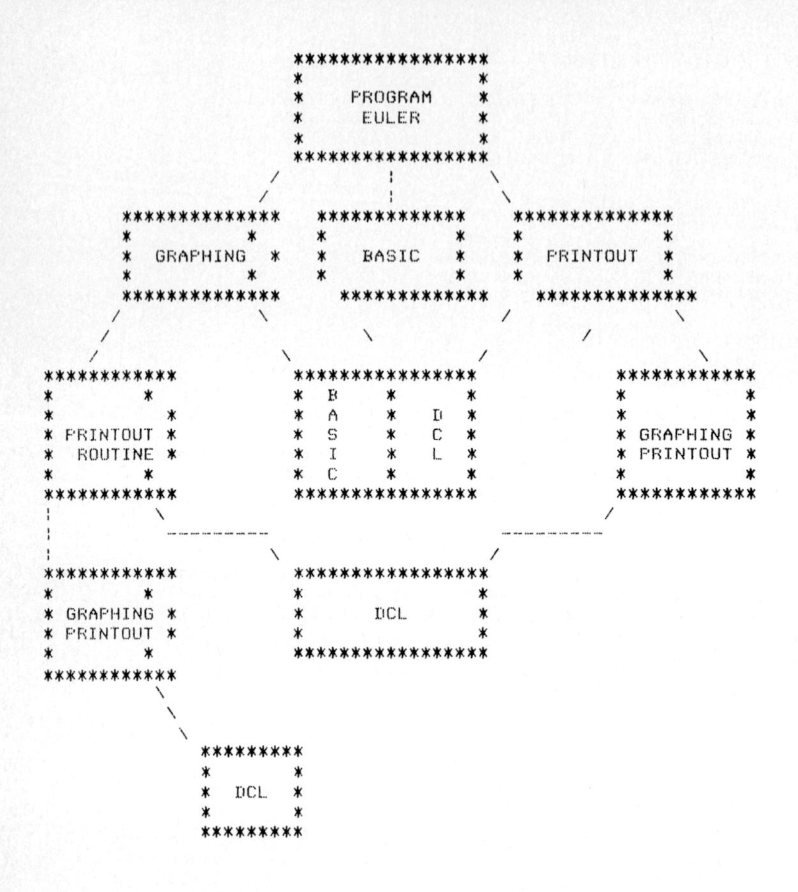

1.4 SCREEN HANDLING

Communication with the user is achieved through use of PRINT and INPUT statements in the BASIC programs.

1.5 SPECIAL PROGRAM INTERFACE

To ease student access to the files in the 'DIFFEQLIB', a command module named 'DEFILES.COM' was created. This module assigns logical names so that the file description is shortened and simplified. It allows the student to access a program in account 'VANIWAARDEN' by simply typing the program name. This also ties in the help library 'MATHHELP.HLB' to the student's account.

The Differential Equations programs all are designed to 'CHAIN' to programs in the 'PRINTGRAF' subdirectory for printouts and graphs. In the subdirectory exists the necessary execution modules.

1.6 ASSEMBLING THE PROGRAMS

In the interactive BASIC mode, a file is automatically compiled and assembled by typing the command 'RUN filespec'. Although the programs that the user accesses directly may be run in this manner, the graphing and printing programs must be previously compiled and linked in order for the 'DIFFEQLIB' programs to 'CHAIN' to them. This need not be done by the user unless a change has been made to one of the graphing or printing programs.

1.7 PROGRAM INSTRUCTIONS

The following is a partial listing of the help file 'HELPFILE.HLP'. This is accessed as a help library under the command 'HELP @MATHHELP'. It contains all the information on the DIFFEQLIB programs, structure, possible errors, and user information.

The help file is organized in a tree-like structure, level 1 being the highest.

1. DIFF_EQ

This help library was created to help you understand and be able to fully utilize the capabilities of the Differential Equations programs.

written by Mike Ely and Joellyn Shull on June 15, 1983
Differential Equations class - Hope College, Holland MI. 49423

2. PRINTOUTS

To obtain a hard copy printout of your data after you have requested it while running the Differential Equations programs, type the following:

```
PRINTOUT
```

This will cause a command module to execute, prompting the user to enter the name to be seen on the banner of the printout. The file PRINTOUT.DAT is renamed to the specified banner name and submitted to the printer. After the printout is made, the file is deleted.

3. /QUEUE

To obtain a printout on a printer at a remote location (not at Durfee Computer Center) type the following:

```
PRINT/QUEUE=queue-name PRINTFILE.DAT
```

where 'queue-name' is the name of the queue at your location. For example, to have a printout made at Vanderwerf Terminal Center, you would type:

```
PRINT/QUEUE=TTAO PRINTFILE.DAT
```

Your printout will now come out on the printer specified.

3. GRAPHING_ROUTINES

The graphing routines which print to the file PRINTFILE.DAT are BESPRINT and DEPRINT. These programs may easily be adjusted to give varying scales and adjustable position of the printing of the graph. To change these values, adjust the constants at the top of the main program.

Caution : When adjusting a printout file remember that the programs chain to compiled routines in Dr. VanIwaarden's account. The 'CHAIN' statements will need to be changed so that they go to an execution file of the program in your account.

Note: The file PRINTFILE.DAT must exist in the user's account before running programs DEPRINT and BESPRINT, since these programs are designed to append an existing file. This is automatically taken care of when program PRINTOUT is run before these routines, which is the normal structure within the program chain (see STRUCTURE for further info).

3. TERMINAL

When you request a printout while running a Differential Equations program, a file called PRINTFILE.DAT is created. This file contains the output which will be sent to a printer with a PRINT command. To see this file on the terminal, type the following:

```
TYPE PRINTFILE.DAT
```

This will cause the contents of the file to be displayed on the screen. The 'NO SCROLL' key in the lower lefthand corner of your keyboard may be used to stop and restart the display for viewing.

3. PRINTOUT_ROUTINE

The printout needed for the Differential Equations assignments is created by a routine called 'PRINTOUT.BAS' found in Dr. VanIwaarden's account. The execution module of this program is 'chained' to by the programs when a printout is asked for.

The program asks for information to be included in the printout that the user must supply. Using this information and the data stored in 'DETEMP.DAT' another file called 'PRINTFILE.DAT' is created, which holds the contents of what will be printed to complete the assignment.

A printout of the graph of the data points may be included in the 'PRINTFILE.DAT' if so requested by the user. This causes 'PRINTOUT.BAS' to chain to one of the graph routines which appends the file 'PRINTFILE.DAT' ('BESPRINT' or 'DEPRINT' depending on the program used).

The user should not have to change program 'PRINTOUT.BAS' to complete the D.E. assignments. If the user does wish to make alterations, programming experience is recommended to correctly adjust 'CHAIN' statements.
(See HELP @DIFFEQHLP DIFF_EQ STRUCTURE for further information)

3. PROBLEMS

Certain intrinsic errors may occur while using the Differential Equations system while trying to obtain printouts. A student account may contain only three versions of a file at any one time. Also, if a file is submitted to the printer and that file is deleted from the student's account before the printout is made, an error will occur when the printer attempts to print a file that no longer exists.

To relieve this problem, the command module PRINTIT.COM is executed when the user types :

```
PRINTOUT
```

This command module renames the file PRINTFILE.DAT to a user specified name, then after printing the file is deleted.

To avoid this problem, the user should name each printout (when prompted by the command module) uniquely, thus avoiding the possibility of the system deleting the file before the printout is made.

2. PROGRAMS

The programs for the Differential Equations class are in Dr. VanIwaarden's account for your use. The list of all the programs available my be seen by typing the following:

```
DIRECTORY [VANIWAARD.DIFFEQLIB]
```

This will show you a listing of all the necessary programs to complete D.E. computer assignments.

3. RUNNING

To run any program in the DIFFEQLIB in Dr. VanIwaarden's account you should first type the following:

```
@[VANIWAARD]DEFILES
```

This routine assigns logical names to your account so that when a program is called (EULER) the computer knows where to find it.

The user should then enter the BASIC environment by typing:

```
BASIC
```

To load a program into memory for use you should type:

```
OLD program-name
```

where 'program-name' is the name of the program you wish to work with. For example, one might type OLD EULER to use program EULER.

Once the program has been loaded into memory, the program may be run by simply typing:

```
RUN
```

after which the computer will run the current program in memory.

Information on making changes to the programs can be found in the CHANGING subtopic of this help library.

IMPORTANT NOTE :

If a graph or printout is requested, the program will be lost in memory. In other words, your changes to the program will no longer be in the program unless you have previously saved it (see SAVING).

3. ACCESSING

To access the programs in the DIFFEQLIB subdirectory of Dr. VanIwaarden's account, two methods may be used. You may access the programs through the logical names assigned by running:

```
  @[VANIWAARD]DEFILES
```

(see RUNNING in this help library). Access to other programs, such as used for printouts and graphing, may be obtained by typing:

```
  [VANIWAARD.DIFFEQLIB.PRINTGRAF]program_name.BAS
```

preceeded by a command such as COPY. An example of this is:

```
  COPY [VANIWAARD.DIFFEQLIB.PRINTGRAF]BESGRAF.BAS GRAPH.BAS
```

which would copy BESGRAF into your account under the name GRAPH.BAS

Caution : To access programs copied or saved under the same name as in Dr. VanIwaarden's account the full program name must be used or the system will default to the program in Dr. VanIwaarden's account. This is caused from using @[VANIWAARD]DEFILES which assigns the program names to be found in his account. For example, if you saved (see SAVING) a program in your account, to be able to reload it in BASIC you would have to type:

```
      OLD program-name.BAS
```

which would then go to your account to find the program.

3. SAVING

In order to save a program once corrections have been made, there are two methods depending on whether the changes where made while in the BASIC environment or in EDIT.

To save a program in the BASIC environment type the following:

```
  SAVE program-name
```

where 'program-name' is the name in your account that you wish the program stored under. The system will use a default type .BAS

To save a program in EDIT depress the CTRL and Z keys together and then type EXIT. This will save the program under the same name you used to enter EDIT, but will increment the version number by 1.

Hint : It is often beneficial to store a program under a name that is different than the program name. In other words, it will be easier to access the program EULER if you store it under EULERA since the computer will not then access Dr. VanIwaarden's account when EULERA is called. If you store a program under the same name, such as EULER, then when you wish to recall it you must use the full program description (EULER.BAS) or the computer will access EULER stored in Dr. VanIwaarden's account. This will save great frustration when deadlines for assignments draw near.

3. CHANGING

Changing the Differential Equations programs is necessary to be able to use the function needed for a particular assignment. There are two methods which can be used to make changes in programs. The line may be retyped while in the BASIC environment, or sections of lines may be altered while in the EDIT mode.

4. BASIC

While in the BASIC environment (type BASIC to enter), after loading the program desired into memory (OLD program-name), the user retypes the line exactly as it is to be in the program. For example, to change the function in program EULER one would type:

```
800    DEF FNF(X,Y) = function
```

where 'function' is the desired function the program should use.

4. EDIT_MODE

To change a program in EDIT is more complicated than in BASIC, yet can be much faster if numerous changes are to be made. To enter the EDIT mode type the following:

```
EDIT/OUTPUT=new-program-name.BAS program-name.BAS
```

where 'new-program-name' is the name you wish the program to be stored under in your account and program-name is the name of the program you wish to edit. (See 'HINT' in SAVING subtopic of this library for suggestions on naming techniques.) Once in EDIT type a 'C' to put you in the screen editor mode. Then proceed to make corrections by moving the curser as necessary, deleting and retyping segments that need to be changed. To leave the screen editor depress the CTRL and Z keys together. Then type 'EXIT' which will return you to DCL.

Note : The editor should only be used by people who have experience using it. To learn screen editing, run the EDTCAI by typing:

```
RUN EDTCAI
```

while in the DCL mode. (You might have to enter 'VINCE' as your name to enter the course since at the time of this writing the course was full)

2. STRUCTURE

The Differential Equations programs are set up in such a way that as many as four programs may be run sequentially under one command, depending on whether printouts or graphs are required. These are not seen by the user and it appears as though only one program were being run.

3. PROGRAMS

The Differential Equations programs can 'chain' to the PRINTOUT and graphing routines if the user desires a printout or graph. If neither is desired, then the system returns to BASIC with the program remaining in memory. If a printout or graph is wanted, the appropriate program is loaded into memory replacing the present program, which is lost if not previously saved in the user's account.(See SAVING.)

3. PRINTOUT

The PRINTOUT routine is chained to by either the Differential Equations programs or a graphing routine. When chained it replaces in memory the calling program.

The PRINTOUT routine builds a file to be printed. It can also chain to a graph-printout routine to add a graph to the printout. If no graph is required the PRINTOUT routine returns the user to DCL so that a printout may be requested. (See PRINTOUTS for information on obtaining printouts.)

3. GRAPHING_ROUTINES

The graphing routines (BESGRAF, DEGRAF) are chained to by the Differential Equations programs to display a graph of the data points on the terminal. These routines may chain either to PRINTOUT if a printout is requested or will return the user to BASIC, with no program stored in memory.

The graph-printout routines (BESPRINT, DEPRINT) are chained to by program PRINTOUT when a printout of the graph is requested. These routines will always return the user to DCL so that a printout may be made.

```
1.8      SAMPLE PROGRAM OUTPUT


      Mike Ely and Joellyn Shull       12345
------------------------------------------------

COMPLETED ON 22-Jun-83      10:59 AM


THESE CHANGES WERE MADE TO PROGRAM 'EULER':


   800  DEF FNF(X,Y) = X**2 - 2


PROBLEM # SAMPLE

       EULER
   ----------------

   X              Y
  ---            ---

-2              0
-1.8             .4
-1.6             .648
-1.4             .76
-1.2             .752
-1               .64
-.8              .44
-.6              .168
-.4             -.16
-.2             -.528001
 .149012E-06    -.920001
 .2             -1.32
 .4             -1.712
 .6             -2.08
 .8             -2.408
 1              -2.68
 1.2            -2.88
 1.4            -2.992
 1.6            -3
 1.8            -2.888
 2              -2.64

MIN. VALUE OF X = -2
MAX. VALUE OF X =  2
MIN. VALUE OF Y = -3
MAX. VALUE OF Y =  .76

Y INCREASES ACROSS THE PAGE TO THE RIGHT.
X INCREASES DOWN THE PAGE.
```

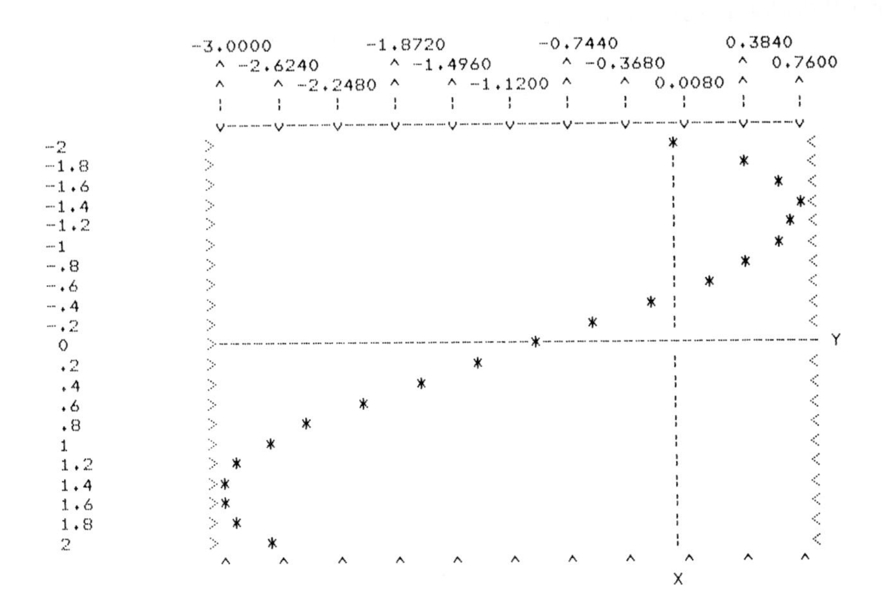

SPACING ACROSS THE PAGE IS EQUAL TO .0752

1.9 PROGRAM ERRORS

This is a partial listing of the help file 'HELPFILE.HLP'.

2. ERRORS

Errors can easily be made while working with the Differential Equations programs. These errors range from entering a function incorrectly to accessing a program or file incorrectly. A description of the error message is given and its probable cause can be found in the subtopics below. Should you encounter an error not found there, consult your instructor.

3. DATA_INPUT

Errors in data input are frequent. There are checks written in the Differential Equations programs to hopefully make erroneous data less of a problem for users. Below is a listing of the most frequently occurring error messages. These can be used to help understand what is causing an error and how to correct it.

Floating point error or overflow:

This error is caused by a value which has gotten out of the computer's range to work with. (For VAX II the range is 1.0E-30 to 1.0E+37). This error is often found when entering a convergence tolerance which is too small for the computer to work to that accuracy. Also, if the function increases quickly and gets out of range, the error will occur.

Data format error:

This error is caused by trying to enter data into a program which is not numeric (such as trying to enter letters for numbers). To cure this simply make sure you enter numeric values.

Illegal argument in LOG:

This error is caused by trying to take the logarithm of a negative or zero value. This is usually caused by incorrect X range or initial values. Check to make sure the function being used will work over the range you specify.

Imaginary square roots:

This error is caused by trying to take the square root of a negative value. This is usually caused by incorrect X range or initial values. Check to make sure the function being used will work over the range you specify.

Division by 0:

This error is caused by trying to divide by a zero value. This is usually caused by incorrect X range or initial values. Check to make sure the function being used will work over the range you specify.

ILLEGAL DATA - explanation of error:

This error is caused by entering data that will not work in the program, such as 0 integrations. Recheck to make sure you enter the correct information for the program.

Note:

Depressing the RETURN key after a comma when entering data to a program causes 0 to be entered. This may cause an error message to appear when a zero value entered is illegal.

3. DEF_FNF

Changing the function definition in each of the programs often causes errors to occur in the programs. The following error messages are ones that could occur from an invalid function definition. Beneath the message is the probable cause and a suggested course of action to correct the difficulty.

Illegal FN redefinition:

This error is caused from trying to define the same function twice in the same program and is most often found when working with system programs such as EULSYS. Check all function definitions to verify that each declares a separate function name.

Arguments don't match:
Too many arguments:
Too few arguments:

This error is caused from defining a function with an argument list that does not match what is written in the program. In other words, the variables between the parenthesis are not the same as those used when the function is called in the program. List out the program to find where the function is called (= FNF()) if you need to see what variables are in the argument list, then make the function definition match this.

Undefined function called:

This error is caused from not having defined a needed function in the program. Check to make sure you have a function definition in your program, and that it is correct.

Syntax error:

This error is caused from entering a line incorrectly in the program, not according to BASIC syntax. The error usually flags where in the line the syntax is incorrect with a '^' symbol underneath the erroneous line. Retype the line with correct BASIC syntax. If you do not know the correct syntax for the statement, ask your instructor.

3. FILE_USAGE

File usage errors occur when the system tries to use a file and is unable to due to various causes. These errors can usually be corrected by using the programs correctly in the right sequence (see STRUCTURE for further information on program chaining and sequence). The errors listed below are those the student could encounter. Along with the error is a probable cause and suggested solution to the error.

Note :
The programs such as PRINTOUT and the graphing routines access the most recent version of the data file DETEMP.DAT (the file with the largest version number).

Not enough data in record:
Too little data in record:

This error is caused from the program trying to input more data from a data file record than in present in the data file. An error of this type is usually found if the wrong program name is specified when requesting a printout to be made. Verify that the program name you entered is the same program name you used to create the data file (the same name as the program you just ran). This error also occurs if the user runs a graphing routine which calls for more data than is present in the data file record. For example, if you create a data file using EULER and then try to graph the data using BESGRAF this error will occur since BESGRAF requires three data coordinates per data record while EULER places only two in the data record.

End of file on device:

This error is caused from trying to run a program such as PRINTOUT or a graphing routine with an empty data file. Verify that the most resent version of the data file has data in it.

File is locked:

This error is caused from someone accessing a file at the same time you wish to. If you are typing out a data file and at the same time on another terminal are trying to run a program requiring the use of this data file the error will occur. Rerun the program again, or run it at a later time when the file is not being used by another user.

Can't find file file or account:

This error occurs when trying to run a program such as PRINTOUT or a graphing routine without the data file DETEMP.DAT being present in your account. Make sure there is at least one version of DETEMP.DAT in your account when trying to run these programs. This error also can occur if you run a graph-printout routine without the file PRINTFILE.DAT in your account, since these two programs (DEPRINT and BESPRINT) are structured to append an existing file.

Protection violation:

This error is caused from trying to access a file without having authorization to do so. This error would occur if the protection on a file where changed to deny access to you. Unless you have changed the protection on the files in your account this error should not occur. If this error occurs when trying to access the programs from the instructor's account, please see him about this.

Appendix B

Computer Programs

This appendix contains the program listings of the numerical methods used in this text. Only the third listed program, called EULER, contains the entire error-control section. In the other programs, the subroutines for error control are not listed. As mentioned in Appendix A, details of these programs are available from the software manufacturer (CONDUIT).

The programs appear in this appendix in alphabetical order by program name, as indicated in Table 1.

Table 1

Program Name	Program Purpose
BESSL0	Calculates and prints values of the two solutions to Bessel's differential equation of order zero.
BESSL1	Calculates and prints values of the two solutions to Bessel's differential equation of order one.
EULER	Calculates an approximate solution to a first-order initial problem by the simple Euler method.
EULSYS	Calculates approximate solutions to a system of first-order initial value problems by the simple Euler method.
HAMMING	Calculates an approximate solution to a first-order initial value problem by Hamming's method.
IMEUL	Calculates an approximate solution to a first-order initial value problem by the improved Euler method.
IMEUL2	Calculates an approximate solution to a second-order initial value problem by the improved Euler method.
IMEUL3	Calculates an approximate solution to a third-order initial value problem by the improved Euler method.
MILNE	Calculates an approximate solution to a first-order initial value problem by Milne's fourth-order method.
MLNS2	Calculates approximate solutions to a system of two first-order initial value problems by Milne's fourth-order method.
RKU4S	Calculates approximate solutions to a system of two first-order initial value problems by the fourth-order Runge-Kutta method.

RUKU3	Calculates an approximate solution to a first-order initial value problem by the third-order Runge-Kutta method.
RUKU4	Calculates an approximate solution to a first-order initial value problem by the fourth-order Runge-Kutta method.
SERSING	Calculates values of two linearly independent solutions of a second-order initial value problem at a regular singular point $x = 0$ by using power-series techniques.
SERSO	Calculates values of the solution of a second-order initial value problem at an ordinary point $x = 0$ by using power-series techniques (when recurrence formula subscripts differ by 2).
SERSO3	Calculates values of the solution of a second-order initial value problem at an ordinary point $x = 0$ by using power-series techniques (when recurrence formula subscripts differ by 3).
YACTUL	Calculates and prints a list of actual x and y values to any initial value problem for which an analytical solution is obtained; used as a comparison tool.

```
100 !              **** PROGRAM BESSLO ****
110 !
120 !       Differential Equations - Hope College
130 !
140 !       program updated for use on VAX II
150 !       by Mike Ely and Joellyn Shull
160 !             on June 15, 1983
170 !
180 !
190 !   THIS IS A SINGLE PURPOSE PROGRAM WHICH WILL CALCULATE
200 !   AND PRINT VALUES OF THE TWO SOLUTIONS TO BESSEL'S
210 !   EQUATION OF ORDER ZERO.
220 !
230 !   THE USER RUNS THE PROGRAM, ENTERING THE LEFT AND RIGHT
240 !   ENDPOINTS OF THE X INTERVAL, THE NUMBER OF POINTS OF
250 !   EVALUATION, AND THE CONVERGENCE TOLERANCE.
260 !
270 !   NOTE: X=0 IS A SINGULAR POINT.
280 !
290 !   COUPLED TO THE PROGRAM IS A GRAPHING ROUTINE ALLOWING
300 !   THE USER TO OPT FOR A PICTORIAL REPRESENTATION OF THE
310 !   DATA WITH BOTH SOLUTIONS ON COMMON AXES.
320 !
330 !
340 !                   VARIABLES
350 !                   ---------
360 !
370 !      LEFT_X_ENDPNT          - LEFT ENDPOINT OF THE X INTERVAL
380 !      RIGHT_X_ENDPNT         - RIGHT ENDPOINT OF THE X INTERVAL
390 !      NUM_EVALUATIONS        - NUMBER OF POINTS OF EVALUATION
400 !      CONVERGENCE_TOL        - CONVERGENCE TOLERANCE
410 !      X_INCREMENT            - X INCREMENT
420 !      I                      - LOOP CONTROL VARIABLE
430 !      PARTIAL_SUM_SOL_1      - PARTIAL SUM OF THE SERIES FOR SOLUTION 1
440 !      PARTIAL_SUM_SOL_2      - PARTIAL SUM OF THE SERIES FOR SOLUTION 2
450 !      SERIES_TERM_SOL_1      - NEWLY COMPUTED SERIES TERM FOR SOLUTION 1
460 !      SERIES_TERM_SOL_2      - NEWLY COMPUTED SERIES TERM FOR SOLUTION 2
470 !      X_SQUARED              - X ** 2 ( X SQUARED )
480 !      SERIES_INDEX           - SERIES INDEX
490 !      HARMONIC_PARTIAL_SUM   - PARTIAL SUM OF HARMONIC SERIES
500 !      ERROR_COUNTER          - COUNTS NUMBER OF ERRORS
510 !      ERROR_NUMBER           - SYSTEM NUMBER OF ERROR
520 !
530 !
540 !  * * * * * * * * * * * * * * * * * * * * * * * * * * * * * * * * *
550 !
```

```
1000     OPEN 'DETEMP.DAT' FOR OUTPUT AS FILE #1
1010 !
1020      PRINT
1030     PRINT 'ENTER LEFT AND RIGHT ENDPOINTS, NUMBER OF EVALUATIONS,'
1040     PRINT 'AND CONVERGENCE TOLERANCE.'
1050 !
1060     ON ERROR GOTO 3000   ! PRINT ERROR DESCRIPTION
1070 !
1080     INPUT LEFT_X_ENDPNT, RIGHT_X_ENDPNT, NUM_EVALUATIONS, CONVERGENCE_TOL
1090 !
1100     GOTO 3200   ! ERROR CHECK FOR INVALID INPUT DATA
1110 !
1120     LET X = LEFT_X_ENDPNT
1130     LET X_INCREMENT = (RIGHT_X_ENDPNT - LEFT_X_ENDPNT) / NUM_EVALUATIONS
1140 !
1150     PRINT
1160     PRINT 'EVALUATIONS:'; NUM_EVALUATIONS
1170     PRINT 'X INCREMENT :'; X_INCREMENT
1180     PRINT
1190     PRINT '  X', '  Y1', '  Y2'
1200     PRINT '------', '-------', '-------'
1210 !
1220     FOR I = 1 TO NUM_EVALUATIONS + 1
1230               LET Y1 = 1
1240        LET SERIES_TERM_SOL_1 = 1
1250        LET HARMONIC_PARTIAL_SUM = 0
1260        LET PARTIAL_SUM_SOL_2 = 0
1270        LET X_SQUARED = X ** 2
1280        LET SERIES_INDEX = 1
1290 !
1300        LET SERIES_TERM_SOL_1 = ( - SERIES_TERM_SOL_1 * X_SQUARED) /  &
               (4 * SERIES_INDEX ** 2)
1310        LET Y1 = Y1 + SERIES_TERM_SOL_1
1320 !
1330        LET SERIES_TERM_SOL_2 = - SERIES_TERM_SOL_1
1340        LET HARMONIC_PARTIAL_SUM = HARMONIC_PARTIAL_SUM + (1 / SERIES_INDEX)
1350        LET PARTIAL_SUM_SOL_2 = PARTIAL_SUM_SOL_2     &
                                   + (SERIES_TERM_SOL_2 * HARMONIC_PARTIAL_SUM)
1360        IF ABS(SERIES_TERM_SOL_1) < CONVERGENCE_TOL THEN 1410
1370 !
1380        LET SERIES_INDEX = SERIES_INDEX + 1
1390        GOTO 1300
1400 !
1410        LET Y2 = Y1 * LOG(X) + PARTIAL_SUM_SOL_2
1420 !
1430        PRINT X, Y1, Y2
1440        PRINT #1, X;',';Y1;',';Y2
1450       LET X = X + X_INCREMENT
1460     NEXT I
1470 !
1480 !
1490  PRINT
1500  PRINT 'Type a 'Y' if you wish these numbers graphed: An 'N' if not...'
1510  INPUT GRAPH$
1520  IF GRAPH$ <> 'Y' THEN 1540
1530  CHAIN '[VANIWAARD.DIFFEQLIB.PRINTGRAF]BESGRAF.EXE'
1540 PRINT 'Type a 'Y' if you want a printout of these numbers; an 'N' if not.'
1550  INPUT P$
1560  IF P$ = 'Y' THEN CHAIN '[VANIWAARD.DIFFEQLIB.PRINTGRAF]PRINTOUT.EXE'
1570 !
1580 !
1590 CLOSE #1
1600 GOSUB 2000   ! DELETE DATA FILES
1610 GOTO 4000   ! END
2000 !
```

```
100 !                  **** PROGRAM BESSL1 ****
110 !
120 !          Differential Equations - Hope College
130 !
140 !
150 !
160 !
170 !
180 !
190 !     THIS IS A SINGLE PURPOSE PROGRAM WHICH WILL CALCULATE
200 !     AND PRINT VALUES OF THE TWO SOLUTIONS TO BESSEL'S
210 !     EQUATION OF ORDER ONE.
220 !
230 !     THE USER RUNS THE PROGRAM, ENTERING THE LEFT AND RIGHT
240 !     ENDPOINTS OF THE X INTERVAL, THE NUMBER OF POINTS OF
250 !     EVALUATION AND THE CONVERGENCE TOLERANCE.
260 !
270 !     NOTE: X=0 IS A SINGULAR POINT
280 !
290 !     COUPLED TO THE PROGRAM IS A GRAPHING ROUTINE ALLOWING
300 !     THE USER TO OPT FOR A PICTORIAL REPRESENTATION OF THE
310 !     DATA WITH BOTH SOLUTIONS ON COMMON AXES.
320 !
330 !
340 !                            VARIABLES
350 !                            ---------
360 !
370 !        LEFT_X_ENDPNT          - LEFT ENDPOINT OF THE X INTERVAL
380 !        RIGHT_X_ENDPNT         - RIGHT ENDPOINT OF THE X INTERVAL
390 !        NUM_EVALUATIONS        - NUMBER OF POINTS OF EVALUATION
400 !        CONVERGENCE_TOL        - CONVERGENCE TOLERANCE
410 !        X_INCREMENT            - X INCREMENT
420 !        I                      - LOOP CONTROL VARIABLE
430 !        PARTIAL_SUM_SOL_1      - PARTIAL SUM OF THE SERIES FOR SOLUTION 1
440 !        PARTIAL_SUM_SOL_2      - PARTIAL SUM OF THE SERIES FOR SOLUTION 2
450 !        SERIES_TERM_SOL_1      - NEWLY COMPUTED SERIES TERM FOR SOLUTION 1
460 !        SERIES_TERM_SOL_2      - NEWLY COMPUTED SERIES TERM FOR SOLUTION 2
470 !        X_SQUARED              - X SQUARED
480 !        SERIES_INDEX           - SERIES INDEX
490 !        HARMONIC_PARTIAL_SUM   - PARTIAL SUM OF HARMONIC SERIES
500 !        ERROR_COUNTER          - COUNTS THE NUMBER OF ERRORS
510 !        ERROR_NUMBER           - SYSTEM NUMBER OF THE ERROR
520 !
530 !
540 !   * * * * * * * * * * * * * * * * * * * * * * * * * * * * * * * * * * *
550 !
1000    OPEN 'DETEMP.DAT'  FOR OUTPUT AS FILE #1
1010 !
1020    PRINT
1030    PRINT 'ENTER LEFT AND RIGHT ENDPOINTS, NUMBER OF EVALUATIONS'
1040    PRINT 'AND CONVERGENCE TOLERANCE.'
1050 !
1060    ON ERROR GOTO 3000  ! PRINT ERROR DESCRIPTION
1070 !
1080    INPUT LEFT_X_ENDPNT, RIGHT_X_ENDPNT, NUM_EVALUATIONS, CONVERGENCE_TOL
1090 !
1100    GOTO 3200  ! ERROR CHECK FOR INVALID DATA
1110 !
1120    LET X = LEFT_X_ENDPNT
1130    LET X_INCREMENT = (RIGHT_X_ENDPNT - LEFT_X_ENDPNT) / NUM_EVALUATIONS
1140 !
1150    PRINT
1160    PRINT 'EVALUATIONS:'; NUM_EVALUATIONS
1170    PRINT 'X INCREMENT :'; X_INCREMENT
1180    PRINT
1190    PRINT '  X', '  Y1', '  Y2'
1200    PRINT '------', '-------', '-------'
1210 !
1220    FOR I =1 TO NUM_EVALUATIONS + 1
1230       LET PARTIAL_SUM_SOL_1 = 1
1240       LET PARTIAL_SUM_SOL_2 = 1
1250       LET SERIES_TERM_SOL_1 = 1
1260       LET SERIES_TERM_SOL_2 = 1
1270       LET X_SQUARED = X ** 2
```

```
1280        LET SERIES_INDEX = 1
1290        LET HARMONIC_PARTIAL_SUM = 0
1300 !
1310        LET SERIES_TERM_SOL_1 = (- SERIES_TERM_SOL_1 * X_SQUARED) /    &
                              (4 * SERIES_INDEX * (SERIES_INDEX + 1))
1320        LET HARMONIC_PARTIAL_SUM = HARMONIC_PARTIAL_SUM + (1 / SERIES_INDEX)
1330        LET SERIES_TERM_SOL_2 = SERIES_TERM_SOL_1 * SERIES_INDEX    &
            * (SERIES_INDEX + 1) * (2 * HARMONIC_PARTIAL_SUM - (1/SERIES_INDEX))
1340 !
1350        LET PARTIAL_SUM_SOL_1 = PARTIAL_SUM_SOL_1 + SERIES_TERM_SOL_1
1360        LET PARTIAL_SUM_SOL_2 = PARTIAL_SUM_SOL_2 - SERIES_TERM_SOL_2
1370 !
1380        IF ABS(SERIES_TERM_SOL_2) < CONVERGENCE_TOL THEN 1420
1390        LET SERIES_INDEX = SERIES_INDEX + 1
1400        GOTO 1310
1410 !
1420        LET Y1 = .5 * X * PARTIAL_SUM_SOL_1
1430        LET Y2 = (-.5) * Y1 * LOG(X) + .5 * (1/X) * PARTIAL_SUM_SOL_2
1440 !
1450        PRINT X,Y1,Y2
1460        PRINT #1,X;",";Y1;",";Y2
1470       LET X = X + X_INCREMENT
1480  NEXT I
1490 !
1500 !
1510 PRINT
1520 PRINT "Type a 'Y' if you wish these numbers graphed: An 'N' if not..."
1530 INPUT GRAPH$
1540 IF GRAPH$ <> 'Y' THEN 1560
1550 CHAIN "[VANIWAARD.DIFFEQLIB.PRINTGRAF]BESGRAF.EXE"
1560 PRINT "Type a 'Y' if you want a printout of these numbers; an 'N' if not."
1570 INPUT P$
1580 IF P$ = 'Y' THEN CHAIN "[VANIWAARD.DIFFEQLIB.PRINTGRAF]PRINTOUT.EXE"
1590 !
1600 !
1610 CLOSE #1
1620 GOSUB 2000   ! DELETE DATA FILES
1630 GOTO 4000   ! END
2000 !
```

```
100 !                   **** PROGRAM EULER ****
110 !
120 !         Differential Equations - Hope College
130 !
140 !
150 !
160 !
170 !
180 !
190 !   THIS PROGRAM USES EULER'S METHOD TO FIND AN APPROXIMATE
200 !   SOLUTION TO A FIRST ORDER INITIAL VALUE PROBLEM:
210 !   Y' = F(X,Y), Y(X0) = Y0.
220 !
230 !   THE USER DEFINES THE DESIRED FUNCTION FNF(X,Y), THEN
240 !   RUNS THE PROGRAM, ENTERING THE LEFT AND RIGHT ENDPOINTS
250 !   OF THE X INTERVAL, THE INITIAL VALUE OF Y, THE NUMBER
260 !   OF INTEGRATIONS, AND THE PRINT STEP SIZE.
270 !
280 !   THE PROGRAM PRINTS THE INITIAL VALUES OF X, Y, STEP SIZE
290 !   AND NUMBER OF INTEGRATIONS.  IT THEN PRINTS A LIST OF
300 !   SOLUTION POINT PAIRS, WHERE X IS INCREMENTED EACH TIME
310 !   BY THE PRINT STEP SIZE, AND Y IS COMPUTED USING EULER'S
320 !   FORMULA.
330 !
340 !   NOTE: FNF(X,Y) DEFINED AT LINE # 800
350 !
360 !
370 !                        VARIABLES
380 !                        ---------
390 !
400 !        FNF                    - FUNCTION DESCRIBING THE DERIVATIVE OF Y
410 !        LEFT_X_ENDPNT          - LEFT ENDPOINT OF THE X-INTERVAL
420 !        RIGHT_X_ENDPNT         - RIGHT ENDPOINT OF THE X-INTERVAL
430 !        Y0                     - INITIAL VALUE OF Y
440 !        NUM_INTEGRATIONS       - NUMBER OF INTEGRATIONS
450 !        PRINT_STEP_SIZE        - COUNTER TO PRINT AFTER M ITERATIONS
460 !        X_INCREMENT            - X INCREMENT
470 !        ERROR_COUNTER          - COUNTS THE NUMBER OF ERRORS
480 !        ERROR_NUMBER           - SYSTEM NUMBER OF AN ERROR
490 !        I                      - LOOP CONTROL VARIABLE
500 !        F                      - VALUE OF FUNCTION AT (X,Y)
510 !
520 !
530 ! * * * * * * * * * * * * * * * * * * * * * * * * * *
540 !
800     DEF FNF(X,Y) = 3*Y - X
1000 !
1010    OPEN 'DETEMP.DAT' FOR OUTPUT AS FILE #1
1020 !
1030    PRINT
1040    PRINT 'ENTER LEFT AND RIGHT ENDPOINTS, INITIAL Y VALUE,'
1050    PRINT 'NUMBER OF INTEGRATIONS AND PRINT STEP SIZE.'
1060 !
1070    ON ERROR GOTO 3000 !ROUTINE TO HANDLE ILLEGAL DATA
1080    INPUT LEFT_X_ENDPNT, RIGHT_X_ENDPNT, Y0, NUM_INTEGRATIONS, &
        PRINT_STEP_SIZE
1090 !
1100    GOTO 3200   ! CHECK FOR ERRORS IN INPUT DATA
1110 !
1120    LET X = LEFT_X_ENDPNT
1130    LET Y = Y0
1140    LET X_INCREMENT = (RIGHT_X_ENDPNT - LEFT_X_ENDPNT)/NUM_INTEGRATIONS
1150 !
1160    PRINT
1170    PRINT 'INTEGRATIONS:'; NUM_INTEGRATIONS
1180    PRINT 'X INCREMENT :'; X_INCREMENT
1190    PRINT
1200    PRINT '  X', '  Y'
1210    PRINT '-----', '-----'
1220    PRINT X, Y
1230    PRINT #1,X;',';Y
1240 !
```

```
1250    FOR I = 1 TO NUM_INTEGRATIONS
1260       LET F = FNF(X,Y)
1270       LET Y = Y + X_INCREMENT * F
1280       LET X = X + X_INCREMENT
1290 !
1300       IF INT(I/PRINT_STEP_SIZE) = I/PRINT_STEP_SIZE THEN 1320
1310       IF I <> NUM_INTEGRATIONS THEN 1340
1320       PRINT X,Y
1330       PRINT #1,X;",";Y
1340    NEXT I
1350 !
1360 !
1370  PRINT
1380  INPUT "Type a 'Y' if you want to see a graph: "; GRAPH$
1400  IF GRAPH$ <> 'Y' THEN 1420
1410    CHAIN "[VANIWAARD.DIFFEQLIB.PRINTGRAF]DEGRAF.EXE"
1420  INPUT "Type a 'Y' if you want a printout file made of the data: "; P$
1440  IF P$ = 'Y' THEN CHAIN "[VANIWAARD.DIFFEQLIB.PRINTGRAF]PRINTOUT.EXE"
1450  CLOSE #1
1460  GOSUB 2000   ! ROUTINE TO DELETE ALL VERSIONS OF DATA FILE
1470 GOTO 4000   ! END
2000 !
2010 !               SUBROUTINE TO DELETE ALL VERSIONS OF "DETEMP.DAT"
2020 !               ------------------------------------------------
2030 !
2040      ON ERROR GOTO 2070
2050       KILL "DETEMP.DAT"
2060      GOTO 2040
2070      ERROR_NUMBER = ERR
2080      IF ERROR_NUMBER = 5 THEN RESUME 2120 !  ALL DATA FILES DELETED
2090      PRINT "ERROR - "; ERT$(ERROR_NUMBER)
2100      PRINT " AT LINE # "; ERL
2110      RESUME 4000  ! GOTO END
2120      RETURN
3000 !
3010 !               ERROR IN INPUT DATA ROUTINE
3020 !               ---------------------------
3030 !
3040      ERROR_NUMBER = ERR
3050      PRINT "ERROR - "; ERT$(ERROR_NUMBER)
3060      PRINT " AT LINE NUMBER ";ERL
3070      ERROR_COUNTER = ERROR_COUNTER + 1
3080      IF ERROR_COUNTER < 5 THEN CLOSE #1 \ RESUME 1010  !ENTER DATA AGAIN
3090      PRINT "ERROR IN INPUT HAS OCCURRED 5 TIMES "
3100      PRINT "** PROGRAM ABORTED **"
3110      RESUME 4000  ! GOTO END
3200 !

3210 !               CHECK FOR EQUAL X ENDPOINTS
3220 !               ---------------------------
3230 !
3240      IF LEFT_X_ENDPNT <> RIGHT_X_ENDPNT THEN GOTO 3410 !  NEXT ERROR CHECK
3250      PRINT "ILLEGAL DATA - LEFT AND RIGHT ENDPOINTS MUST NOT BE EQUAL"
3260      GOTO 3670 !COUNT NUMBER OF ERRORS - IF 5 ERRORS ABORT PROGRAM
3400 !
3410 !               CHECK FOR NONPOSITIVE NUMBER OF INTEGRATIONS
3420 !               --------------------------------------------
3430 !
3440      IF NUM_INTEGRATIONS > 0 THEN GOTO 3610 !  NEXT ERROR CHECK
3450      PRINT "ILLEGAL DATA - NUMBER OF INTEGRATIONS MUST BE POSITIVE"
3460      GOTO 3670 !COUNT NUMBER OF ERRORS - IF 5 ABORT PROGRAM
3600 !
3610 !               CHECK FOR NONPOSITIVE PRINT STEP SIZE
3620 !               -------------------------------------
3630 !
3640      IF PRINT_STEP_SIZE > 0 THEN GOTO 1120  ! NO ERRORS FOUND
3650      PRINT "ILLEGAL DATA - PRINT STEP SIZE MUST BE POSITIVE"
3660 !
3670      ERROR_COUNTER = ERROR_COUNTER + 1
3680      IF ERROR_COUNTER < 5 THEN GOTO 1020  !ENTER DATA AGAIN
3690      PRINT "ERROR IN INPUT HAS OCCURRED 5 TIMES"
3700      PRINT "** PROGRAM ABORTED **"
3800  GOTO 1450 !CLOSE AND DELETE DATA FILES - END PROGRAM
4000  END
```

```
100 !                **** PROGRAM EULSYS ****
110 !
120 !         Differential Equations - Hope College
130 !
140 !
150 !
160 !
170 !
180 !
190 !  THIS PROGRAM USES EULER'S METHOD TO SOLVE A SYSTEM OF
200 !  TWO FIRST ORDER DIFFERENTIAL EQUATIONS WITH INITIAL
210 !  CONDITIONS: Y' = F(X,Y,Z), Z' = G(X,Y,Z), Y(XO) = YO,
220 !  Z(XO) = ZO.
230 !
240 !  THE USER DEFINES THE TWO FUNCTIONS, FNF(X,Y,Z) AND
250 !  FNG(X,Y,Z), RUNS THE PROGRAM AND ENTERS THE LEFT AND
260 !  RIGHT ENDPOINTS OF THE X INTERVAL, INITIAL VALUES OF
270 !  Y AND Z, THE NUMBER OF INTEGRATIONS AND THE PRINT STEP
280 !  SIZE.
290 !
300 !  THE PROGRAM PRINTS THE INITIAL VALUES, THEN PRINTS A
310 !  LIST OF X, Y, AND Z VALUES, WHERE X IS INCREMENTED BY
320 !  THE PRINT STEP SIZE, AND Y AND Z ARE COMPUTED USING
330 !  EULER'S FORMULA.
340 !
350 !
360 !  NOTE: FNF(X,Y,Z) DEFINED AT LINE # 800
370 !        FNG(X,Y,Z) DEFINED AT LINE # 810
380 !
390 !
400 !                    VARIABLES
410 !                    ---------
420 !
430 !      FNF                 - FUNCTION DESCRIBING THE DERIVATIVE OF Y
440 !      FNG                 - FUNCTION DESCRIBING THE DERIVATIVE OF Z
450 !      LEFT_X_ENDPNT       - LEFT ENDPOINT OF THE X INTERVAL
460 !      RIGHT_X_ENDPNT      - RIGHT ENDPOINT OF THE X INTERVAL
470 !      YO                  - INITIAL VALUE OF Y
480 !      ZO                  - INITIAL VALUE OF Z
490 !      NUM_INTEGRATIONS    - NUMBER OF INTEGRATIONS
500 !      PRINT_STEP_SIZE     - PRINT STEP SIZE
510 !      X_INCREMENT         - X INCREMENT
520 !      ERROR_COUNTER       - COUNTS THE NUMBER OF ERRORS
530 !      ERROR_NUMBER        - SYSTEM NUMBER OF THE ERROR
540 !      I                   - LOOP VARIABLE
550 !      F                   - VALUE OF FUNCTION FNF AT (X,Y,Z)
560 !      G                   - VALUE OF FUNCTION FNG AT (X,Y,Z)
570 !
580 !
590 ! * * * * * * * * * * * * * * * * * * * * * * * * * * * * *
600 !
800     DEF FNF(X,Y,Z) = Z
810     DEF FNG(X,Y,Z) = 3*Y + X
1000 !
1010    OPEN 'DETEMP.DAT' FOR OUTPUT AS FILE #1
1020 !
1030    PRINT
1040    PRINT 'ENTER LEFT AND RIGHT ENDPOINTS OF THE X INTERVAL,'
1050    PRINT 'INITIAL VALUES OF Y AND Z, NUMBER OF INTEGRATIONS'
1060    PRINT 'AND THE PRINT STEP SIZE.'
1070 !
1080     ON ERROR GOTO 3000  ! ROUTINE TO HANDLE ILLEGAL DATA
1090     INPUT LEFT_X_ENDPNT, RIGHT_X_ENDPNT, YO, ZO, NUM_INTEGRATIONS, &
              PRINT_STEP_SIZE
1100 !
1110    GOTO 3200   ! CHECK FOR ERRORS IN INPUT DATA
1120 !
1130    LET X = LEFT_X_ENDPNT
1140    LET Y = YO
1150    LET Z = ZO
1160    LET X_INCREMENT = (RIGHT_X_ENDPNT - LEFT_X_ENDPNT)/NUM_INTEGRATIONS
1170 !
```

```
1180     PRINT
1190     PRINT 'INTEGRATIONS:'; NUM_INTEGRATIONS
1200     PRINT 'X INCREMENT :'; X_INCREMENT
1210     PRINT
1220     PRINT '  X', '  Y', '  Z'
1230     PRINT '------', '------', '------'
1240 !
1250     PRINT X, Y, Z
1260     PRINT #1,X;',';Y;',';Z
1270 !
1280     FOR I= 1 TO NUM_INTEGRATIONS
1290                LET F = FNF(X,Y,Z)
1300        LET G = FNG(X,Y,Z)
1310 !
1320        LET Y = Y + X_INCREMENT * F
1330        LET Z = Z + X_INCREMENT * G
1340        LET X = X + X_INCREMENT
1350 !
1360        IF INT(I/PRINT_STEP_SIZE) = I/PRINT_STEP_SIZE THEN 1380
1370        IF I <> NUM_INTEGRATIONS THEN 1400
1380        PRINT X,Y,Z
1390        PRINT #1,X;',';Y;',';Z
1400     NEXT I
1410 !
1420 !
1430 PRINT
1440 PRINT
1450 INPUT 'Type a 'Y' if you want to see a graph: ';GRAPH$
1470 IF GRAPH$ <> 'Y' THEN 1490
1480    CHAIN '[VANIWAARD.DIFFEQLIB.PRINTGRAF]BESGRAF.EXE'
1490 INPUT 'Type a 'Y' if you want a printout file made of the data: '; P$
1510 IF P$ = 'Y' THEN CHAIN '[VANIWAARD.DIFFEQLIB.PRINTGRAF]PRINTOUT.EXE'
1520 CLOSE #1
1530 GOSUB 2000   ! ROUTINE TO DELETE ALL VERSIONS OF DATA FILE
1540 GOTO 4000   ! END
2000 !
```

```
100 !              **** PROGRAM HAMMING ****
110 !
120 !       Differential Equations - Hope College
130 !
140 !
150 !
160 !
170 !
180 !
190 !  THIS PROGRAM USES HAMMING'S METHOD TO APPROXIMATE THE
200 !  SOLUTION OF A FIRST ORDER INITIAL VALUE PROBLEM:
210 !  Y' = F(X,Y), Y(X0) = Y0.
220 !
230 !  THE USER DEFINES THE DESIRED FUNCTION, FNF(X,Y), THEN
240 !  RUNS THE PROGRAM, ENTERING THE LEFT AND RIGHT ENDPOINTS
250 !  OF THE X INTERVAL, INITIAL VALUES OF Y, THE NEXT THREE
260 !  'STARTING' VALUES OF Y: Y1, Y2, Y3, (OBTAINED BY SOME
270 !  OTHER METHOD), AND THE NUMBER OF INTEGRATIONS.
280 !
290 !  THIS PROGRAM PRINTS THE INITIAL VALUES OF THE
300 !  FUNCTION AT THE 'STARTING' VALUES Y(1) THRU Y(4),
310 !  THEN USING HAMMING'S PREDICTOR AND CORRECTOR FORMULAE,
320 !  COMPUTES AND PRINTS THE APPROXIMATE SOLUTION POINTS.
330 !
340 !  NOTE: FNF(X,Y) DEFINED AT LINE # 800
350 !
360 !
370 !                    VARIABLES
380 !                    ---------
390 !
400 !       FNF                - FUNCTION DESCRIBING THE DERIVATIVE OF Y
410 !       Y()                - Y VALUES, Y(1)-Y(4) ARE 'INITIAL' VALUES
420 !       X()                - X VALUES
430 !       LEFT_X_ENDPNT      - LEFT ENDPOINT OF THE X INTERVAL
440 !       RIGHT_X_ENDPNT     - RIGHT ENDPOINT OF THE X INTERVAL
450 !       NUM_INTEGRATIONS   - NUMBER OF INTEGRATIONS
460 !       X_INCREMENT        - X INCREMENT
470 !       F()                - VALUE OF FUNCTION AT (X(I),Y(I))
480 !       I                  - LOOP CONTROL VARIABLE
490 !       Y_PREDICTED        - PREDICTED VALUE OF Y
500 !       VALUE_FNF          - VALUE OF FUNCTION AT (X(I),Y_PREDICTED)
510 !       Y(I)               - CORRECTED VALUE OF Y
520 !       ERROR_COUNTER      - COUNTS THE NUMBER OF ERRORS
530 !       ERROR_NUMBER       - SYSTEM NUMBER OF AN ERROR
540 !
550 !
560 ! * * * * * * * * * * * * * * * * * * * * * * * * * * * * *
570 !
580 !
800     DEF FNF(X,Y) = 2 * Y - X
1000 !
1010     DIMENSION X(100), Y(100)
1020     DIMENSION F(4)
1030 !
1040    OPEN 'DETEMP.DAT' FOR OUTPUT AS FILE #1
1050 !
1060    PRINT
1070    PRINT 'ENTER LEFT AND RIGHT ENDPOINTS OF THE X INTERVAL,'
1080    PRINT 'INITIAL VALUE OF Y, THE NEXT THREE 'STARTING' VALUES'
1090    PRINT 'OF Y: Y1, Y2, Y3, AND NUMBER OF INTEGRATIONS.'
1100 !
1110     ON ERROR GOTO 3000 ! BAD DATA INPUT
1120     INPUT LEFT_X_ENDPNT, RIGHT_X_ENDPNT, Y(1), Y(2), Y(3), Y(4),  &
             NUM_INTEGRATIONS
1130      GOTO 3200  ! CHECK FOR ERRORS IN INPUT DATA
1140 !
1150    LET X_INCREMENT = (RIGHT_X_ENDPNT - LEFT_X_ENDPNT)/NUM_INTEGRATIONS
1160 !
1170    PRINT
1180    PRINT 'INTEGRATIONS:'; NUM_INTEGRATIONS
1190    PRINT 'X INCREMENT :'; X_INCREMENT
```

```
1200     PRINT
1210     PRINT '  X', '  Y'
1220     PRINT '------', '------'
1230     FOR I = 1 TO 4
1240               LET X(I) = LEFT_X_ENDPNT + (I-1) * X_INCREMENT
1250        LET F(I) = FNF(X(I),Y(I))
1260        PRINT X(I),Y(I)
1270        PRINT #1,X(I);',';Y(I)
1280     NEXT I
1290     FOR I = 5 TO NUM_INTEGRATIONS + 1
1300        LET X(I) = X(I-1) + X_INCREMENT
1310        LET Y_PREDICTED = Y(I-4) + (4*X_INCREMENT/3) * (2*F(4)-F(3) + 2*F(2))
1320        LET VALUE_FNF = FNF(X(I),Y_PREDICTED)
1330        LET Y(I) = .125*(9*Y(I-1) - Y(I-3))+(.375*X_INCREMENT)*(VALUE_FNF  &
              + 2 * F(4) - F(3))
1340        PRINT X(I),Y(I)
1350        PRINT #1,X(I);',';Y(I)
1360        FOR J = 1 TO 3
1370          LET F(J) = F(J+1)
1380        NEXT J
1390        LET  F(4) = FNF(X(I),Y(I))
1400     NEXT I
1410 !
1420 !
1430 PRINT
1440 INPUT 'Type a 'Y' if you want to see a graph: '; GRAPH$
1460 IF GRAPH$ <> 'Y' THEN 1480
1470 CHAIN '[VANIWAARD.DIFFEQLIB.PRINTGRAF]DEGRAF.EXE'
1480 INPUT 'Type a 'Y' if you want a printout made of the data: ';PRINT$
1500 IF PRINT$ = 'Y' THEN CHAIN '[VANIWAARD.DIFFEQLIB.PRINTGRAF]PRINTOUT.EXE'
1510 CLOSE #1
1520 GOSUB 2000  !ROUTINE TO DELETE DATA FILES
1530 GOTO 4000  ! END
2000 !
```

```
100 !              **** PROGRAM IMEUL ****
110 !
120 !        Differential Equations - Hope Collge
130 !
140 !
150 !
160 !
170 !
180 !
190 !   THIS PROGRAM USES THE IMPROVED EULER TECHNIQUE TO
200 !   APPROXIMATE A SOLUTION TO A FIRST ORDER INITIAL VALUE
210 !   PROBLEM: Y' = F(X,Y), Y(XO) = YO.
220 !
230 !   THE USER DEFINES THE DESIRED FUNCTION FNF(X,Y), AND
240 !   RUNS THE PROGRAM, ENTERING THE LEFT AND RIGHT ENDPOINTS
250 !   OF THE X INTERVAL, INITIAL VALUE OF Y, NUMBER OF
260 !   INTEGRATIONS, AND THE PRINT STEP SIZE.
270 !
280 !   THE PROGRAM PRINTS THE INITIAL VALUES, THEN A LIST OF
290 !   X AND Y VALUES WHERE X IS INCREMENTED BY THE PRINT STEP
300 !   SIZE AND Y IS COMPUTED USING THE IMPROVED EULER
310 !   FORMULA.
320 !
330 !   NOTE: FNF(X,Y) IS DEFINED AT LINE # 800
340 !
350 !
360 !                    VARIABLES
370 !                    ---------
380 !
390 !        FNF                - FUNCTION DESCRIBING THE DERIVATIVE OF Y
400 !        LEFT_X_ENDPNT      - LEFT ENDPOINT OF THE X INTERVAL
410 !        RIGHT_X_ENDPNT     - RIGHT ENDPOINT OF THE X INTERVAL
420 !        YO                 - INITIAL VALUE OF Y
430 !        NUM_INTEGRATIONS   - NUMBER OF INTEGRATIONS
440 !        PRINT_STEP_SIZE    - PRINT STEP SIZE
450 !        X_INCREMENT        - X INCREMENT
460 !        I                  - LOOP VARIABLE
470 !        FO                 - VALUE OF FUNCTION AT (X,Y)
480 !        Y_PREDICTED        - PREDICTOR VALUE OF Y
490 !        VALUE_FNF          - VALUE OF FUNCTION AT (X,Y_PREDICTED)
500 !        ERROR_COUNTER      - COUNTS THE NUMBER OF ERRORS MADE
510 !        ERROR_NUMBER       - SYSTEM NUMBER OF THE ERROR
520 !
530 !
540 ! * * * * * * * * * * * * * * * * * * * * * * * * * * * * *
550 !
800     DEF FNF(X,Y) = X**2 + Y
1000 !
1010    OPEN 'DETEMP.DAT' FOR OUTPUT AS  FILE #1
1020 !
1030    PRINT
1040    PRINT 'ENTER LEFT AND RIGHT ENDPOINTS OF THE X INTERVAL,'
1050    PRINT 'INITIAL Y VALUE, NUMBER OF INTEGRATIONS,'
1060     PRINT 'AND PRINT STEP SIZE.'
1070 !
1080     ON ERROR GOTO 3000 !PRINT ERROR DESCRIPTION
1090     INPUT LEFT_X_ENDPNT, RIGHT_X_ENDPNT, YO, NUM_INTEGRATIONS, &
         PRINT_STEP_SIZE
1100     GOTO 3200   ! CHECK FOR ERRORS IN INPUT DATA
1110 !
1120    LET X = LEFT_X_ENDPNT
1130    LET Y = YO
1140    LET X_INCREMENT = (RIGHT_X_ENDPNT - LEFT_X_ENDPNT)/NUM_INTEGRATIONS
1150 !
1160    PRINT
1170    PRINT 'INTEGRATIONS:'; NUM_INTEGRATIONS
1180    PRINT 'X INCREMENT :'; X_INCREMENT
1190    PRINT
1200    PRINT '  X', '  Y'
1210    PRINT '-----', '-----'
1220    PRINT X, Y
```

```
1230    PRINT #1,X;",";Y
1240 !
1250    FOR I = 1 TO NUM_INTEGRATIONS
1260              LET F0 = FNF(X,Y)
1270       LET Y_PREDICTED = Y + X_INCREMENT * F0
1280       LET X = X + X_INCREMENT
1290       LET VALUE_FNF = FNF(X,Y_PREDICTED)
1300       LET Y = Y + .5 * X_INCREMENT * (F0 + VALUE_FNF)
1310       IF INT(I/PRINT_STEP_SIZE) = I/PRINT_STEP_SIZE THEN 1330
1320       IF I <> NUM_INTEGRATIONS THEN 1350
1330       PRINT X,Y
1340       PRINT #1,X;",";Y
1350    NEXT I
1360 !
1370 !
1380 PRINT
1390 INPUT "Type a 'Y' if you want to see a graph: ";GRAPH$
1410 IF GRAPH$ <> 'Y' THEN 1430
1420 CHAIN "[VANIWAARD.DIFFEQLIB.PRINTGRAF]DEGRAF.EXE"
1430 INPUT "Type a 'Y' if you want a printout file made of the data: ";PRINT$
1450 IF PRINT$ = 'Y' THEN CHAIN "[VANIWAARD.DIFFEQLIB.PRINTGRAF]PRINTOUT.EXE"
1460 CLOSE #1
1470 GOSUB 2000   ! ROUTINE TO DELETE DATA FILES
1480 GOTO 4000   ! END
2000 !
```

```
100 !              **** PROGRAM IMEUL2 ****
110 !
120 !        Differential  Equations - Hope College
130 !
140 !
150 !
160 !
170 !
180 !
190 !  THIS PROGRAM USES THE IMPROVED EULER TECHNIQUE TO
200 !  APPROXIMATE A SOLUTION TO A SECOND ORDER DIFFERENTIAL
210 !  EQUATION WITH INITIAL CONDITIONS: Y'' = F(X,Y,Y'),
220 !  Y(XO) = YO, Y'(XO) = VO.
230 !
240 !  THE USER DEFINES THE DESIRED FUNCTION, FNF(X,Y,V), RUNS
250 !  THE PROGRAM AND ENTERS THE LEFT AND RIGHT ENDPOINTS OF
260 !  THE X INTERVAL, INITIAL VALUES OF Y AND Y'(NOTE V=Y'),
270 !  NUMBER OF INTEGRATIONS AND THE PRINT STEP SIZE.
280 !
290 !  THE PROGRAM PRINTS THE INITIAL VALUES, INCREMENTS X BY
300 !  THE PRINT STEP SIZE, COMPUTES AND LISTS X, Y AND Y' AT EACH
310 !  X VALUE USING THE IMPROVED EULER FORMULAE.
320 !
330 !  NOTE: FNF(X,Y,V) IS DEFINED AT LINE # 800
340 !
350 !
360 !                    VARIABLES
370 !                    ---------
380 !
390 !       FNF                  - FUNCTION DESCRIBING THE SECOND DERIVATIVE OF Y
400 !       LEFT_X_ENDPNT        - LEFT ENDPOINT OF THE X INTERVAL
410 !       RIGHT_X_ENDPNT       - RIGHT ENDPOINT OF THE X INTERVAL
420 !       YO                   - INITIAL VALUE OF Y
430 !       VO                   - INITIAL VALUE OF V
440 !       NUM_INTEGRATIONS     - NUMBER OF INTEGRATIONS
450 !       PRINT_STEP_SIZE      - PRINT STEP SIZE
460 !       X_INCREMENT          - X INCREMENT
470 !       ERROR_COUNTER        - COUNTS THE NUMBER OF ERRORS
480 !       ERROR_NUMBER         - SYSTEM NUMBER OF AN ERROR
490 !       I                    - LOOP VARIABLE
500 !       F                    - VALUE OF FUNCTION AT (X,Y,V)
510 !       Y_PREDICTED          - PREDICTOR VALUE OF Y
520 !       V_PREDICTED          - PREDICTOR VALUE OF V
530 !
540 !
550 ! * * * * * * * * * * * * * * * * * * * * * * * * * * * * *
560 !
800     DEF FNF(X,Y,V) = X - Y**2*V
1000 !
1010    OPEN 'DETEMP.DAT' FOR OUTPUT AS FILE #1
1020 !
1030    PRINT
1040    PRINT 'ENTER LEFT AND RIGHT ENDPOINTS OF THE X INTERVAL,'
1050    PRINT 'INITIAL VALUES OF Y AND Y', NUMBER OF INTEGRATIONS'
1060    PRINT 'AND PRINT STEP SIZE.'
1070 !
1080     ON ERROR GOTO 3000  ! PRINT ERROR DESCRIPTION
1090     INPUT LEFT_X_ENDPNT, RIGHT_X_ENDPNT, YO, VO, &
              NUM_INTEGRATIONS, PRINT_STEP_SIZE
1100     GOTO 3200  ! CHECK FOR ERRORS IN INPUT DATA
1110 !
1120    LET X_INCREMENT = (RIGHT_X_ENDPNT - LEFT_X_ENDPNT)/NUM_INTEGRATIONS
1130    LET X = LEFT_X_ENDPNT
1140    LET Y = YO
1150    LET V = VO
1160 !
1170    PRINT
1180    PRINT 'INTEGRATIONS:'; NUM_INTEGRATIONS
1190    PRINT 'X INCREMENT :'; X_INCREMENT
1200    PRINT
1210    PRINT '  X', '  Y', '  Y''
```

```
1220     PRINT '------', '------', '-------'
1230     PRINT X, Y, V
1240     PRINT #1,X;',';Y;',';V
1250 !
1260     FOR I = 1 TO NUM_INTEGRATIONS
1270               LET Y_PREDICTED = Y + X_INCREMENT * V
1280        LET F = FNF(X,Y,V)
1290        LET V_PREDICTED = V + X_INCREMENT * F
1300        LET X = LEFT_X_ENDPNT + X_INCREMENT * I
1310        LET Y = Y + .5 * X_INCREMENT * (V + V_PREDICTED)
1320        LET V = V + .5 * X_INCREMENT * (F + FNF(X,Y_PREDICTED,V_PREDICTED))
1330 !
1340        IF INT(I/PRINT_STEP_SIZE) = I/PRINT_STEP_SIZE THEN 1360
1350        IF I <> NUM_INTEGRATIONS THEN 1380
1360        PRINT X,Y,V
1370        PRINT #1,X;',';Y;',';V
1380     NEXT I
1390 !
1400 !
1410 PRINT
1420 INPUT 'Type a 'Y' if you want to see a graph: ';GRAPH$
1440 IF GRAPH$ <> 'Y' THEN 1460
1450 CHAIN '[VANIWAARD.DIFFEQLIB.PRINTGRAF]DEGRAF.EXE'
1460 INPUT 'Type a 'Y' if you want a printout made of the data: '; PRINT$
1480 IF PRINT$ = 'Y' THEN CHAIN '[VANIWAARD.DIFFEQLIB.PRINTGRAF]PRINTOUT.EXE'
1490 CLOSE #1
1500 GOSUB 2000   ! ROUTINE TO DELETE DATA FILES
1510 GOTO 4000   ! GOTO END
2000 !
```

```
100 !            **** PROGRAM IMEUL3 ****
110 !
120 !      Differential Equations - Hope College
130 !
140 !
150 !
160 !
170 !
180 !
190 !  THIS PROGRAM USES THE IMPROVED EULER TECHNIQUE TO
200 !  APPROXIMATE A SOLUTION TO THE THIRD ORDER DIFFERENTIAL
210 !  EQUATION WITH INITIAL CONDITIONS: Y''' =  F(X,Y,Y',Y''),
220 !  Y(XO) = YO, Y'(XO) = VO, Y''(XO) = WO.
230 !
240 !  THE USER DEFINES THE DESIRED FUNCTION FNF(X,Y,V,W),
250 !  RUNS THE PROGRAM AND ENTERS THE LEFT AND RIGHT END-
260 !  POINTS OF THE X INTERVAL, INITIAL VALUES OF Y, Y',
270 !  Y'' (NOTE V = Y', W = Y''), NUMBER OF INTEGRATIONS AND
280 !  THE PRINT STEP SIZE.
290 !
300 !  THE PROGRAM PRINTS THE INITIAL VALUES, INCREMENTS X
310 !  BY THE PRINT STEP SIZE, AND COMPUTES AND LISTS X, Y, Y',
320 !  Y'' AT EACH X VALUE USING THE IMPROVED EULER FORMULAE.
330 !
340 !
350 !                  VARIABLES
360 !                  ---------
370 !
380 !       LEFT_X_ENDPNT      - LEFT ENDPOINT OF THE X INTERVAL
390 !       RIGHT_X_ENDPNT     - RIGHT ENDPOINT OF THE X INTERVAL
400 !       YO                 - INITIAL VALUE OF Y
410 !       VO                 - INITIAL VALUE OF V
420 !       WO                 - INITIAL VALUE OF W
430 !       NUM_INTEGRATIONS   - NUMBER OF INTEGRATIONS
440 !       PRINT_STEP_SIZE    - PRINT STEP SIZE
450 !       X_INCREMENT        - X INCREMENT
460 !       J                  - LOOP CONTROL VARIABLE
470 !       Y_PREDICTOR        - PREDICTOR VALUE OF Y
480 !       V_PREDICTOR        - PREDICTOR VALUE OF V
490 !       W_PREDICTOR        - PREDICTOR VALUE OF W
500 !       F                  - VALUE OF FUNCTION AT (X,Y,V,W)
510 !       ERROR_COUNTER      - COUNTS NUMBER OF ERRORS
520 !       ERROR_NUMBER       - SYSTEM NUMBER OF AN ERROR
530 !
540 !
550 ! * * * * * * * * * * * * * * * * * * * * * * * * * * * * * *
560 !
800     DEF FNF(X,Y,V,W) = -1 * Y * W
1000 !
1010    OPEN 'DETEMP.DAT' FOR  OUTPUT  AS FILE #1
1020 !
1030    PRINT
1040    PRINT 'ENTER LEFT AND RIGHT ENDPOINTS, INITIAL VALUES OF '
1050    PRINT "Y, Y', Y'', NUMBER OF INTEGRATIONS, AND PRINT STEP SIZE."
1060 !
1070            ON ERROR GOTO 3000   ! PRINT ERROR DESCIPTION
1080       INPUT LEFT_X_ENDPNT, RIGHT_X_ENDPNT, YO, VO, WO, &
                NUM_INTEGRATIONS, PRINT_STEP_SIZE
1090            GOTO 3200   ! CHECK FOR ERRORS IN INPUT DATA
1100 !
1110    LET X_INCREMENT = (RIGHT_X_ENDPNT - LEFT_X_ENDPNT)/NUM_INTEGRATIONS
1120    LET X = LEFT_X_ENDPNT
1130    LET Y = YO
1140    LET V = VO
1150    LET W = WO
1160 !
1170    PRINT
1180    PRINT 'INTEGRATIONS:'; NUM_INTEGRATIONS
1190    PRINT 'X INCREMENT :'; X_INCREMENT
1200    PRINT
1210    PRINT '  X', '  Y', '  Y'', '  Y'''
```

```
1220     PRINT '------', '------', '-------', '-------'
1230     PRINT X, Y, V, W
1240     PRINT #1,X;',';Y;',';V;',';W
1250 !
1260     FOR I = 1 TO NUM_INTEGRATIONS
1270              LET F = FNF(X,Y,V,W)
1280        LET Y_PREDICTOR = Y + X_INCREMENT * V
1290        LET V_PREDICTOR = V + X_INCREMENT * W
1300        LET W_PREDICTOR = W + X_INCREMENT * F
1310        LET X = X + X_INCREMENT
1320        LET Y = Y + .5 * X_INCREMENT * (V + V_PREDICTOR)
1330        LET V = V + .5 * X_INCREMENT * (W + W_PREDICTOR)
1340        LET W = W + .5 * X_INCREMENT * (F + FNF(X,Y_PREDICTOR,  &
                V_PREDICTOR,W_PREDICTOR))
1350 !
1360        IF INT(I/PRINT_STEP_SIZE) = I/PRINT_STEP_SIZE THEN 1380
1370        IF I <> NUM_INTEGRATIONS THEN 1400
1380        PRINT X,Y,V,W
1390        PRINT #1,X;',';Y;',';V;',';W
1400     NEXT I
1410 !
1420 !
1430 PRINT
1440 INPUT 'Type a 'Y' if you want to see a graph: '; GRAPH$
1460 IF GRAPH$ <> 'Y' THEN 1480
1470 CHAIN '[VANIWAARD.DIFFEQLIB.PRINTGRAF]DEGRAF.EXE'
1480 INPUT 'Type a 'Y' if you want a printout file made of the data: '; PRINT$
1500 IF PRINT$ = 'Y' THEN CHAIN '[VANIWAARD.DIFFEQLIB.PRINTGRAF]PRINTOUT.EXE'
1510 CLOSE #1
1520 GOSUB 2000  ! ROUTINE TO DELETE DATA FILES
1530 GOTO 4000   ! END
2000 !
```

```
100 !                   **** PROGRAM MILNE ****
110 !
120 !       Differential Equations - Hope College
130 !
140 !
150 !
160 !
170 !
180 !
190 !  THIS PROGRAM USES MILNE'S FOURTH ORDER METHOD FOR
200 !  APPROXIMATING THE SOLUTION OF A FIRST ORDER INITIAL
210 !  VALUE PROBLEM: Y' = F(X,Y), Y(X0) = Y0.
220 !
230 !  AFTER DEFINING THE DESIRED FUNCTION, FNF(X,Y), THE USER
240 !  RUNS THE PROGRAM, ENTERING THE LEFT AND RIGHT ENDPOINTS
250 !  OF THE X INTERVAL,  THE INITIAL VALUE OF Y, THE NEXT
260 !  THREE 'STARTING' VALUES OF Y: Y1, Y2, Y3, (OBTAINED BY
270 !  SOME OTHER METHOD),  AND THE NUMBER OF INTEGRATIONS.
280 !
290 !  THE PROGRAM FIRST LISTS INITIAL VALUES OF THE FUNCTION
300 !  FOR Y(1) THRU Y(4). IT THEN COMPUTES AND PRINTS
310 !  SUCCESSIVE APPROXIMATIONS OF Y USING MILNE'S FORMULAE.
320 !
330 !
340 !                  VARIABLES
350 !                  ---------
360 !
370 !       FNF                 - FUNCTION DESCRIBING THE DERIVATIVE OF Y
380 !       Y()                 - Y VALUES, Y(1)-Y(4) ARE 'INITIAL' VALUES
390 !       X()                 - X VALUES
400 !       LEFT_X_ENDPNT       - LEFT ENDPOINT OF THE X INTERVAL
410 !       RIGHT_X_ENDPNT      - RIGHT ENDPOINT OF THE X INTERVAL
420 !       NUM_INTEGRATIONS    - NUMBER OF INTEGRATIONS
430 !       X_INCREMENT         - X INCREMENT
440 !       F()                 - VALUE OF FUNCTION AT (X(I),Y(I))
450 !       I                   - LOOP VARIABLE
460 !       Y_PREDICTED         - PREDICTED VALUE OF Y
470 !       VALUE_FNF           - VALUE OF FUNCTION AT (X(I),Y_PREDICTED)
480 !       Y(I)                - CORRECTED VALUE OF Y
490 !       ERROR_COUNTER       - COUNTS NUMBER OF ERRORS
500 !       ERROR_NUMBER        - SYSTEM NUMBER OF AN ERROR
510 !
520 !
530 !
540 ! * * * * * * * * * * * * * * * * * * * * * * * * * * * * *
550 !
560     DIMENSION X(100), Y(100)
570     DIMENSION F(4)
580 !
800     DEF FNF(X,Y) = 2*Y - X
1000 !
1010    OPEN 'DETEMP.DAT' FOR OUTPUT AS FILE #1
1020 !
1030    PRINT
1040    PRINT 'ENTER LEFT AND RIGHT ENDPOINTS OF THE X INTERVAL,'
1050    PRINT 'INITIAL VALUE OF Y, THE NEXT THREE 'STARTING' VALUES'
1060    PRINT 'OF Y: Y1, Y2, Y3, AND THE NUMBER OF INTEGRATIONS.'
1070 !
1080     ON ERROR GOTO 3000  !PRINT ERROR DESCRIPTION
1090      INPUT LEFT_X_ENDPNT, RIGHT_X_ENDPNT,Y(1), Y(2), Y(3), Y(4), &
              NUM_INTEGRATIONS
1100     GOTO 3200  ! CHECK FOR ERRORS IN INPUT DATA
1110 !
1120    LET X_INCREMENT = (RIGHT_X_ENDPNT - LEFT_X_ENDPNT)/NUM_INTEGRATIONS
1130 !
1140    PRINT
1150    PRINT 'INTEGRATIONS:'; NUM_INTEGRATIONS
1160    PRINT 'X INCREMENT :'; X_INCREMENT
1170    PRINT
1180    PRINT '  X', '  Y'
1190    PRINT '-----', '-----'
```

```
1200      FOR I = 1 TO 4
1210         LET X(I) = LEFT_X_ENDPNT + (I-1) * X_INCREMENT
1220         LET F(I) = FNF(X(I),Y(I))
1230         PRINT X(I),Y(I)
1240         PRINT #1,X(I);",";Y(I)
1250      NEXT I
1260      FOR I = 5 TO NUM_INTEGRATIONS + 1
1270        LET X(I) = X(I-1) + X_INCREMENT
1280        LET Y_PREDICTED = Y(I-4) + (4*X_INCREMENT /3)*(2*F(4) - F(3) + 2*F(2))
1290        LET VALUE_FNF = FNF(X(I),Y_PREDICTED)
1300        LET Y(I) = Y(I-2) + (X_INCREMENT /3)*(F(3) + 4*F(4)+ VALUE_FNF)
1310        PRINT X(I),Y(I)
1320        PRINT #1,X(I);",";Y(I)
1330        FOR J = 1 TO 3
1340          LET F(J) = F(J+1)
1350        NEXT J
1360        LET  F(4) = FNF(X(I),Y(I))
1370      NEXT I
1380 !
1390 !
1400 PRINT
1410 INPUT "Type a 'Y' if you want to see a graph: "; GRAPH$
1430 IF GRAPH$ <> 'Y' THEN 1450
1440 CHAIN "[VANIWAARD.DIFFEQLIB.PRINTGRAF]DEGRAF.EXE"
1450 INPUT "Type a 'Y' if you want a printout file made of this data: "; PRINT$
1470 IF PRINT$ = 'Y' THEN CHAIN "[VANIWAARD.DIFFEQLIB.PRINTGRAF]PRINTOUT.EXE"
1480 CLOSE #1
1490 GOSUB 2000   ! DELETE DATA FILES
1500 GOTO 4000   ! END
2000 !
```

```
100 !             **** PROGRAM MLNS2 ****
110 !
120 !     Differential Equations - Hope College
130 !
140 !
150 !
160 !
170 !
180 !
190 !  THIS PROGRAM USES MILNE'S METHOD TO SOLVE A SYSTEM OF
200 !  TWO FIRST ORDER DIFFERENTIAL EQUATIONS WITH INITIAL
210 !  CONDITIONS: Y' = F(X,Y,Z), Z' = G(X,Y,Z), Y(X0) = Y0,
220 !  Z(X0) = Z0.
230 !
240 !  THE USER DEFINES THE TWO FUNCTIONS FNF(X,Y,Z) AND
250 !  FNG(X,Y,Z), THEN RUNS THE PROGRAM, ENTERING THE LEFT
260 !  AND RIGHT ENDPOINTS OF THE X INTERVAL, INITIAL VALUE OF
270 !  Y AND THE NEXT THREE 'STARTING' VALUES OF Y: Y1, Y2, Y3
280 !  (OBTAINED BY SOME OTHER METHOD), INITIAL VALUE OF Z AND
290 !  THE NEXT THREE 'STARTING' VALUES OF Z: Z1, Z2, Z3
300 !  (OBTAINED BY SOME OTHER METHOD), NUMBER OF INTEGRATIONS.
310 !
320 !  THE PROGRAM FIRST LISTS THE VALUES OF THE FUNCTIONS
330 !  AT THE INITIAL AND 'STARTING' VALUES OF Y AND Z. IT
340 !  THEN COMPUTES AND LISTS FROM THE PREDICTOR AND
350 !  CORRECTOR FORMULAE THE SUBSEQUENT SOLUTION VALUES OF
360 !  THE SYSTEM USING MILNE'S FORMULAE.
370 !
380 !
390 !                    VARIABLES
400 !                    ----------
410 !
420 !     FNF              - FUNCTION DESCRIBING THE DERIVATIVE OF Y
430 !     FNG              - FUNCTION DESCRIBING THE DERIVATIVE OF Z
440 !     X()              - X VALUES
450 !     Y()              - Y VALUES, Y(1)-Y(4) ARE 'INITIAL' VALUES
460 !     Z()              - Z VALUES, Z(1)-Z(4) ARE 'INITIAL' VALUES
470 !     LEFT_X_ENDPNT    - LEFT ENDPOINT OF THE X INTERVAL
480 !     RIGHT_X_ENDPNT   - RIGHT ENDPOINT OF THE X INTERVAL
490 !     NUM_INTEGRATIONS - NUMBER OF INTEGRATIONS
500 !     X_INCREMENT      - X INCREMENT
510 !     I                - LOOP CONTROL VARIABLE
520 !     Y_PREDICTED      - PREDICTED VALUE OF Y
530 !     Z_PREDICTED      - PREDICTED VALUE OF Z
540 !     VALUE_FNF        - VALUE OF FNF AT (X(I),Y_PREDICTED,Z_PREDICTED)
550 !     VALUE_FNG        - VALUE OF FNG AT (X(I),Y_PREDICTED,Z_PREDICTED)
560 !     Y(I)             - CORRECTED VALUE OF Y
570 !     Z(I)             - CORRECTED VALUE OF Z
580 !     ERROR_COUNTER    - COUNTS THE NUMBER OF ERRORS
590 !     ERROR_NUMBER     - SYSTEM NUMBER OF THE ERROR
600 !
610 !
620 ! * * * * * * * * * * * * * * * * * * * * * * * * * * * * * *
630 !
640 !
800     DEF FNF(X,Y,Z)= Z
810     DEF FNG(X,Y,Z) = 5*Z - 4*Y
1000 !
1010    DIMENSION X(100), Y(100), Z(100)
1020    DIMENSION F(4),G(4)
1030 !
1040    OPEN 'DETEMP.DAT' FOR OUTPUT AS FILE #1
1050 !
1060    PRINT
1070    PRINT 'ENTER LEFT AND RIGHT ENDPOINTS OF THE X INTERVAL,'
1080    PRINT 'AND THE NUMBER OF INTEGRATIONS'
1090 !
1100     ON ERROR GOTO 3000  ! PRINT ERROR DESCRIPTION
1110      INPUT LEFT_X_ENDPNT, RIGHT_X_ENDPNT, NUM_INTEGRATIONS
1120            GOTO 3200  ! CHECK FOR ERRORS IN INPUT DATA
1130 !
```

```
1140     PRINT 'NOW ENTER THE INITIAL VALUE OF Y AND THE NEXT THREE'
1150     PRINT "'STARTING' VALUES OF Y: Y1,Y2,Y3."
1160 !
1170     INPUT Y(1), Y(2), Y(3), Y(4)
1180 !
1190     PRINT 'NOW ENTER THE INITIAL VALUE OF Z AND THE NEXT THREE'
1200     PRINT "'STARTING' VALUES OF Z: Z1,Z2,Z3."
1210 !
1220     INPUT Z(1), Z(2), Z(3), Z(4)
1230 !
1240     LET X_INCREMENT = (RIGHT_X_ENDPNT - LEFT_X_ENDPNT)/NUM_INTEGRATIONS
1250 !
1260     PRINT
1270     PRINT 'INTEGRATIONS:'; NUM_INTEGRATIONS
1280     PRINT "X INCREMENT :"; X_INCREMENT
1290     PRINT
1300     PRINT '  X', '  Y', '  Z'
1310     PRINT '------', '------', '------'
1320 !
1330     FOR I = 1 TO 4
1340        LET X(I) = LEFT_X_ENDPNT + (I-1) * X_INCREMENT
1350        LET F(I) = FNF(X(I),Y(I),Z(I))
1360        LET G(I) = FNG(X(I),Y(I),Z(I))
1370        PRINT X(I),Y(I),Z(I)
1380        PRINT #1,X(I);",";Y(I);",";Z(I)
1390     NEXT I
1400     FOR I = 5 TO NUM_INTEGRATIONS + 1
1410       LET X(I) = X(I-1) + X_INCREMENT
1420       LET Y_PREDICTED = Y(I-4) + (4*X_INCREMENT/3)*(2*F(4) - F(3) + 2*F(2))
1430       LET Z_PREDICTED = Z(I-4)+(4*X_INCREMENT/3)*(2*G(4)-G(3)+2*G(2))
1440     !
1450       LET VALUE_FNF = FNF(X(I),Y_PREDICTED,Z_PREDICTED)
1460       LET VALUE_FNG = FNG(X(I),Y_PREDICTED,Z_PREDICTED)
1470     !
1480       LET Y(I) = Y(I-2) + (X_INCREMENT/3)*(F(3) + 4*F(4) + VALUE_FNF)
1490       LET Z(I) = Z(I-2) + (X_INCREMENT/3)*(G(3) + 4*G(4) + VALUE_FNG)
1500     !
1510       PRINT X(I),Y(I),Z(I)
1520       PRINT #1,X(I);",";Y(I);",";Z(I)
1530     !
1540       FOR J= 1 TO 3
1550         LET F(J)= F(J+1)
1560         LET G(J)= G(J+1)
1570       NEXT J
1580       LET F(4) = FNF(X(I),Y(I),Z(I))
1590       LET G(4) = FNG(X(I),Y(I),Z(I))
1600     NEXT I
1610 !
1620 !
1630 PRINT
1640 INPUT "Type a 'Y' if you want to see a graph: "; GRAPH$
1660 IF GRAPH$ <> 'Y' THEN 1680
1670 CHAIN '[VANIWAARD.DIFFEQLIB.PRINTGRAF]BESGRAF.EXE'
1680 INPUT "Type a 'Y' if you want a printout file made of the data: "; P$
1700 IF P$ = 'Y' THEN CHAIN '[VANIWAARD.DIFFEQLIB.PRINTGRAF]PRINTOUT.EXE'
1710 CLOSE #1
1720 GOSUB 2000   ! DELETE DATA FILES
1730 GOTO 4000   ! END
2000 !
```

```
100 !              **** PROGRAM RKU4S ****
110 !
120 !       Differential Equations - Hope College
130 !
140 !
150 !
160 !
170 !
180 !
190 !  THIS PROGRAM USES THE FOURTH ORDER RUNGE-KUTTA METHOD
200 !  TO SOLVE A SYSTEM OF TWO FIRST ORDER DIFFERENTIAL
210 !  EQUATIONS WITH INITIAL CONDITIONS: Y' = F(X,Y,Z),
220 !  Z' = G(X,Y,Z), Y(XO) = YO, Z(XO) = ZO.
230 !
240 !  THE USER DEFINES THE TWO FUNCTIONS, FNF(X,Y,Z) AND
250 !  FNG(X,Y,Z), THEN RUNS THE PROGRAM, ENTERING THE LEFT
260 !  AND RIGHT ENDPOINTS OF THE X INTERVAL, INITIAL VALUES
270 !  OF Y AND Z, NUMBER OF INTEGRATIONS AND THE PRINT
280 !  STEP SIZE.
290 !
300 !  THE PROGRAM PRINTS THE INITIAL VALUES, THEN LISTS X, Y,
310 !  AND Z VALUES, WHERE X IS INCREMENTED BY THE PRINT STEP
320 !  SIZE, AND Y AND Z ARE COMPUTED USING THE FOURTH ORDER
330 !  RUNGE-KUTTA FORMULAE.
340 !
350 !  NOTE: FNF(X,Y,Z) IS DEFINED AT LINE # 800
360 !        FNG(X,Y,Z) IS DEFINED AT LINE # 810
370 !
380 !
390 !                  VARIABLES
400 !                  ---------
410 !
420 !       FNF                - FUNCTION DESCRIBING THE DERIVATIVE OF Y
430 !       FNG                - FUNCTION DESCRIBING THE DERIVATIVE OF Z
440 !       LEFT_X-ENDPNT      - LEFT ENDPOINT OF THE X INTERVAL
450 !       RIGHT_X_ENDPNT     - RIGHT ENDPOINT OF THE X INTERVAL
460 !       YO                 - INITIAL VALUE OF Y
470 !       ZO                 - INITIAL VALUE OF Z
480 !       NUM_INTEGRATIONS   - NUMBER OF INTEGRATIONS
490 !       PRINT_STEP_SIZE    - PRINT STEP SIZE
500 !       X_INCREMENT        - X INCREMENT
510 !       ERROR_COUNTER      - COUNTS THE NUMBER OF ERRORS
520 !       ERROR_NUMBER       - SYSTEM NUMBER OF AN ERROR
530 !       I                  - LOOP VARIABLE
540 !       K1                 - RUNGE-KUTTA FUNCTION COMPUTATION FORMULAE
550 !                           USED IN COMPUTING THE NEW VALUE OF Y
560 !       K2                 -    .      .      .      .
570 !       K3                 -    .      .      .      .
580 !       K4                 -    .      .      .      .
590 !       L1                 - RUNGE-KUTTA FUNCTION COMPUTATION FORMULAE
600 !                           USED IN COMPUTING THE NEW VALUE OF Z
610 !       L2                 -    .      .      .      .
620 !       L3                 -    .      .      .      .
630 !       L4                 -    .      .      .      .
640 !
650 !
660 ! * * * * * * * * * * * * * * * * * * * * * * * * * * * * * * * * * * *
670 !
800     DEF FNF(X,Y,Z) = Z
810     DEF FNG(X,Y,Z) = 3*Y + X
1000 !
1010    OPEN 'DETEMP.DAT' FOR OUTPUT AS FILE #1
1020 !
1030    PRINT
1040    PRINT 'ENTER LEFT AND RIGHT ENDPOINTS OF THE X INTERVAL,'
1050    PRINT 'INITIAL VALUES OF Y AND Z, NUMBER OF INTEGRATIONS'
1060    PRINT 'AND THE PRINT STEP SIZE.'
1070 !
1080          ON ERROR GOTO 3000   ! PRINT ERROR MESSAGE
1090      INPUT LEFT_X_ENDPNT, RIGHT_X_ENDPNT, YO, ZO, NUM_INTEGRATIONS,  &
               PRINT_STEP_SIZE
```

```
1100             GOTO 3200  ! CHECK FOR ERRORS IN INPUT DATA
1110 !
1120     LET X = LEFT_X_ENDPNT
1130     LET Y = Y0
1140     LET Z = Z0
1150 !
1160     LET X_INCREMENT = (RIGHT_X_ENDPNT - LEFT_X_ENDPNT)/NUM_INTEGRATIONS
1170 !
1180     PRINT
1190     PRINT "INTEGRATIONS:"; NUM_INTEGRATIONS
1200     PRINT "X INCREMENT :"; X_INCREMENT
1210     PRINT
1220     PRINT "  X", "  Y", "  Z"
1230     PRINT "------", "------", "------"
1240 !
1250     PRINT X, Y, Z
1260     PRINT #1,X;",";Y;",";Z
1270 !
1280     FOR I =1 TO NUM_INTEGRATIONS
1290        LET K1 = X_INCREMENT * FNF(X,Y,Z)
1300        LET L1 = X_INCREMENT * FNG(X,Y,Z)
1310       !
1320        LET K2 = X_INCREMENT * (FNF(X+.5*X_INCREMENT, Y+.5*K1, Z+.5*L1))
1330        LET L2 = X_INCREMENT * (FNG(X+.5*X_INCREMENT, Y+.5*K1, Z+.5*L1))
1340       !
1350        LET K3 = X_INCREMENT * (FNF(X+.5*X_INCREMENT, Y+.5*K2, Z+.5*L2))
1360       LET L3 = X_INCREMENT * (FNG(X+.5*X_INCREMENT, Y+.5*K2, Z+.5*L2))
1370      !
1380       LET K4 = X_INCREMENT * (FNF(X+X_INCREMENT, Y+K3, Z+L3))
1390       LET L4 = X_INCREMENT * (FNG(X+X_INCREMENT, Y+K3, Z+L3))
1400      !
1410       LET Y = Y+(1/6)*(K1+2*K2+2*K3+K4)
1420       LET Z = Z+(1/6)*(L1+2*L2+2*L3+L4)
1430      !
1440       LET X = X + X_INCREMENT
1450      !
1460       IF I/PRINT_STEP_SIZE = INT(I/PRINT_STEP_SIZE) THEN 1480
1470         IF I<>NUM_INTEGRATIONS THEN 1500
1480           PRINT X,Y,Z
1490           PRINT #1,X;",";Y;",";Z
1500     NEXT I
1510 !
1520 !
1530 PRINT
1540 INPUT "Type a 'Y' if you want to see a graph: ";GRAPH$
1560 IF GRAPH$ <> 'Y' THEN 1600
1570 ! 'NOTE:  TO GRAPH Y VS. Z OR X VS. Z INSTEAD OF X VS. Y,'
1580 ! 'CHANGE ST. # 900 & 1130 TO PRINT APPROPRIATE VALUES.'
1590 CHAIN "[VANIWAARD.DIFFEQLIB.PRINTGRAF]BESGRAF.EXE"
1600 INPUT "Type a 'Y' if you want a printout file made of the data: ";P$
1620 IF P$ = 'Y' THEN CHAIN "[VANIWAARD.DIFFEQLIB.PRINTGRAF]PRINTOUT.EXE"
1630 !
1640 CLOSE #1
1650 GOSUB 2000  ! DELETE DATA FILES
1660 GOTO 4000  ! END
2000 !
```

```
100 !              **** PROGRAM RUKU3 ****
110 !
120 !        Differential Equations - Hope College
130 !
140 !
150 !
160 !
170 !
180 !
190 !  THIS PROGRAM USES THE THIRD ORDER RUNGE-KUTTA METHOD FOR
200 !  APPROXIMATING A SOLUTION TO A FIRST ORDER INITIAL VALUE
210 !  PROBLEM: Y' = F(X,Y), Y(X0) = Y0.
220 !
230 !  THE USER DEFINES THE DESIRED FUNCTION, FNF(X,Y), THEN
240 !  RUNS THE PROGRAM, ENTERING THE LEFT AND RIGHT ENDPOINTS
250 !  OF THE X INTERVAL, THE INITIAL VALUE OF Y, THE NUMBER OF
260 !  INTEGRATIONS AND THE PRINT STEP SIZE.
270 !
280 !  THE PROGRAM PRINTS THE GIVEN INITIAL VALUES, INCREMENTS
290 !  X, CALCULATES AND LISTS THE APPROXIMATE Y VALUES USING
300 !  THE THIRD ORDER RUNGE-KUTTA FORMULAE.
310 !
320 !  NOTE: FNF(X,Y) IS DEFINED AT LINE # 800
330 !
340 !
350 !                    VARIABLES
360 !                    ---------
370 !
380 !       FNF                 - FUNCTION DESCRIBING THE DERIVATIVE OF Y
390 !       LEFT_X_ENDPNT       - LEFT ENDPOINT OF THE X INTERVAL
400 !       RIGHT_X_ENDPNT      - RIGHT ENDPOINT OF THE X INTERVAL
410 !       Y0                  - INITIAL VALUE OF Y
420 !       NUM_INTEGRATIONS    - NUMBER OF INTEGRATIONS
430 !       PRINT_STEP_SIZE     - PRINT STEP SIZE
440 !       X_INCREMENT         - X INCREMENT
450 !       ERROR_COUNTER       - COUNTS THE NUMBER OF ERRORS
460 !       ERROR_NUMBER        - SYSTEM NUMBER OF AN ERROR
470 !       I                   - LOOP VARIABLE
480 !       F0                  - RUNGE-KUTTA FUNCTION COMPUTATION FORMULAE
490 !                             USED IN COMPUTING THE NEW Y VALUE
500 !       F1                  -   .     .       .          .         .
510 !       F2                  -   .     .       .          .         .
520 !
530 !
540 ! * * * * * * * * * * * * * * * * * * * * * * * * * * * * * * * * * * *
550 !
800     DEF FNF(X,Y) = 2*Y - X
1000 !
1010    OPEN 'DETEMP.DAT' FOR OUTPUT AS FILE #1
1020 !
1030    PRINT
1040    PRINT 'ENTER LEFT AND RIGHT ENDPOINTS, INITIAL Y VALUE,'
1050    PRINT 'NUMBER OF INTEGRATIONS AND PRINT STEP SIZE.'
1060 !
1070            ON ERROR GOTO 3000   ! PRINT ERROR DESCRIPTIONS
1080      INPUT LEFT_X_ENDPNT, RIGHT_X_ENDPNT, Y0, NUM_INTEGRATIONS,    &
          PRINT_STEP_SIZE
1090            GOTO 3200   ! CHECK FOR ERRORS IN INPUT DATA
1100 !
1110    LET X = LEFT_X_ENDPNT
1120    LET Y = Y0
1130    LET X_INCREMENT = (RIGHT_X_ENDPNT - LEFT_X_ENDPNT)/NUM_INTEGRATIONS
1140 !
1150    PRINT
1160    PRINT 'INTEGRATIONS:'; NUM_INTEGRATIONS
1170    PRINT 'X INCREMENT :'; X_INCREMENT
1180    PRINT
1190    PRINT '  X', '  Y'
1200    PRINT '-----', '-----'
1210    PRINT X, Y
1220    PRINT #1,X;',';Y
1230 !
```

```
1240     FOR I = 1 TO NUM_INTEGRATIONS
1250        LET F0 = FNF(X,Y) * X_INCREMENT
1260        LET F1 = FNF(X + .5 * X_INCREMENT, Y + .5 * F0) * X_INCREMENT
1270        LET F2 = FNF(X + X_INCREMENT, Y - F0 + 2 * F1) * X_INCREMENT
1280        LET Y = Y + (1/6) * (F0 + 4 * F1 + F2)
1290        LET X = X + X_INCREMENT
1300     !
1310        IF I/PRINT_STEP_SIZE = INT(I/PRINT_STEP_SIZE) THEN 1330
1320          IF I <> NUM_INTEGRATIONS THEN 1350
1330            PRINT X,Y
1340                   PRINT #1,X;",";Y
1350     NEXT I
1360 !
1370 !
1380 PRINT
1390 INPUT "Type a 'Y' if you want to see a graph: ";GRAPH$
1410 IF GRAPH$ <> 'Y' THEN 1430
1420 CHAIN "[VANIWAARD.DIFFEQLIB.PRINTGRAF]DEGRAF.EXE"
1430 INPUT "Type a 'Y' if you want a printout file made of the data: ";P$
1450 IF P$ = 'Y' THEN CHAIN "[VANIWAARD.DIFFEQLIB.PRINTGRAF]PRINTOUT.EXE"
1460 !
1470 CLOSE #1
1480 GOSUB 2000   ! DELETE DATA FILES
1490 GOTO 4000   ! END
2000 !
```

```
100 !                 **** PROGRAM RUKU4 ****
110 !
120 !        Differential Equations - Hope College
130 !
140 !
150 !
160 !
170 !
180 !
190 !  THIS PROGRAM USES THE FOURTH ORDER RUNGE-KUTTA METHOD FOR
200 !  APPROXIMATING A SOLUTION TO A FIRST ORDER INITIAL VALUE
210 !  PROBLEM: Y' = F(X,Y), Y(XO) = YO.
220 !
230 !  THE USER DEFINES THE DESIRED FUNCTION FNF(X,Y), THEN
240 !  RUNS THE PROGRAM, ENTERING THE LEFT AND RIGHT ENDPOINTS
250 !  OF THE X INTERVAL, INITIAL VALUE OF Y, NUMBER OF INTE-
260 !  GRATIONS, AND THE PRINT STEP SIZE.
270 !
280 !  THE PROGRAM PRINTS THE INITIAL VALUES, INCREMENTS X BY
290 !  THE STEP SIZE, THEN CALCULATES AND LISTS THE APPROXIMATE
300 !  Y VALUES USING THE FOURTH ORDER RUNGE-KUTTA FORMULAE.
310 !
320 !  NOTE: FNF(X,Y) IS DEFINED AT LINE # 800
330 !
340 !
350 !                    VARIABLES
360 !                    ---------
370 !
380 !       FNF                 - FUNCTION DESCRIBING THE DERIVATIVE OF Y
390 !       LEFT_X_ENDPNT       - LEFT ENDPOINT OF THE X INTERVAL
400 !       RIGHT_X_ENDPNT      - RIGHT ENDPOINT OF THE X INTERVAL
410 !       YO                  - INITIAL VALUE OF Y
420 !       NUM_INTEGRATIONS    - NUMBER OF INTEGRATIONS
430 !       PRINT_STEP_SIZE     - PRINT STEP SIZE
440 !       X_INCREMENT         - X INCREMENT
450 !       ERROR_COUNTER       - COUNTS THE NUMBER OF ERRORS
460 !       ERROR_NUMBER        - SYSTEM NUMBER OF AN ERROR
470 !       I                   - LOOP CONTROL VARIABLE
480 !       F0                  - RUNGE-KUTTA FUNCTION COMPUTATION FORMULAE
490 !                             USED IN COMPUTING THE NEW Y VALUE
500 !       F1                  -   .     .      .        .       .
510 !       F2                  -   .     .      .        .       .
520 !       F3                  -   .     .      .        .       .
530 !
540 !
550 ! * * * * * * * * * * * * * * * * * * * * * * * * * * * * * * * * * *
560 !
800      DEF FNF(X,Y) = 2*Y - X
1000 !
1010 !
1020     OPEN 'DETEMP.DAT' FOR OUTPUT AS FILE #1
1030 !
1040     PRINT
1050     PRINT 'ENTER LEFT AND RIGHT ENDPOINTS, INITIAL Y VALUE,'
1060     PRINT 'NUMBER OF INTEGRATIONS AND PRINT STEP SIZE.'
1070 !
1080     ON ERROR GOTO 3000  ! PRINT ERROR DESCRIPTION
1090      INPUT LEFT_X_ENDPNT, RIGHT_X_ENDPNT, YO, NUM_INTEGRATIONS,  &
                PRINT_STEP_SIZE
1100      GOTO  3200  ! CHECK FOR ERRORS IN INPUT DATA
1110 !
1120     LET X = LEFT_X_ENDPNT
1130     LET Y = YO
1140     LET X_INCREMENT = (RIGHT_X_ENDPNT - LEFT_X_ENDPNT)/NUM_INTEGRATIONS
1150 !
1160     PRINT
1170     PRINT 'INTEGRATIONS:'; NUM_INTEGRATIONS
1180     PRINT 'X INCREMENT :'; X_INCREMENT
1190     PRINT
1200     PRINT '  X', '  Y'
1210     PRINT '------', '------'
```

```
1220    PRINT X, Y
1230    PRINT #1,X;",";Y
1240 !
1250    FOR I = 1 TO NUM_INTEGRATIONS
1260              LET F0 = FNF(X,Y)
1270       LET F1 = FNF(X + .5 * X_INCREMENT, Y + .5 * X_INCREMENT * F0)
1280       LET F2 = FNF(X + .5 * X_INCREMENT, Y + .5 * X_INCREMENT * F1)
1290       LET F3 = FNF(X + X_INCREMENT, Y + X_INCREMENT * F2)
1300       LET Y = Y + (1/6) * X_INCREMENT * (F0 + 2 * F1 + 2 * F2 + F3)
1310       LET X = X + X_INCREMENT
1320    !
1330      IF INT(I/PRINT_STEP_SIZE) = I/PRINT_STEP_SIZE THEN 1350
1340      IF I <> NUM_INTEGRATIONS THEN 1370
1350      PRINT X, Y
1360      PRINT #1,X;",";Y
1370    NEXT I
1380 !
1390 !
1400 PRINT
1410 INPUT "Type a 'Y' if you want to see a graph: "; GRAPH$
1430 IF GRAPH$ <> 'Y' THEN 1450
1440   CHAIN "[VANIWAARD.DIFFEQLIB.PRINTGRAF]DEGRAF.EXE"
1450 INPUT "Type a 'Y' if you want a printout file made of the data: ";P$
1470 IF P$ = 'Y' THEN CHAIN "[VANIWAARD.DIFFEQLIB.PRINTGRAF]PRINTOUT.EXE"
1480 !
1490 CLOSE #1
1500 GOSUB 2000   ! DELETE DATA FILES
1510 GOTO 4000   ! END
2000 !
```

```
100 !           **** PROGRAM SERSING ****
110 !
120 !       Differential Equations - Hope College
130 !
140 !
150 !
160 !
170 !
180 !
190 !   THIS PROGRAM COMPUTES VALUES OF THE TWO LINEARLY
200 !   INDEPENDENT SOLUTIONS OF A SECOND ORDER INITIAL VALUE
210 !   PROBLEM AT THE REGULAR SINGULAR POINT X = 0 USING
220 !   THE POWER SERIES TECHNIQUE. IT IS DESIGNED TO BE USED
230 !   WITH RECURRENCE FORMULAS WHOSE SERIES COEFFICIENT
240 !   SUBSCRIPTS DIFFER BY ONE.
250 !
260 !   THE USER RUNS THE PROGRAM, ENTERING THE LEFT AND
270 !   RIGHT ENDPOINTS OF THE X INTERVAL, NUMBER OF
280 !   EVALUATIONS, CONVERGENCE TOLERANCE AND ONE ROOT OF
290 !   THE INDICIAL EQUATION.
300 !
310 !   NOTE: RECURRENCE FORMULA, FNF(N,An_1), IS DEFINED AT LINE # 800
320 !
330 !   THE PROGRAM INITIALIZES VALUES FOR USE IN THE RECURRENCE
340 !   FORMULA FOR THE INDICIAL ROOT AND BEGINS COMPUTING
350 !   THE SUCCESSIVE TERMS OF THE SERIES. WHEN THE TOLERANCE
360 !   LIMIT HAS BEEN REACHED, THE VALUES OF X AND Y ARE PRINTED
370 !   AND THE NEXT VALUE OF X IS USED. IF THE TOLERANCE HAS
380 !   NOT BEEN REACHED, THE NEXT TERM OF THE SERIES IS
390 !   COMPUTED AND ADDED. WHEN THE COMPLETE LIST OF (X,Y)
400 !   VALUES HAS BEEN PRINTED AND OPTIONALLY GRAPHED, THE USER
410 !   MUST RUN THE PROGRAM AGAIN, THIS TIME ENTERING THE
420 !   OTHER INDICIAL ROOT.
430 !
440 !
450 !                  VARIABLES
460 !                  ---------
470 !
480 !         LEFT_X_ENDPNT      - LEFT ENDPOINT OF THE X INTERVAL
490 !         RIGHT_X_ENDPNT     - RIGHT ENDPOINT OF THE X INTERVAL
500 !         NUM_EVALUATIONS    - NUMBER OF EVALUATIONS
510 !         CONVERGENCE_TOL    - CONVERGENCE TOLERANCE
520 !         INDICIAL_ROOT      - ROOT OF THE INDICIAL EQUATION
530 !         X_INCREMENT        - X INCREMENT
540 !         ERROR_COUNTER      - COUNTS THE NUMBER OF ERRORS
550 !         ERROR_NUMBER       - SYSTEM NUMBER OF AN  ERROR
560 !         I                  - LOOP CONTROL VARIABLE
570 !         An_2               - A SUB (N-2) IN THE RECURRENCE FORMULA
580 !         An_1               - A SUB (N-1) IN THE RECURRENCE FORMULA
590 !         N                  - INDEX IN THE RECURRENCE FORMULA
600 !         X_TO_THE_N         - X TO THE N POWER
610 !         NEXT_TERM          - NEXT COMPUTED TERM OF THE SERIES
620 !         A0                 - INITIAL TERM OF RECURRENCE FORMULA
630 !
640 !
650 ! * * * * * * * * * * * * * * * * * * * * * * * * * * * * *
660 !
670 !
800      DEF FNF(N,An_1,INDICIAL_ROOT) = -An_1 /(2*N + 2*INDICIAL_ROOT - 1)
1000 !
1010      OPEN 'DETEMP.DAT' FOR OUTPUT AS FILE #1
1020 !
1030     PRINT
1040     PRINT 'ENTER LEFT AND RIGHT ENDPOINTS OF THE X INTERVAL,'
1050     PRINT 'NUMBER OF EVALUATIONS, THE CONVERGENCE TOLERANCE,'
1060     PRINT 'AND ONE INDICIAL ROOT.'
1070 !
1080             ON ERROR GOTO 3000   ! PRINT ERROR MESSAGE
1090      INPUT LEFT_X_ENDPNT, RIGHT_X_ENDPNT, NUM_EVALUATIONS,   &
          CONVERGENCE_TOL, INDICIAL_ROOT
1100     GOTO 3200   ! CHECK FOR ERRORS IN DATA
1110 !
```

```
1120    LET X = LEFT_X_ENDPNT
1130    LET X_INCREMENT = (RIGHT_X_ENDPNT - LEFT_X_ENDPNT)/NUM_EVALUATIONS
1140 !
1150    PRINT
1160    PRINT 'EVALUATIONS :'; NUM_EVALUATIONS
1170    PRINT 'X INCREMENT :'; X_INCREMENT
1180    PRINT
1190    PRINT '  X', '  Y'
1200    PRINT '------', '------'
1210    FOR I = 1 TO NUM_EVALUATIONS + 1
1220                    LET A0 = 1
1230                    LET Y1 = X ** INDICIAL_ROOT
1240                    LET Y2 = A0
1250                    LET N = 1
1260                    LET An_1 = A0
1270                    LET X_TO_THE_N = X
1280 !
1290     LET An_2 = FNF(N,An_1,INDICIAL_ROOT)
1300 !
1310                    LET NEXT_TERM = An_2 * X_TO_THE_N
1320                    LET Y2 = Y2 + NEXT_TERM
1330                    IF ABS(NEXT_TERM) < CONVERGENCE_TOL THEN 1400
1340 !
1350            LET N = N + 1
1360            LET An_1 = An_2
1370            LET X_TO_THE_N = X_TO_THE_N * X
1380            GOTO 1290
1390 !
1400            LET Y = Y1 * Y2
1410 !
1420            PRINT X,Y
1430     PRINT #1, X;',';Y
1440            LET X = X + X_INCREMENT
1450    NEXT I
1460 !
1470 !
1480 PRINT
1490 INPUT 'Type a 'Y' if you want to see a graph: '; GRAPH$
1510 IF GRAPH$ <> 'Y' THEN 1530
1520 CHAIN '[VANIWAARD.DIFFEQLIB.PRINTGRAF]DEGRAF.EXE'
1530 INPUT 'Type a 'Y' if you want a printout file made of the data: ';P$
1550 IF P$ = 'Y' THEN CHAIN '[VANIWAARD.DIFFEQLIB.PRINTGRAF]PRINTOUT.EXE'
1560 !
1570 !
1580 CLOSE #1
1590 GOSUB 2000   ! DELETE DATA FILES
1600 GOTO 4000   ! END
2000 !
```

```
100 !              **** PROGRAM SERSO ****
110 !
120 !      Differential Equations - Hope College
130 !
140 !
150 !
160 !
170 !
180 !
190 !  THIS PROGRAM COMPUTES VALUES OF THE SOLUTION TO A SECOND
200 !  ORDER INITIAL VALUE PROBLEM AT THE ORDINARY POINT X = 0
210 !  USING THE POWER SERIES TECHNIQUE.  IT IS DESIGNED TO BE
220 !  USED WITH RECURRENCE FORMULAS WHOSE SERIES COEFFICIENTS
230 !  DIFFER BY 2.
240 !
250 !  THE USER RUNS THE PROGRAM, ENTERING THE LEFT
260 !  AND RIGHT ENDPOINTS OF THE X INTERVAL,  THE NUMBER OF
270 !  EVALUATIONS, INITIAL VALUES OF Y AND Y', AND THE
280 !  CONVERGENCE TOLERANCE.
290 !
300 !  NOTE: RECURRENCE FORMULA, FNF(N,An_1,An_2), IS DEFINED
310 !        AT LINE # 800
320 !
330 !  THE PROGRAM INITIALIZES VALUES FOR USE IN THE RECURRENCE
340 !  FORMULA AND STARTS COMPUTING THE INFINITE SERIES. WHEN
350 !  THE TOLERANCE LIMIT HAS BEEN REACHED, THE VALUES OF X
360 !  AND Y ARE PRINTED AND THE NEXT POINT OF THE X INTERVAL
370 !  IS USED. IF THE TOLERANCE HAS NOT BEEN REACHED, THE NEXT
380 !  TERM OF THE SERIES IS COMPUTED AND ADDED.
390 !
400 !
410 !                  VARIABLES
420 !                  ---------
430 !
440 !       LEFT_X_ENDPNT     - LEFT ENDPOINT OF THE X INTERVAL
450 !       RIGHT_X_ENDPNT    - RIGHT ENDPOINT OF THE X INTERVAL
460 !       NUM_EVALUATIONS   - NUMBER OF EVALUATIONS
470 !       Y0                - INITIAL VALUE OF Y
480 !       Y1                - INITIAL VALUE OF Y'
490 !       CONVERGENCE_TOL   - CONVERGENCE TOLERANCE
500 !       X_INCREMENT       - X INCREMENT
510 !       I                 - LOOP CONTROL VARIABLE
520 !       An                - THE PRESENT A SUB (N) IN THE FORMULA
530 !       An_2              - A SUB (N-2) IN THE RECURRENCE FORMULA
540 !       An_1              - A SUB (N-1) IN THE RECURRENCE FORMULA
550 !       N                 - INDEX IN THE RECURRENCE FORMULA
560 !       X_TO_THE_N        - VARIOUS POWERS OF X
570 !       NEXT_TERM         - NEXT COMPUTED TERM OF THE SERIES
580 !       ERROR_COUNTER     - COUNTS THE NUMBER OF ERRORS
590 !       ERROR_NUMBER      - SYSTEM NUMBER OF AN ERROR
600 !
610 !
620 !
630 ! * * * * * * * * * * * * * * * * * * * * * * * * * * * * * * * * *
640 !
800     DEF FNF(N,An_1,An_2) = ((N*N-7*N+16)*An_2+(N-1)*An_1)/(-4*N*(N-1))
1000 !
1010    OPEN 'DETEMP.DAT' FOR OUTPUT AS FILE #1
1020 !
1030    PRINT
1040    PRINT 'ENTER LEFT AND RIGHT ENDPOINT OF THE X INTERVAL,'
1050    PRINT 'NUMBER OF EVALUATIONS, INITIAL VALUES OF Y AND Y',' 
1060    PRINT 'AND THE CONVERGENCE TOLERANCE.'
1070 !
1080            ON ERROR GOTO 3000   ! PRINT ERROR DESCRIPTION
1090     INPUT LEFT_X_ENDPNT, RIGHT_X_ENDPNT, NUM_EVALUATIONS,   &
               Y0, Y1, CONVERGENCE_TOL
1100            GOTO 3200   ! CHECK FOR ERRORS IN DATA INPUT
1110 !
1120    LET X = LEFT_X_ENDPNT
1130    LET X_INCREMENT = (RIGHT_X_ENDPNT - LEFT_X_ENDPNT)/NUM_EVALUATIONS
```

```
1140 !
1150     PRINT
1160     PRINT "EVALUATIONS :"; NUM_EVALUATIONS
1170     PRINT "X INCREMENT :"; X_INCREMENT
1180     PRINT
1190     PRINT "  X", "  Y"
1200     PRINT "------", "------"
1210     FOR I =1 TO NUM_EVALUATIONS+1
1220         LET An_2 = Y0
1230         LET An_1 = Y1
1240         LET Y = Y0 + Y1 * X
1250         LET N = 2
1260         LET X_TO_THE_N = X * X
1270 !
1280          LET An = FNF(N, An_1, An_2)
1290 !
1300         LET NEXT_TERM = An * X_TO_THE_N
1310         LET Y = Y + NEXT_TERM
1320         IF ABS(NEXT_TERM) < CONVERGENCE_TOL THEN 1400
1330 !
1340         LET N = N + 1
1350         LET An_2 = An_1
1360         LET An_1 = An
1370         LET X_TO_THE_N = X_TO_THE_N * X
1380         GOTO 1280
1390 !
1400         PRINT X,Y
1410         PRINT #1,X;",";Y
1420         LET X = X + X_INCREMENT
1430     NEXT I
1440 !
1450 !
1460 PRINT
1470 INPUT "Type a 'Y' if you want to see a graph: "; GRAPH$
1490 IF GRAPH$ <> 'Y'  THEN 1510
1500 CHAIN "[VANIWAARD.DIFFEQLIB.PRINTGRAF]DEGRAF.EXE"
1510 INPUT "Type a 'Y' if you want a printout file made of the data: "; P$
1530 IF P$ = 'Y' THEN CHAIN "[VANIWAARD.DIFFEQLIB.PRINTGRAF]PRINTOUT.EXE"
1540 !
1550 !
1560 CLOSE #1
1570 GOSUB 2000   ! DELETE DATA FILES
1580 GOTO 4000   ! END
2000 !
```

```
100 !               **** PROGRAM SERSO3 ****
110 !
120 !        Differential Equations - Hope College
130 !
140 !
150 !
160 !
170 !
180 !
190 !  THIS PROGRAM COMPUTES VALUES OF THE SOLUTION TO A SECOND
200 !  ORDER INITIAL VALUE PROBLEM AT THE ORDINARY POINT X = 0,
210 !  USING THE POWER SERIES TECHNIQUE. IT IS DESIGNED TO BE
220 !  USED WITH RECURRENCE FORMULAS WHOSE SERIES COEFFICIENT
230 !  SUBSCRIPTS DIFFER BY THREE.
240 !
250 !  THE USER RUNS THE PROGRAM, ENTERING THE LEFT AND RIGHT
260 !  ENDPOINTS OF THE X INTERVAL, THE NUMBER OF EVALUATIONS,
270 !  THE INITIAL VALUES OF Y, Y', Y'', AND THE CONVERGENCE
280 !  TOLERANCE.
290 !
300 !  NOTE: RECURRENCE FORMULA, FNF(N,An_1,An_2,An_3), IS DEFINED
310 !        AT LINE # 800
320 !
330 !  THE PROGRAM INITIALIZES VALUES FOR USE IN THE RECURRENCE
340 !  FORMULA AND STARTS COMPUTING THE INFINITE SERIES. WHEN
350 !  THE TOLERANCE LIMIT HAS BEEN REACHED, THE VALUES OF X
360 !  AND Y ARE PRINTED AND THE NEXT POINT OF THE X INTERVAL
370 !  IS USED. IF THE TOLERANCE HAS NOT BEEN REACHED, THE NEXT
380 !  TERM OF THE SERIES IS COMPUTED AND ADDED.
390 !
400 !
410 !                    VARIABLES
420 !                    ---------
430 !
440 !       LEFT_X_ENDPNT     - LEFT ENDPOINT OF THE X INTERVAL
450 !       RIGHT_X_ENDPNT    - RIGHT ENDPOINT OF THE X INTERVAL
460 !       NUM_EVALUATIONS   - NUMBER OF EVALUATIONS
470 !       Y0                - INITIAL VALUE OF Y
480 !       Y1                - INITIAL VALUE OF Y'
490 !       Y2                - INITIAL VALUE OF Y''
500 !       CONVERGENCE_TOL   - CONVERGENCE TOLERANCE
510 !       X_INCREMENT       - X INCREMENT
520 !       ERROR_COUNTER     - COUNTS THE NUMBER OF ERRORS
530 !       ERROR_NUMBER      - SYSTEM NUMBER OF AN ERROR
540 !       I                 - LOOP CONTROL VARIABLE
550 !       An_3              - A SUB (N-3) IN THE RECURRENCE FORMULA
560 !       An_2              - A SUB (N-2) IN THE RECURRENCE FORMULA
570 !       An_1              - A SUB (N-1) IN THE RECURRENCE FORMULA
580 !       An                - THE PRESENT A SUB (N) IN THE RECURRENCE FORMULA
590 !       N                 - INDEX IN THE RECURRENCE FORMULA
600 !       X_TO_THE_N        - VARIOUS POWERS OF X
610 !       NEXT_TERM         - NEXT COMPUTED TERM OF THE SERIES
620 !
630 !
640 ! * * * * * * * * * * * * * * * * * * * * * * * * * * * * *
650 !
800     DEF FNF(N,An_1,An_2,An_3) = ((N-2)*N*An_2 + An_3) / (N*(N-1))
1000 !
1010    OPEN "DETEMP.DAT" FOR OUTPUT AS FILE #1
1020 !
1030    PRINT
1040    PRINT "ENTER LEFT AND RIGHT ENDPOINTS OF THE X INTERVAL,"
1050    PRINT "NUMBER OF EVALUATIONS, INITIAL VALUES OF Y, Y', Y'',"
1060    PRINT "AND THE CONVERGENCE TOLERANCE."
1070 !
1080     ON ERROR GOTO 3000  ! PRINT ERROR DESCRIPTION
1090     INPUT LEFT_X_ENDPNT, RIGHT_X_ENDPNT, NUM_EVALUATIONS,  &
              Y0, Y1, Y2, CONVERGENCE_TOL
1100     GOTO 3200  ! CHECK FOR ERRORS IN DATA INPUT
1110 !
1120    LET X = LEFT_X_ENDPNT
```

```
1130     LET X_INCREMENT = (RIGHT_X_ENDPNT - LEFT_X_ENDPNT)/NUM_EVALUATIONS
1140 !
1150     PRINT
1160     PRINT "EVALUATIONS :"; NUM_EVALUATIONS
1170     PRINT "X INCREMENT :"; X_INCREMENT
1180     PRINT
1190     PRINT "  X", "  Y"
1200     PRINT "------", "------"
1210     FOR I = 1 TO NUM_EVALUATIONS + 1
1220         LET An_3 = Y0
1230         LET An_2 = Y1
1240         LET An_1 = Y2
1250         LET Y = Y0 + Y1 * X + Y2 * X * X
1260         LET N = 3
1270         LET X_TO_THE_N = X ** 3
1280     !
1290         LET An = FNF(N, An_1, An_2, An_3)
1300     !
1310         LET NEXT_TERM = An * X_TO_THE_N
1320         LET Y = Y + NEXT_TERM
1330         IF ABS(NEXT_TERM) < CONVERGENCE_TOL THEN 1420
1340     !
1350         LET N = N + 1
1360         LET An_3 = An_2
1370         LET An_2 = An_1
1380         LET An_1 = An
1390         X_TO_THE_N = X_TO_THE_N * X
1400         GOTO 1290
1410     !
1420         PRINT X,Y
1430         PRINT #1,X;",";Y
1440         LET X = X + X_INCREMENT
1450     NEXT I
1460 !
1470 !
1480 PRINT
1490 PRINT "Type a 'Y' if you want to see a graph: ";
1500 INPUT GRAPH$
1510 IF GRAPH$  <> 'Y' THEN 1530
1520 CHAIN "[VANIWAARD.DIFFEQLIB.PRINTGRAF]DEGRAF.EXE"
1530 PRINT "Type a 'Y' if you want a printout file made of the data: ";
1540 INPUT P$
1550 IF P$ = 'Y' THEN CHAIN "[VANIWAARD.DIFFEQLIB.PRINTGRAF]PRINTOUT.EXE"
1560 !
1570 !
1580 CLOSE #1
1590 GOSUB 2000  ! DELETE DATA FILES
1600 GOTO 4000  ! END
2000 !
```

```
100 !                    **** PROGRAM YACTUL ****
110 !
120 !          Differential Equations - Hope College
130 !
140 !
150 !
160 !
170 !
180 !
190 !  FOR PURPOSES OF COMPARING THE ACCURACY OF AN APPROXIMATE
200 !  SOLUTION TO A DIFFERENTIAL EQUATION WITH ITS ANALYTICAL
210 !  SOLUTION, THIS PROGRAM PRINTS A LIST OF ACTUAL X AND
220 !  Y VALUES TO COMPARE WITH THE APPROXIMATIONS.
230 !
240 !  THE USER THEN RUNS THE PROGRAM, ENTERING THE LEFT AND
250 !  RIGHT ENDPOINTS OF THE X INTERVAL, AND THE NUMBER OF
260 !  EVALUATIONS DESIRED.
270 !
280 !  THE PROGRAM COMPUTES THE VALUE OF Y, PRINTS THE
290 !  X, Y PAIR, INCREMENTS X AND LOOPS.
300 !
310 !  NOTE: FNF(X,Y) IS DEFINED AT LINE # 800
320 !
330 !                  VARIABLES
340 !                  ---------
350 !
360 !       LEFT_X_ENDPNT     - LEFT ENDPOINT OF THE X INTERVAL
370 !       RIGHT_X_ENDPNT    - RIGHT ENDPOINT OF THE X INTERVAL
380 !       NUM_EVALUATIONS   - NUMBER OF EVALUATIONS
390 !       X_INCREMENT       - X INCREMENT
400 !       I                 - LOOP VARIABLE
410 !       ERROR_COUNTER     - COUNTS THE NUMBER OF ERRORS
420 !       ERROR_NUMBER      - SYSTEM NUMBER OF AN ERROR
430 !
440 !
450 !
460 ! * * * * * * * * * * * * * * * * * * * * * * * * * * * * * * *
470 !
800     DEF FNF(X)= 8*EXP(X-1) - X**2 - 2*X - 2
1000 !
1010    OPEN 'DETEMP.DAT' FOR  OUTPUT AS FILE #1
1020 !
1030    PRINT
1040    PRINT 'ENTER LEFT AND RIGHT ENDPOINTS, NUMBER OF EVALUATIONS.'
1050 !
1060    ON ERROR GOTO 3000  !ILLEGAL DATA INPUT
1070    INPUT LEFT_X_ENDPNT, RIGHT_X_ENDPNT, NUM_EVALUATIONS
1080    GOTO 3200  ! ERROR CHECK FOR INVALID DATA
1090 !
1100    LET X_INCREMENT = (RIGHT_X_ENDPNT-LEFT_X_ENDPNT)/NUM_EVALUATIONS
1110    LET X = LEFT_X_ENDPNT
1120 !
1130    PRINT
1140    PRINT '  X', '  Y'
1150    PRINT '------', '------'
1160 !
1170    FOR I = 1 TO NUM_EVALUATIONS+1
1180        LET Y = FNF(X)
1190        PRINT X,Y
1200        PRINT #1,X;',';Y
1210        LET X = X + X_INCREMENT
1220    NEXT I
1230 !
1240 PRINT
1250 INPUT 'Type a 'Y' if you want to see a graph: '; GRAPH$
1270 IF GRAPH$ <> 'Y' THEN 1290
1280 CHAIN '[VANIWAARD.DIFFEQLIB.PRINTGRAF]DEGRAF.EXE'
1290 PRINT 'Type a 'Y' if you want a printout file made of the data: ';
1300 INPUT P$
1310 IF P$ = 'Y' THEN CHAIN '[VANIWAARD.DIFFEQLIB.PRINTGRAF]PRINTOUT.EXE'
1320 !
1330 !
1340  CLOSE #1
1350  GOSUB 2000  ! DELETE DATA FILES
1360  GOTO 4000  ! END
2000 !
```

Answers: Selected Exercises, Problems, and Applications Problems

Section 1.1

1. (a) Second order, linear, variable coefficients, homogeneous.
   (b) First order, nonlinear, constant coefficients, nonhomogeneous.
   (c) Third order, linear, constant coefficients, homogeneous.
   (d) Second order, nonlinear, variable coefficients, nonhomogeneous.
   (e) First order, nonlinear, variable coefficients, nonhomogeneous..
   (f) Second order, nonlinear, variable coefficients, nonhomogeneous.
   (g) First order, nonlinear, constant coefficients, nonhomogeneous.
   (h) First order, nonlinear, variable coefficients, nonhomogeneous.
   (i) Second order, linear, variable coefficients, nonhomogeneous.
   (j) Fourth order, linear, constant coefficients, nonhomogeneous.

9. (a) $k = 1, k = 3$

   (b) $k = -\frac{1}{2}, k = -1$

   (c) $k = 1, k = -2$
   (d) $k = -1$

15. $C_1 = \frac{1}{2}, C_2 = -\frac{1}{2}$

Section 1.2

1. $y = x^4 + C$

3. $y = \frac{1}{2}(e^{2x} + 3)$

5. $y = -\cos x + 2x + 2$

7. $y = 2x - \frac{x^2}{2} + C$

9. $y = \frac{x^5}{20} + 2x^2 + 2x - 1$

11. $y = -\frac{1}{6}\cos 3t - \frac{7}{6}$

13. $y = \frac{1}{4}\ln|x| + \frac{1}{8}\ln|x+2| + \frac{1}{8}\ln|x-2| - \frac{1}{8}\ln 3$

15. $(1+x^2)^2 = Ke^t$

17. $y = \frac{C_1}{2}x^2 + C_2x + C_3$

19. $y = \sec x$

21. \$187.25

23. $p = 0.03(27 + 14t - t^2)^{1/2} + 0.144$; 7 hours

Section 1.3

1. $y = Cx^3$

3. $\frac{1}{2}y^{-2} - \frac{1}{y} = \frac{-1}{x} + C$

5. $x = \frac{6(1 - e^{[(1/10)\ln 2]t}}{1 - 2e^{[(1/10)\ln 2]t}}$

7. $y = C \sec x$

9. $s = C\left(\frac{t - 2s}{t}\right)^{3/2}$

11. $y = x\ln|Cx|$

13. $y = \frac{t^2}{4t+1} + C$

15. $v^2 = 1 - \frac{K}{x}$

17. $y = \sqrt{3}\,x^{-1/2}$

19. $y = \text{Arccot}\,\frac{(1+e^x)^3}{C}$

21. $y^2 = -x^3 + 17$

25. (a) $t = \frac{1}{k}\ln\frac{4}{3} + 3$

 (b) $x = \frac{a[akt_1(1-n) - 1]}{akt_1(1-n) - n}$

27. $x = 64$

29. $v = \dfrac{10 - 10e^{-0.3t}}{3}$

31. 11.55 years

33. $\phi(x) = \left(\dfrac{C_1}{K+1}\right)x + C_2$

Section 1.4

1. $y = x\ln|x| + Cx$

3. $y^2 = 2Cx + C^2$

5. $y = \dfrac{Cx}{1 - Cx}$

7. $y = \ln|x| + 1 + Cx$

9. $x = \dfrac{y^3}{2} + Cy$

11. $\dfrac{-1}{xy} = y^2 + C$

13. $Cy^2 - y = -\dfrac{1}{2}x^{-2}$

15. $\dfrac{1}{2}\ln|x^2 + y^2| - \text{Arctan}\,\dfrac{y}{x} = C$

Section 1.5

1. $x^2 + xy + y^2 = C$

3. $xy - \dfrac{y^4}{4} = C$

5. Not exact

7. $x^2 - \dfrac{x}{y} + 3\ln|y| = C$

9. $\cos x \cos y = C$

11. $y^2 = (x-1)^3 + C$

13. $x^2 \cos y - e^x = C$

15. $xy^2 + \dfrac{x^3}{3} = 12$

17. $t^2y - \frac{y^2}{2} = C$

19. $x(x^2 - 3 + t) + t(2 + t) = C$

21. $xe^y + ye^x = 1$

23. (a) $v = -4$; $y = Cx^2$

(b) $v = \frac{1}{2}$; $e^{3x^2}\sin 3y = C$

25. (a) $\frac{y^2}{2} - \frac{y}{x} + 2x = C$

(b) $y = \frac{2x}{C - x^2}$

(c) $e^{y/x} + y^2 = C$

(d) $x^5y + x^4y^2 + \frac{x^4}{4} = C$

Section 1.6

1. $y = e^x(x + C)$

3. $y = 2x^{-1} + Cx^{1/2}$

5. $y = \frac{1}{3}(1 + x^2) + \frac{C}{(1 + x^2)^{1/2}}$

7. $y = \frac{x - 1}{x} + \frac{C}{xe^x}$

9. $y = -x - 2 + Ce^x$

11. $y = \frac{-\cos x}{\cot x} + \frac{C}{\cot x}$

13. $y = (\ln|x| + 1 + Cx)^{-1}$

15. $y = \left(\frac{1}{3}e^x + e^{x^2/4}C\right)^2$

17. $y = \left(Cx^{2/3} - \frac{3}{7}x^3\right)^{-3}$

19. $v = \left(-\frac{1}{2} + Ce^{-u^2}\right)^{-1}$

21. $x^2 = -2y^2\ln|y| + Cy^2$

23. $y = \left(\dfrac{1}{1 + Ce^{3x}}\right)^{1/3}$ or $y = \dfrac{e^{-x}}{(e^{-3x} + C)^{1/3}}$

25. $y = \dfrac{5}{x + 1} - \dfrac{1}{e^x(x + 1)} = \dfrac{5 - e^{-x}}{x + 1}$

27. $y = (x + C) \cos x$

29. $x = 200 - 150e^{-0.9}$

31. $i = \dfrac{e}{R} + Ce^{-(R/L)t}$

33. $n(t) = \dfrac{n_0(ag - bj)e^{(ag - bj)t}}{n_0bje^{(ag - bj)t} + ag - bj(1 + n_0)}$

Section 1.7

1. $y = x^2 + 2x + C$

3. $e^x y - 2xe^x + 2e^x = C$

5. $x^2 + y^2 = C$

7. $y = -\dfrac{5}{4} - x + Ce^x$

9. $y^2 = -\dfrac{x^2}{5} + \dfrac{2}{5}C$

11. $y^2 = \ln|Cx|$

13. $y^3 = -x^3 + 3C$

15. $\sin y = x + C$

17. $x^2 + y^2 = C$

19. $x^2 + \dfrac{y^2}{2} = C$

23. $x^2 + y^2 = C$

25. $\int (y^2 - 1)^{-1/2}\, dy = x + C$

Section 1.8

1. $y = \dfrac{2\sqrt{2}}{3}(x + C_1)^{3/2} + C_2$

3. $y = C_2 + \dfrac{1}{4}\ln|4x - C_1|$

5. $y = C_1\left(x - \frac{1}{3}x^3\right) + C_2$

7. $y = \frac{x^3}{3} + C_1\frac{x^2}{2} + C_2$

9. $\ln|y| + C_1 y = -x + C_2$

11. $y = \sqrt{x^2 - 1} - \sqrt{3}$

13. $y = -\frac{1}{2}\ln|1 - C_1 e^{2x}| + C_2$

17. $y'' = \left[\frac{1 + (y')^2}{y}\right]^{3/2}$

19. $H = \frac{4}{35} - 0.0343e^{-0.35t}$

Section 1.9

1. $\frac{25}{4}$ feet; $\frac{625}{256}$ feet

3. $v = 20 - 20e^{-t/300}$; $v_t = 20$

5. 1060.3 feet

Miscellaneous Exercises for Chapter 1

1. $y^2 + 2xy = C$

2. $s = \frac{3}{3 - t^3}$

3. $y = \frac{2\,e^{x^2}}{2C - e^{x^2}}$

4. $kt^2 = 2\ln|Cv|$

5. $4x^2 = y(C - x^4)$

6. $x^2 + y^2 = Cxy^2$

7. $y = x^2\ln|x| - x^2 + (2 - 2\ln 2)x$

8. $e^y = Cx^2(y + 3)^3$

9. $y - x = C(y + x)^3$

10. $r = C\sec\theta$

11. $y = 3x - \frac{3}{2} + Ce^{-2x}$

12. $x^2 \tan y - 3xy^2 + 2y = K$

13. $y = C(\sec x + \tan x)$

14. $y = \dfrac{x^2}{x^3 + C}$

15. $x - y = Cxy^2$

16. $y^2 = C - e^{x^2}$

17. $y = \dfrac{1}{5}x^6 + Cx$

18. $y = \dfrac{Cx}{(x + 1)^3}$

19. $y = \dfrac{1}{2}\ln|K - 2e^x|$

20. $y^2 = C^2x^2 + 1 = 1 - \dfrac{1}{4}x^2$

21. $v^2 = v_0^2 + 2g(x - x_0)$

22. $\dfrac{y^2}{2x^2} - \dfrac{2x}{y} + \ln|x| = C$

23. $y = \dfrac{1}{4}x - \dfrac{1}{48} + Ce^{4x}$

24. $y = [\ln|C - \sin\sqrt{x}|]^{-1}$

25. $3x \cos y - \cos(x^2) = K$

26. $y = \dfrac{x + C}{\sec x + \tan x}$

27. $y^3 = C(y^2 - x^2)$

28. $y^2 - x^2e^{y/x} = C$

29. $r^2 = 2k^2 \sin^2\theta + C$

30. $i = \dfrac{E}{R}(1 - e^{-(R/L)t})$

31. $y = -\dfrac{C}{2}\cos^2 x + K \sin^2 x$

32. $y^2 = \dfrac{x^2}{Cx - 1}$

33. $y = (x^3 - x)[\ln|x| + C]$

34. $y = Ce^{x^2/4} - 1$

35. $y = (\ln|x| - \frac{1}{2}x + 1)^{-1}$

36. $x^2 + 2xy + 3y^2 = 48$

37. $y = \dfrac{\ln|x|}{2} + \dfrac{C}{\ln|x|}$

38. $y = Cx - x \sin\dfrac{1}{x}$

39. $y = \dfrac{(x+1)^5}{2} + C(x+1)^3$

40. $y = \frac{1}{3}x^{5/2} - \frac{1}{4}x^{3/2} + Cx^{-1/2}$

41. $xy^3 = x^2y + C$

42. $x + C = \displaystyle\int \frac{dy}{\cos^{-1}(1-y)}$

43. $e^y = 1 - Ce^{-x}$

44. $y = \dfrac{1 + Ce^{-x}}{x^2}$

45. $C^2x^2 - 2Cy - 1 = 0$

46. $y = \sin x + C_1 x + C_2$

47. $y = 2e^{2x} - 2x - 1$

48. $y^3 + e^y = \sin x + 1$

49. $y = (x^2 + 1)\left[\text{Arctan}\, x + \dfrac{\pi}{4}\right]$

50. $y = \sin\left(e^x + \dfrac{\pi}{6} - 1\right)$

51. $p = \dfrac{\alpha SI}{\beta} + \left(p_0 - \dfrac{\alpha SI}{\beta}\right)e^{-\beta t}$

52. $L = L_0 e^{-K\sigma\zeta t}$

54. $y = \dfrac{C^2x^2}{3} + \dfrac{K}{x}$

55. $T_2(t) = 50e^{-It/v} + T_1(t)$

56. $L = Ke^{at/3\eta}$

57. ≈ 2.87 grams

59. $A = 6.57 \times 10^{13}$

60. $v = 15.9$ ft/sec

61. General n: $\dfrac{r^{1-n}}{1-n} = k\ln|s| + C$

$n = 0$: $r = k\ln|s| + C$
$n = 1$: $r = Cs^k$
$n = 2$: $r = \dfrac{-1}{k\ln|s| + C}$

62. (a) $n(1000 - n) = 9900e^{9t}$
(b) 1000 schools

63. (a) $T = T_1 e^{\mu\theta}$
(b) $50e^{\pi/10}$, $50e^{6\pi/10}$
(c) $9000e^{-2\pi}$

64. 4179 years

65. $C = \dfrac{[(1-N)r \cdot 10^6 + s]e^{(1-N)r(t-t_0)} - s}{(1-N)r}$

Problems for Chapter 2

1. (a) $y = 2x^{-1/3}$, YACTUL: $y(2) = 1.58740$
(b) Using EULER and $h = 0.1$: $Y(2) = 1.56911$
Using EULER and $h = 0.05$: $Y(2) = 1.57842$
(c) Using EULER and $h = 0.001$: $Y(2) = 1.58722$
(d) Using IMEUL and $h = 0.01$: $Y(2) = 1.58740$

2. (a) $y = 2\tan(2x)$, YACTUL: $y\left(\dfrac{1}{2}\right) = 3.11482$

(b) Using EULER and $h = 0.1$: $Y\left(\dfrac{1}{2}\right) = 2.58826$

(c) Using IMEUL and $h = 0.1$: $Y\left(\dfrac{1}{2}\right) = 3.08968$

(d) Using IMEUL and $h = 0.01$: $Y\left(\dfrac{1}{2}\right) = 3.11451$

3. (a) Using EULER and $h = 0.01$: $Y\left(\frac{\pi}{2}\right) = 0.990739$

   Using EULER and $h = 0.01$: $Y(\pi) = 2.69608$

   Using EULER and $h = 0.01$: $Y\left(\frac{3\pi}{2}\right) = 0.96041$

   Using EULER and $h = 0.01$: $Y(2\pi) = 0.361865$

   (b) Using IMEUL and $h = 0.01$: $Y\left(\frac{\pi}{2}\right) = 0.999959$

   Using IMEUL and $h = 0.01$: $Y(\pi) = 2.71814$

   Using IMEUL and $h = 0.01$: $Y\left(\frac{3\pi}{2}\right) = 1.00029$

   Using IMEUL and $h = 0.01$: $Y(2\pi) = 0.367867$

   (c) $y = e^{-\cos x}$

   Using YACTUL: $y\left(\frac{\pi}{2}\right) = 1$

   Using YACTUL: $y(\pi) = 2.71828$

   Using YACTUL: $y\left(\frac{3\pi}{2}\right) = 1.00003$

   Using YACTUL: $y(2\pi) = 0.367879$

   (d) Percent error using EULER = 1.635%
   Percent error using IMEUL = 0.004%

4. (a) Using EULER and $h = 0.1$: $Y(1) = 1.73593$
   Using EULER and $h = 0.05$: $Y(1) = 1.78971$
   Using EULER and $h = 0.01$: $Y(1) = 1.83279$

   (b) Using IMEUL and $h = 0.1$: $Y(1) = 1.8277$
   Using IMEUL and $h = 0.05$: $Y(1) = 1.83763$
   Using IMEUL and $h = 0.01$: $Y(1) = 1.84277$

   (c) Using IMEUL and $h = 0.001$; $Y(1) = 1.84327$

5. (a) Using EULER and $h = 0.1$: $Y(2) = 0.246956$
   Using EULER and $h = 0.05$: $Y(2) = 0.230508$
   Using EULER and $h = 0.01$: $Y(2) = 0.217522$

   (b) Using IMEUL and $h = 0.1$: $Y(2) = 0.213005$
   Using IMEUL and $h = 0.05$: $Y(2) = 0.213985$
   Using IMEUL and $h = 0.01$: $Y(2) = 0.214286$

   (c) Using EULER and $h = 0.001$: $Y(2) = 0.21461$
   Using IMEUL and $h = 0.001$: $Y(2) = 0.214288$

6. $y = \sqrt{2x + 1}$, YACTUL: $y(2) = 2.23607$
   (a) Using EULER and $h = 0.1$: $Y(2) = 2.50471$
   Using EULER and $h = 0.01$: $Y(2) = 2.27104$
   (b) Using IMEUL and $h = 0.1$: $Y(2) = 2.27072$
   Using IMEUL and $h = 0.01$: $Y(2) = 2.23643$

7. (a) Using EULER and $h = 0.0005$: $Y(1) = 0.256926$
   (b) Using IMEUL and $h = 0.001$: $Y(1) = 0.257196$

8. (a) Using EULER and $h = 0.0002$: $Y(0.2) = 0.93056$
   (b) Using IMEUL and $h = 0.0002$: $Y(0.2) = 0.930455$

9. Using IMEUL with $B = 1$ and $h = 0.002$: $Y(1) = 16.6984$

Applications Problems for Chapter 2

A–1. Using IMEUL with $h = 0.01$, $P(5) = 4015.61$

A–2. (a) Using IMEUL and $h = 1.0$: $Y(100) = 76{,}596.8$
   Using IMEUL and $h = 0.01$: $Y(100) = 76{,}612.2$
   (c) $K_1 = 2.2 \times 10^{-2}$
   (d) Using IMEUL and $h = 1.0$: $Y(2050) = 240, 422$

A–3. (a) Using IMEUL and $h = 0.1$: $T(4) = 90.3987$
   (b) Using YACTUL: $T(4) = 90.3836$

A–4. (a) Using IMEUL and $h = 0.01$ with $P_0 = 25$: $P(15) = 125$
   Using IMEUL and $h = 0.01$ with $P_0 = 75$: $P(15) = 125$
   (b) Using YACTUL with $P_0 = 25$: $P(15) = 125$
   Using YACTUL with $P_0 = 75$: $P(15) = 125$

A–5. (a) $h = 0.01$ yields $x = 300$ using IMEUL when $t = 12.67$

A–6. (b) Using IMEUL and $h = 0.01$: $Y = 50$ when $t = 1.69$
   Using IMEULL and $h = 0.01$: $Y = 80$ when $t = 8.95$
   (c) Using IMEUL and $h = 0.01$: $Y = 50$ when $t = 1.145$; at $t = 10$, $Y = 77.7676$; maximum Y is 77.7778

A–7. (a) Using IMEUL and $P(0) = 111$: $P(25) = 289.381$
   Using IMEUL and $P(0) = 110$: $P(8.5) = 10.137$
   (b) Using IMEUL and $P(0) = 250$: $P(25) = 210.191$
   Using IMEUL and $P(0) = 300$: $P(25) = 211.348$
   (c) 4.433 fish remain after 20 days.

A–8. (a) $r(t) = 40e^{-.02t}$, $r(0) = 40$
   Using YACTUL: $r(5) = 36.1935$
   (b) Using EULER and $h = 0.05$: $r(5) = 36.1917$
   Using IMEUL and $h = 0.05$: $r(5) = 36.1935$

A–9. (a) Using EULER and $h = 1.0$: $Y(100) = 0.753769$
(b) Using IMEUL and $h = 1.0$: $Y(100) = 0.749962$
(c) Using YACTUL: $y(100) = 0.75$

A–10. When $t = 10.1$, height $= 0$.

A–11. (a) Using YACTUL: $y(50) = 3000$ employed; 33.1% were women.
(b) Using IMEUL and $h = 1.0$: $Y(50) = 993.73$

A–12. (a) $r = 15.5\,(1 - e^{-0.000368p})$
Using YACTUL: $r(5000) = 13.0383$
(b) Using EULER: $r(5000) = 13.1223$
Using IMEUL: $r(5000) = 13.0383$

A–13. Convert: $p = 2160\ \text{lb/ft}^2$, $\rho = 0.0023769\ \text{slugs/ft}^3$. Then $p = (2.16783 - 0.281962 \cdot 10^{-5} \cdot h)^{7/2}$. With RUKU4, $p(100{,}000) = 9.2107\ \text{lb/in}^2$; with YACTUL, $p(100{,}000) = 9.2106\ \text{lb/in}^2$.

A–14. (b) Using IMEUL with $b = -\frac{1}{2}$ and $h = 0.01$: $Y(3) = 0.05343$
Using IMEUL with $b = 0$ and $h = 0.01$: $Y(3) = 0.04979$
Using IMEUL with $b = \frac{1}{2}$ and $h = 0.01$: $Y(3) = 0.046398$

A–15. (a) Using EULER with $h = 0.01$ and $u(0) = 320°\text{F}$:
$U(1.623) = 120°\text{F}$
Using IMEUL with $h = 0.01$ and $u(0) = 320°\text{F}$:
$U(1.633) = 120°\text{F}$

Section 3.2

1. General: $y(x) = C_1 e^{-4x} + C_2 e^{3x}$

   Specific: $y(x) = \frac{5}{7} e^{-4x} + \frac{2}{7} e^{3x}$

3. $y = C_1 x + C_2 x e^{2/x}$

5. $y = C_1 + C_2 e^{3x} + C_3 e^{-3x}$

Section 3.3

1. $y = C_1 e^x + C_2 e^{6x} + 2$

3. $y = C_1 e^{3x/2} + C_2 e^{-2x} + \frac{1}{3} e^{3x} + \frac{1}{6} x + \frac{1}{36}$

5. $y = C_1 e^{-2x} + C_2 e^{-x} + e^x + 2$

7. $y = C_1 e^{x/2} + C_2 e^{5x} - \frac{2}{5} e^{3x} - 2$

9. $y_p = x$

11. $y_p = 2e^x$

13. $y_p = 2 + 3x + x^2$

15. $y = C_1 e^{3x} + C_2 e^{-3x} + \frac{1}{9} - \frac{1}{3} x$

17. $y = C_1 e^{2x} + C_2 e^{-2x} - \frac{3}{4} x$

19. $y = C_1 e^{-2x} + C_2 e^{-x} + \frac{5}{6} e^x + 2x - \frac{9}{2}$

Section 3.4

1. (a) $\cos 4x + i \sin 4x$
   (b) $e^{2x}\cos 3x - ie^{2x}\sin 3x$
   (c) $e^{-2x}\cos x + ie^{-2x}\sin x$
   (d) $e^{-3x}\cos 5x - ie^{-3x}\sin 5x$
   (e) $e^{x/2}\cos \frac{x}{2} + ie^{x/2}\sin \frac{x}{2}$
   (f) $e^{(2a/5)x}\cos\left(-\frac{ax}{5}\right) + ie^{(2a/5)x}\sin\left(-\frac{ax}{5}\right)$

3. $u_1(x) = \cos 5x$, $u_2(x) = \sin 5x$

5. $u_1(x) = e^{-x/4}\cos\left(\frac{\sqrt{7}}{4} x\right)$, $u_2(x) = e^{-x/4}\sin\left(\frac{\sqrt{7}}{4} x\right)$

9. $\cos i = 1 - \frac{i^2}{2} + \frac{i^4}{24} - \frac{i^6}{720} + \cdots \simeq 1.54305$

Section 3.5

1. $r^2 - 6r + 5 = 0$; $r_1 = 5$, $r_2 = 1$

3. $3r^2 + 13r - 10 = 0$; $r_1 = \frac{2}{3}$, $r_2 = -5$

5. $r^3 - 7r^2 + 12r = 0$; $r_1 = 0$, $r_2 = 4$, $r_3 = 3$

7. $u_1 = e^{-5x}$, $u_2 = e^{2x}$; $y = C_1 e^{-5x} + C_2 e^{2x}$

9. $u_1 = e^{-(4/3)x}$, $u_2 = e^{2x}$; $y = C_1 e^{-(4/3)x} + C_2 e^{2x}$

11. $u_1 = e^{-7x}$, $u_2 = e^{6x}$; $y = C_1 e^{-7x} + C_2 e^{6x}$

Section 3.7

1. $y = C_1 e^{3x} + C_2 e^x$
3. $y = e^{-(5/2)x} - \dfrac{1}{2} x e^{-(5/2)x}$
5. $y = C_1 e^{(3/5)x} + C_2 e^{-(3/5)x}$
7. $y = C_1 e^{-x} + C_2 e^x + C_3 + C_4 e^{-2x}$
9. $y = -\dfrac{2}{3} + \dfrac{2}{3} e^{3x}$
11. $y = 2e^{-(5/2)x} + \dfrac{15}{2} x e^{-(5/2)x}$
13. $y^{(iv)} - 7y''' + 13y'' + 3y' - 18y = 0$
15. $x = C_1 e^{-t/2} + C_2 e^{0.7t} + C_3 e^{-0.7t}$

Section 3.8

1. $y = C_1 e^{-3x} + C_2 x e^{-3x}$
3. $y = C_1 e^{-x} + C_2 e^x + C_3 x e^x$
5. $y = C_1 e^x + C_2 x e^x + C_3 x^2 e^x$
7. $y = C_1 e^{-x} \cos(\sqrt{2}x) + C_2 e^{-x} \sin(\sqrt{2}x)$
9. $y = C_1 e^{-3x} + C_2 e^{[(3+\sqrt{5})/2]x} + C_3 e^{[(3-\sqrt{5})/2]x} + 2$
11. $y = C_1 e^{\sqrt{3}x} + C_2 x e^{\sqrt{3}x} + C_3 e^{-\sqrt{3}x} + C_4 x e^{-\sqrt{3}x}$
13. $y = C_1 e^{2x} + C_2 e^{-2x} + C_3 x^2 + C_4 x + C_5$
15. $y = 0.544 e^{[(1+\sqrt{13})/2]x} - 0.344 e^{[(1-\sqrt{13})/2]x}$
17. $y = \dfrac{6}{5} e^x - \dfrac{1}{5} \cos 2x - \dfrac{3}{5} \sin 2x$
19. $y = e^{-2x} \cos x + 2e^{-2x} \sin x$
21. $y = -2e^{-x/2} + 2e^x$
23. $\omega = \dfrac{\pi}{2}$ at $t = \dfrac{\pi}{8}$
25. $y = \dfrac{1}{2} \cos 7.905x$; amplitude $= \dfrac{1}{2}$, depth $= 0.512$ ft.
27. $x = C_1 e^{[(r/2)+(1/2)\sqrt{r^2+(4r\beta/b)}]t} + C_2 e^{[(r/2)-(1/2)\sqrt{r^2+(4r\beta/b)}]t} + \dfrac{K}{\beta}$

Section 3.9

1. (a) $y_p = \dfrac{-30}{100}\sin 2x$ using e^{2ix}

   (b) $y_p = \dfrac{40}{289}\cos 4x - \dfrac{75}{289}\sin 4x$ using e^{4ix}

   (c) $y_p = \dfrac{308}{8845}\cos 6x + \dfrac{216}{8845}\sin 6x$ using e^{6ix}

3. $y = C_1e^{-3x} + C_2e^{4x} + \left(\dfrac{-13+i}{170}\right)e^{ix}$

4. $y = C_1e^{-3x} + C_2e^{4x} - \dfrac{13}{170}\sin x - \dfrac{1}{170}\cos x$

5. (a) $y = C_1e^{-5x} + C_2e^{3x} - \dfrac{64}{260}\cos x + \dfrac{8}{260}\sin x$

   (b) $y = C_1e^{-5x} + C_2e^{3x} - \dfrac{8}{260}\cos x - \dfrac{64}{260}\sin x$

7. $y = C_1e^{-3x} + C_2e^{-2x} + \dfrac{8}{5}\cos x + \dfrac{8}{5}\sin x$

9. $y = C_1e^{x/2} + C_2e^{-4x} - \dfrac{6}{170}\cos 2x + \dfrac{7}{170}\sin 2x$

11. $y = C_1e^{-(2/3)x} + C_2e^{x} + \dfrac{1}{4} - \dfrac{1}{2}x - \dfrac{10}{13}\cos x - \dfrac{2}{13}\sin x$

13. $y = C_1e^{-x/2} + C_2xe^{-x/2} - 4 + x - \dfrac{36}{1369}\cos 3x - \dfrac{105}{1369}\sin 3x$

15. $y = C_1\cos x + C_2\sin x - \dfrac{1}{2}x + \dfrac{1}{5}e^{2x} - \dfrac{4}{3}\cos 2x$

17. $y = C_1e^{(-2/3)x} + C_2xe^{-(2/3)x} + \dfrac{\alpha}{4} + \dfrac{\beta}{64}e^{2x} - \dfrac{221\gamma}{52441}\cos 5x$

    $+ \dfrac{60\gamma}{52441}\sin 5x$

19. $y = C_1e^{4x} + C_2e^{-3x} + \dfrac{1}{4} - \dfrac{\sqrt{2}}{170}(6\cos x + 7\sin x)$

Section 3.10

1. $y = C_1e^{-3x} + C_2e^{2x} + \frac{1}{7}e^{4x}$

3. $y = C_1e^{2x} + C_2e^{x} + \frac{7}{4} + \frac{3}{2}x + \frac{1}{2}x^2 + 2xe^x$

5. $y = C_1e^{-(3/2)x}\cos\left(\frac{\sqrt{11}}{2}x\right) + C_2e^{-(3/2)x}\sin\left(\frac{\sqrt{11}}{2}x\right) - \frac{8}{97}\sin 3x$

   $-\frac{18}{97}\cos 3x$

7. $y = C_1e^{(3/2)x}\cos\left(\frac{\sqrt{3}}{2}x\right) + C_2e^{(3/2)x}\sin\left(\frac{\sqrt{3}}{2}x\right) + e^x + xe^x$

9. $y = C_1 + C_2e^{-2x} - 2x$

11. $y = C_1\cos 2x + C_2\sin 2x - \frac{1}{42}\sin 5x - \frac{1}{6}\sin x$

13. $y = C_1\cos x + C_2\sin x + C_3e^x + C_4e^{-x} + \frac{1}{2}xe^x$

15. $y = C_1\cos x + C_2\sin x - x^2\cos x + x\sin x - \frac{1}{3}\cos 2x$

17. $y = C_1 + C_2x + C_3e^{4x} - \frac{65}{64}x^2 - \frac{1}{48}x^3 - \frac{1}{48}x^4$

19. $y = C_1\cos x + C_2\sin x + \frac{3}{4}x^2\cos x + \frac{3}{4}x\sin x$

    $y = 2\cos x + \sin x + \frac{3}{4}x^2\cos x + \frac{3}{4}x\sin x$

21. $y = \frac{-57}{504}e^{4x} - \frac{454}{504}e^{-3x} + \frac{1}{72} + \frac{1}{6}x$

23. $y = 2.9915e^{5x} - 16.9848xe^{5x} + 0.00842\cos x + 0.0148x\cos x$

    $+ 0.0125\sin x + 0.0355x\sin x$

25. $y = \frac{17}{64} - \frac{1}{64}e^{4x} + \frac{1}{16}x + \frac{1}{8}x^2$

27. $x(t) = C_1\cos 8\sqrt{3}t + C_2\sin 8\sqrt{3}t - \frac{1}{192}\sin 8\sqrt{6}t$

Section 3.11

1. $y = C_1x^2 + C_2x^{-2}$
3. $y = C_1x^2 + C_2x + x^2 \ln x$
5. $y = C_1x^3 + C_2x^2$
7. $y = C_1x^3\cos(2\ln x) + C_2x^3\sin(2\ln x)$
9. $y = C_1x^3 + C_2x^{-1} - \frac{2}{9}x^2 - \frac{1}{3}x^2 \ln x$
11. $y = C_1x^2 + C_2x^{-1}$
13. $y = C_1x^{-1} + C_2x^{-2} + C_3x^4$
15. $y = C_1x + C_2x^{(1/3 + 1/3\sqrt{31})} + C_3x^{(1/3 - 1/3\sqrt{31})} - \frac{2}{9}x^{-2}$
17. $y = C_1x + C_2x\int \frac{e^{x^2}}{x^2}\,dx + 2$
19. $y = C_1 + C_2r^{-3}$
21. $y = C_1 + C_2x^2 - 3x - \frac{1}{2}x^2 \ln x$
23. $u = \frac{u_2r_1^{-1} - u_1r_2^{-1}}{r_1^{-1} - r_2^{-1}} + \left(\frac{u_1 - u_2}{r_1^{-1} - r_2^{-1}}\right)r^{-1}$
25. $y = C_1e^x + C_2e^x\left[\ln x - x + \frac{x^2}{2\cdot 2!} - \frac{x^3}{3\cdot 3!} + \cdots\right]$
27. $y = C_1e^{x^2} + C_2 + \frac{1}{4}e^{x^2}(x^2 - 1)$
29. $y = C_1x - C_2\left(\frac{1}{x+1}\right) + x^2 - \frac{x^2}{x+1}\left(\frac{2x+3}{6}\right)$
31. $y = C_1e^x + C_2x^2 + (x^3 - 3x^2)e^x$

Section 3.12

1. $y = C_1\cos x + C_2\sin x - x\cos x + (\sin x)\ln|\sin x|$
3. $y = C_1\cos bx + C_2\sin bx - \left(\frac{1}{b^2}\cos bx\right)\ln|\sec bx| + \left(\frac{1}{b}x\right)\sin bx$
5. $y = C_1e^x + C_2xe^x - e^x\ln|1 - x|$

7. $y = C_1\cos 3x + C_2\sin 3x - \left(\frac{1}{9}\cos 3x\right)\ln|\sec 3x + \tan 3x|$

$+ \left(\frac{1}{9}\sin 3x\right)\ln|\csc 3x - \cot 3x|$

9. $y = C_1 + C_2e^{-x} + \frac{1}{4}e^x - \frac{1}{2}xe^{-x}$

11. $y = C_1 + C_2\ln|x| + \frac{1}{4}x^2 + x$

13. $y = C_1e^x + C_2e^{-x} - \frac{1}{2} + \frac{1}{2}e^x\ln|1 + e^{-x}| - \frac{1}{2}e^{-x}\ln|1 + e^x|$

15. $y = C_1e^{-4x} + C_2e^{2x} + e^{2x}\ln|x|$

17. $y = C_1e^x + C_2xe^x + C_3x^2e^x + e^x\left(\frac{-11}{2x} - \frac{3}{x}\ln|x|\right)$

19. $y = C_1\cos\frac{1}{3}x + C_2\sin\frac{1}{3}x - 2 + \left(\sin\frac{x}{3}\right)\ln\left|\sec\frac{x}{3} + \tan\frac{x}{3}\right|$

21. $y = C_1e^{x/2} + C_2xe^{x/2} + \left(\frac{1}{8}x^2e^{x/2}\right)\ln|x| - \frac{3}{16}x^2e^{x/2}$

23. $y = C_1\cos 2x + C_2\sin 2x + (\sin 2x)\ln|\csc 2x - \cot 2x|$

25. $y = C_1 + C_2\cos 2x + C_3\sin 2x + \frac{1}{8}\ln|\sec 2x + \tan 2x| - \frac{1}{4}x\cos 2x$

$- \left(\frac{1}{8}\sin 2x\right)\ln|\sec 2x|$

27. $y = C_1 + C_2e^{-kt/m} + \frac{mg}{2k}t - \frac{m^2g}{k^2}$

Miscellaneous Exercises for Chapter 3

1. $y = C_1\cos x + C_2\sin x + e^x$
2. $y = C_1x + C_2x^{-1}$
3. $y = C_1 + C_2\cos x + C_3x\cos x + C_4\sin x + C_5x\sin x$
4. $y = C_1 + C_2x + C_3e^x + C_4xe^x$
5. $y = C_1e^{x/2} + C_2e^{2x}$
6. $z = C_1x^{-2} + C_2x^{-1}$

7. $y = C_1e^x\cos 2x + C_2e^x\sin 2x + \frac{1}{5}$

8. $y = C_1e^x + C_2e^{-x} - \frac{1}{2} + \frac{1}{10}\cos 2x$

9. $y = C_1e^{-x} + C_2xe^{-x} + \frac{1}{2}x^2e^{-x} + \frac{1}{4}e^x$

10. $y = C_2\left(1 + \frac{1}{x}\right) + C_1\left(x + 1 - \frac{1}{x} - 2\left(\frac{x+1}{x}\right)\ln(x+1)\right)$

11. $C_1\cos\sqrt{2}x + C_2\sin\sqrt{2}x + x\cos x + 2\sin x$

12. $y = C_1 + C_2e^{3x} + C_3e^{-3x} + C_4e^x + C_5e^{-x}$

13. $y = K_2x^3 + C_2x^2 + C_3x + C_4 + \frac{x^3}{6}\ln x - \frac{1}{18}x^3$

14. $y = C_1e^{-x} + C_2e^{x/2}\cos\frac{\sqrt{3}}{2}x + C_3e^{x/2}\sin\frac{\sqrt{3}}{2}x$

15. $y = C_1e^{-x} + C_2\cos 2x + C_3\sin 2x - \frac{1}{10}x\cos 2x - \frac{1}{20}x\sin 2x$

16. $y = C_1e^{2x} + C_2e^x + [(\ln 2)^2 - 3\ln 2 + 2]^{-1}\cdot 2^x$

17. $y = C_1\cos x + C_2\sin x + 4\sin\frac{x}{2}$

18. $y = C_1xe^{-x^2} + C_2e^{-x^2}$

19. $y = C_2\sin x$

20. $y = C_1e^x + C_2e^{2x} + C_3e^{4x}$

21. $y = \left(\frac{34}{100} - \frac{3}{10}x\right)\cos x + \left(\frac{12}{100} + \frac{1}{10}x\right)\sin x + C_1e^{-2x} + C_2e^{-x}$

22. $y = C_1 + C_2e^{-x} + \frac{1}{4}e^x + \frac{1}{2}xe^{-x}$

23. $y = C_1\cos 2x + C_2\sin 2x + \frac{1}{8}x\sin 2x - \frac{1}{24}\cos 4x$

24. $y^5 = -8\left(\frac{x+2}{x-3}\right)$

25. $y = C_1\cos x + C_2\sin x + \frac{1}{4}x\cos x + \frac{1}{4}x^2\sin x + \frac{1}{2}e^x$

26. $y = C_2x + C_1xe^{1/x}$

27. $y = \pm 1$

28. $y = C_1 e^{-x} + C_2 e^{x/2} \cos \dfrac{\sqrt{7}}{2} x + C_3 e^{x/2} \sin \dfrac{\sqrt{7}}{2} x + \dfrac{1}{12} e^{-x} \sin 2x$

29. $y = \dfrac{1}{4} e^{x} - \dfrac{1}{4} e^{-x} - \dfrac{1}{2} \sin x$

30. $y = C_1 x + C_2 x^2 + C_3 x^3$

31. $y = K_1 \cos 2x + K_2 \sin 2x - \dfrac{1}{4} + \dfrac{1}{4} (\sin 2x) \ln|\sec 2x + \tan 2x|$

32. $y = \dfrac{66}{29} e^{-2x} \cos 3x + \dfrac{85}{87} e^{-2x} \sin 3x - \dfrac{8}{29} \cos 2x + \dfrac{9}{29} \sin 2x$

33. $y = C_1 x^2 \cos(3 \ln x) + C_2 x^2 \sin(3 \ln x)$

34. $y = 2e^{-2x} + 9xe^{-2x}$

35. $y = 3e^{-(1/3)x} \cos \dfrac{2}{3} x + \dfrac{3}{2} e^{-(1/3)x} \sin \dfrac{2}{3} x + 1$

36. $y = C_1 e^{3x} + C_2 e^{2x} + \dfrac{2}{3} + \dfrac{1}{5} x$

37. $y = C_1(x^2 - 1) + C_2 x$

38. $y = C_2 e^{c_1 x}$

39. $y = e^{-2x} \cos x + C_2 e^{-2x} \sin x + \left(\dfrac{1}{10} - \dfrac{1}{6} x\right) e^{x}$

40. $y = C_1 e^{x} + C_2 e^{-x} + C_3 \cos x + C_4 \sin x$

41. $y_p = (A_0 x^3 + A_1 x^4 + A_2 x^5) e^{2x} = \dfrac{1}{10} x^5 e^{2x}$

42. $y = \dfrac{14}{5} e^{-3x} + \dfrac{1}{5} e^{2x} + x$

43. $y = C_1 e^{-2x} + C_2 x e^{-2x} + \dfrac{1}{4} e^{2x} - \dfrac{1}{4} + \dfrac{1}{4} x$

44. $y = C_1 e^{x} + C_2 x e^{x} + C_3 x^2 e^{x} + x^3 e^{x}$

45. $y = C_1(e^{(2+\sqrt{6})x}) + C_2(e^{(2-\sqrt{6})x}) - 120 + 54x - 12x^2 + 2x^3$

46. $y = K_1 \cos ax - \dfrac{1}{2a^2} \tan^2 ax \cos ax + K_2 \sin ax + \dfrac{1}{a^2} \tan ax \sin ax$

47. $y = C_1 + C_2 x^2 - 2 \ln x$

48. $y = C_1 + C_2x + C_3e^x + 2\cos x - 2\sin x$

49. $y = C_1e^{-ax} + C_2e^{ax/2}\cos\frac{\sqrt{3}}{2}ax + C_3e^{ax/2}\sin\frac{\sqrt{3}}{2}ax$

50. $y = C_1e^{3x} + C_2e^{2x} + C_3e^{-4x} + C_4e^{-x}$

51. $u(r) = C_1r^m + C_2r^{-m}$

52. (a) $y_1 = b_0\left(x - \frac{k^2}{2}x^2 + \frac{k^4}{2!3!}x^3 - \frac{k^6}{3!4!}x^4 + \frac{k^8}{4!5!}x^5 - \cdots\right)$

53. $x(t) = \frac{6}{\pi}\sin\frac{4\pi}{3}t$

54. $x(t) = \frac{1}{6}\cos(4\sqrt{6})t$

55. $y = \sqrt{6}\,R\sec\left(\frac{1}{2}\sqrt{\frac{6g}{R}}\right)x$

56. $y(t) = \frac{1}{2}e^{-4t} + 2te^{-4t}$

57. Yes, at $t = \frac{1}{10}\ln\frac{28}{3}$

58. $s(t) = \frac{16}{3}t^2$

Problems for Chapter 4

1. (a) $y = \frac{9}{4}e^{\sqrt{2}x} + \frac{9}{4}e^{-\sqrt{2}x} - \frac{15}{2}$

   (d) Using IMEUL2 and $h = 0.1$: $Y(1) = 2.26552$; YACTUL: $y(1) = 2.30183$
   Using IMEUL2 and $h = 0.05$: $Y(1) = 2.29217$
   Using IMEUL2 and $h = 0.01$: $Y(1) = 2.30142$

2. (c) With $h = 0.1$, $Y(1) = 1.32519$
   With $h = 0.02$, $Y(1) = 1.32935$
   With $h = 0.01$, $Y(1) = 1.32949$

3. With $h = 0.02$, $Y(-1) = 3.12314$, $Y(1) = 3.15053$
   With $h = 0.01$, $Y(-1) = 3.12318$, $Y(1) = 3.15091$

4. With $h = 0.02$, $Y(2) = -0.049345$
   With $h = 0.01$, $Y(2) = -0.0485147$
   With $h = 0.002$, $Y(2) = -0.0482487$
   With $h = 0.0005$, $Y(2) = -0.0482367$

5. With $h = 0.01$, $Y(2) = 5.64506$
   With $h = 0.002$, $Y(2) = 5.64399$
   With $h = 0.001$, $Y(2) = 5.64396$

6. (a) $y = e^{-x} + 2\sin x$ gives $y(7) = 1.31489$
   (b) With $h = 0.1$, $Y(7) = 1.334$
   With $h = 0.01$, $Y(7) = 1.31506$
   With $h = 0.005$, $Y(7) = 1.31493$

7. With $h = 0.01$, $Y(2) = 6.81543$
   With $h = 0.005$, $Y(2) = 6.81551$

8. With $h = 0.01$, $Y(4) = -12.1975$
   With $h = 0.005$, $Y(4) = -12.1935$
   With $h = 0.001$, $Y(4) = -12.1915$

Applications Problems for Chapter 4

A–1. (a) Using $h = 0.0001$, at the x-value 0.0004 the intercept is made at $y = 0.215276$.
(b) Using $h = 0.0001$, at the x-value 0.0001 the intercept is made at $y = 0.414332$. (Theoretically no intercept is possible.)

A–2. (a) Using $h = 0.5$, $Y(10) = -0.021527$
(b) Using $h = 0.1$, $Y(10) = -0.0206711$
(c) Using $h = 0.1$, $Y(80) = 0.09628$
The period is approximately 37 and the maximum amplitude is 0.2.

A–3. Using $h = 0.01$, $Y(2) = 1.78811$
Using $h = 0.001$, $Y(2) = 1.78807$

A–4. (a) $\theta(t) = \dfrac{1}{4}\cos 4t$

(b) $\theta\left(\dfrac{\pi}{4}\right) = -0.249976$; also $\theta(2\pi) = 0.2488$

(c) With $h = 0.001$, $\theta\left(\dfrac{\pi}{4}\right) = -0.98149$. From $\theta(t) = \cos 4t$,

$\theta\left(\dfrac{\pi}{4}\right) = -1.$

(d) If $L = 4$, $\theta(t) = \dfrac{1}{4}\cos\sqrt{8}t$; the period is 2.2214.

With IMEUL2, $\theta(1.12) = -0.249976$ and
$\theta(2.221) = 0.249918$

A–5. (a) $x(8) = 0.22507 \times 10^{-5}$
(b) $X(8) = 0.225314 \times 10^{-5}$
(c) $X(5) = 1.17649$; the period is approximately 1.57.

A–6. (a) With $h = 0.01$, $Y(1.5) = 0.182054$; max value occurs at $x = 0.86$.
(b) With $h = 0.001$, $Y(1) = 0.0745183$; max value occurs at $x = 0.54$.

A–7. For $n = 2$, $Y(10) = -1.12329$
For $n = 3$, $Y(10) = -0.0913887$
For $n = 4$, $Y(10) = 0.058858$
For $n = 5$, $Y(10) = 0.169991$
YACTUL, $y(10) = 0.170664$

A–8. (a) Using $h = 0.01$, $X(1.57) = 0.33356$; the period is $\frac{\pi}{4}$.
(b) Using $h = 0.01$, $X(1.57) = 0.0050853$
(c) $\beta = 4$; using $h = 0.01$, $X(1) = -0.68979 \times 10^{-3}$. On $[0, 1]$ no oscillation occurs.
(d) $\beta = 3$; using $h = 0.01$, $X(0.5) = -0.0313$, $X(1) = 0.00293$. We have oscillation on $[0, 1]$. The value for critical damping on $[0, 1]$ is in $3 < \beta < 4$.
(e) Oscillation of amplitude approximately 1.11 occurs.
(f) Oscillation of increasing magnitude results; the period is approximately 0.8.
(g) $Y(1) = 0.6813$, $Y(3) = 0.2189$, $Y(5) = -1.6480$, $Y(6) = 0.28272$, $Y(6.3) = 0.020282$. The graph shows the interesting properties.

A–9. Using IMEUL2 gives $Y(20) = 13.3337$, $Y'(20) = 0.8889$,
$\delta = 13.3333\cdots$ $\theta_b = 0.8888\cdots$

A–10. With $h = 0.1$, $Y(8) = 1.23035$, $V(8) = -0.973007$; the phase-plane plot is a clockwise spiral curve that closes.

Problems for Chapter 5

1. YACTUL: $Y(1) = 3.29179$ using $y = \frac{3}{4}e^{2x} - \frac{7}{4} - \frac{1}{2}x$

   EULER: $Y(1) = 2.3938$
   IMEUL: $Y(1) = 3.22847$
   RUKU3: $Y(1) = 3.28864$
   RUKU4: $Y(1) = 3.29167$

2. YACTUL: $Y(2) = 2.34726$ using $y = x^{-2}(xe^x - e^x + 2)$
   IMEUL: $Y(2) = 2.35049$

RUKU4: $Y(2) = 2.34727$
MILNE: $Y(2) = 2.34722$
HAMMING: $Y(2) = 2.34711$

3. (a) $y = x^3 - 4x$
   (b) $Y(-3) = -15$, $Y(-2) = 0$, $Y(-1) = 3$, $Y(0) = 0$, $Y(1) = -3$, $Y(2) = 0$, $Y(3) = 15$
   (c) Results are same as part (b).

4. RUKU4: $Y(1) = -0.694943$
   (b) MILNE: $Y(1) = -0.694978$
   HAMMING: $Y(1) = -0.695075$

5. From RUKU4: $Y(2) = 16.8026$; MILNE: $Y(2) = 15.9181$ using $h = 0.1$. The analytical solution is $y = \sqrt{2}(\tan\sqrt{2}x + 1.802)$, giving $y(2) = 17.2159$.

6.

$Y(1)$ with $h = 0.1$	Program	$Y(1)$ with $h = 0.01$
1.67515	IMEUL	1.67819
1.67826	RUKU4	1.67822
1.67826	MILNE	1.67822

7. RUKU4: $Y(1)$ with $h = 0.1$ is 2.16297; $Y(1)$ with $h = 0.05$ is 2.16297
   HAMMING: $Y(1)$ with $h = 0.05$ is 2.16298

8. RUKU4: $Y(2) = 2.43401$
   MILNE: $Y(2) = 2.43392$

9. RUKU4: $Y(2) = 1.87298$, $Y(2.5) = 1.91588$, $Y(3) = 2$, $Y(3.5) = 2.11249$, $Y(4) = 2.245$

10. RUKU4: $Y(2) = 1.5764$

11. YACTUL: $Y(2) = 11.7781$; MILNE: $Y(2) = 11.7781$; HAMMING: $Y(2) = 11.7782$; RUKU4: $Y(2) = 11.7788$; ADAMS-MOULTON: $Y(2) = 11.7788$

12. The analytical solution is $x^2 + \left(y - \frac{5}{4}\right)^2 = \left(\frac{5}{4}\right)^2$. If we begin at $(1, 2)$ on this circle, we cannot use x values outside of $-\frac{5}{4} \le x \le \frac{5}{4}$.

13. The solution is $y = e^{\sin(x^4)}$, which oscillates rapidly.

Applications Problems for Chapter 5

A–1. (a) $y' = -0.22y$, $y(0) = 10$ produces $Y(6.30) = 2.50074$ (25% of original value).
   (b) $Y(10.46) = 1.00138$ (10% of original value).

A–2. (a) $y' + \frac{1}{4}y = 0$, $y(0) = 20$

(b) $y(4) = 7.35759$, $y(8) = 2.70671$
(c) RUKU4: $Y(4) = 7.35759$, $Y(8) = 2.70671$

A–3. (b) $y(0.5)[3/(3 + x)]^2$, which gives $y(5) = 0.0703125$
(c) Using RUKU4 and MILNE, $Y(5) = 0.0703125$
(d) RUKU4: $Y(5) = 0.164396 \times 10^{-4}$; MILNE: $Y(5) = 0.654923 \times 10^{-4}$

A–4. $Y(-0.2) = 0.0422574$; $Y(0.6) = 0.257741$

A–5. $P(0) = 3900$ (in thousands) $= P(1790)$; $P(1830) = 13017$; $P(1870) = 39135$; $P(1910) = 91603$; $P(1950) = 148400$

A–6. $C(10) = 29.9954$

A–7. $\psi(2) = 1.40366$; $\psi(4) = 1.40373$

A–8. (a) $y = 4$ EXP[LOG(40) EXP($-0.43x$)], giving $y(3) = 11.0423$, $y(9) = 4.31993$, $y(16) = 4.0152$, $y(20) = 4.00272$
(b) RUKU4: $Y(3) = 11.0423$, $Y(6) = 5.28999$
IMEUL: $Y(3) = 11.0427$, $Y(6) = 5.29005$
(c) The limiting value is 4.

Section 6.1

1. $y = 2 + 2x + \sum_{3}^{\infty} \frac{3}{(n-1)!} x^{n-1}$

3. $y = Ce^{\text{Arctan}\, x}$

5. $y = 2 + \frac{3}{2}x^2 + \frac{3}{8}x^4 + \frac{1}{16}x^6 + \cdots = 3e^{x^2/2} - 1$

Section 6.2

1. (a) All points except $x = 0$.
(b) All points ordinary.
(c) All points except $x = 0$ and $x = -1$.
(d) All points except $x = 0$.

3. $y = 3x^2 - 1$; valid in $(-1, 1)$

5. $$y = a_0\left[1 + \sum_{n=1}^{\infty} \frac{1\cdot 4\cdot 7\cdots(3n-2)}{(3n)!} x^{3n}\right] + a_1\left[x + \sum_{n=1}^{\infty} \frac{2\cdot 5\cdot 8\cdots(3n-1)}{(3n+1)!} x^{3n+1}\right]$$

7. $a_{n+3} = \dfrac{-a_n(n+1)}{(n+3)(n+2)}, \qquad a_2 = 0$

9. $y = a_0(1 - x^2) + a_1\left(x - \dfrac{1}{3}x^3 - \dfrac{1}{3\cdot 5}x^5 - \dfrac{1\cdot 3}{3\cdot 5\cdot 7}x^7 - \cdots\right)$

11. $y = a_0\left(1 - \dfrac{2}{3!}x^3 + \dfrac{16}{6!}x^6 - \dfrac{384}{9!}x^9 + \cdots\right)$

$$+ a_1\left(x - \frac{6}{4!}x^4 + \frac{36}{7!}x^7 + \cdots\right) + a_2\left(x^2 - \frac{16}{5!}x^5 + \frac{32}{8!}x^8 \cdots\right)$$

13. #5 converges on all reals.
#6 converges on $-1 < x < 1$.
#7 converges on all reals.
#8 converges on $-1 < x < 1$.
#9 converges on $-1 < x < 1$.
#10 converges on all reals.
#11 converges on all reals.
#12 converges on $-\sqrt{8} < x < \sqrt{8}$.

17. $y = a_0\left[1 + \displaystyle\sum_{n=1}^{\infty} \frac{2^n(-k/2)(-k/2+2)\cdots(-k/2+2n-2)x^{2n}}{(2n)!}\right]$

$$+ a_1\left[x + \sum_{n=1}^{\infty} \frac{2^n(1-k/2)(1-k/2+2)\cdots(1-k/2+2n-2)x^{2n+1}}{(2n+1)!}\right]$$

Section 6.3

1. $y = 1 + 0\cdot x + \dfrac{3}{2!}x^2 + \dfrac{0}{3!}x^3 + \dfrac{18}{24}x^4 + \cdots$

3. $y = 2 + 2x + \dfrac{5}{2}x^2 + x^3 + \dfrac{7}{12}x^4 + \dfrac{1}{6}x^5 + \cdots$

5. $y = 2 + \dfrac{1}{2}x^2 + \dfrac{4\ln 2}{6}x^3 + \dfrac{3}{24}x^4 + \cdots$

7. $y = 0 + 0\cdot x + \dfrac{2}{2!}x^2 + \dfrac{0}{3!}x^3 + \dfrac{0}{4!}x^4 + \dfrac{0}{5!}x^5 + \dfrac{-24}{6!}x^6 + \cdots$

9. $y = 3 - 5(x+1) + \dfrac{4}{2!}(x+1)^2 - \dfrac{12}{3!}(x+1)^3 + \dfrac{12}{4!}(x+1)^4 + \cdots$

Section 6.4

1. (a) $x = 0$ is a regular singular point.
(b) $x = 0$ is a regular singular point.

(c) $x = 0$ is an irregular singular point.
(d) $x = 0$ is an irregular singular point.
(e) $x = 0$ is a regular singular point.
(f) $x = 0$ is a regular singular point.
(g) $x = 0$ is a regular singular point.
$x = -1$ is an irregular singular point.

3. $$y = a_0\left(\sum_{n=0}^{\infty} \frac{2^{2n}}{(2n)!} x^n\right) + b_0 x^{1/2}\left(\sum_{n=0}^{\infty} \frac{2^{2n+1}(n+1)}{(2n+2)!} x^n\right)$$
$$= a_0(\cosh 2\sqrt{x}) + b_0\left(\frac{1}{2}\sinh 2\sqrt{x}\right)$$

5. $$a_n = \left(\frac{n+s-3}{2n+2s-3}\right) a_{n-1};\ s_1 = 0,\ s_2 = \frac{3}{2}$$

7. $$y = a_0\left(1 + \sum_{n=1}^{\infty} \frac{(n+1)x^{2n}}{7\cdot 17\cdot 27\cdots(10n-3)}\right)$$
$$+ b_0 x^{3/5}\left(\sum_{n=0}^{\infty} \frac{10n+13}{13\cdot 5^n\cdot 10^n n!} x^{2n}\right)$$

9. $$a_n = \frac{-2a_{n-1}}{25(n+s)(n+s-1)+4};\ s_1 = \frac{1}{5},\ s_2 = \frac{4}{5}$$

11. $$a_n = \left(\frac{n+s+2}{n+s}\right) a_{n-2};\ s_1 = 0,\ s_2 = -1$$

13. $$y = \frac{a_0}{x} + \frac{b_0}{1+x^2}$$

15. $$y = a_0 x\left(1 + \frac{1}{5}x - \frac{2}{35}x^2 + \frac{22}{945}x^3 - \cdots\right)$$
$$+ b_0 x^{-1/2}\left(1 - \frac{19}{4}x - \frac{209}{32}x^2 + \frac{1045}{1152}x^3 - \cdots\right)$$

25. $$y = 1 - \frac{1}{2^2}x^2 + \frac{1}{2^2\cdot 4^2}x^4 - \frac{1}{2^2\cdot 4^2\cdot 6^2}x^6$$
$$+ \cdots + (-1)^n \frac{1}{2^{2n}(n!)^2} x^{2n} + \cdots$$

27. $L_0(x) = 1, \qquad L_1(x) = 1 - x, \qquad L_2(x) = x^2 - 4x + 2$

29. $$a_n = \frac{-a_{n-2}}{(n+s)(n+s)(n+s-1)(n+s+1)},\quad s_1 = 0,\quad s_2 = 0,\quad s_3 = 1,$$
$s_4 = -1$

Miscellaneous Exercises for Chapter 6

1. Regular Singular Point: $x = \frac{1}{2}$; Irregular Singular Point: None; Ordinary Points: All real numbers except $x = \frac{1}{2}$.
2. R.S.P.: $x = 0$, $x = -4$; I.S.P: None; O.P.: All reals except $x = 0$ and -4.
3. All real numbers ($\mathbb{R}$) are ordinary points.
4. R.S.P.: $x = \pm 1$; I.S.P.: None; O.P.: $\mathbb{R} - \{1, -1\}$
5. R.S.P.: $x = \frac{1}{2}$, $x = -2$; I.S.P.: None; O.P.: $\mathbb{R} - \left\{\frac{1}{2}, -2\right\}$
6. R.S.P.: None; I.S.P.: $x = 0$; O.P.: $\mathbb{R} - \{0\}$
7. R.S.P.: $x = 0$; I.S.P.: None; O.P.: $\mathbb{R} - \{0\}$
8. R.S.P.: $x = -1$, $x = -2$; I.S.P.: None; O.P.: $\mathbb{R} - \{-1, -2\}$
9. R.S.P.: $x = 0$, $x = -4$; I.S.P.: None; O.P.: $\mathbb{R} - \{0, -4\}$
10. R.S.P.: $x = 0$, $x = 2$; I.S.P.: None; O.P.: $\mathbb{R} - \{0, 2\}$
11. $y = 1 - \frac{1}{2!}x^2 + \frac{2}{4!}x^4 - \frac{10}{6!}x^6 + \cdots$
12. $y = x - \frac{1}{9}x^3 + \frac{1}{54}x^4 - \frac{1}{2430}x^6 + \cdots$
13. $y = 1 + 3x + 4x^2 + \frac{13}{6}x^3 + \cdots$
14. $y = 1 + \frac{1}{2!}(x-1)^2 + \frac{2}{4!}(x-1)^4 - \frac{1}{5!}(x-1)^5 + \cdots$
15. $y = 2x + 2x^2 + \frac{1}{3}x^4 + \frac{2}{5}x^5 + \cdots$
16. $y = 1 + \frac{1}{4}x - \frac{3}{32}x^2 + \frac{7}{128}x^3 + \cdots$
17. $y(x) = 1 + 2x^2 + \frac{3}{2}x^4 + \cdots + \frac{3^{n-1}}{2^{n-2}n!}x^{2n} + \cdots$
18. $y(x) = 2 + x^2 - \frac{1}{3}x^3 + \frac{1}{4}x^4 - \frac{1}{15}x^5 + \cdots$

19. $y(x) = a_0\left(\frac{1}{x} - \frac{1}{2}\right) + a_1 x^2\left[1 - \frac{2}{4}x + \frac{2\cdot 3}{1\cdot 4\cdot 2\cdot 5}x^2 - \cdots + (-1)^n \frac{3!(n+1)!}{(n+3)!n!}x^n + \cdots\right]$

20. $y(x) = a_0\left(1 + \frac{x^2}{2} - \frac{x^4}{2\cdot 4} + \frac{1\cdot 3\, x^6}{2\cdot 4\cdot 6} - \cdots\right) + a_1 x = a_0\sqrt{1 + x^2} \quad a_1 x$

21. $y = a_1 x + a_2 x^2$

22. $y = a_0\left(1 + x + \frac{x^2}{2}\right) + b_0 e^x$

23. $y = A\left(1 + \frac{x^2}{3!} + \frac{x^4}{5!} + \cdots\right) = A\frac{\sinh x}{x}$

24. $y = \frac{C_1}{x} + \frac{C_2 e^{x^2}}{x}$

25. $y = a_0 x^{-1} + a_1(1 + x^2)$

26. $s_1 = \frac{1}{2}, s_2 = \frac{1}{2}$

27. $s_1 = 1, s_2 = -1$

28. Roots: $s_1 = 1, s_2 = -\frac{2}{3}$; recurrence formula: $a_1 = \frac{a_0}{3s + 5}$;

$$a_n = \frac{a_{n-1}}{3s + 3n + 2} + \frac{a_{n-2}}{(s + n - 1)(3s + 3n + 2)}$$

29. Roots: $s_1 = -1, s_2 = \frac{1}{4}$; recurrence formula:

$$a_n = \frac{-(n + s - 1)(n + s - 2)a_{n-1}}{(4n + 4s - 1)(n + s + 1)}$$

30. Roots: $s_1 = -1, s_2 = \frac{1}{3}$; recurrence formula:

$$a_n = \frac{(n + s - 1)a_{n-1} - 2a_{n-2}}{(3n + 3s - 1)(n + s + 1)}$$

PROBLEMS FOR CHAPTER 7

1. (a) Recurrence formula is $a_n = \frac{-2n + 3}{n}a_{n-1} - \frac{2}{n(n-1)}a_{n-2}$.

   (b) At $x = 0.25$, Y(SERSO) = 1.17087, y(actual) = 1.17130.
   At $x = 0.50$, Y(SERSO) = 1.22415, y(actual) = 1.24453.

2. Recurrence formula is $a_n = \dfrac{a_{n-2}}{2(n-1)}$; $Y(-1) = 0.308271$, $Y(1) = 2.87632$.

3. R.F. is $a_n = -\dfrac{n}{2(n-1)} a_{n-2}$; $Y(-1.3) = -0.278566$, $Y(1.3) = 0.75972$.

4. $y = 1 + x - x^2 - \frac{3}{4}x^3 + \frac{2}{3}x^4 + \frac{15}{64}x^5 - \frac{6}{15}x^6 - \frac{105}{768}x^7 + \frac{8}{35}x^8 + \cdots$; good results for small x.

5. R.F. is $a_n = \dfrac{n-3}{n} a_{n-2}$; $Y(-1) = 4$, $Y(0.25) = 1.1875$, $Y(1) = -2$. From a Taylor series solution we get $Y(0.25) = 1.18652$, $Y(0.50) = 0.234375$.

6. If $t = x + 3$, then $a_n = \dfrac{2n-1}{n(n-1)} a_{n-2}$. Modify SERSO and run on $(-2, 2)$, obtaining $Y(-5) = 7$, $Y(-1) = 7$.

7. R.F. is $a_n = \dfrac{2a_{n-2} - a_{n-3}}{n(n-1)}$ with $a_2 = a_0$. With SERSO3 we get $Y(0.10) = 3.15992$, $Y(0.20) = 3.44004$, $Y(0.50) = 5.02735$. Using a series approximation we get, $Y(0.10) = 3.12987$, $Y(0.20) = 3.31928$, $Y(0.50) = 4.25$.

8. R.F. is $a_n = \dfrac{a_{n-3}}{n}$. With SERSO3, $Y(-1) = -0.08333$, $Y(0.5) = 1.55729$, $Y(1) = 2.58333$. With a Taylor series approximation, $Y(0.5) = 1.55833$, $Y(1) = 2.61667$.

9. R.F. is $a_n = \dfrac{-a_{n-1}}{n+s+2}$. For $s_1 = 1$, $Y(0.5) = 0.443264$, $Y(2) = 1.297$; YACTUL gives $Y(0.5) = 0.443262$, $Y(2) = 1.297$. For $s_2 = -2$, SERSING give $Y(0.5) = 2.42612$, $Y(2) = 0.0338341$; YACTUL gives $Y(0.5) = 2.42612$, $Y(2) = 0.0338338$.

10. R.F. is $a_n = \dfrac{-7(n+s-1)a_{n-1}}{(2n+2s-3)(n+s-3)}$. For $s_1 = \frac{3}{2}$, $Y(0.5) = -0.26116$, $Y(1.1) = -1.05298$. For $s_2 = 3$, $Y(0.5) = 0.01292$, $Y(1.1) = -0.003831$.

11. R.F. is $a_n = \dfrac{-a_{n-1}}{(4n+4s-1)(2n+2s+1)}$. For $s_1 = \frac{1}{4}$, $Y(0.5) = 0.8112$, $Y(1.1) = 0.9456$. For $s_2 = -\frac{1}{2}$, $Y(0.5) = 1.0694$, $Y(1.1) = 0.45731$.

12. For BESSL0, zeros are at 2.40483 and 5.52007. For BESSL1, zeros are at 3.83168 and 7.01559.

Section 8.3

1. Converges to $\frac{\pi}{2}$.

3. Converges to $\frac{1}{3}$.

13. $\frac{3}{2}e^{3x} - \frac{1}{2}e^{x}$

15. $2e^{x} - 2$

19. $\frac{2}{s}e^{-4s} - \frac{3}{s}e^{-7s}$

Section 8.4

1. $\frac{6as^2 - 2a^3}{(s^2 + a^2)^3}$

3. $\frac{1}{2(s-a)} - \frac{1}{2(s+a)} = \frac{a}{s^2 - a^2}$

5. $\frac{s^2 + a^2}{(s^2 - a^2)^2}$

11. $\frac{4x^6}{6!} + 2e^{3x}$

13. $\frac{24s(s^2 - 1)}{(s^2 + 1)^4}$

15. $\frac{-16 + (s+1)^2}{[(s+1)^2 + 16]^2}$

17. $\frac{\sqrt{\pi}}{2}(s-3)^{-3/2}$

Section 8.5

1. $y = e^{-t} - e^{-2t}$

3. $x = \frac{16}{5}e^{5t} - \frac{1}{5}$

5. $y = -\frac{1}{2}(e^{x} + \cos x + \sin x)$

7. $y = -\frac{1}{4} + \frac{1}{2}x + \frac{1}{4}e^{-2x} + e^{-x}$

9. $y = 2e^x\cos 2x + \frac{1}{2}e^x\sin 2x$

11. $y = 3\cos 2x - \frac{3}{2}\sin 2x + 3$

13. $y = \frac{3}{4}e^x + \frac{1}{2}e^{-x} - \frac{1}{4}e^{-3x}$

15. $y = \frac{9}{4} + \frac{1}{2}x + \frac{5}{4}e^{2x} - e^x$

17. $y = \frac{1}{2}\sin x - \frac{1}{2}\cos x + 2e^{-x} - 2$

19. $y = e^x\left(\frac{x^3}{6} - \frac{3}{2}x^2 + x\right)$

21. $y = e^{-x}\cos x + 2e^{-x}\sin x$

23. $x(t) = -\frac{1}{4}e^{-3t} + \frac{2}{5}e^{-2t} - \frac{3}{20}e^{-t}\cos 2t - \frac{1}{20}e^{-t}\sin 2t$

Section 8.6

1. (a) $\frac{1}{a^2}(ax + e^{-ax} - 1)$

   (b) $\frac{1}{6}(\sin 3x + 3x\cos 3x)$

   (c) $\frac{1}{a^2}(1 - \cos ax)$

3. (a) $\frac{3}{s^2(s^2 + 9)}$

   (b) $\frac{2}{s^3(s - 1)}$

5. $I(t) = \frac{3}{100}\sin t - \frac{1}{100}\cos t - \frac{1}{25}e^{-t/2} + \frac{1}{20}e^{-t}$

7. $\frac{1}{840}$

Section 8.7

1. $y = (1 - \cos t) - (1 - \cos(t - 4))H_4(t)$

3. $y = 8(1 - \cos t) + [(t - 2) - \sin(t - 2)]H_2(t)$

5. $\dfrac{1}{s^2}\tanh\dfrac{\pi s}{2}$

7. $\dfrac{\int_0^{\pi} e^{-sx}f(x)dx}{1 - e^{-\pi s}}$

Miscellaneous Exercises for Chapter 8

1. $5\cos\sqrt{8}x$

2. $\dfrac{2\beta s}{(s^2 + \beta^2)^2}$

3. $6\cos 2x + \sin 2x$

4. $H_4(x)$

5. $(x + 1)e^{2x}$

6. $x(e^x + e^{-x}) - 1(e^x - e^{-x})$

7. $e^{-3x}\cos 2x$

8. $\dfrac{1}{2}(3e^{-3(x-2)} - e^{-(x-2)})H_2(x)$

9. $\dfrac{2}{s(s^2 + 4)}$

10. $7e^{-x} + 9e^{3x} - 4xe^{-x}$

11. $\dfrac{x^2}{2}e^{-x}$

12. $\dfrac{1}{2}e^{-3x}\sin 2x$

13. $y = te^{-t}$

14. $y = e^{-t} - e^{-2t}$

15. $y = -4e^{3t}$

16. $y = e^{-t}(\cos 2t + 3\sin 2t)$

17. $y = \frac{1}{3}\sin t - \frac{1}{6}\sin 2t$

18. $y = -2(e^{-0.005t})\cos t - 0.01\sin t + 2$

19. $y = 3t - 3 + \frac{3}{2}(\cos t - \sin t + e^{-t})$

20. $y = (x^2 + x)e^{-x}$

21. $y = \frac{1}{3}e^{-x}\sin 3x$

22. $y = 5\cos\frac{\sqrt{7}}{2}x + \frac{2\sqrt{7}}{7}\sin\frac{\sqrt{7}}{2}x$

23. $y = \cos\alpha x$

24. $i(t) = -1 + \cos 4t - H_3(t)[1 - \cos 4(t - 3)]$

25. $\frac{2}{s}(2e^{-s} - e^{-3s})$

26. $\frac{e^{-3s}(1 + 3s)}{s^2}$

27. $\frac{1}{s^2}[1 - 2e^{-2s} + e^{-4s}]$

28. $y = \frac{1}{4}x\sin 2x + \cos 2x - \frac{3}{2}\sin 2x$

29. $\frac{1}{s}\left(\frac{1}{1 - e^{-s}}\right)$

30. (a) Unit step function
(b) Forward sawtooth
(c) Backward sawtooth

(d) $\mathscr{L}[K(t)] = \frac{1}{s}\left(\frac{e^{-s}}{1 - e^{-s}}\right)$, $\mathscr{L}[t - K(t - 1)] = \frac{1 - e^{-s} - se^{-s}}{s^2(1 - e^{-s})}$

$\mathscr{L}[K(t) - t] = \frac{s - 1 + e^{-s}}{s^2(1 - e^{-s})}$

31. $\frac{a}{s^2 + a^2}$

32. $\frac{s}{s^2 + 1}$

33. $\frac{\pi}{2} - \text{Arctan } s$ and $\frac{\pi}{2} - \text{Arctan } \frac{s}{2}$

35. (a) $\frac{s}{(s^2+1)^2}$ (b) $\frac{s}{s^2(s^2+1)}$ (c) $\frac{1}{(s+1)^2(s^2+1)}$ (d) $\frac{2}{s^2(s^2+4)}$

36. (a) $\frac{1}{4}x + \frac{3}{8}\sin 2x$ (b) $\frac{1}{2}e^{-2x} + \frac{1}{2}e^{2x} + 2xe^{2x}$

(c) $\frac{4}{3}\sin 2x - \frac{2}{3}\sin x$ (d) $\frac{1}{2}(\sin x + x\cos x)$

39. (a) $-\frac{s^2+a^2}{(s^2-a^2)^2}$ (b) $\frac{-2as}{(s^2-a^2)^2}$

Section 9.2

1. (a) 2
   (b) 2
   (c) 4
   (d) 4
   (e) 2

3. $x(t) = C_1e^{2t} + C_2te^{2t}$
   $y(t) = C_1e^{2t} + C_2te^{2t} - C_2e^{2t}$

5. $r(t) = C_1e^{t} + C_2e^{2t}$
   $s(t) = C_1e^{t} + 3C_2e^{2t}$

7. $x_1(t) = C_1e^{3t/2}\cos\frac{\sqrt{2}}{2}t + C_2e^{3t/2}\sin\frac{\sqrt{2}}{2}t - \frac{18}{11}t - \frac{62}{121}$

$$x_2(t) = \left(C_1 + \frac{\sqrt{2}}{2}C_2\right)e^{3t/2}\cos\frac{\sqrt{2}}{2}t + \left(C_2 - \frac{\sqrt{2}}{2}C_1\right)e^{3t/2}\sin\frac{\sqrt{2}}{2}t - \frac{2}{11}t - \frac{46}{121}$$

9. $x = C_1e^{t} + \left(C_3 + \frac{1}{2}C_4\right)e^{t}\cos t + \left(C_4 - \frac{1}{2}C_3\right)e^{t}\sin t$

$y = C_1e^{t} + C_3e^{t}\cos t + C_4e^{t}\sin t$

$z = C_1e^{t} + \frac{1}{2}C_3e^{t}\cos t + \frac{1}{2}C_4e^{t}\sin t$

11. $y = C_1e^{x} + C_2xe^{x} - \frac{3}{25}\cos 2x - \frac{4}{25}\sin 2x$

$z = C_3e^{-x} + \left(C_1 + \frac{1}{2}C_2\right)e^{x} + C_2xe^{x} - \frac{8}{25}\cos 2x - \frac{4}{25}\sin 2x$

13. $x = \dfrac{aC_1e^t}{1 + C_1e^t}, \qquad y = C_2(1 + C_1e^t)^{1/10}$

15. $x = -E^{-1/2}(C_1e^\tau - C_2e^{-\tau})$
$y = C_1e^\tau + C_2e^{-\tau}$

17. $x_1 = C_1 + C_2e^{-(1/v_1 + 1/v_2)t}$
$x_2 = C_1 - C_2\dfrac{v_1}{v_2}e^{-(1/v_1 + 1/v_2)t}$

19. $E(t) = \dfrac{2\sqrt{2}}{3}e^{-t}\sin\sqrt{2}t + \dfrac{1}{3}e^{-t}\cos\sqrt{2}t + \dfrac{2}{3}$

$F(t) = -\dfrac{\sqrt{2}}{6}e^{-t}\sin\sqrt{2}t + \dfrac{2}{3}e^{-t}\cos\sqrt{2}t + \dfrac{1}{3}$

21. $C_1 = c_1\cos\dfrac{\sqrt{2}}{10}t + c_2\sin\dfrac{\sqrt{2}}{10}t + \dfrac{1}{2}\cdot 10^{-4}$

$C_2 = (c_1 - \sqrt{2}\,c_2)\cos\dfrac{\sqrt{2}}{10}t + (c_2 + \sqrt{2}\,c_1)\sin\dfrac{\sqrt{2}}{10}t + \dfrac{5}{2}\cdot 10^{-4}$

Section 9.3

1. $x = \dfrac{2}{5}e^{4t} + \dfrac{3}{5}e^{-t}$

$y = -\dfrac{3}{5}e^{4t} + \dfrac{3}{5}e^{-t}$

3. $x = -\dfrac{1}{2} + 3e^t - \dfrac{3}{2}e^{4t}$

$y = -2e^t + 2e^{4t}$

5. $y = z = e^x - 1$

7. $y = e^{-2x}$

$z = \dfrac{4}{3}e^{2x}$

9. $y = \dfrac{94}{99}e^{-t} + \dfrac{1}{10001}\left(\dfrac{10}{99}e^{-100t} - 9496\cos t + 9506\sin t\right)$

Section 9.4

1. $q = 10^{-3}\left[1 - e^{-500t}\left(\cos 1000t + \dfrac{1}{2}\sin 1000t\right)\right]$ coulombs

Miscellaneous Exercises for Chapter 9

1. $y = t^2 - 1, \qquad z = t^2 + 2t + 1$

2. $y = C_1e^{2x} + C_2e^{-x} - 3e^x + 5, \qquad z = C_1e^{2x} + \frac{1}{2}C_2e^{-x} - \frac{5}{2}e^x + 3$

3. $x = -e^{-2t}\cos t + 3e^{-2t}\sin t + 8, \qquad y = 5e^{-2t}\cos t - 5e^{-2t}\sin t - 10$

4. $x = e^{-t}(2\cos t + 3\sin t) + \frac{1}{2}(t - 1), \qquad y = e^{-t}(-\sin t - 5\cos t) + \frac{1}{2}$

5. $x_1 = -e^t + 2e^{2t}, \qquad x_2 = -e^t + 6e^{2t}$

6. $x = C_1e^{2t}(\cos t - \sin t) + C_2e^{2t}(\cos t + \sin t)$
   $y = -C_1e^{2t}\cos t - C_2e^{2t}\sin t$

7. $x_1 = -3e^t + e^{5t}, \qquad x_2 = 3e^t + 3e^{5t}$

8. $x = (1 - t)e^{2t} + \frac{1}{2}t + \frac{1}{4}, \qquad y = (2 - t)e^{2t} + t^2 + \frac{3}{2}t + \frac{1}{4}$

9. $x = 4e^{8t} - 6t - 3,\ y = 6e^{8t} + 3t + 2$

10. $x = C_1,\ y = C_1$

11. $y = \left(-\frac{1}{2}C_1\ln t + C_2\right)\cdot t + \frac{C_3}{t} - 2\ln t - 2$

    $z = \left(C_1 - C_2 + \frac{1}{2}C_1\ln t\right)\cdot t - \frac{C_3}{t} + \ln t + 1$

12. $y = \sqrt{2}\cos\left(x^2 - 1 + \frac{\pi}{4}\right), \qquad z = \sqrt{2}\sin\left(x^2 - 1 + \frac{\pi}{4}\right)$

13. $x = 3C_1e^t + 3C_2e^{-t} + C_3\cos t + C_4\sin t$
    $y = C_1e^t + C_2e^{-t} + C_3\cos t + C_4\sin t$

14. $y = 4e^{-3x} - 3e^{-7x}, \qquad z = 6e^{-3x} - 6e^{-7x}$

15. $y = 2 + 2e^t$
    $x = -2 + (2t + 1)e^t$
    $z = -2 + (2t + 3)e^t$

16. $x = C_1t^2 + C_2t^{-1}$
    $y = C_1t^2 + C_3t^{-1}$
    $z = C_1t^2 - (C_2 + C_3)t^{-1}$

Problems for Chapter 10

1. (b) $Y(1) = -18.8777,\ Z(1) = -128.829$
   (c) Analytical solution is $y = -\frac{1}{5}e^{6x} + \frac{6}{5}e^x,\ y(1) = -77.4239.$

2. (a) Using RKU4S with $h = 0.1$ gives $Y(1) = -77.1062$.
   (b) Using MLNS2 with $h = 0.1$ gives $Y(1) = -76.9174$.

3. EULSYS with $h = 0.1$ gives $Y(1) = -0.477325$, $Z(1) = -2.23822$.
   EULSYS with $h = 0.05$ gives $Y(1) = -0.453019$, $Z(1) = -2.01491$.
   Analytical solution is $y = \cos 2x$ and $y(1) = -0.416147$; $y'(1) = z(1) = -1.81859$.

4. RKU4S with $h = 0.1$ gives $Y(1) = -0.416121$, $Z(1) = -1.81861$.
   RKU4S with $h = 0.05$ gives $Y(1) = -0.416145$, $Z(1) = -1.8186$.
   Analytical results are $y(1) = -0.416147$, $z(1) = -1.81859$.
   MLNS2 with $h = 0.1$ gives $Y(1) = -0.416153$, $Z(1) = -1.81864$.
   MLNS2 with $h = 0.05$ gives $Y(1) = -0.416148$, $Z(1) = -1.81860$.

5. Using modified EULSYS with $h = 0.1$ gives $Y(1) = 1.33122$, $Z(1) = 0.553194$, $W(1) = -0.0181616$.

6. Using modified RKU4S with $h = 0.1$ gives $Y(1) = 1.33118$, $Z(1) = 0.486667$, $W(1) = -0.068952$.

7. Using RKU4S with $h = 0.1$ gives $Y(2) = 2.51639$, $Z(2) = 3.54784$. Then with MLNS2 we get $Y(2) = 2.51621$, $Z(2) = 3.5477$.

Applications Problems for Chapter 10

A–1. (a) $Q(t) = e^{-t}\cos 7t + \frac{1}{7}e^{-t}\sin 7t$. First zero is at 0.2445.
   (b) RKU4S: $Y(1) = 0.312171$, $Z(1) = -1.72531$
   (c) MLNS2: $Y(1) = 0.312583$, $Z(1) = -1.73112$

A–2. (a) $\theta(1.57) = -0.0047$, $\theta(3.14) = 0.000217$, $\theta(6.28) = 0.44 \times 10^{-6}$
   (b) $\theta(1.57) = -0.0045$, $\theta(3.14) = 0.00020$, $\theta(6.28) = 0.375 \times 10^{-6}$

A–3. (a) $P(t) = e^{-(t/2)}\cos\frac{\sqrt{3}}{6}t + 2\sqrt{3}\,e^{-(t/2)}\sin\frac{\sqrt{3}}{6}t + 4$

   (b)

x	Y(YACTUL)	Y(IMEUL2)	Y(RKU4S)
0	5	5	5
6	4.16224	4.16746	4.16158
12	3.99493	3.9996	3.99499
23	4.00002	3.9999	4.00002

A–4. (a) Critical values: $w_c = 6$, $y_c = 10$
   (b) EULSYS: $Y(2) = 5.969$, $W(2) = 0.0396$; $Y(4) = -0.000389$, $W(4) = 0.3035$; $Y(5) = -0.13024$, $W(5) = 0$.
   (c) Elliptic plots, with y becoming extinct around $t = 3.75$.

A–5. (a) $Y(5) = 2.3506$, $Z(5) = 7.666$
   (b) $Y(5) = 18.496$, $Z(5) = 21.504$

(c) $Y(5) = 120$, $Z(5) = 0.11 \times 10^{-6}$
(d) $Y(5) = 0.44 \times 10^{-9}$, $Z(5) = 199.9$

A–6. Y returns to its orignal value at approximately 1.6×10^{-2} seconds.

A–7. X decreases from 2000 to $X(0) = 411.158$. Y varies from $Y(0) = 0$ to $Y_{max}(2.5) = 795.37$ to $Y(10) = 583.134$.

A–8. RKU4S: $i_1(2) = 0.2262$, $i_2(2) = 0.9636$; $i_1(4) = 1.2849$, $i_2(4) = -1.221$
YACTUL: $i_1(2) = 0.2263$, $i_2(2) = 0.9634$; $i_1(4) = 1.2852$, $i_2(4) = -1.2217$

A–9. RKU4S: $i_1(8) = 0.481294$, $i_2(8) = 0.329423$, $i_3(8) = 0.151871$

A–10. MLNS2: $Y(3) = 1.83809$, $Z(3) = 1.69235$

A–12. The period on x_1 is approximately 8.52 seconds. When $\beta = 6.4$, the period is approximately 2.54 seconds.

A–13. $H(t)$ rises to a maximum value of 4000 at approximately 17.5 and then oscillates.

A–14.
$$\frac{dX}{d\tau} = X\left(1 - \frac{X}{10}\right) - \frac{XY}{X + 1}$$
$$\frac{dY}{d\tau} = \frac{1}{6}Y\left(1 - \frac{Y}{X}\right)$$

The value of X decreases to a minimum value of 0.7 at $\tau = 7.5$ and then increases to a maximum of 6.94 at $\tau = 18$. Then it slowly oscillates to smaller values.

A–15. Under these parameter values, m_1 decreases from $m_1(0) = 40$ to $m_1(2) = 4.42718$ while $m_2(0) = 6$ increases to a maximal value of 21.43 at $t = 1.6$

A–16. The same revised RKU4S program from Problem A–11 may be used.

A–17. The values of F oscillate with period approximately 0.75.

A–18. (a) $T = 0.1575$, $b_0 = -0.02025$

A–19. (a) $\phi(-15) = 0.000024$, $\phi(0) = 0.5$, $\phi(15) = 1$, $\theta(-15) = \theta(15) = 0$, $\theta(0) = 0.1767$, $\phi(-10) = 0.0008486$, $\theta(-10) = 0.000599545$
(b) RKU4S: $\phi(0) = 0.5$, $\theta(0) = 0.1767$

A–20. The value of n_1 rises rapidly to 802 at $t = 13$, falls to 0.147 at $t = 24$, rises to 607.75 at $t = 39$ and then to 2.54 at $t = 50$. The valuc of n_2 first falls to 0.384 at $t = 4$, rises to 1472 at $t = 16$, continues to oscillate and ends at $t = 50$ at value 361.

Index

A 5
B 6
C 7
D 8
E 9
F 0
G 1
H 2
I 3
J 4

Laplace Transforms

$f(x)$	$\mathscr{L}[f(x)]$	$f(x)$	$\mathscr{L}[f(x)]$
1	$\frac{1}{s}$	$e^{-ax}\sin(bx)$	$\frac{b}{(s+a)^2+b^2}$
x	$\frac{1}{s^2}$	$e^{-ax}\cos(bx)$	$\frac{s+a}{(s+a)^2+b^2}$
x^n	$\frac{n!}{s^{n+1}}$	xe^{-ax}	$\frac{1}{(s+a)^2}$
e^{-ax}	$\frac{1}{s+a}$	$y'(x)$	$s\mathscr{L}[y(x)] - y(0)$
$\sin(ax)$	$\frac{a}{s^2+a^2}$	$y''(x)$	$s^2\mathscr{L}[y(x)] - sy(0) - y'(0)$
$\cos(ax)$	$\frac{s}{s^2+a^2}$	$y^{(n)}(x)$ $n = 1, 2, 3, \ldots$	$s^n\mathscr{L}[y(x)] - s^{n-1}y(0) - \cdots - y^{(n-1)}(0)$
$x\sin(ax)$	$\frac{2as}{(s^2+a^2)^2}$	$e^{ax}f(x)$	$F(s-a)$ where $\mathscr{L}[f(x)] = F(s)$
$x\cos(ax)$	$\frac{s^2-a^2}{(s^2+a^2)^2}$	$x^n f(x)$ $n = 1, 2, 3, \ldots$	$(-1)^n F^{(n)}(s)$ where $\mathscr{L}[f(x)] = F(s)$